BIONANOPHOTONICS

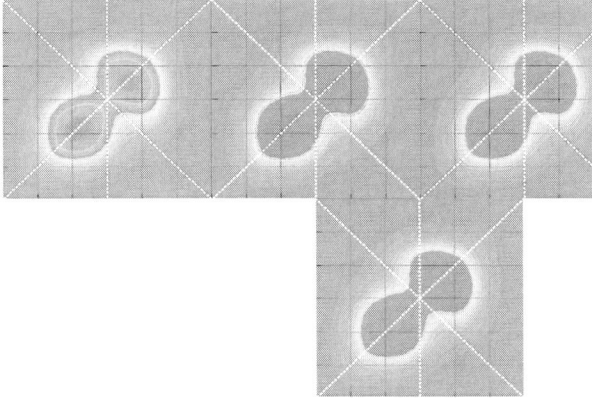

BIONANOPHOTONICS

AN INTRODUCTORY TEXTBOOK

Shuichi Kinoshita

PAN STANFORD PUBLISHING

Published by

Pan Stanford Publishing Pte. Ltd.
Penthouse Level, Suntec Tower 3
8 Temasek Boulevard
Singapore 038988

Email: editorial@panstanford.com
Web: www.panstanford.com

British Library Cataloguing-in-Publication Data
A catalogue record for this book is available from the British Library.

Bionanophotonics: An Introductory Textbook

ISBN 978-981-4364-71-3 (Hardcover)
ISBN 978-981-4364-72-0 (eBook)

Printed in the USA

Contents

Preface

Structural color must be one of the most marvelous arts that nature has ever created during the long history of the universe. Humanity has been mesmerized by their beauty and has become enslaved by their wonderfulness. After the development of modern science in the 19th century, research studies to explore their mysteries began to flourish. However, the fine structures that contributed to it were so small that they were far beyond the resolution of optical microscopes at that time. When the electron microscope was invented, studies on structural color progressed quickly and various sophisticated structures were discovered one after another. In spite of that, the structures thus observed were so complicated that their true optical functions were still behind the veil.

At the end of the 20th century, the concept of photonics was constructed and a new research field to control light with the aid of fine structures progressed rapidly. On the other hand, owing to the development of various techniques of microscopic observations and manipulations, a new field of nanotechnology surged like an avalanche, which enabled us to manipulate and process even on an atomic or molecular level. Research on structural colors also matched these flows and developed quickly. This was partly because fine structures created by nature were themselves just preemptions of photonics and also because research and development in industries became energetic to search for new coloring materials.

Five years ago, I published a book titled *Structural Colors in the Realm of Nature* with World Scientific Publishing. In this book, I tried to summarize how structural colors are distributed in nature, what elaborate structures they exhibit, and how they have been studied from a scientific standpoint. I tried to summarize these points by biological species. This work was actually very hard because I had to gather a lot of material, and then to classify and summarize it from species to species. Hence, it was quite difficult, at that time, to add detailed optical processes related with them, which I thought very important to truly understand their optical functions and biological roles.

By good fortune, I got a splendid opportunity to write a new book concerning structural colors, which was proposed by Pan Stanford Publishing. I decided to write the book, focusing mainly on the fundamental optical processes related with structural colors. I had the intention to make this book largely different from ordinary books on optics. So, I have organized it so that only the basic optical processes that are deeply connected with structural colors will become the main flow of this book. Yet, it should contain ample examples of natural products that contribute to coloring and also contain the recent technological advances in photonics, which are rooted in nature. As a result, I believe this book will occupy a special position among similar books, and will be devoted to scientific research on the natural structural colors and technological research on photonics. In this sense, it completely differs from my previous book that has a rather biological basis. Thus, I recommend this book to be used in conjunction with the previous one, in which more complete examples are explored in nature.

In this book, I have picked up the following topics: light as electromagnetic wave, interference and diffraction of light, photonic crystal, light scattering and our recent work using electromagnetic field simulation. Although the topics themselves are commonplace, I have attempted to describe them as accurate as possible. For example, I have shown the derivations of formulas carefully and minutely so that the readers can follow them only with their eyes. In addition, I have added many footnotes to make the readers pay attention to their importance. I hope these attempts will be helpful for students or beginners to study by themselves. I dare to pick up the electromagnetic field simulation in the final chapter, though the work itself is not completed yet. This is because I intend to give an example of how the interaction of light with real structures is actually analyzed, which may give a hint to solve more complicated problems.

Furthermore, contrary to the ordinary books on optics, this book contains a section on how man detects and processes information on color and also on how the detected colors are expressed quantitatively. The structure and mechanism of animal vision are largely indebted to Prof. S. Kawamura, who suggested that I add interesting research on visual information processing within the retina. I would like to express my deep gratitude to him and also hope that the readers will get interested in this fascinating research area. Since I have placed the whole part of this book on the fundamental optical processes, the book itself takes the form of a textbook for students and beginners. I have added several exercises in each chapter and also several experimental methods that will be easily performed in laboratory, which I hope will help the readers understand the physical basis of this mysterious world of nature.

Before ending the preface, I would like to express my gratitude to Dr. S. Yoshioka, Dr. D. Zhu, Dr. E. Lee, Prof. H. Ghiradella, Prof. A. Saito, Prof. T. Hariyama, Prof. N. Oshima, and many collaborating students in my laboratory. Finally, I express my sincere gratitude to my wife, Sachiko, and to my parents for their continuous encouragement and support. I also thank the publisher for giving me an opportunity to write this book.

Shuichi Kinoshita
Spring 2013

Chapter 1

Introduction

1.1 Bionanophotonics

1.1.1 *Bionanophotonics and Structure-Based Colors*

Nowadays, our daily life is becoming more and more colorful. Paintings of cars and various household electric appliances are much more colorful than before, the display of a television set provides us stimulating colors, and wrapping paper and decorations are so beautiful. Especially, we are often anxious about the brilliantly colored coatings that change their colors with viewing angles. One may think such coatings were not present in old days. However, our ancestors skillfully used such strange colors to decorate craftwork. Plumes using peacock feathers, mother-of-pearl inlay work, opal, and Tamamushi-no-zushi (Japanese miniature shrine whose pillars are decorated with the jewel beetle's wings) are typical examples. In these decorations, natural products were used without further modifications, while recent rapidly growing nanotechnology can make it possible to mass-produce these decorations.

In nature, a tremendous number of orders are generated spontaneously, which cause strikingly brilliant colors owing to

Bionanophotonics: An Introductory Textbook
Shuichi Kinoshita
Copyright © 2013 Pan Stanford Publishing Pte. Ltd.
ISBN 978-981-4364-71-3 (Hardcover), 978-981-4364-72-0 (eBook)
www.panstanford.com

elaborate structures furnished in them. They sometimes reflect surprisingly intense light with a specific color, while in other cases they prohibit any reflection of light. These are natural consequences of complicated interactions between light and microstructures through purely physical processes and are fully utilized in the biological world as signals for courtship, thread, concealing, mating and so on. Such colors are generally called *structural colors*, which are considered one of the main directions of bionanophotonics [Simon (1971); Fox (1976); Srinivasarao (1999); Parker (2000); Vukusic and Sambles (2003); Parker (2003); Kinoshita and Yoshioka (2005b); Kinoshita (2005); Kinoshita and Yoshioka (2005c); Berthier (2007); Kinoshita *et al.* (2008); Kinoshita (2008)].

Another aspect of bionanophotonics is based on the idea to freely handle light. Such examples that take root in natural products are moth-eye structure and opal-type photonic crystal. The moth-eye structure consists of regularly arranged projections with a height of ~250 nm and was first discovered on the corneal surface of the moth. This structure was considered to have a function of anti-reflection, which was soon found as an excellent device to reduce surface reflection for wide wavelength and angular ranges.

Opal is well known as a jewel made of silica, which is contained in sedimentary rocks. Its microstructure was clarified using an electron microscope and was found to be a typical photonic crystal made of silica spheres with diameters of 150–350 nm that were arranged in face-centered-cubic (fcc) form. This discovery shortly blossomed and became an important source of the three-dimensional (3D) photonic crystal, which is difficult to fabricate under lithographic technique even now. Its porous structure becomes a template to fabricate inverse opal after infiltrating materials of a higher refractive index and then removing the silica spheres. This structure guarantees the presence of complete photonic band gap, within which light cannot even exist as in case of an electron in the energy gap in semiconductors.

Thus, the now flourishing photonic technology considerably takes root in nature and particularly coloring mechanisms are too many to say that they are of a biological origin. In the following chapters, we will show the optical fundamentals of these natural devices in detail. We will also show how these natural devices

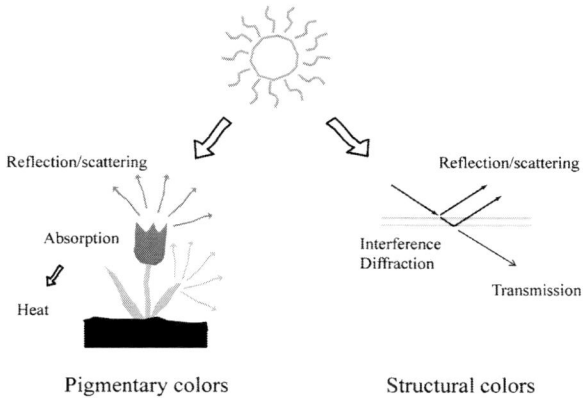

Figure 1.1 Pigmentary colors and structural colors (illustrated by Dr. S. Yoshioka).

manifest their functions through physical operation of light and how animals utilize these devices.

1.1.2 *Difference between Structural and Pigmentary Colors*

The scientific definition of structural color has not yet been settled and its characteristics are often explained as compared with pigmentary color. Pigmentary colors basically originate from atoms or molecules composing the pigment. We usually call them *dyes* when they are soluble in solvent and are normally made of organic materials, while they are called *pigments* when they are insoluble and are often made of inorganic materials.

When matter is illuminated by light, a molecule composing the matter absorbs light and is excited. Since this transition occurs when the energy of light matches the energy needed for the transition from the ground to the excited state, light having a particular energy will be lost by absorption. Usually, the energies of the ground and excited states are broadened owing to the interaction with the surroundings so that light absorption occurs with a narrow or wide energy range depending on the matter employed. Anyway, light having other energies then transmits or reflects, and then comes into our eyes. That is exactly the pigmentary color that we perceive. The molecule in the excited state shortly loses energy by making a transition to the ground state through the emission of light or the release of thermal energy to the surroundings. Thus, it can be said that pigmentary color is the remaining color that is not consumed in the matter. The color in this case is anyway caused by the exchange of energies between light and electron.

The other is a case where light with a certain energy range is reflected, scattered, and deflected not to reach the eyes. In this case, no electronic transition in molecule is necessary and coloration takes place mainly under the presence of a structure, which causes a purely physical operation of light to remove a specific energy range of light. This is the reason why it is called structural color. Since light does not essentially lose energy, the coloration based on this mechanism is sometimes said to be economical and also environmentally friendly as compared with pigments involving heavy metals.

1.1.3 *Optical Phenomena Causing Structural Colors*

In order to reflect, scatter and deflect light using a structure, it is necessary to prepare a special device that functions differently with the color or wavelength of light. The optical functions that are different with the wavelength of light are usually categorized into the following three: The first one is a case where the refractive index of matter constituting a structure differs with the wavelength of light. This is called *refractive index dispersion* and is deeply related to the location of the absorption spectral band in the matter. We can take a prism as an example for this case. When light is incident on a prism, it is refracted and deflected at the apex. It will be soon found that the light is dispersed according to the wavelength of light and shows so-called prismatic colors. This is because the refractive indices are different with wavelengths and light waves with different colors are deflected to different directions. In general, the refractive index for blue is higher than that of red so that the blue light wave is more largely deflected at the apex. In a similar manner, the coloration related to the polarization of light can be classified into this category. One will find this type of coloration when a transparent plate having slight optical anisotropy[a] is sandwiched between two sheet polarizers whose polarization directions are crossed with each other (see Exp. II in Chapter 2). Thus, refractive index dispersion can be considered as a cause of structural colors, though the structure itself is macroscopic.

As the second function depending on the wavelength of light, we will show a phenomenon related to the superposition of waves. This phenomenon is known as interference and diffraction of light, and is deeply associated with the wave nature of light. The colorations due to this function are most abundantly distributed in the natural world. The principle of interference or diffraction is simply explained if we consider light is a kind of wave. Consider a case where two waves expressed by a sine function propagate to the same direction with the same amplitude and wavelength. When the two waves are superposed in phase, the amplitude will be double, while they will completely disappear when they are out of phase.

On the contrary, when the two waves have different wavelengths, the absolute enhancement or disappearance will not occur and only the amplitude will oscillate with time, which is known as *beat*. Hence, if we take a temporal average of the light intensity, which will be performed finally in our eyes or at the detectors in camera and video camera, the effect of the superposition cannot be eventually detected. Thus, the enhancement and disappearance of waves can become an origin of structural colors, if we prepare a special device that can enhance the waves only in a particular wavelength range. The examples categorized into this mechanism are thin-film and multilayer interference, diffraction grating and photonic crystal, which are illustrated in Fig. 1.2. The structural colors widely distributed in the natural world mostly belong to this category.

The last one is a phenomenon related to the scattering of light. This function comes from a fact that the scattering efficiency is dependent on the wavelength of light. In general, if the scatters are small enough as compared with the wavelength of light, the blue is preferably scattered, which is considered to be the origin of the blue sky (see Fig. 1.2).

---

[a]A cellophane adhesive tape and a transparent set square become the good samples for this experiment.

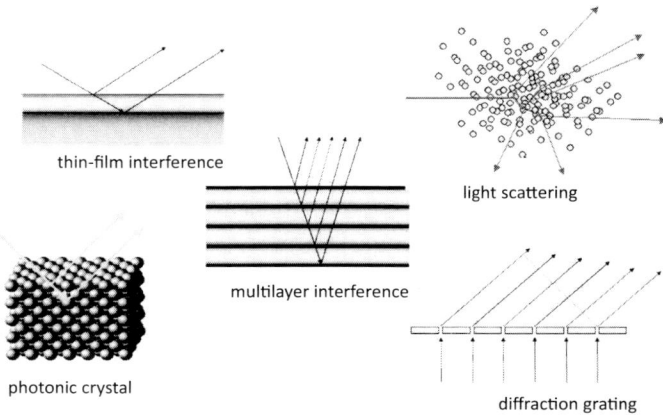

Figure 1.2   Various mechanisms of structural colors.

On the other hand, structural colors are classified from a different perspective. The peculiar color due to structural color is often expressed by a word *iridescence*, which originally means "rainbow" in Latin, but is usually used in case the color changes with viewing angles like prismatic colors. In contrast, the color not showing iridescence is called *non-iridescence*. For example, thin-film interference is generally iridescent, while light scattering is usually non-iridescent but of a structural origin. Recently, non-iridescent color has attracted considerable attention of physicists and biologists, because it appears even when the structure responsible for the color is seemingly irregular, which are commonly distributed in birds and insects.

Though we have exemplified various mechanisms of structural colors, it is rare that these mechanisms themselves work to display colors. In the natural world, these mechanisms are somehow combined with each other and also with the other macroscopic structures so that their original mechanisms will be largely modified to produce a new visual effect. In some case, structural colors are intended to cause special effects in combination with visual and cognitive processes in the observer. The wisdom of nature built up during a long evolutional period now supplies a tremendous amount of devices in photonics, which started only fifteen years ago. At present, we can give only a few examples of the natural materials, which are actually used in the field of photonics. Thus, most of the natural devices remain untouched from a viewpoint of science and technology. In this sense, the research fields of bionanophotonics have just opened and will be endlessly extending in future.

1.2 Historical Review

1.2.1 *Dawn of Bionanophotonic Research (−1930s)*

The study of bionanophonics originated from the research on structure-based colors. Probably, the oldest scientific description on structural colors appeared in *Micrographia* written by Hooke [Hooke (1665)]. In this book, he described the microscopic observation of the brilliant feathers of the peacock and the duck, and found that their colors were destroyed by a drop of water. He speculated that alternate layers of thin plate and air might strongly reflect light. Newton described in his book *Opticks* that the colors of iridescent peacock arose from the thinness of a transparent part of feathers [Newton (1704)]. In spite of these pioneering works on structural colors, the further scientific development must have waited for the establishment of electromagnetic theory.

The study deserving special mention in the 19th century was that performed by Tyndall. When minute particles were involved in liquid and the liquid was observed at right angles to the incident illumination, he found that bluish color was perceived to the eye and in addition, the scattered light was polarized. This phenomenon is now well known as *Tyndall phenomenon* and its color is sometimes called *Tyndall blue*. Stimulated by this discovery, Lord Rayleigh proved that the blue sky originates from the light scattering due to minute particles in the atmosphere, whose efficiency was inversely proportional to the fourth power of the wavelength of light [Rayleigh (1871a,b)]. The scattering according to such a mechanism is now called *Rayleigh scattering*. On the other hand, Brewster described in his book *A Treatise on Optics* which the mother-of-pearl showed the characteristics of diffraction grating, whose casting also showed the same characteristics [Brewster (1845)]. A botanist Reinitzer discovered a strange melting phenomenon when a mixture of cholesterol and benzoic acid was heated [Reinitzer (1888)]. In the next year, a physicist Lehmann analyzed this phenomenon and found a new phase that showed birefringence and anisotropy in spite of liquid-like appearance [Lehmann (1889)]. This material was later called *liquid crystal*, and particularly that discovered by Reinitzer was called *cholesteric liquid crystal*.

In the middle of the 18th century, Maxwell established the mathematical expressions, so-called *Maxwell equations*, by which various electromagnetic phenomena reported so far were treated in a unified way [Maxwell (1873)]. From his equations, an electromagnetic wave was predicted to be emitted by an oscillating current, which was later confirmed experimentally by Hertz [Hertz (1887)]. Thus, it was established that light was a kind of electromagnetic wave. After that, the fundamental properties of light such as reflection, refraction, interference and diffraction could be quantitatively treated, and the study on structural colors approached its first climax.

However, there arose a significant conflict between two hypotheses concerning the mechanisms of structural colors: One was *surface-color*, which was proposed by Walter and was thought to originate from the reflection at a surface involving pigments [Walter (1895)]. The other was *structure-color* that originated from a purely physical operation of light [Rayleigh (1888a,b, 1919)]. These two hypotheses drove the world of physics into two at that time. Walter explained the variation of colors with varying incident angle as due to the change of polarization in reflection at the absorption band edge. The idea of surface-color was then succeeded by Michelson, who conducted experiments of reflectivity on seemingly metallic samples such as the *Morpho* butterfly, and described that they resembled the surface reflection from a very thin surface layer involving dye [Michelson (1911)]. He also reported circularly polarized reflection from a golden scarab.

Lord Rayleigh, on the other hand, derived a formula to express the reflection properties from a regularly stratified medium using electromagnetic theory [Rayleigh (1917)], and considered it as the origin of colors of twin crystals, old decomposed glass and probably those of some beetles and butterflies. He surveyed the studies

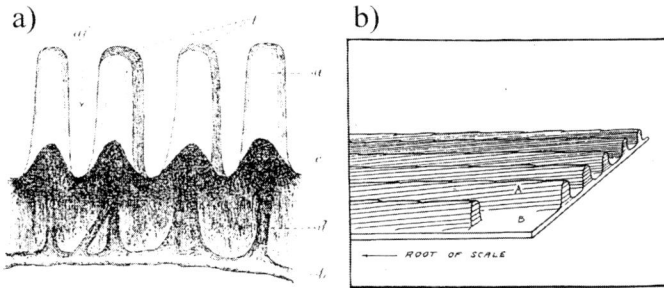

Figure 1.3 Sketches of the *Morpho* butterfly scale illustrated by (a) Onslow [Onslow (1923)] and (b) Mason [Mason (1927a)].

performed so far, and described that the brilliant colors, which varied strongly with varying incident angle, were not due to the ordinary operation of dyes, but came from structure-color [Rayleigh (1919)].

Many experimental works were performed, on an optical microscopic level, to clarify the relationship between brilliant colorations and microstructures at the surface of iridescent, metallic, and whitish materials [Onslow (1923); Süffert (1924); Merritt (1925); Mason (1923a,b, 1926, 1927a,b); Rayleigh (1923a,b,c,d, 1930)]. Onslow observed more than 50 iridescent animals to settle the conflict between these hypotheses (see Fig. 1.3a). Merritt measured the reflection spectra of tempered steel and the *Morpho* butterfly, and interpreted in terms of thin-layer interference. In 1923–27, Mason published a series of papers on various types of color-producing structures in animals and supported the interference theory (see Fig. 1.3b). Thus, the theory based on the interference of light gradually increases its momentum, which finally kept away the interest of physicists. Independently of the tide in Europe, an Indian physicist Raman investigated the origin of the iridescent colors found in shells and birds through optical measurements [Raman (1934a,b,c)].

The theories that deserve special mention in this period are those developed by Lord Rayleigh. As described above, he developed the mathematical expressions for Rayleigh scattering [Rayleigh (1871a,b)], for reflection at a surface having a gradually changing refractive index [Rayleigh (1880)], and for multilayer reflection [Rayleigh (1917)]. On the other hand, a German scientist Mie showed a general expression for so-called *Mie scattering* in his paper [Mie (1908)].

1.2.2 *Growth Period (1940s–1980s)*

The complete understanding of the microstructures that contributed to the structure-color came after the invention of electron microscope. In 1939, the first attempt was made to clarify the mechanism of non-iridescent blue coloring in the bird feathers. Frank and Ruska applied the first marketable electron microscope, which was developed in this year, and found spongy structure consisting of keratin and air on the inner wall of the medullary cell of blue feather of a bird pitta [Frank and Ruska (1939)]. Anderson and Richards [Anderson and Richards (1942)], and Gentil [Gentil (1942)] investigated the scales of famous *Morpho* butterflies. Their observations revealed a surprisingly complicated structure on a tiny scale of the butterfly wing, which accelerated the structural study on a nanometer scale.

Many biologists attempted to elucidate the structures causing iridescence and accumulated the enormous data. Sophisticated microstructure was reported by Greenewalt on the feathers of humming birds, which had a form of pancake and piled in several layers within a barbule of the feather [Greenewalt *et al.* (1960a,b); Greenewalt (1960)]. Durrer investigated the feather of peacock and found two-dimensional lattice structure made of melanin granules [Durrer (1962)]. He advanced the microscopic observations on more than 100 bird species and classified their microstructures according to the shape of melanin granule and its arrangement [Durrer (1977)].

Structural coloration due to helicoidal structure analogous to cholesteric liquid crystal was found in insect's outer layers by Neville *et al.* They measured the reflection properties of various scarab beetles and found most of them showed high reflectivity for left-circular polarization [Neville and Caveney (1969)]. Particularly a golden scalab, reported before by Michelson [Michelson (1911)], showed a special structure that allowed the reflection for both left-

and right-circular polarizations. Beautiful multilayered structure was found in a kind of jewel beetle by Durrer [Durrer and Villiger (1972)]. This study led to the detailed structural study on tiger beetles by Schultz and Rankin in the 1980s [Schultz and Rankin (1985a)]. They also reported the development of multilayer structure before and after ecdysis [Schultz and Rankin (1985b)].

The sophisticated structure found in the *Morpho* butterfly was further investigated by Lippert and Gentil [Lippert and Gentil (1959)] and Ghiradella [Ghiradella (1974, 1984)], and the regular lamellar structure was found to be present on each ridge running in parallel on a butterfly scale. Ghiradella performed extensive studies on the development of UV-reflecting scale of the alfalfa butterfly, which showed a structure analogous to the *Morpho* butterfly [Ghiradella (1974)]. She found that cuticulin deposition took place nearly equidistantly in a flattened scale-forming cell, which grew with increasing height and narrowing, showed buckling first at the top, which grew into a lamellar structure that was followed by filling the protein epicuticle. Further, she investigated the structural origin of colors in butterfly scales and classified them into several types, which contributed to the specific appearance of the butterfly wing [Ghiradella (1991, 1998)].

The structure of opal was investigated by Sanders using an electron microscope, and spherical particles of amorphous silica with the diameter of 150–350 nm were found to be arranged in fcc crystalline form [Jones *et al.* (1964); Sanders (1964)]. Later, Darragh *et al.* showed that each sphere consisted of fine spheres of 30–40 nm diameter, which were arranged in a way of double or triple shell [Darragh *et al.* (1964)]. The first photonic crystal in living thing was discovered by Morris in a wing of a small green butterfly, *Callophrys rubi*, which was reported as air-filled spheres arranged on a simple cubic lattice [Morris (1975)]. Similar photonic crystals were discovered one after another since then. Ghiradella and Radican also reported the electron microscopic observation on the same butterfly [Ghiradella and Radican (1976)]. Ghiradella also performed the developmental study on a butterfly scale equipped with photonic crystal [Ghiradella (1989)] and reported a photonic crystal in a scale of weevil [Ghiradella (1984)].

On the other hand, Bernhard *et al.* discovered regular projections on corneal surface of moth [Bernhard and Miller (1962); Bernhard *et al.* (1965); Bernard and Miller (1968); Bernhard *et al.* (1968)], which was later called *moth-eye structure* and became one of the important bases for photonics. The moth-eye structure in insects was considered to have a function of anti-reflection to incident light in order to hide themselves against predators and also to increase the transmission of light into the eyes. Strangely enough, the moth-eye structure was found to be distributed among only limited species that were even distant in a genealogical chart, which arouse various hypotheses [Bernhard *et al.* (1970)].

At around the same time, Kawaguchi reported that guanine crystal platelets were involved in the outer layer of the fish skin [Kawaguti (1965)]. These platelets were found to have a function of highly reflecting multilayer, which were extensively studied by Denton and Land [Denton and Land (1967, 1971)]. Such structures were later found widely in integuments and eyes of animals, and are now known as *animal reflector* [Land (1972); Parker (2000)]. The variation of color in the iridescent cell of fish to the ambient illumination was investigated by Lythgoe and Shand, which was thought to be due to a change in the interval between adjacent regularly arranged platelets [Lythgoe and Shand (1982)].

These biological investigations stimulated the attempts to mimic such microstructures. The first attempt to fabricate the moth-eye structure was made in 1973 by Clapham and Hutley, who used a holographic photoresist method to obtain the 2D pattern with a pitch of 210 nm [Clapham and Hutley (1973)]. The method to fabricate synthetic opal was proposed in 1964 and came into the market in 1972. Together with the attempts to fabricate the structures, the methods for analyses to elucidate the optical properties also progressed little by little. Huxley presented a paper to calculate the multilayer reflection [Huxley (1968)], which was the improvement of the work reported by Lord Rayleigh [Rayleigh (1917)]. The transfer matrix method was described in the famous book *Principles of Optics*, by Born and Wolf, and became a general method to calculate the multilayer interference [Born and Wolf (1959)]. The Mie theory originally applied to a spherical dielectric particle was extended to a cylinder with infinite length, as described in van de Hulst's book *Light Scattering by Small Particles* [van de Hulst (1957)], and further extended to concentric spheres and infinite cylinders.

1.2.3 *Developing Period (1990s–)*

In spite of the above progresses in growth period, the physical interpretation of structural colors did not proceed much, since Lord Rayleigh proposed a theory of the multilayer interference. Only recently, structural colors have been a subject of extensive studies because their applications have been rapidly growing in many industrial fields related to vision such as painting, automobile, cosmetics, display and textile. It has been soon noticed that simple multilayer interference no longer reproduces actual appearance. Furthermore, recent research studies have revealed that even a very simple structure in nature has a surprisingly multiple function, which by far exceeds our speculation.

The primary motivation of this progress was provided by a couple of papers published in 1987 by Yablonovitch and Jones, respectively. In this paper, Yablonovitch predicted that 3D array of periodic dielectric scatterers possessed a photonic band gap, within which spontaneous emission would be strongly inhibited [Yablonovitch (1987)]. He aimed to fit the energy of the photonic band gap with that of electronic band edge to inhibit the radiative recombination in semiconductor, which would lead to a significant effect in solid-state physics and laser technology. On the other hand, John hypothesized that 3D photonic crystal with moderate disorder would provide a predictable and systematic observation of strong photon localization even within nondissipative material [John (1987)]. These somewhat specific predictions, however, strongly stimulated the research field of photonics and accelerated a trend to search for new photonic materials and to fabricate 3D photonic crystals.

Research studies on structural colors have also accelerated explosively. This acceleration is supported on the one hand by the advance in measurement technologies and on the other hand by the development of a variety of calculation methods. In the former, various apparatuses for optical measurement has been constructed on a laboratory level, and has also come into the market, which enables us to measure BRDF (bidirectional reflectance distribution function) in order to characterize the visual appearance of matter. In the latter, various calculation methods such as FDTD (finite-difference time-domain), Pendry's transfer matrix method, and RCWA (rigorous coupled wave analysis) are available on a personal computer level. These environmental improvements in addition to the demands for new photonic materials have greatly promoted the search for the natural photonic materials and contributed to the development of a new technological field of "biomimetics".

Although various living things assuming structural colors have been investigated in the last ten years, traditionally the most important subject will be to clarify the mechanism of the *Morpho* blue. Thus, let us pursue the recent research on the *Morpho* butterfly as a typical example. After a long period of biological investigations, the detailed optical measurement was performed by Tabata *et al.* [Tabata *et al.* (1996)]. They measured the optical properties of the *Morpho* wings using an angle- and wavelength-resolved reflection spectrometer and analyzed the high reflectivity of the wing using a multilayer interference model. Later, they mimicked the multilayer found in a *Morpho* scale and fabricated new structurally colored fiber, in which 61 alternate layers of nylon 6 and polyester were incorporated [Iohara *et al.* (2000)].

The measurement of the angle-resolved transmission and reflection properties on a single scale level was reported by Vukusic *et al.* [Vukusic *et al.* (1999)]. They found that the reflectivity from a scale amounted to 70% and the reflection pattern was divided into two distinct lobes. In the transmission side, they found clear diffraction spots, that was reasonable by considering the spacing of the ridges. On the other hand, our group noticed the irregular structures inherent to the natural products and insisted that the interplay between regularity and irregularity was extremely important to understand the strongly blue and yet remarkably diffusive nature of the *Morpho* wing [Kinoshita *et al.* (2002a,b)]. Berthier *et al.* performed the optical measurement on the wings of totally 14 *Morpho* species and classified them according to the size of cover scale, the number and tilt angle of lamellae and the BRDF pattern obtained by scanning the direction of detection hemispherically over a wing [Berthier *et al.* (2003, 2006)].

The FDTD algorithm is an effective method to calculate the scattering problem due to a complicated structure. Plattner employed this method to extract the scattering feature using model *Morpho* structures and found that a strong extinction of specular reflection occurs when alternate lamellar structure was employed [Plattner (2004)]. Zhu *et al.* employed nonstandard FDTD algorithm to calculate the electromagnetic fields using an intact scale structure directly obtained from a transmission electron micrograph and found that densely distributed ridges had a function to uniformize the reflection intensity irrespectively of incident polarizations [Zhu *et al.* (2009)]. Further, Kambe *et al.* proved experimentally that the alternate lamellar structure contributed to the retroreflection of blue colors [Kambe *et al.* (2011)]. Thus, we can say that the study on the *Morpho* butterfly has come into the world of exact science.

Similar interesting topics are biological photonic crystal and seemingly irregular network structure found in birds and insects. As for the former, it has been recently clarified that biological photonic crystals are widely distributed in animal world such as butterflies, weevil, long-horn, birds, sea mouse, jellyfish and so on. However, it is soon clear that these are not complete and always include domain and disorder in themselves, which may function to diffuse the color and reflection direction to accommodate to their struggles for existence. Recently, it is noticed that these photonic structures can be well expressed by a gyroid structure [Michielsen and Stavenga (2008)], which is deeply connected with their forming processes and the detailed experiment and analyses are now in progress.

As for the latter, light scattering has been believed as a cause of non-iridescent bluish colors in a wide variety of birds and dragonflies, whose microstructures were characterized by random network of fibers or random packing of spheres. Prum *et al.* conducted a spatial Fourier transformation of the TEM image of animals and found a clear ring structure around the origin in wave vector space [Prum *et al.* (1998)]. If the random structure with a wide variety of sizes were distributed, a Gaussian-like distribution would be obtained around the origin. The presence of a ring structure indicated the presence of short-range order hidden in seemingly random structure. Various optical, X-ray and computational studies are now being performed to clarify their optical characteristics and formation mechanism.

Seeing from an industrial aspect, we will find that the idea of photonics gives a revolutionary development to handle light by various newly developed devices such as waveguide, memory, switching, display and so on. In the field of structural colorations, various pigments have appeared, which usually contain flakes with the surfaces coated by metal and metal oxide to increase the reflectivity of light and the variability of color appearance using the mechanisms of thin-layer interference and diffraction grating. Synthetic opal will be fabricated first as a jewel but in later times, as ideal 3D photonic materials after transforming into inverse opal. Moth-eye structure is one of the most promising method to reduce the surface reflection for solar panel.

Thus, during the long history of bionanophotonics, we can say that a period from the end of the 19th century to the beginning of the 20th century corresponds to the dawn of bionanophotonics, while a period from a time of the invention of electron microscope to a time when papers on photonics were published, does to a growth period and a developing period begins after that. Although from now on, new natural photonic materials will be discovered one after another and the applications of such natural products to nanophotonic devices will be rapidly growing, the present author deeply feels the necessity to reconsider the principle of such natural photonic materials and their physical and/or biological meanings from a purely scientific basis. The present book is aimed for such students and researchers, who want to learn the bionanophotonics from the fundamentals.

Chapter 2

Light and Color

2.1 Fundamental Properties of Light

2.1.1 *What Is Light?*

Bionanophotonics is a consequence of complicated interaction between light and minute structure. Before entering into its details, it will be meaningful to build up an image of what light is. First we have to create an image of the wave nature of light. We sometimes notice that the sun's ray peeps through wisps of cloud. In such a case, we feel that light is a kind of ray that travels straight through a space. When light illuminates a matter of a size of 1 μm or less, or reflects from a thin layer, however, the wave nature of light will definitely come out. For example, when light passes through a narrow hole, the light does not travel straight, but its direction will be slightly uncertain, the degree of which is dependent on the size of a hole. When the hole is small enough, the uncertainty of the propagation direction will be prominent, whereas it will not be so when light passes through a large hole. This phenomenon is called *diffraction of light.*

On the other hand, if two monochromatic light beams with the same wavelengths and the same propagation directions are super-

Bionanophotonics: An Introductory Textbook
Shuichi Kinoshita
Copyright © 2013 Pan Stanford Publishing Pte. Ltd.
ISBN 978-981-4364-71-3 (Hardcover), 978-981-4364-72-0 (eBook)
www.panstanford.com

posed with each other, they will cause *interference phenomenon* resulting in the increase or decrease of the total intensity. The colorings noticed in soap bubble and oil film on the surface of water are just due to the effect of interference. The interference of light is frequently employed to enhance a certain color specifically under the presence of microstructures such as multilayer and photonic crystal. Another example for the wave nature of light is *polarization of light*, which directly expresses the oscillatory nature of light and indicates the direction of oscillation with respect to the propagation direction. All these phenomena are explainable fully in terms that light is a wave. Thus, we unconsciously come across the wave nature of light in our daily life.

Light as a wave can be compared conveniently with sound. The sound is also a kind of wave that propagates by vibrating the atmosphere or materials. We can hear the sounds whose frequencies approximately lie between 20 Hz and 15000 Hz, where Hz (hertz) is the SI unit to express a frequency of vibration per second. When the frequency of sound is high, we perceive it as a high-pitched sound, while when it is low, we do as a low-pitched one. Thus, we can say that the variation in the frequency generates a variety of sound pitches, which results in voice and music in the end. Above the highest limit of the frequency, we cannot hear the sound any more and such a sound is called *supersonic wave*, while when it is too low, we cannot also hear it as a sound but can feel as a vibration. Such a sound is called ultra *low-frequency sound*.

The same is true for light. Namely, we can perceive light when its frequency ranges from 3.9×10^{14} to 7.9×10^{14} Hz (Note its frequency is incomparably higher than that of sound). When the frequency of light is higher, we feel blue, while when it is lower, we feel red (see Fig. 2.1). Thus, we perceive the difference of color as that of the frequency. If the frequency becomes higher than the limit, we cannot see the color any more. The light in this frequency is called *ultraviolet ray*. Although we cannot see the ultraviolet rays, they destroy chemicals within our body so that we can actually feel its existence. When the frequency becomes much higher, such kinds of light are called *X-ray* ($10^{18} \sim 10^{20}$ Hz) and γ-*ray* (10^{20} Hz$\sim$). Though these rays are far from feeling light, they also belong to light.

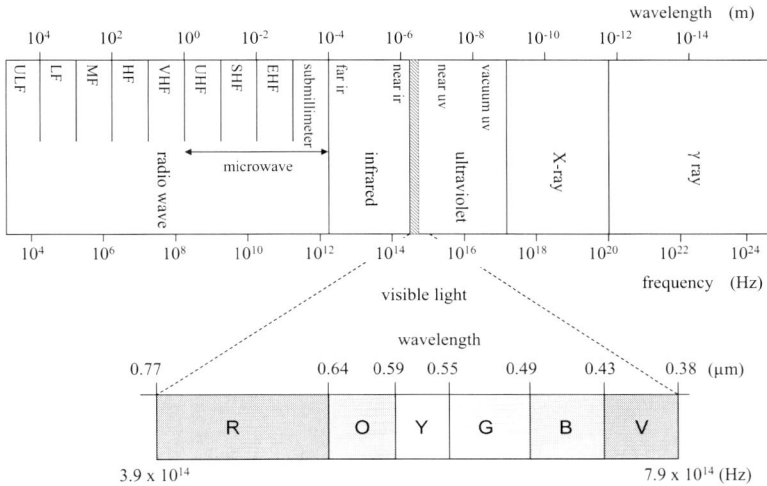

Figure 2.1   Wavelength and frequency of electromagnetic wave.

On the other hand, when the frequency of light becomes lower than the limit, we cannot see it as well. Such kind of light is called *infrared ray*. In a similar way as in ultraviolet ray, we feel its existence by warming our body through the absorption of infrared ray. When the frequency of light becomes much lower, such a wave is called *microwave* ($10^9 \sim 10^{12}$ Hz) and is used to heat up food in a microwave oven. The light with much lower frequency is generally called *radio wave* and is used as broadcasting and telecommunication. Thus, if we overview the light with its frequency as in Fig. 2.1, we notice light is variously called with its frequency and is utilized in various ways. We usually name all of these waves generically as *electromagnetic wave* rather than light.

Let us examine how light is connected with electromagnetic properties. For this purpose, it is convenient to consider a generation process of light. Let us consider an atom as a source of light. An atom consists of a nucleus and electrons surrounding it. For simplicity, we consider a case that a single electron of a negative charge revolves about a nucleus of a positive charge. The electron is assumed to behave like an electron cloud surrounding the nucleus as is shown in Fig. 2.2a. Since the centers of mass of both particles

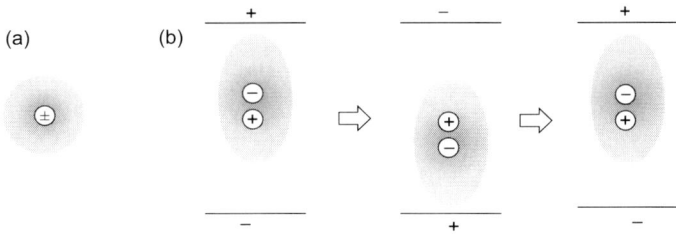

Figure 2.2 (a) An atom and (b) that with an oscillating electric field applied to it. The atom consists of a nucleus and an electron cloud, whose centers of mass are coincident with each other. When an oscillating electric field is applied, the electron cloud oscillates following the electric field, while a nucleus remains unmoved, which makes an oscillating dipole.

originally coincide with each other, they seem to be neutral when we see them from far away. Let us apply an oscillating electric field to this atom. The frequency of the oscillation is so fast that the heavy nucleus does not follow it and remains unmoved, while the light electron can follow the oscillation. Thus, the center of mass of each particle will deviate slightly and hence the positive and negative charges will be separated in space and with time as shown in Fig. 2.2b. We call a state with separated charges as *polarization* and a pair of these charges as *electric dipole* or simply *dipole*. Since the applied electric field oscillates with its direction changing alternately, the direction of a dipole[a] also changes following its oscillation.

If we observe such an oscillating dipole from far away, what will happen? Instead of simply observing, we place a positive point charge and consider a force operating on this change. It is easily understood that the positive charge on the nucleus gives a repulsive force on the point charge, while the negative charge on the center of mass of the electron cloud gives an attractive force. Then, according to a rule of the composition of forces, the direction of the force applied on the point charge is obtained as that nearly perpendicular to the direction of observation (see Fig. 2.3a). If the distance between the dipole and the point charge is far enough, the direction of the resultant force is regarded to be completely perpendicular to the

---

[a]The direction of a dipole is defined as a position of a positive charge measured from that of a negative charge.

(a)

polarization   electric field

(b)

current   magnetic field

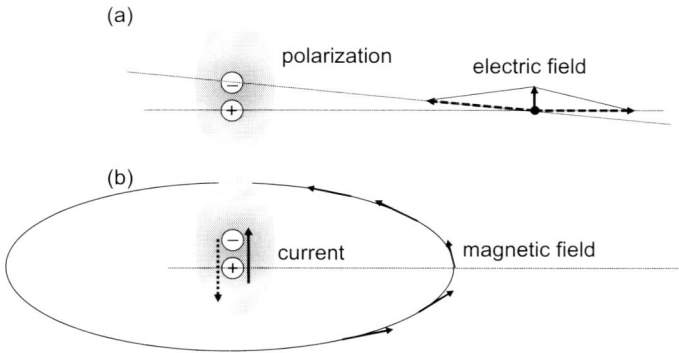

Figure 2.3   Oscillating dipole that generates the electric and magnetic fields.

direction of observation. This means the oscillating electric field is actually generated at a point far from the dipole, whose direction is perpendicular to the direction of observation. This is true for an arbitrary point when it is only placed far from the dipole, but the magnitude of the fields will be maximized when the directions of the dipole and observation make an angle of 90° and will decrease with decreasing angle according to a sine function (see Fig. 2.4a).

Since the electron cloud repeatedly moves up and down around the nucleus, the electric current is inevitably generated. This current gives a magnetic field at the same point, which is directed along a circumference shown in Fig. 2.3b and is again perpendicular to the direction of observation. It is also perpendicular to that of the electric field. Thus, the oscillating dipole creates the oscillating electric and magnetic fields simultaneously[a].

When the electric and magnetic fields thus generated at various points oscillate in synchronism with the oscillation of the

---

[a]It should be noted that this intuitive explanation is actually not correct when we use it to explain the origin of light. This is because the electric and magnetic fields are assumed to be induced instantaneously even at a distant place. This assumption corresponds to a limit of $c \to \infty$ in Eqs. (6.28) and (6.26), where c is the light velocity in vacuum. Actually, the electromagnetic wave is resulted from a fact that these fields should propagate with the light velocity and hence corresponds to the terms that are nonzero only when c is finite and also inversely proportional to the distance from the oscillating dipole. However, this explanation will help us to understand how the oscillatory electric and magnetic fields are generated by the oscillatory motion of a charged particle.

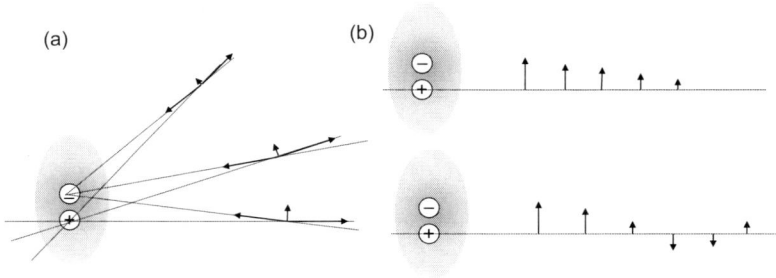

Figure 2.4 (a) Electric fields induced at various positions. (b) The cases where electric field is induced instantaneously at every position (upper) and with a time delay determined by the light velocity (lower) .

dipole, these fields cannot propagate as a wave and only vary their amplitudes according to the Coulomb's law (see the upper illustration in Fig. 2.4b). In order to let these fields propagate like a wave, it is necessary to assume that the operation of the Coulomb's law is anyway time-delayed and should propagate with the speed of light. In such a case, both the electric and magnetic fields will oscillate with the time delay dependent on the distance from the dipole and will propagate as a wave (see the lower illustration in Fig. 2.4b). This is exactly the light. Thus, the light is a wave accompanying the oscillating electric and magnetic fields, that is, electromagnetic field.

Then, the next question will be aroused. What is an origin of the external electric field that oscillates with such a high frequency. Normally, there are two origins for that. One is that the electric field of light itself is the origin, and the other is a case that an electron cloud spontaneously oscillates without applying external field. The electric field of the light so rapidly oscillates that it can also oscillate the electron cloud of an atom and forces to emit another electromagnetic field. This is the case of *light scattering* and *stimulated emission*. On the other hand, light emission due to the spontaneous oscillation of the electron cloud is called *spontaneous emission*, which is deeply connected with the quantum nature of light. However, we will make no mention of it because it is clearly beyond the scope of this book.

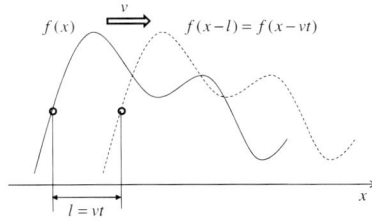

Figure 2.5 Mathematical expression for a moving wave.

2.1.2 *Mathematical Expression for Wave*

First, let us express a wave using a mathematical formula. The essential feature of wave is that it propagates in a space without changing its shape. Consider a wave expressed by a function $f(x)$ as shown in Fig. 2.5. If we transfer $f(x)$ without changing its shape to a positive direction along the x-axis by an amount of l, then the transferred function should be expressed as $f(x - l)$. When a wave propagates toward a positive direction of the x axis with a velocity of v, l should be time-dependent and is described as $l = vt$. Thus, the wave propagating toward the positive direction is simply expressed as $f(x - vt)$. In a similar manner, a wave propagating toward the negative direction is obtained by putting $v \rightarrow -v$ as $f(x+vt)$. These are general mathematical expressions for waves in one-dimensional (1D) space.

If we use a cosine function instead of f, a cosine wave propagating toward the positive direction will be expressed as $\cos k(x - vt)$, where k is a constant. The constant k is called *wave number* and can be determined easily. A cosine wave is a periodic function with a period of 2π. On the other hand, periods of a wave in space and time are called *wavelength* λ, and *period* T, respectively. Thus, if we put $t = 0$, then the relation $k\lambda = 2\pi$ is easily derived, which gives $k = 2\pi/\lambda$. Similarly, if we put $x = 0$, then the relation $kvT = 2\pi$ is derived and then $k = 2\pi/(vT)$ is obtained. Since T is expressed as $T = 1/v$ using the *frequency* v, the above relation is rewritten as $k = 2\pi v/v = \omega/v$, where $\omega = 2\pi v$ is called *angular frequency*. Thus, the cosine wave is generally expressed as $\cos(kx - \omega t)$. Furthermore, in case of light wave, if the velocity in

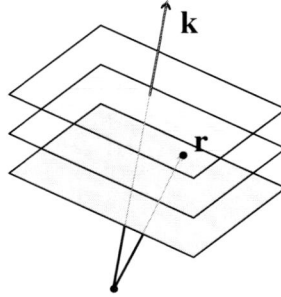

Figure 2.6   Propagation of wave in three-dimensional space.

vacuum is put as c, then the ratio of the velocity in vacuum to that in the medium is expressed as $c/v \equiv n$, where n is called *refractive index*.

Next, we regard $f(x - vt)$ as a function of two variables of x and t, and put $y(x, t) = f(x - vt)$. Then, a partial derivative with respect to x becomes $\partial y(x, t)/\partial x = f'(x - vt)$ and that with respect to t becomes $\partial y(x, t)/\partial t = -vf'(x - vt)$, where $f'(u) = \mathrm{d}f(u)/\mathrm{d}u$. The second partial derivative with respect to x becomes $\partial^2 y(x, t)/\partial x^2 = f''(x - vt)$ and that with respect to t becomes $\partial^2 y(x, t)/\partial t^2 = v^2 f''(x - vt)$, where $f''(u) = \mathrm{d}^2 f(u)/\mathrm{d}u^2$. Combining these relations by eliminating $f''(x - vt)$, we obtain

$$\frac{\partial^2 y}{\partial t^2} = v^2 \frac{\partial^2 y}{\partial x^2}, \tag{2.1}$$

which corresponds to a *wave equation* in 1D system, and y is called *wave function*. Completely the same equation is derived for $f(x+vt)$. Thus, both $f(x - vt)$ and $f(x + vt)$ are two independent particular solutions for the wave equation of Eq. (2.1).

Consider that x in a wave function $y(x, t)$ can be expressed in terms of $x = \mathbf{e}_x \cdot \mathbf{r}$, where $\mathbf{e}_x$ and $\mathbf{r}$ are a unit vector directing to the x axis and a position vector in three-dimensional (3D) space. Then, it is easy to extend the wave function in 1D system into that in 3D system. For example, a cosine wave expressed as $\cos(kx - \omega t) = \cos(k\mathbf{e}_x \cdot \mathbf{r} - \omega t)$ is extended to $\cos(\mathbf{k} \cdot \mathbf{r} - \omega t)$, where the propagation direction of wave is represented by a vector $\mathbf{k} \equiv k_x \mathbf{e}_x + k_y \mathbf{e}_y + k_z \mathbf{e}_z$. The meaning of this expression is easy to understand, if one sees Fig. 2.6. The inner product of $\mathbf{k} \cdot \mathbf{r}$ becomes constant, if a vector $\mathbf{r}$ moves within a plane perpendicular to $\mathbf{k}$. If we put $t = 0$, $\mathbf{k} \cdot \mathbf{r} = 2m\pi$, that is, $k\mathbf{e_k} \cdot \mathbf{r} = 2m\pi$, determines the periodicity of the wave, where $\mathbf{e_k}$ is a unit vector along the vector $\mathbf{k}$ and m is an integer. Hence putting $\mathbf{e_k} \cdot \mathbf{r} = \lambda$, we obtain the relation $k\lambda = 2\pi$ for $m = 1$, which leads to the same relation as above, i.e. $k = 2\pi/\lambda$. Accordingly, as in the 1D case, the cosine wave in 3D space is generally expressed as

$$y(\mathbf{r}, t) = y_0 \cos(\mathbf{k} \cdot \mathbf{r} - \omega t), \qquad (2.2)$$

where y_0 is an amplitude of the wave and $\mathbf{k}$ is called *wave vector*. Thus, in 3D space, a plane-like wave (called *plane wave*) propagates along the direction of $\mathbf{k}$ with a spatial period of λ.

Let us construct a wave equation in 3D space as has been done in 1D space. Consider a wave function expressed by $g(\mathbf{k} \cdot \mathbf{r} - \omega t)$ after the example of $\cos(\mathbf{k} \cdot \mathbf{r} - \omega t)$. If we spatially differentiate the wave function in 3D space using a differential operator $\nabla = (\partial/\partial x, \partial/\partial y, \partial/\partial z)$, the first derivative of the wave function becomes $\nabla g = \mathbf{k}g'$, while that with respect to time becomes $\partial g/\partial t = -\omega g'$, where we put $g' = dg(u)/du$. In a similar manner, the second derivative becomes $\nabla^2 g = \nabla \cdot \mathbf{k}g' = \mathbf{k}^2 g''$, while that with respect to time becomes $\partial^2 g/\partial t^2 = \omega^2 g''$. Thus, eliminating g'' from the two relations, we obtain

$$\frac{\partial^2 g}{\partial t^2} = v^2 \nabla^2 g, \qquad (2.3)$$

where we put $v^2 = \omega^2/\mathbf{k}^2$.

2.1.3 Complex Representation of Wave

Since a wave actually exists in a real world, the amplitude of a wave should be, of course, expressed by a real function of time and position. However, it is often convenient to express its amplitude as a complex function, because the mathematical calculation becomes considerably simplified. It is a custom to perform the calculation in a complex form and then to calculate the real part to obtain a real picture of wave. Here we will follow this way of calculation. Since a wave should be anyway expressed by a real function, it should be noted that there is a rule when one treats the wave in a complex form.

When a wave expressed in a real form is converted into that in a complex form, the following Euler's formula[a] is normally used,

$$e^{i\theta} = \cos\theta + i\sin\theta. \tag{2.4}$$

By using this formula, it is easy to express sin and cos functions in a complex form as

$$\sin\theta = \frac{e^{i\theta} - e^{-i\theta}}{2i}, \tag{2.5}$$

$$\cos\theta = \frac{e^{i\theta} + e^{-i\theta}}{2}. \tag{2.6}$$

When a wave in 1D space is described as $y(x, t) = y_0 \cos(kx - \omega t)$, the corresponding complex form is given as

$$\tilde{y}(x, t) = y_0 e^{i(kx - \omega t)}, \tag{2.7}$$

where we put a symbol $\tilde{y}$, which indicates that a function y is expressed in a complex form[b]. When a wave in a complex form is converted to that in a real form, it is easy to calculate the real part of the complex form such that $y(x, t) = \text{Re}\{\tilde{y}(x, t)\}$ or to calculate $y(x, t) = (1/2)\{\tilde{y}(x, t) + \tilde{y}^*(x, t)\}$, where $\tilde{y}^*(x, t)$ is a complex conjugate of $\tilde{y}(x, t)$.

When we calculate the interference and diffraction phenomena, it is quite convenient to use a complex form for the calculation. However, it should be noted that this rule cannot be applied, in principle, to the product of the amplitudes of waves. For example, if we calculate a square of the amplitude for a wave function given by $\tilde{y}(x, t) = y_0 \exp[i(kx - \omega t)]$, the following two methods will come to our mind intuitively: One is to simply calculate its square as $\tilde{y}^2(x, t)$, and the other is to calculate $\tilde{y}(x, t)\tilde{y}^*(x, t)$. However, neither of them gives the correct result. In fact, the former gives $y_0^2 \exp[2i(kx - \omega t)]$, which results in $y_0^2 \cos\{2(kx - \omega t)\}$ in a real form, and the latter

---

[a]The Euler's formula is easily proved by comparing the terms on both sides after expanding to the Taylor series. It is also proved by differentiating the relation $f(\theta) = (\cos\theta - i\sin\theta)\exp[i\theta]$ with respect to θ, which leads to $f'(\theta) = 0$, meaning $f(\theta)$ should be unity since $f(0) = 1$. Another proof is given by putting $g(\theta) = \cos\theta + i\sin\theta$ and differentiating the both sides, which gives $g'(\theta) = ig(\theta)$. Solving this differential equation immediately gives the Euler's formula by putting the condition $g(0) = 1$.

[b]In this book, a symbol ~ is added when a wave is treated generally, whereas in case of the electromagnetic wave, it is normally omitted.

simply gives y_0^2, both of which are completely different from the correct answer of $y_0^2 \cos^2(kx - \omega t)$.

However, in case of light, a special treatment is possible. This is mainly because the frequency of light is extremely high so that any detector can follow the oscillation directly. Thus, the operation of time averaging is anyway necessary. We often employ a concept of *cycle-averaging* in such a case. We consider two waves oscillating with the same angular frequency ω with the same wave number k, which are expressed by

$$y_1(x, t) = y_1 \cos(kx - \omega t + \phi_1), \qquad (2.8)$$
$$y_2(x, t) = y_2 \cos(kx - \omega t + \phi_2), \qquad (2.9)$$

where ϕ_1 and ϕ_2 are initial phases for their oscillations. The cycle-averaged product of the amplitudes of the two waves is expressed in terms of an average during one cycle of the oscillation and is defined as

$$\overline{y_1(x, t)y_2(x, t)} \equiv \frac{1}{T} \int_0^T dt\, y_1(x, t)y_2(x, t)$$

$$= \frac{\omega}{2\pi} \int_0^{2\pi/\omega} dt\, y_1 y_2 \cos(kx - \omega t + \phi_1) \cos(kx - \omega t + \phi_2)$$

$$= \frac{\omega}{2\pi} \int_0^{2\pi/\omega} dt\, y_1 y_2 \frac{1}{2} \{\cos[2(kx - \omega t) + \phi_1 + \phi_2] + \cos(\phi_1 - \phi_2)\}$$

$$= \frac{\omega}{2\pi} \cdot y_1 y_2 \cdot \frac{1}{2}\frac{2\pi}{\omega} \cos(\phi_1 - \phi_2) = \frac{1}{2} y_1 y_2 \cos(\phi_1 - \phi_2), \qquad (2.10)$$

where $\overline{A}$ means that A should be cycle-averaged and $T = 2\pi/\omega$ is a period of the oscillation.

On the other hand, the corresponding wave functions in the complex representation are expressed as

$$\tilde{y}_1(x, t) = y_1 \exp[i(kx - \omega t + \phi_1)], \qquad (2.11)$$
$$\tilde{y}_2(x, t) = y_2 \exp[i(kx - \omega t + \phi_2)]. \qquad (2.12)$$

If we calculate the following quantity

$$\frac{1}{2} \text{Re}\{\tilde{y}_1(x, t)\tilde{y}_2^*(x, t)\} = \frac{1}{2} y_1 y_2 \text{Re}\{e^{i(\phi_1 - \phi_2)}\}$$

$$= \frac{1}{2} y_1 y_2 \cos(\phi_1 - \phi_2),$$

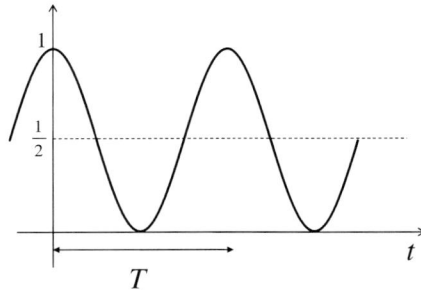

Figure 2.7 Principle of cycle averaging.

we notice that completely the same result is obtained as above. Thus, for the cycle-averaging, the following relation generally holds

$$\overline{y_1(x, t)y_2(x, t)} = \frac{1}{2}\text{Re}\{\tilde{y}_1(x, t)\tilde{y}_2^*(x, t)\}. \tag{2.13}$$

The factor of $1/2$ can be easily explained if we see Fig. 2.7, in which the average of $\cos^2(kx - \omega t)$ during one period is shown to give the factor of $1/2$.

2.1.4 *Superposition of Waves and Fourier Transformation*

It is generally known that waves obey the superposition principle, which indicates that waves can be superposed without any deformation. The essential point of this principle is based on a fact that a wave equation expressed by

$$\nabla^2 u = \frac{1}{v^2}\frac{\partial^2 u}{\partial t^2}, \tag{2.14}$$

is linear with respect to its solution. That is, if u and v are solutions of this wave equation, then their sum $u + v$ also becomes its solution. This is really a simple result, but will cause various peculiar phenomena through interference and diffraction of light when light interacts with microstructures. The detailed descriptions for these optical phenomena will appear later. Here, we will only describe a simple but truly important, mathematical treatment concerning the superposition of waves, which is known as Fourier transformation.

Two classes of Fourier transformation are known. One is discrete Fourier transformation applied to a finite region and the other is

Fourier transformation applied to an infinite region. The former is used for the analysis for a finite region or for a periodic function and is useful for actual computation, while the latter is more general, which is, in principle, applicable to an arbitrary function, as long as the mathematical restriction is permitted. Here, we will confine ourselves to the latter case.

The Fourier transformation is defined as follows: Consider a function $f(t)$ expressed by a sum of many oscillations with angular frequencies ω's with complex amplitudes $F(\omega)$'s. Then, $f(t)$ will be described as

$$f(t) = \int_{-\infty}^{\infty} d\omega \, F(\omega) e^{-i\omega t}. \tag{2.15}$$

This mathematical operation is called *Fourier transformation*. Thus, the superposition of oscillations with various angular frequencies and complex amplitudes creates a function $f(t)$. When an arbitrary function $f(t)$ is given, the corresponding amplitudes $F(\omega)$ is obtained as follows: The both sides of the above relation are first multiplied by $\exp[i\omega't]$ and then integrated with respect to t, which results in

$$\int_{-\infty}^{\infty} dt \, f(t) e^{i\omega' t} = \int_{-\infty}^{\infty} dt \int_{-\infty}^{\infty} d\omega \, F(\omega) e^{-i\omega t} e^{i\omega' t}$$

$$= 2\pi \int_{-\infty}^{\infty} d\omega \, F(\omega) \delta(\omega - \omega')$$

$$= 2\pi F(\omega').$$

Thus, by changing $\omega' \to \omega$,

$$F(\omega) = \frac{1}{2\pi} \int_{-\infty}^{\infty} dt \, f(t) e^{i\omega t}, \tag{2.16}$$

is obtained, which is called *inverse Fourier transformation*. Here, we have used the definition of a delta function

$$\delta(\omega) = \frac{1}{2\pi} \int_{-\infty}^{\infty} dt \, e^{i\omega t}, \tag{2.17}$$

and its properties

$$\int_{-\infty}^{\infty} d\omega \, F(\omega) \delta(\omega - \omega') = F(\omega'). \tag{2.18}$$

The similar expression is possible for a spatial Fourier transformation in 3D space as

$$g(\mathbf{r}) = \int d\mathbf{k} \, G(\mathbf{k}) e^{i\mathbf{k} \cdot \mathbf{r}}. \tag{2.19}$$

and its inverse Fourier transformation is defined as

$$G(\mathbf{k}) = \left(\frac{1}{2\pi}\right)^3 \int d\mathbf{r}\, g(\mathbf{r}) e^{-i\mathbf{k}\cdot\mathbf{r}}, \qquad (2.20)$$

where the integrations should be performed over the whole wave vector and spatial ranges, respectively.

2.2 Light as an Electromagnetic Wave

2.2.1 *Maxwell Equations and Wave Equations*

It was the 19th century when scientists reached a conclusion that light was a kind of electromagnetic wave. This was brought about by two great scientists who founded the basis of electromagnetic theory. The most important work was done by a Scottish physicist and mathematician, James Clerk Maxwell, who predicted the existence of electromagnetic wave through the famous Maxwell equations, while the other was done by a German physicist, Heinrich Rudolf Hertz, who performed an experiment to confirm that electromagnetic wave was actually emitted from electric oscillation.

Maxwell equations for electric and magnetic fields in medium are generally summarized as follows:

$$\nabla \times \mathbf{E} = -\frac{\partial \mathbf{B}}{\partial t}, \qquad (2.21)$$

$$\nabla \times \mathbf{H} = \mathbf{j} + \frac{\partial \mathbf{D}}{\partial t}, \qquad (2.22)$$

$$\nabla \cdot \mathbf{D} = \rho, \qquad (2.23)$$

$$\nabla \cdot \mathbf{B} = 0, \qquad (2.24)$$

with $\mathbf{D} = \epsilon\mathbf{E} = \epsilon_0\mathbf{E} + \mathbf{P}$ and $\mathbf{B} = \mu\mathbf{H} = \mu_0\mathbf{H} + \mathbf{M}$, where $\mathbf{E}$, $\mathbf{D}$, $\mathbf{H}$, $\mathbf{B}$, $\mathbf{P}$ and $\mathbf{M}$ are *electric field, electric displacement, magnetic field, magnetic flux density, polarization* and *magnetization vectors*, respectively. $\mathbf{j}$ and ρ are *current* and *charge densities*, both of which will be set at zero when a dielectric insulator without true charge is employed as a medium. ϵ, ϵ_0, μ and μ_0 are the *permittivities* of the medium and vacuum, and the *magnetic permeability* of the medium and vacuum, respectively.

For an uniform and isotropic insulator not having true charge, the above equations are further reduced to

$$\nabla \times \mathbf{E} = -\frac{\partial \mathbf{B}}{\partial t}, \tag{2.25}$$

$$\nabla \times \mathbf{B} = \epsilon\mu\frac{\partial \mathbf{E}}{\partial t}, \tag{2.26}$$

$$\nabla \cdot \mathbf{E} = 0, \tag{2.27}$$

$$\nabla \cdot \mathbf{B} = 0. \tag{2.28}$$

Employing these equations, we first operate $\nabla\times$ from the left of Eq. (2.25) and then insert Eq. (2.26) into it:

$$\nabla \times (\nabla \times \mathbf{E}) = -\frac{\partial}{\partial t}\nabla \times \mathbf{B}$$

$$= -\epsilon\mu\frac{\partial^2 \mathbf{E}}{\partial t^2}.$$

Using the vector relation

$$\nabla \times (\nabla \times \mathbf{E}) = \nabla(\nabla \cdot \mathbf{E}) - \nabla^2\mathbf{E}, \tag{2.29}$$

and Eq. (2.27), we obtain a partial differential equation:

$$\nabla^2\mathbf{E} = \epsilon\mu\frac{\partial^2 \mathbf{E}}{\partial t^2}. \tag{2.30}$$

In a similar manner, operating $\nabla\times$ from the left of Eq. (2.26) and then inserting Eq. (2.25) into it, we obtain the following relation:

$$\nabla \times (\nabla \times \mathbf{B}) = -\epsilon\mu\frac{\partial^2 \mathbf{B}}{\partial t^2}.$$

Using the same vector relation and Eq. (2.28), we obtain the completely equivalent equation corresponding to the magnetic flux density:

$$\nabla^2\mathbf{B} = \epsilon\mu\frac{\partial^2 \mathbf{B}}{\partial t^2}. \tag{2.31}$$

It should be noted that these wave equations have completely the same form as that derived before (Eq. (2.3)), if we put $v^2 = 1/(\epsilon\mu)$. In vacuum, the permittivity and permeability take values of ϵ_0 and μ_0, respectively. Thus, by considering that v is the velocity of the wave, the light velocity in vacuum is expressed as $c^2 = 1/(\epsilon_0\mu_0)$, that is, $c = 1/\sqrt{\epsilon_0\mu_0}$. Further, the ratio of the velocity in vacuum and medium gives the refractive index: $n \equiv c/v = \sqrt{\epsilon\mu/(\epsilon_0\mu_0)}$. In a usual case, the function of magnetic field working on the medium is sufficiently small compared with that of electric field. Hence, it is a custom to put $\mu \to \mu_0$ so that the function of the magnetic field on the medium will not be considered explicitly[a].

As is described in Sec. 2.1.2, particular solutions for wave equations of Eqs. (2.30) and (2.31) are expressed as plane waves and are described as

$$\mathbf{E} = \mathbf{E}_0 \exp[i(\mathbf{k} \cdot \mathbf{r} - \omega t)], \tag{2.32}$$

$$\mathbf{B} = \mathbf{B}_0 \exp[i(\mathbf{k} \cdot \mathbf{r} - \omega t)], \tag{2.33}$$

which means that a plane wave propagates toward the direction of $\mathbf{k}$ with the angular frequency of ω. The necessary condition for these expressions to be solutions of Eqs. (2.30) and (2.31) is obtained by inserting them into the wave functions. In this case, it is convenient to use the following rules applicable to a plane wave: $\nabla \times \to i\mathbf{k} \times$ and $\partial/\partial t \to -i\omega$. Thus, we obtain a relation $k^2 = \omega^2 \epsilon \mu$ from Eq. (2.30) or Eq. (2.31), which leads to $k = \omega/v$.

The electric and magnetic fields thus given are connected with each other owing to the Maxwell equations. By using the above rules, it is easy to derive the Maxwell equations for a propagating electromagnetic wave from Eqs. (2.25)–(2.28) as

$$i\mathbf{k} \times \mathbf{E} = i\omega\mathbf{B}, \tag{2.34}$$

$$i\mathbf{k} \times \mathbf{B} = -i\epsilon\mu\omega\mathbf{E}, \tag{2.35}$$

$$i\mathbf{k} \cdot \mathbf{E} = 0, \tag{2.36}$$

$$i\mathbf{k} \cdot \mathbf{B} = 0. \tag{2.37}$$

---

[a]Consider a case where an electromagnetic field is applied to an electron. The Lorentz force in this case is given as $\mathbf{F} = q(\mathbf{E} + \mathbf{v} \times \mathbf{B})$, where q is a charge of electron. Thus, the magnitudes of the electric and magnetic forces become qE_0 and qvB_0, respectively. From the relation of Eq. (2.38), $qvB_0 = qvE_0/c$ holds in vacuum. Hence, the ratio of the electric and magnetic contributions is expressed by a factor of c/v. Since $c \sim 3 \times 10^8$ (m/s) and $v \sim 10^6$ (m/s) for a Bohr electron, c/v amounts to ~ 300 and hence the magnetic force does not give a large contribution in a usual case. In the sense that the magnetic contribution will not be considered explicitly, μ is often replaced by μ_0. This approximation is actually valid because the relative permeability defined as μ/μ_0 is known to be quite close to unity for paramagnetic and diamagnetic materials, e.g. $\mu/\mu_0 = 0.999991$ for water and 0.999984 for graphite. However, it should be noted that for ferromagnetic materials, it normally becomes much larger than unity, e.g. for iron, $\mu/\mu_0 \sim 150$ is known as the initial static permeability.

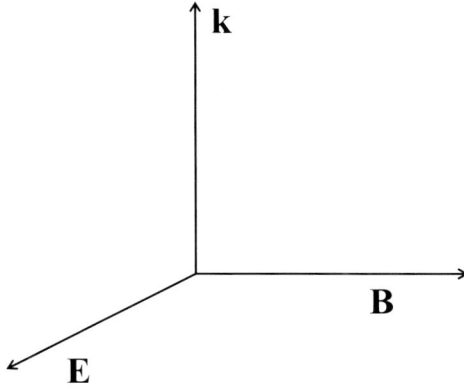

Figure 2.8 Relations among the wave vector and the electric and magnetic flux density vectors in isotropic medium.

From these equations and the feature of direct product, it is found that **k**, **E** and **B** are all orthogonal to each other (see Fig. 2.8), and **E** and **B** simultaneously oscillate in space and time, if **k**, and hence $\epsilon\mu$, is real. The relation between their magnitudes is derived from the relations $kE_0 = \omega B_0$ and $kB_0 = \epsilon\mu\omega E_0$, which leads to

$$E_0/B_0 = (\epsilon\mu)^{-1/2} = v, \tag{2.38}$$

where $|\mathbf{E}_0| = E_0$ and $|\mathbf{B}_0| = B_0$.

2.2.2 *Complex Refractive Index*

When a medium has movable charges, the Maxwell equation of Eq. (2.22) should be used instead of Eq. (2.26). Thus, the fundamental equations in this case become

$$\nabla \times \mathbf{E} = -\frac{\partial \mathbf{B}}{\partial t},$$

$$\nabla \times \mathbf{H} = \mathbf{j} + \frac{\partial \mathbf{D}}{\partial t}.$$

In case of monochromatic light wave traveling within a uniform medium, the wave function will be expressed by $\mathbf{E} = \mathbf{E}_0 \exp[i(\mathbf{k} \cdot \mathbf{r} - \omega t)]$ and the right-hand side of the second equation becomes

$$\mathbf{j} + \frac{\partial \mathbf{D}}{\partial t} = \mathbf{j} - i\epsilon\omega\mathbf{E} = (\sigma - i\epsilon\omega)\mathbf{E}$$

$$= -i\omega \left(\epsilon + i\frac{\sigma}{\omega} \right) \mathbf{E}, \tag{2.39}$$

where we have used Ohm's law of $\mathbf{j} = \sigma\mathbf{E}$ with σ electric conductivity. If we replace $\epsilon + i\sigma/\omega$ with ϵ in the last relation, newly defined permittivity ϵ becomes a complex number and is called *complex permittivity*.

Similar expression is possible when a polarization induced by oscillating external electric field does not oscillate simultaneously but delays somehow or other. Many reasons for the delay are known, because the polarizations are generated through various mechanisms such as orientations of molecules, ionic movements, atomic displacements within a molecule, electronic displacement and so on. In case of light in a visible region, the major reason for the delay comes from the electronic displacement near resonance because of the loss of energy through the interaction with the light field.

When the oscillation of the polarization is anyhow delayed as compared with that of the applied electric field, the polarization will be described as

$$\mathbf{P}(t) = \epsilon_0 \int_{-\infty}^{t} dt' \, \chi(t - t')\mathbf{E}(t'), \tag{2.40}$$

where $\chi(t)$ is *electric susceptibility* and a fact that the upper limit of the integration is described as t ensures the *causality*[a]. For an oscillating electric field expressed as $\mathbf{E}(t) = \mathbf{E}(\omega)\exp[-i\omega t]$, $\mathbf{P}(t)$ should have a form of $\mathbf{P}(\omega)\exp[-i\omega t]$ and Eq. (2.40) becomes

$$\mathbf{P}(\omega)e^{-i\omega t} = \epsilon_0 \int_{-\infty}^{t} dt' \, \chi(t - t')e^{-i\omega t'}\mathbf{E}(\omega)$$

$$\equiv \epsilon_0\chi(\omega)\mathbf{E}(\omega)e^{-i\omega t}, \tag{2.41}$$

---

[a]Causality assures that a polarization is induced only by electric field applied in the past. Sometimes $\chi(t)$ is expressed as $\chi(t) = 2\chi_\infty\delta(t) + \chi'(t)$, where the first term in the right-hand side expresses the instantaneous response, which does not obey the causality, with $\delta(t)$ a delta function, while the second term obeys the causality. The first term is considered to come from nonresonance electronic response and gives a contribution to the real part of complex admittance.

where we put

$$\int_{-\infty}^{t} dt' \, \chi(t-t') e^{-i\omega t'} = \int_{0}^{\infty} dt' \, \chi(t') e^{i\omega t'} e^{-i\omega t} \equiv \chi(\omega) e^{-i\omega t}, \quad (2.42)$$

which is called *complex admittance*.

Since the left- and right-hand sides of Eq. (2.40) should be real, $\chi(t)$ is also a real function of time. Hence,

$$\chi(\omega) = \int_{0}^{\infty} dt' \, \chi(t') e^{i\omega t'},$$

is a complex function and the relation $\chi^*(\omega) = \int_{0}^{\infty} dt' \, \chi(t')$ $\exp[-i\omega t'] = \chi(-\omega)$ holds generally. Further, the electric displacement vector under the application of an oscillating electric field is also expressed as $\mathbf{D}(\omega) \exp[-i\omega t]$ and hence $\mathbf{D}(\omega) = \epsilon_0 \mathbf{E}(\omega) +$ $\mathbf{P}(\omega) = \epsilon(\omega)\mathbf{E}(\omega)$ holds naturally, where $\epsilon(\omega) = \epsilon_0(1 + \chi(\omega))$ with the same relation as $\epsilon^*(\omega) = \epsilon(-\omega)$.

In either case, ϵ becomes a complex number with a function of ω. The refractive index in such a case is derived in a usual way as $n = \sqrt{\epsilon\mu/(\epsilon_0\mu_0)}$. Thus, the refractive index becomes also a complex number and will be expressed as[a]

$$n(\omega) = \sqrt{\epsilon(\omega)\mu/(\epsilon_0\mu_0)} \equiv \eta(\omega) + i\kappa(\omega), \quad (2.43)$$

which anyway becomes a function of ω. Inserting this relation into a 1D wave function of $\mathbf{E}(x, t) = \mathbf{E}_0 \exp[i(kx - \omega t)]$ with $k = n\omega/c$, we obtain

$$\mathbf{E}(x, t) = \mathbf{E}_0 \exp[i\omega(\eta x/c - t)] \exp[-\kappa\omega x/c].$$

As described in the next section, since the intensity of light is proportional to the square of the amplitude, the intensity of light propagating in a medium with a complex refractive index becomes

$$I \propto |\mathbf{E}_0|^2 \exp[-2\kappa\omega x/c] \equiv |\mathbf{E}_0|^2 \exp[-\alpha x],$$

which means that the light intensity decreases exponentially with a characteristic length of $1/\alpha$, where $\alpha \equiv 2\kappa\omega/c$ is called *absorption*

---

[a]When a complex representation of a wave is defined as $\exp[-i(\mathbf{k} \cdot \mathbf{r} - \omega t)]$, the corresponding refractive index should be defined as $\eta(\omega) - i\kappa(\omega)$.

coefficient[a]. Thus, the imaginary part of the refractive index is directly connected with the absorption of light.

2.2.3 Intensity of Light

The intensity of light I is defined using a Poynting vector

$$\mathbf{S} = \mathbf{E} \times \mathbf{H}. \tag{2.44}$$

The vector product of two oscillating quantities, for instance $\mathbf{A}(t)$ and $\mathbf{B}(t)$, with the same frequencies, which are on the order of light frequency, is usually evaluated by its cycle average as

$$\overline{\mathbf{A}(t) \times \mathbf{B}(t)} = \frac{1}{T} \int_0^T dt\, \mathbf{A}(t) \times \mathbf{B}(t) = \frac{1}{2}\mathrm{Re}\{\tilde{\mathbf{A}}(t) \times \tilde{\mathbf{B}}^*(t)\}, \tag{2.45}$$

where $T = 2\pi/\omega$ is a period of the oscillation. Putting $\mathbf{E} = \mathbf{E}_0 \exp[i(\mathbf{k} \cdot \mathbf{r} - \omega t)]$ and using Eq. (2.34), we obtain $\mathbf{H} = (\omega\mu)^{-1}\mathbf{k} \times \mathbf{E}_0 \exp[i(\mathbf{k} \cdot \mathbf{r} - \omega t)]$. This relation assures that the electric and magnetic fields are orthogonal to each other and hence the cycle-averaged intensity of light is expressed as

$$I \equiv \overline{|\mathbf{S}(t)|} = \overline{|\mathbf{E}(t) \times \mathbf{H}(t)|} = \frac{1}{2}E_0 H_0$$

$$= \frac{1}{2}\left(\frac{\epsilon}{\mu}\right)^{\frac{1}{2}} E_0^2 = \frac{1}{2}\sqrt{\frac{\epsilon/\mu}{\epsilon_0/\mu_0}}\epsilon_0 c E_0^2 = \frac{1}{2}\epsilon_0 m c E_0^2 \rightarrow \frac{1}{2}\epsilon_0 n c E_0^2, \tag{2.46}$$

where $E_0 \equiv |\mathbf{E}_0|$ and $H_0 \equiv (\omega\mu)^{-1}|\mathbf{k} \times \mathbf{E}_0|$ are the amplitudes of the electric and magnetic fields, respectively, and we put $m \equiv \sqrt{(\epsilon/\mu)/(\epsilon_0/\mu_0)} = n\mu_0/\mu$. Further, we have used the relation $k/(\omega\mu) = \sqrt{\epsilon\mu}/\mu = \sqrt{\epsilon/\mu}$. It should be noted that $|\cdots|$ in this case is used as the magnitude of a vector. The last result is obtained by putting $\mu \rightarrow \mu_0$ and hence $m \rightarrow n$.

---

[a]The absorption coefficients are variously defined. When the definition given here is employed, it is directly connected with *absorbance* or *optical density* (*OD*) as in the following: When light travels over a pass length of x, its intensity decreases according to a relation of $I(x) = I(0)\exp[-\alpha x]$, which leads to $\log_{10}[I(x)/I(0)] = -\alpha x \log_{10} e$. The absorbance A is defined as $A = -\log_{10}[I(x)/I(0)]$, which results in $A = \alpha x \log_{10} e = (2\kappa\omega x/c)\log_{10} e$. On the other hand, *molar extinction coefficient* ϵ is defined as $I(l) = I(0)10^{-\epsilon C l}$, where C and l are the concentration of absorber in molarity and a pass length in cm unit. Thus, ϵ is derived as $\epsilon = \alpha \log_{10} e/C = 2\kappa\omega \log_{10} e/(cC)$.

The light intensity in an absorbing medium is obtained in a similar way to that in a transparent medium. Since the electromagnetic properties in the absorbing medium are determined only by putting ϵ as a complex number, the Maxwell equations hold generally. The Maxwell equations of Eqs. (2.34) and (2.35) in this case become

$$\mathbf{H} = \sqrt{\frac{\epsilon}{\mu}} \mathbf{e}_k \times \mathbf{E}, \tag{2.47}$$

$$\mathbf{E} = -\sqrt{\frac{\mu}{\epsilon}} \mathbf{e}_k \times \mathbf{H}, \tag{2.48}$$

where $\mathbf{e}_k$ is a unit vector directing to a vector $\mathbf{k}$. These relations show that the electric and magnetic vectors in the absorbing medium are still orthogonal to each other, but the oscillating phases of both fields differ owing to the presence of the complex permittivity, ϵ. This is because if we put $\epsilon = |\epsilon| \exp[i\phi]$, $\mathbf{E} \propto -\sqrt{\mu/|\epsilon|}\mathbf{e}_k \times \mathbf{H} \exp[-i\phi/2]$, where ϕ is obtained from $\tan\phi = \epsilon''/\epsilon'$ with $\epsilon = \epsilon' + i\epsilon''$.

Since the amplitudes of electromagnetic fields changes with the propagation, we put

$$\mathbf{E} = \mathbf{E}_0 \exp[i(\mathbf{k} \cdot \mathbf{r} - \omega t)] \equiv \mathbf{E}(\mathbf{r}) \exp[-i\omega t],$$
$$\mathbf{H} = \mathbf{H}_0 \exp[i(\mathbf{k} \cdot \mathbf{r} - \omega t)] \equiv \mathbf{H}(\mathbf{r}) \exp[-i\omega t].$$

The intensity of light is obtained as a cycle-averaged Poynting vector such that

$$
\begin{aligned}
I &= \overline{|\mathbf{E} \times \mathbf{H}|} \\
&= \frac{1}{2}\mathrm{Re}\{|\mathbf{E}(\mathbf{r}) \times \mathbf{H}^*(\mathbf{r})|\} \\
&= \frac{1}{2}\mathrm{Re}\left\{ \left| \mathbf{E}(\mathbf{r}) \times \left(\sqrt{\frac{\epsilon^*}{\mu}}\mathbf{e}_k \times \mathbf{E}^*(\mathbf{r}) \right) \right| \right\} \\
&= \frac{1}{2}\mathrm{Re}\left\{ \left| \sqrt{\frac{\epsilon^*}{\mu}} \left\{ \mathbf{e}_k(\mathbf{E}(\mathbf{r}) \cdot \mathbf{E}^*(\mathbf{r})) - \mathbf{E}^*(\mathbf{r})(\mathbf{e}_k \cdot \mathbf{E}(\mathbf{r})) \right\} \right| \right\} \\
&= \frac{1}{2}\mathrm{Re}\left\{ \sqrt{\frac{\epsilon^*}{\mu}}(\mathbf{E}(\mathbf{r}) \cdot \mathbf{E}^*(\mathbf{r})) \right\},
\end{aligned}
$$

where we have used a vector relation of $\mathbf{A} \times (\mathbf{B} \times \mathbf{C}) = \mathbf{B}(\mathbf{A} \cdot \mathbf{C}) - \mathbf{C}(\mathbf{A} \cdot \mathbf{B})$ and relations $\mathbf{e}_k \cdot \mathbf{E} = 0$ and $|\mathbf{e}_k| = 1$. Using $\sqrt{\epsilon\mu/(\epsilon_0\mu_0)} =$

$n = \eta + i\kappa$, we can further simplify the relation as

$$I = \frac{1}{2}\mathrm{Re}\left\{n^* \frac{\sqrt{\epsilon_0\mu_0}}{\mu}(\mathbf{E}(\mathbf{r}) \cdot \mathbf{E}^*(\mathbf{r}))\right\}$$

$$= \frac{1}{2}\mathrm{Re}\left\{n^* \frac{\epsilon_0 c\mu_0}{\mu}(\mathbf{E}(\mathbf{r}) \cdot \mathbf{E}^*(\mathbf{r}))\right\}$$

$$= \frac{1}{2}\eta\frac{\epsilon_0 c\mu_0}{\mu}(\mathbf{E}(\mathbf{r}) \cdot \mathbf{E}^*(\mathbf{r}))$$

$$\rightarrow \frac{1}{2}\eta\epsilon_0 c E_0^2 e^{-2(\kappa\omega/c)\mathbf{e}_k\cdot\mathbf{r}},$$

where the last relation is obtained by putting $\mu \rightarrow \mu_0$. It is immediately understood that the light intensity in the absorbing medium is related only to the real part of the refractive index, while the imaginary part contributes to the exponential decay of the intensity which is dependent on the position.

2.2.4 Fresnel's Law

2.2.4.1 Derivation of Fresnel's law

When a plane wave of light is incident on an interface between two isotropic media with different refractive indices, it will be reflected or refracted at the interface. The behavior of the light wave in such a case is summarized as Fresnel's law. Consider a case that a light wave is incident on an interface between two isotopic media designated as 1 and 2. We put the y axis to be vertical to the interface, while the xz plane lies in accord with the interface, as shown in Fig. 2.9. The light wave is assumed to be incident within the xy plane. We consider a rectangular region A enclosed by four straight lines of $x = x_1, x_2$, and $y = y_1, y_2$, and apply Eq. (2.25) to this area. Then, the integration of the z-component of Eq. (2.25) over the area A results in

$$\int_{x_1}^{x_2} dx \int_{y_1}^{y_2} dy (\nabla \times \mathbf{E})_z = -\int_{x_1}^{x_2} dx \int_{y_1}^{y_2} dy\, \mu \frac{\partial H_z}{\partial t}, \qquad (2.49)$$

and the left-hand side of this relation reduces to

$$\int_{x_1}^{x_2} dx \int_{y_1}^{y_2} dy \left(\frac{\partial E_y}{\partial x} - \frac{\partial E_x}{\partial y}\right)$$

$$= \int_{y_1}^{y_2} dy [(E_y)_{x=x_2} - (E_y)_{x=x_1}] - \int_{x_1}^{x_2} dx [(E_x)_{y=y_2} - (E_x)_{y=y_1}].$$

$$(2.50)$$

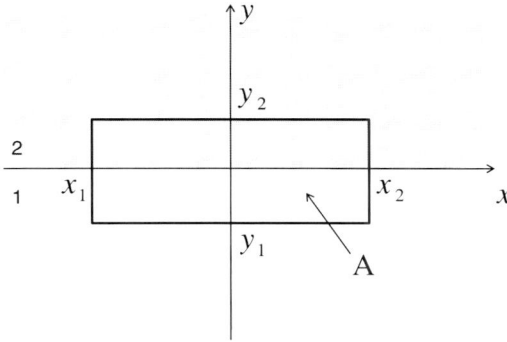

Figure 2.9 Calculation of the boundary conditions at an interface between two isotropic media, 1 and 2, with different refractive indices.

Let y_1 and y_2 approach zero, then the right-hand side of Eq. (2.49) and the first term in the right-hand side of Eq. (2.50) become zero. Thus, a relation

$$- \int_{x_1}^{x_2} dx \left[(E_x)_{y=+0} - (E_x)_{y=-0} \right] = 0, \qquad (2.51)$$

generally holds for arbitrary values of x_1 and x_2. This is only possible when the relation $[(E_x)_{y=+0} - (E_x)_{y=-0}] = 0$ holds generally, and hence $(E_x)_{y=+0} = (E_x)_{y=-0}$. In other words, in case light reflection and refraction take place at an interface perpendicular to y axis, E_x should be always continuous at the interface. In a similar manner, the continuity conditions are also proved to hold for E_z, H_x, and H_z (Exercise (1)).

By using this principle, various rules concerning the reflection and refraction of light at an interface are derived. Consider first that a plane wave of light is incident on an interface between two kinds of isotropic media with the refractive indices of n_1 and n_2, as shown in Fig. 2.10, which is then reflected and refracted at the interface. We describe the incident, refracted and reflected waves in complex forms as

$$\mathbf{E}_1 = \mathbf{E}_{10} \exp[i(\mathbf{k}_1 \cdot \mathbf{r} - \omega_1 t)], \qquad (2.52)$$

$$\mathbf{E}_2 = \mathbf{E}_{20} \exp[i(\mathbf{k}_2 \cdot \mathbf{r} - \omega_2 t)], \qquad (2.53)$$

$$\mathbf{E}_r = \mathbf{E}_{r0} \exp[i(\mathbf{k}_r \cdot \mathbf{r} - \omega_r t)], \qquad (2.54)$$

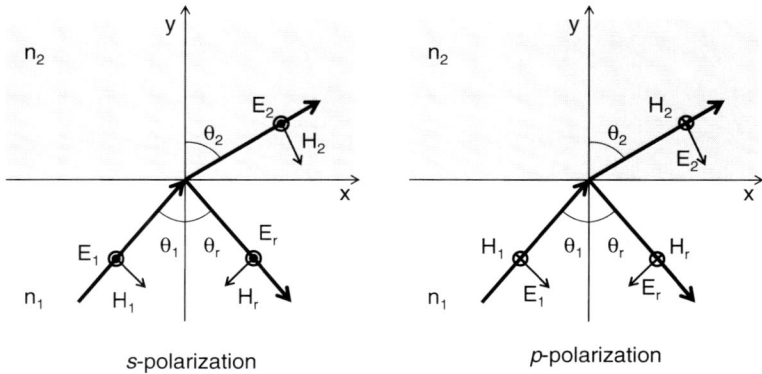

Figure 2.10 Definitions of electric and magnetic fields in cases of reflection and refraction of light at the interface. (left) *s*- and (right) *p*-polarization.

where we assume the wave vectors and angular frequencies of these waves are different from each other for a while, and discriminate them by suffices of 1, 2 and r, respectively.

Consider an arbitrary point on the interface corresponding to the xz plane, which is expressed by a position vector of $\mathbf{r}_0$. Since a plane wave is incident on a flat interface, it is natural to consider that the phase relationships between these three waves should be independent of the position. For example, the phase difference between the incident and reflected waves at a time t and a position $\mathbf{r}_0$ is generally expressed as

$$(\mathbf{k}_1 \cdot \mathbf{r}_0 - \omega_1 t) - (\mathbf{k}_r \cdot \mathbf{r}_0 - \omega_r t) = (\mathbf{k}_1 - \mathbf{k}_r) \cdot \mathbf{r}_0 - (\omega_1 - \omega_r)t. \quad (2.55)$$

When the phase difference becomes independent of time and position, $\mathbf{k}_1 \cdot \mathbf{r}_0 = \mathbf{k}_r \cdot \mathbf{r}_0$ and $\omega_1 = \omega_r$ should hold naturally.

In a similar manner, the following relations hold generally:

$$\omega_1 = \omega_r = \omega_2, \quad (2.56)$$

$$\mathbf{k}_1 \cdot \mathbf{r}_0 = \mathbf{k}_r \cdot \mathbf{r}_0 = \mathbf{k}_2 \cdot \mathbf{r}_0. \quad (2.57)$$

The first relation assures the conservation of angular frequency and we put simply $\omega_1 = \omega_r = \omega_2 \equiv \omega$. On the other hand, the important laws can be derived from the second relation. When the position vector $\mathbf{r}_0$ lies along the z-axis and $\mathbf{k}_1$ does within the xy plane, the left-hand side of Eq. (2.57) becomes zero, i.e. $\mathbf{k}_1 \cdot \mathbf{r}_0 = 0$ and hence

$\mathbf{k}_r \cdot \mathbf{r}_0 = \mathbf{k}_2 \cdot \mathbf{r}_0 = 0$, which gives that $\mathbf{k}_r$ and $\mathbf{k}_2$ also lie within the xy-plane. Thus, the incidence, reflection and refraction of light generally take place within the same plane, which is called *plane of incidence*.

When we assume that $\mathbf{r}_0$ lies along the x-axis, Eq. (2.57) leads to

$$k_{1x} = k_{rx} = k_{2x}. \tag{2.58}$$

Using the refractive indices of the media, we will express the magnitudes of the wave vectors as $k_1 = k_r = n_1\omega/c$ and $k_2 = n_2\omega/c$. Further, if the angles of incidence, reflection and refraction are defined as θ_1, θ_r and θ_2, as shown in Fig. 2.10, the relation $k_{1x} = k_{rx}$ gives $k_1 \sin\theta_1 = k_r \sin\theta_r$, which leads to $\theta_1 = \theta_r$. This relation is called *law of reflection*. On the other hand, $k_{1x} = k_{2x}$ leads to a well-known relation,

$$n_1 \sin\theta_1 = n_2 \sin\theta_2, \tag{2.59}$$

which is called *law of refraction*, or *Snell's law*.

Now, we will derive the reflectivity and transmittance when light is incident on an interface, using the continuity conditions derived above. Before doing that, we should again note that the relation between the electric and magnetic fields is describable as

$$\mathbf{H}_i = \frac{1}{\omega\mu_i} \mathbf{k}_i \times \mathbf{E}_i, \tag{2.60}$$

where $i = 1, r$ and 2.

Consider first a case that the electric field is perpendicular to the plane of incidence, which is called s-polarization (see Fig. 2.10a). Putting $A_1 = E_{10z}$, $A_r = E_{r0z}$ and $A_2 = E_{20z}$, and using the continuity conditions for E_z and H_x, we can derive the following relations:

$$A_1 + A_r = A_2, \tag{2.61}$$

$$\left(k_{1y}A_1 + k_{ry}A_r\right)/\mu_1 = k_{2y}A_2/\mu_2. \tag{2.62}$$

Further, inserting $k_{1y} = (n_1\omega/c)\cos\theta_1$, $k_{2y} = (n_2\omega/c)\cos\theta_2$, $k_{ry} = -(n_1\omega/c)\cos\theta_r$ into Eq. (2.62), we obtain $m_1(A_1 - A_r)\cos\theta_1 = m_2 A_2 \cos\theta_2$, where $m_i = n_i\mu_0/\mu_i = \sqrt{(\epsilon_i/\mu_i)/(\epsilon_0/\mu_0)}$ with $i = 1, 2$. Eliminating A_2 or A_r from Eq. (2.61) and this relation, we obtain the amplitude reflectivity and transmittance for s-polarization as

$$r_s = \frac{A_r}{A_1} = \frac{m_1\cos\theta_1 - m_2\cos\theta_2}{m_1\cos\theta_1 + m_2\cos\theta_2} \rightarrow \frac{n_1\cos\theta_1 - n_2\cos\theta_2}{n_1\cos\theta_1 + n_2\cos\theta_2}, \tag{2.63}$$

$$t_s = \frac{A_2}{A_1} = \frac{2m_1 \cos\theta_1}{m_1 \cos\theta_1 + m_2 \cos\theta_2} \rightarrow \frac{2n_1 \cos\theta_1}{n_1 \cos\theta_1 + n_2 \cos\theta_2}, \quad (2.64)$$

where the last relations are obtained by putting $\mu_i \rightarrow \mu_0$ and hence $m_i \rightarrow n_i$ with $i = 1, 2$.

In a similar manner, the amplitude reflectivity and transmittance for a case that the electric field is parallel to a plane of incidence (p-polarization) can be derived as follows (see Fig. 2.10b). Putting $A_1 = |\mathbf{E}_{10}|$, $A_r = |\mathbf{E}_{r0}|$ and $A_2 = |\mathbf{E}_{20}|$, and using the continuity conditions for E_x and H_z, we can derive the following relations:

$$m_1(A_1 + A_r) = m_2 A_2, \quad (2.65)$$

$$A_1 \cos\theta_1 - A_r \cos\theta_r = A_2 \cos\theta_2, \quad (2.66)$$

which lead to

$$r_p = \frac{A_r}{A_1} = \frac{m_2 \cos\theta_1 - m_1 \cos\theta_2}{m_2 \cos\theta_1 + m_1 \cos\theta_2} \rightarrow \frac{n_2 \cos\theta_1 - n_1 \cos\theta_2}{n_2 \cos\theta_1 + n_1 \cos\theta_2}, \quad (2.67)$$

$$t_p = \frac{A_2}{A_1} = \frac{2m_1 \cos\theta_1}{m_2 \cos\theta_1 + m_1 \cos\theta_2} \rightarrow \frac{2n_1 \cos\theta_1}{n_2 \cos\theta_1 + n_1 \cos\theta_2}. \quad (2.68)$$

From these relations, the following general relations are derived for the amplitude reflectivity and transmittance from a medium i to j, which are denoted as r_{ij} and t_{ij}:

$$r_{ij} = -r_{ji}, \quad r_{ij}^2 + t_{ij}t_{ji} = 1. \quad (2.69)$$

This shows that light incident from a medium of a lower refractive index to higher one, changes its sign, meaning that the phase change occurs by $180°$, while it does not in an opposite case.

Under normal incidence, these relations are reduced to

$$r_s = \frac{A_r}{A_1} = \frac{m_1 - m_2}{m_1 + m_2} \rightarrow \frac{n_1 - n_2}{n_1 + n_2}, \quad (2.70)$$

$$t_s = \frac{A_2}{A_1} = \frac{2m_1}{m_1 + m_2} \rightarrow \frac{2n_1}{n_1 + n_2}, \quad (2.71)$$

under a geometry defined for s-polarization[a].

The power reflectivity and transmittance are calculated from a pointing vector in each medium and are given as[b]

---

[a]Note that the reflectivity for p-polarization gives an opposite sign because the definition of the electric field for the reflected light is opposite.

[b]A factor of $\cos\theta_2 / \cos\theta_1$ comes by evaluating the light intensity in a unit area at the interface.

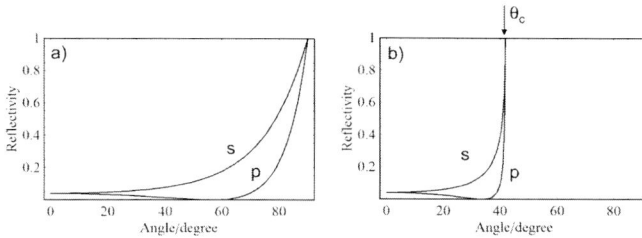

Figure 2.11 Reflectivity for s- and p-polarized incidence with changing incidence angle at an interface between two isotropic media with (a) $n_1 = 1.0$ and $n_2 = 1.5$, and (b) $n_1 = 1.5$ and $n_2 = 1.0$. θ_c is a critical angle for total internal reflection.

$$R = |r_{s,p}|^2 \quad \text{and} \quad T = \frac{m_2 \cos \theta_2}{m_1 \cos \theta_1} |t_{s,p}|^2, \qquad (2.72)$$

and the energy conservation is confirmed by the relation (Exercise (2))

$$R + T = 1. \qquad (2.73)$$

Figure 2.11 shows a typical power reflectance against incident angle when light is incident from a medium having the lower refractive index to that having the higher one, and that in an opposite case. In the former case, the reflectance gradually increases with increasing incidence angle for s-polarization, while the reflectivity initially decreases and shows a minimum around 60° for p-polarization. In either case, the reflectance approaches unity in the vicinity of 90°. On the other hand, in the case light incident from the higher refractive index to the lower one, the reflectivity reaches a maximum at the incidence angle of ~40° (indicated as θ_c in the figure) for both s- and p-polarizations, although the general trend is similar to that for s-polarization when the incidence angle is small. Above this angle, the reflectivity is eventually unity independently of the incidence angle. In the following, we will explain these two phenomena in detail.

2.2.4.2 Brewster's angle

At first, we will explain the origin of the reflection minimum appearing only in p-polarization. For this purpose, employing the approximation of $\mu \to \mu_0$ and using the Snell's law of $n_1 \sin \theta_1 =$

$n_2 \sin \theta_2$, we will transform Eq. (2.67) into

$$
\begin{aligned}
r_p &= \frac{\sin \theta_1 \cos \theta_1 - \sin \theta_2 \cos \theta_2}{\sin \theta_1 \cos \theta_1 + \sin \theta_2 \cos \theta_2} = \frac{\sin 2\theta_1 - \sin 2\theta_2}{\sin 2\theta_1 + \sin 2\theta_2} \\
&= \frac{2 \cos(\theta_1 + \theta_2) \sin(\theta_1 - \theta_2)}{2 \sin(\theta_1 + \theta_2) \cos(\theta_1 - \theta_2)} = \frac{\tan(\theta_1 - \theta_2)}{\tan(\theta_1 + \theta_2)},
\end{aligned}
\tag{2.74}
$$

which shows that the complete depletion of reflection occurs when the relation $\theta_1 - \theta_2 = 0$ or $\theta_1 + \theta_2 = \pi/2$ is satisfied.

The former case leads to a trivial result that the two media have the same refractive indices. On the other hand, the latter case leads to a special condition for the angle of incidence. Let us denote this angle as θ_{1B} and the corresponding refraction angle θ_{2B}. Since $\sin \theta_{2B} = \sin(\pi/2 - \theta_{1B}) = \cos \theta_{1B}$ and further from the Snell's law of $\sin \theta_{2B} = (n_1/n_2) \sin \theta_{1B}$, the following relation is easily derived:

$$
\tan \theta_{1B} = \frac{n_2}{n_1}. \tag{2.75}
$$

The angle of incidence satisfying this condition is called *Brewster's angle*, which is understandable in terms that the direction of the dipole induced by the incident light within the medium coincides with that of the reflection. The extremely different behavior found between *s*- and *p*-polarizations around the Brewster's angle is often used to generate a polarized state from unpolarized light or to avoid reflection loss in laser technology.

2.2.4.3 Total internal reflection

From Fig. 2.11, one will notice that a quite different behavior is seen at large incidence angles of $\theta_1 > \theta_c$ for $n_1 > n_2$, which is characterized by the complete reflection in a wide angular range. This is called *total internal reflection*, whose mechanism is explained as follows: From the Snell's law of $\sin \theta_2 = (n_1/n_2) \sin \theta_1 \equiv \alpha$ with $n_1 > n_2$, it is expected that $\alpha > 1$ will occur when the angle of incidence becomes larger. In this case, the Snell's law does not hold within a range of real number. The limit of this criterion for the incidence angle is given as

$$
\sin \theta_c = n_2/n_1, \tag{2.76}
$$

where θ_c is called *critical angle*. Above the critical angle, the relation $\sin \theta_2 = \alpha > 1$ holds and θ_2 should be expressed by a complex number.

Since $\cos\theta_2 = \pm\sqrt{1-\alpha^2} = \pm i\sqrt{\alpha^2-1}$, inserting it into the wave function (s-polarization, for example) within a medium 2, we obtain

$$
\begin{aligned}
E_{2z} &= A_2 e^{i(\mathbf{k}_2 \cdot \mathbf{r} - \omega t)} \\
&= A_2 e^{i(k_2 x \sin\theta_2 + k_2 y \cos\theta_2 - \omega t)} \\
&= A_2 e^{i(k_2 x \alpha - \omega t)} e^{\mp k_2 y \sqrt{\alpha^2-1}}.
\end{aligned}
$$

Since an exponential factor including y in the right-hand side will diverse in the limit of $y \to \infty$ when a plus sign is chosen, it is clear that only a minus sign is physically relevant. Thus,

$$
E_{2z} = A_2 e^{i(k_2 x \alpha - \omega t)} e^{-k_2 y \sqrt{\alpha^2-1}}. \tag{2.77}
$$

From this relation, it is understood that above the critical angle ($\alpha > 1$), a light wave in a medium 2 propagates along the x axis, which is parallel to the interface, having a larger wave vector by a factor of α, while along the y axis, the light wave does not propagate and decreases exponentially with a characteristic length of $d \equiv 1/(2k_2\sqrt{\alpha^2-1}) = \lambda/(4\pi n_2\sqrt{\alpha^2-1})$, which is obtained by taking a square of the absolute value of Eq. (2.77), i.e. $|E_{2z}|^2 \propto \exp[-2k_2 y\sqrt{\alpha^2-1}]$. Such a wave that cannot propagate and only exists in the vicinity of the interface is called *evanescent wave* and the length d is called *penetration length*.

If we insert the relation $\cos\theta_2 = i\sqrt{\alpha^2-1}$ into the Fresnel's law of Eqs. (2.63) and (2.67) after putting $\mu \to \mu_0$, we obtain

$$
r_s = \frac{n_1 \cos\theta_1 - i n_2 \sqrt{\alpha^2-1}}{n_1 \cos\theta_1 + i n_2 \sqrt{\alpha^2-1}},
$$

$$
r_p = \frac{n_2 \cos\theta_1 - i n_1 \sqrt{\alpha^2-1}}{n_2 \cos\theta_1 + i n_1 \sqrt{\alpha^2-1}}.
$$

In either case, the formula has a form of $(a - bi)/(a + bi)$ and is transformed into $(\sqrt{a^2+b^2} \exp[-i\phi])/(\sqrt{a^2+b^2} \exp[i\phi]) = \exp[-2i\phi]$, where ϕ is given as $\phi_s = \tan^{-1}\{(n_2\sqrt{\alpha^2-1})/(n_1 \cos\theta_1)\}$ for s-polarization and $\phi_p = \tan^{-1}\{(n_1\sqrt{\alpha^2-1})/(n_2 \cos\theta_1)\}$ for p-polarization. Thus, above the critical angle, the magnitude of the reflectivity becomes unity, while the phase changes during the reflection, which is dependent on the polarization.

The phase change during the process of total internal reflection is called *Goos-Hänchen effect*. Taking account of a fact that a light wave

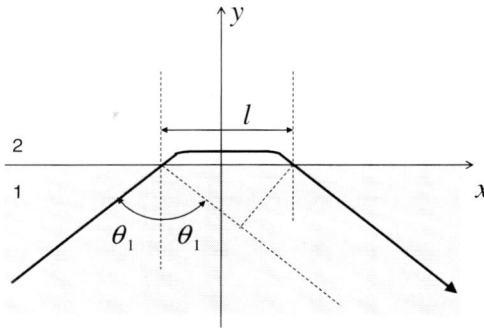

Figure 2.12 Goos-Hänchen effect with the Goos-Hänchen length *l*.

in the medium 2 propagates along the interface, we can speculate that the light wave propagates for a while within the medium 2 and then is directed back into the medium 1, as shown in Fig. 2.12. If the phase change is caused by this short-length propagation in the medium 2, we can estimate the length *l* through a geometrical inspection. From Fig. 2.12, *l* is obtained from the phase change of 2ϕ, which corresponds to $k_1 l \sin \theta_1$ and leads to $l_j = 2\phi_j/(k_1 \sin \theta_1) = \lambda \phi_j/(\pi \sin \theta_1)$ with $j = s, p$. That is, the magnitude of the spatial shift of the light beam is on the order of the wavelength of light and is called *Goos-Hänchen length*.

In Fig. 2.13, we plot the penetration length, phase change and Goos-Hänchen length in case of the refractive indices of 1.5 and 1.0 for the media 1 and 2, respectively. It is remarkable that both of the phase change and Goos-Hänchen length increase with increasing incidence angle, but the increasing rate is more prominent for *p*-polarization. Finally, the results for *s*- and *p*-polarizations agree with each other. On the other hand, the penetration length initially rapidly and then gradually decreases with increasing incidence angle in a sub-100 nm range.

The difference of the phase changes for the *s*- and *p*-polarizations is used to generate circular polarization in *Fresnel rhomb*, the principle of which is shown in Fig. 2.14. As shown in Fig. 2.13b, the difference of the phase changes for the *s*- and *p*-polarizations in case of the reflection between the media having the refractive indices of 1.5 and 1.0 is less than 50° so that a light wave needs to

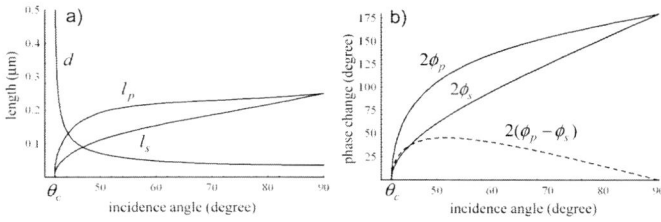

Figure 2.13 (a) Penetration depth d and the Goos-Hänchen lengths l_s and l_p for s- and p-polarizations, respectively. (b) Phase changes $2\phi_s$ and $2\phi_p$ that occur during the total internal reflection for s- and p-polarizations, and their difference $2(\phi_p - \phi_s)$. The refractive indices of medium 1 and 2 are set at 1.5 and 1.0, respectively.

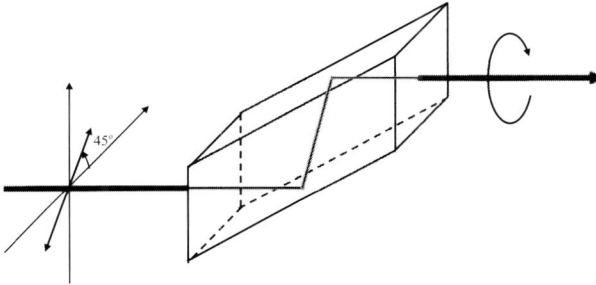

Figure 2.14 Schematic view of Fresnel rhomb, which converts a linearly polarized light beam into circularly polarized one.

reflect at least twice at the interfaces to obtain the phase difference of 90°, which corresponds to circular polarization (see Sec. 2.2.5; Exercise (3)). It is possible to increase the phase difference to 180° by combining two sets of Fresnel rhombs in series, as shown in the photographs of Fig. 2.14. This produces linear polarization whose direction is orthogonal to the incident one or produces inversely rotating circular polarization in case of circularly polarized incidence.

2.2.4.4 Reflection at a surface of absorbing medium

Next, we investigate the light reflection at a surface of an absorbing medium. We assume that the refractive index of medium 2 is complex, while that of medium 1 is real. Even in this case, the continuity conditions for the electric and magnetic fields at the interface, e.g. Eqs. (2.61) and (2.62) for s-polarization, should be satisfied. Thus, the Fresnel's laws of Eqs. (2.63) and (2.67) associated with the reflectivities for s- and p-polarizations are also applicable.

However, the relation Eq. (2.58) should be slightly modified, because k_{1x} and k_{rx} are real, while k_{2x} is complex. Putting $k_{2x} = k'_{2x} + i k''_{2x}$, we rewrite the relation as follows:

$$k_{1x} = k_{rx} = k'_{2x},$$
$$k''_{2x} = 0.$$

The second relation indicates that the imaginary part of the wave vector in the medium 2 has no x component. Further, putting $n_2 = \eta_2 + i\kappa_2$, we obtain from the Snell's law

$$\sin\theta_2 = \frac{n_1}{\eta_2 + i\kappa_2} \sin\theta_1, \tag{2.78}$$

which leads to

$$\cos\theta_2 = \sqrt{1 - \left(\frac{n_1}{\eta_2 + i\kappa_2}\right)^2 \sin^2\theta_1}$$

$$= \sqrt{1 - \frac{n_1^2(\eta_2^2 - \kappa_2^2)}{(\eta_2^2 + \kappa_2^2)^2} \sin^2\theta_1 + i\frac{2n_1^2\eta_2\kappa_2}{(\eta_2^2 + \kappa_2^2)^2} \sin^2\theta_1} \equiv q e^{i\gamma}, \tag{2.79}$$

where we put the magnitude and phase of $\cos\theta_2$ as q and γ, respectively.

Then,

$$q^2 \cos 2\gamma = 1 - \frac{n_1^2(\eta_2^2 - \kappa_2^2)}{(\eta_2^2 + \kappa_2^2)^2} \sin^2\theta_1,$$

$$q^2 \sin 2\gamma = \frac{2n_1^2\eta_2\kappa_2}{(\eta_2^2 + \kappa_2^2)^2} \sin^2\theta_1.$$

Inserting these relation into a wave function in the medium 2 of

$$E_{2z} = A_2 \exp[i\{k_2(x \sin\theta_2 + y \cos\theta_2) - \omega t\}],$$

for s-polarization as an example, and calculating the first half of the exponent, we obtain

$$ik_2(x \sin\theta_2 + y \cos\theta_2)$$

$$= i(\eta_2 + i\kappa_2)k_0 \left\{\frac{n_1}{\eta_2 + i\kappa_2} x \sin\theta_1 + yq(\cos\gamma + i\sin\gamma)\right\}$$

$$= ik_0 \{n_1 x \sin\theta_1 + yq(\eta_2 \cos\gamma - \kappa_2 \sin\gamma)\}$$

$$- k_0 yq(\kappa_2 \cos\gamma + \eta_2 \sin\gamma),$$

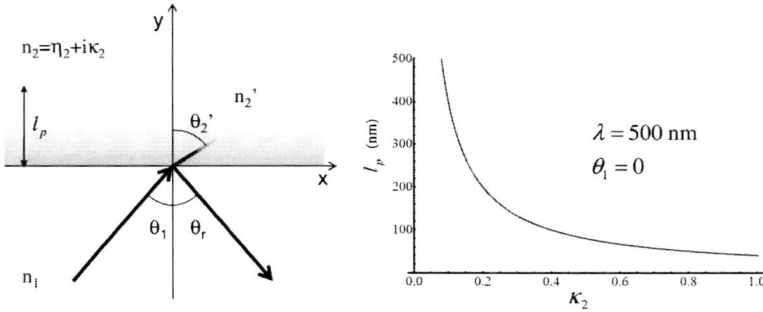

Figure 2.15 Schematic illustration of light reflection at an interface between non-absorbing 1 and absorbing medium 2. The light wave in the absorbing medium propagates along a refraction angle of θ_2' with an apparent refractive index of n_2' and then decreases with a penetration length of l_p. The right figure shows the dependence of the penetration length on the imaginary part of the complex refractive index of medium 2. The wavelength and angle of incidence are set at 500 nm and $0°$, respectively.

with $k_0 = \omega/c$. The light wave propagates toward the direction determined by

$$\tan \theta_2' = n_1 \sin \theta_1 / \{q(\eta_2 \cos \gamma - \kappa_2 \sin \gamma)\},$$

while its intensity decreases exponentially, perpendicularly to the interface, with a penetration length of $l_p = 1/\{2k_0 q(\kappa_2 \cos \gamma + \eta_2 \sin \gamma)\}$ independently of the directions of propagation and polarization. As shown in Fig. 2.15, the penetration length becomes smaller with increasing imaginary part of the refractive index. Thus, the electric field in the medium 2 is expressed as

$$E_{2z} = A_2 \exp\left[i\{n_2' k_0 (x \sin \theta_2' + y \cos \theta_2') - \omega t\}\right] \exp[-y/(2l_p)],$$
$$(2.80)$$

with $n_2' = \sqrt{(n_1 \sin \theta_1)^2 + q^2(\eta_2 \cos \gamma - \kappa_2 \sin \gamma)^2}$. We plot the apparent refractive index, n_2', and the angle of refraction, θ_2', in Fig. 2.16, where we put the real part of the refractive index of 1.5 with the incidence angle of 45°. It is clear that with increasing imaginary part of the refractive index, the deviations from their original values become prominent, which appear as the increase of the apparent refractive index and the decrease of the angle of refraction.

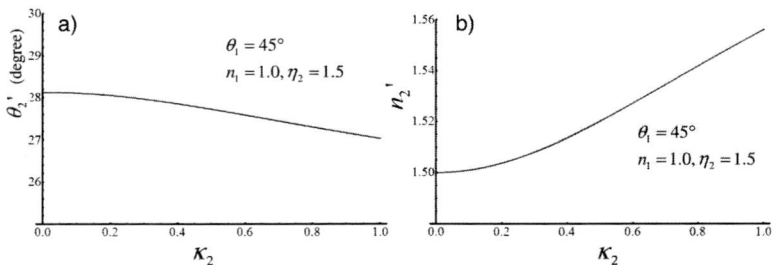

Figure 2.16 (a) Refraction angle and (b) apparent refractive index for a propagating light wave in the absorbing medium. n_1, η_2 and θ_1 are set at 1.0, 1.5 and 45°, respectively. Note $\theta_2' \rightarrow 28.125°$ and $n_2' \rightarrow 1.5$ for $\kappa_2 \rightarrow 0$.

The Fresnel's law in this case becomes

$$r_s = \frac{n_1 \cos\theta_1 - (\eta_2 + i\kappa_2)\cos\theta_2}{n_1 \cos\theta_1 + (\eta_2 + i\kappa_2)\cos\theta_2},$$

$$r_p = \frac{(\eta + i\kappa_2)\cos\theta_1 - n_1 \cos\theta_2}{(\eta + i\kappa_2)\cos\theta_1 + n_1 \cos\theta_2},$$

where $\cos\theta_2$ is given in Eq. (2.79).

In case of ideal metal, where the refractive index is expressed by a pure imaginary of $n_2 = i\kappa_2$, the Fresnel's law becomes

$$r_s = \frac{n_1 \cos\theta_1 - i\kappa_2 \cos\theta_2}{n_1 \cos\theta_1 + i\kappa_2 \cos\theta_2},$$

$$r_p = \frac{i\kappa_2 \cos\theta_1 - n_1 \cos\theta_2}{i\kappa_2 \cos\theta_1 + n_1 \cos\theta_2},$$

where $\cos\theta_2$ in this case is given as

$$\cos\theta_2 = \sqrt{1 + (n_1/\kappa_2)^2 \sin^2\theta_1}.$$

Thus, $\cos\theta_2$ becomes real and the Fresnel's law is anyway expressed as $(a - bi)/(a + bi)$ or $(-a + bi)/(a + bi)$ as in the case of total internal reflection. Therefore, the same discussion is applicable and gives the complete reflection with the phase change. In contrast to the total internal reflection, the complete reflection does not depend on the incident angle, while the phase change depends both on the incident angle and polarization.

2.2.5 *Polarizations*

Light is a wave accompanying electric and magnetic fields that oscillating at a high frequency. In isotropic medium, both the oscillating electric and magnetic fields are perpendicular to each other and also to the direction of the propagation. It is a custom to take the direction of polarization as that of the electric field. Since in the research field of nanophotonics, the polarization of light plays an important role, here we will summarize its characteristics.

Consider a plane wave of light propagating along the z axis of the Cartesian coordinate fixed in a space so that the electric field oscillates within the xy plane. We denote unit vectors directed to positive directions of the x and y axes as $\mathbf{e}_x$ and $\mathbf{e}_y$, respectively. Then, the electric field of the light wave is generally written as

$$\mathbf{E}(z, t) = E_x(z, t)\mathbf{e}_x + E_y(z, t)\mathbf{e}_y, \qquad (2.81)$$

where $E_x(z, t)$ and $E_y(z, t)$ are the components of the electric fields along the x and y axes, respectively, which are generally described as

$$E_x(z, t) = A_x \exp[i(kz - \omega t + \phi_x)], \qquad (2.82)$$
$$E_y(z, t) = A_y \exp[i(kz - \omega t + \phi_y)]. \qquad (2.83)$$

Dividing Eq. (2.82) by A_x and Eq. (2.83) by A_y, and adding or subtracting with each other, we obtain the following expressions in a form of real representation[a]:

$$\frac{E_x}{A_x} + \frac{E_y}{A_y} = \cos(kz - \omega t + \phi_x) + \cos(kz - \omega t + \phi_y)$$

$$= 2\cos\left(kz - \omega t + \frac{\phi_x + \phi_y}{2}\right)\cos\left(\frac{\phi_x - \phi_y}{2}\right), \quad (2.84)$$

$$\frac{E_x}{A_x} - \frac{E_y}{A_y} = \cos(kz - \omega t + \phi_x) - \cos(kz - \omega t + \phi_y)$$

$$= -2\sin\left(kz - \omega t + \frac{\phi_x + \phi_y}{2}\right)\sin\left(\frac{\phi_x - \phi_y}{2}\right). \quad (2.85)$$

By putting $\phi \equiv \phi_y - \phi_x$, and adding after squaring and dividing a appropriate factor for each formula, these relations lead to

$$\frac{1}{\cos^2(\phi/2)}\left(\frac{E_x}{A_x} + \frac{E_y}{A_y}\right)^2 + \frac{1}{\sin^2(\phi/2)}\left(\frac{E_x}{A_x} - \frac{E_y}{A_y}\right)^2 = 4,$$

---

[a]The real representation is employed because we will calculate a product of the electric fields.

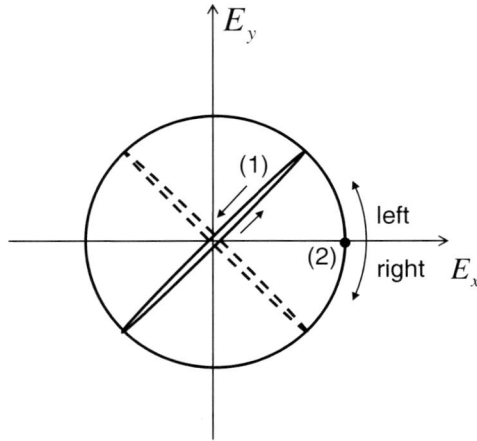

Figure 2.17 (1) Linear and (2) circular polarizations. "right" and "left" indicate right-handed and left-handed circular polarizations.

and hence

$$\sin^2(\phi/2)\left(\frac{E_x}{A_x}+\frac{E_y}{A_y}\right)^2+\cos^2(\phi/2)\left(\frac{E_x}{A_x}-\frac{E_y}{A_y}\right)^2=4\sin^2\frac{\phi}{2}\cos^2\frac{\phi}{2}.$$

Further transforming it, we finally obtain

$$\left(\frac{E_x}{A_x}\right)^2+\left(\frac{E_y}{A_y}\right)^2-2\left(\frac{E_x}{A_x}\right)\left(\frac{E_y}{A_y}\right)\cos\phi=\sin^2\phi. \qquad (2.86)$$

Various polarization states of light are derivable from this expression. At first, we should say that a locus of the electric field is generally expressed by an ellipse and is called *elliptical polarization*. There are the following special cases in elliptical polarization with *m* an integer:

Case 1 $\phi = m\pi$

In this case, Eq. (2.86) is reduced to

$$\left(\frac{E_x}{A_x}\right)^2+\left(\frac{E_y}{A_y}\right)^2\pm2\left(\frac{E_x}{A_x}\right)\left(\frac{E_y}{A_y}\right)=0, \qquad (2.87)$$

and hence

$$\left(\frac{E_x}{A_x}\pm\frac{E_y}{A_y}\right)^2=0. \qquad (2.88)$$

Thus, the loci within the $E_x E_y$ plane in this case are expressed by two line segments, which are symmetric with respect to E_y axis (see Fig. 2.17). We call the polarization state in this case as *linear polarization*.

Case 2 $A_x = A_y \equiv A$ and $\phi = (m + 1/2)\pi$

Equation (2.86) in this case is reduced to a simple form of

$$\left(\frac{E_x}{A}\right)^2 + \left(\frac{E_y}{A}\right)^2 = 1, \tag{2.89}$$

which expresses a circle and the polarization state expressed by this formula is called *circular polarization*. When $\phi = \pm\pi/2$, putting $\phi_x = 0$, we rewrite the electric fields as

$$E_x = A\cos(kz - \omega t),$$
$$E_y = A\cos(kz - \omega t \pm \pi/2).$$

Putting $z = 0$, we see that the electric field draws a circular orbit counterclockwise for $\phi = \pi/2$ and clockwise for $\phi = -\pi/2$ with increasing t, as shown in Fig. 2.17, which correspond to *left-handed (counterclockwise)* and *right-handed (clockwise) circular polarizations*, respectively. In complex representation, the electric fields are expressed as

$$E_x = A e^{i(kz - \omega t)},$$
$$E_y = A e^{i(kz - \omega t \pm \pi/2)}.$$

Since $E_y = E_x \exp(\pm i\pi/2)$, these expressions are sometimes rewritten as

$$E_y = \begin{pmatrix} +i \\ -i \end{pmatrix} E_x. \begin{array}{l} \cdots \text{left,} \\ \cdots \text{right.} \end{array} \tag{2.90}$$

When the phase difference ϕ changes, the polarization state of light variously changes according to the value of ϕ. In Fig. 2.18, we show such an example in case of $A_x = A_y$. When $\phi = 0$, the polarization state is expressed by linear polarization, whose direction is inclined by $45°$ from the x axis, while in case of $\phi = \pi/2$, it gives the left-handed circular polarization. Between these states, the elliptical polarization appears. With further increasing

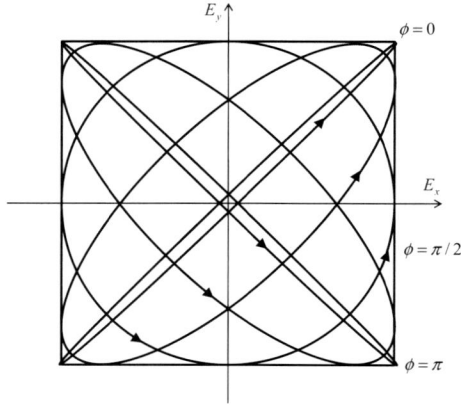

Figure 2.18 Polarization states appearing with changing the phase difference ϕ.

ϕ, the linear polarization state again appears with the polarization direction perpendicular to the case of $\phi = 0$. Thereafter, the polarization state changes to the right-handed circular polarization and comes back to the initial linear polarization. Thus, the polarization state is periodic with respect to the change of ϕ with the period of 2π.

Let us consider what happens when a linearly polarized light wave passes through a thin plate of crystal with the thickness of d, as shown in Fig. 2.19. The crystal is assumed to be placed so that the crystallographic axes are coincident with those of the Cartesian coordinate. We set the propagation direction of light along the z axis. Further, it is assumed that the crystal is optically anisotropic: i.e. the refractive indices are different for the electric fields oscillating along the x and y axes, which are denoted as n_x and n_y, respectively[a], as shown in Fig. 2.19. We set the origin of the coordinate on the incident surface of the crystal. If the electric field of the incident light is linearly polarized, whose polarization direction is inclined by an angle of θ from the x axis, the x and y components of the electric

---

[a]We consider here an optically uniaxial crystal. For example, we adjust the c-axis to agree with the x or y axis.

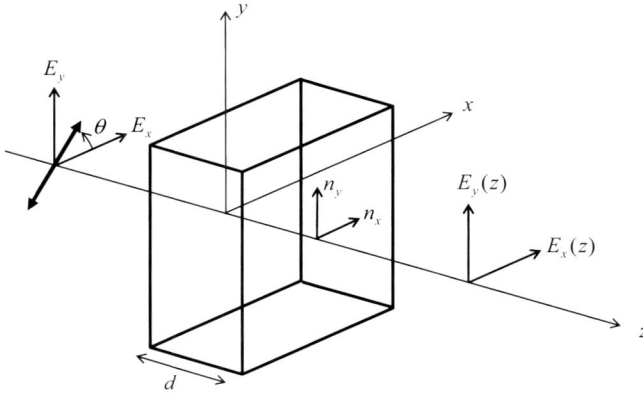

Figure 2.19 Linearly polarized light passing through a plate of anisotropic crystal with the thickness of d.

field of the incident light will be generally expressed as

$$E_x(z) = E_0 \cos \theta \; e^{i(kz-\omega t)},$$
$$E_y(z) = E_0 \sin \theta \; e^{i(kz-\omega t)},$$

for $z < 0$ with $k = n_0 \omega/c$, where E_0 and n_0 are the amplitude of the electric field and the refractive index of the surrounding medium, respectively.

On the other hand, those at $z = d$ will be

$$E_x(d) = E_0 \cos \theta \; e^{i(n_x \omega d/c - \omega t)},$$
$$E_y(d) = E_0 \sin \theta \; e^{i(n_y \omega d/c - \omega t)},$$

respectively, if the reflection loss at the crystal surface is negligible. Then, the electric field of the light wave passing through the anisotropic crystal becomes

$$E_x(z) = E_0 \cos \theta \; e^{i\{k(z-d)+n_x\omega d/c - \omega t\}}$$
$$= E_0 \cos \theta \; e^{i\{kz-\omega t+(n_x - n_0)\omega d/c\}}$$
$$= E_0 \cos \theta \; e^{i(kz-\omega t+\phi_x)},$$

and

$$E_y(z) = E_0 \sin \theta \; e^{i\{k(z-d)+n_y\omega d/c - \omega t\}}$$
$$= E_0 \sin \theta \; e^{i\{kz-\omega t+(n_y - n_0)\omega d/c\}}$$
$$= E_0 \sin \theta \; e^{i(kz-\omega t+\phi_y)},$$

where we put $\phi_{x,y} = (n_{x,y} - n_0)\omega d/c$. These expressions completely agree with what we have initially assumed in Eqs. (2.82) and (2.83). Thus, the polarization state of the light passing through an anisotropic crystal is essentially determined by the phase difference, $\phi = \phi_y - \phi_x = (n_y - n_x)\omega d/c$.

For example, if we generate a circularly polarized light wave using an anisotropic crystal, we prepare a crystal whose thickness is determined by a relation $\phi = \pm\pi/2$ and the initial polarization direction is set so as to satisfy $E_0 \sin\theta = E_0 \cos\theta$. The former condition for the plus sign leads to $(n_y - n_x)\omega d/c = 2\pi(n_y - n_x)d/\lambda = \pi/2$, which reduces to $d = \lambda/\{4(n_y - n_x)\}$ in case of $n_y > n_x$, while for the minus sign, the crystal should be rotated by $90°$ so that n_x and n_y will be exchanged. On the other hand, the latter condition gives a simple result of $\theta = 45°$.

In case of quartz, which belongs to the hexagonal system, the refractive indices are known to be $n_1 = n_2 = n_3 = 1.5462$ and $n_c = 1.5553$ at 18°C for the wavelength of 546.1 nm. Thus, if we prepare a thin plate of the crystal that is cut in parallel to the c axis and direct the c axis along the y axis, we can obtain a plate producing circular polarization when d is set at $d = 0.5461/\{4 * (1.5553 - 1.5462)\} \approx 15$ (μm). Such a plate generating circular polarization is normally called $\lambda/4$ *plate* (Another good example to cause an elliptical polarization is found when linearly polarized light is reflected at a metal surface. See Exercise (4).)

On the contrary, completely different case appears when the polarization direction is unsettled in time and changes randomly. We call such a state of light as *unpolarized light* or *unpolarized state*. In order to express the polarization states including unpolarized one, a set of *Stokes parameters* is often employed. For z-propagating light wave, a Stokes vector S is defined as

$$S = \begin{pmatrix} I \\ Q \\ U \\ V \end{pmatrix} = \begin{pmatrix} \langle|E_x|^2\rangle + \langle|E_y|^2\rangle \\ \langle|E_x|^2\rangle - \langle|E_y|^2\rangle \\ 2\mathrm{Re}\langle E_x^* E_y\rangle \\ -2\mathrm{Im}\langle E_x^* E_y\rangle \end{pmatrix}. \qquad (2.91)$$

Let us investigate the nature of the Stokes vector by setting $A_x = A\cos\theta$ and $A_y = A\sin\theta$. In this case, the electric fields expressed by

$$E_x = A \cos\theta \cdot \exp[i(kz - \omega t + \phi_x)], \qquad (2.92)$$
$$E_y = A \sin\theta \cdot \exp[i(kz - \omega t + \phi_y)], \qquad (2.93)$$

give the following Stokes parameters:

$$I = A^2,$$
$$Q = A^2 \cos 2\theta,$$
$$U = A^2 \sin 2\theta \cos\phi,$$
$$V = -A^2 \sin 2\theta \sin\phi,$$

where $\phi = \phi_y - \phi_x$.

For linear polarization, where $\phi = m\pi$, the S vector becomes

$$S = A^2(1, \cos 2\theta, \pm\sin 2\theta, 0), \qquad (2.94)$$

and for circular polarization, where we put $A_x = A_y = A/\sqrt{2}$ with $\phi = (m + 1/2)\pi$, it becomes

$$S = A^2(1, 0, 0, \pm 1). \qquad (2.95)$$

On the other hand, the components of the electric field of unpolarized light are formally expressed as

$$E_x = A_{ux} \exp[i(kz - \omega t + \phi_x(t))], \qquad (2.96)$$
$$E_y = A_{uy} \exp[i(kz - \omega t + \phi_y(t))], \qquad (2.97)$$

where $\phi_{x,y}(t)$ are random functions of time and then their time averages satisfy the relations $\langle \exp[i\phi_{x,y}(t)] \rangle = 0$. The amplitudes A_{ux} and A_{uy} are also random functions of time, which vary slowly as compared with the phase change. Thus, we put $\langle A_{ux}^2 \rangle = \langle A_{uy}^2 \rangle \equiv (1/2)\langle A_u^2 \rangle$ with $\langle A_{ux} \rangle = \langle A_{uy} \rangle = 0$ and $\langle A_{ux} A_{uy} \rangle = \langle A_{ux} \rangle \langle A_{uy} \rangle = 0$. The Stokes vector in this case is expressed as

$$S = \langle A_u^2 \rangle (1, 0, 0, 0). \qquad (2.98)$$

With mixing polarized and unpolarized light waves, the components of the electric field are generally expressed as

$$E_x = A_p \cos\theta \cdot \exp[i(k_z z - \omega t + \phi_{x0})]$$
$$+ A_{ux} \exp[i(k_z z - \omega t + \phi_{x1}(t))], \qquad (2.99)$$
$$E_y = A_p \sin\theta \cdot \exp[i(k_z z - \omega t + \phi_{y0})]$$
$$+ A_{uy} \exp[i(k_z z - \omega t + \phi_{y1}(t))], \qquad (2.100)$$

where ϕ_{x0} and ϕ_{y0} are the phases of the polarized component for the x- and y-component, while $\phi_{x1}(t)$ and $\phi_{y1}(t)$ are those of unpolarized one, which are assumed to vary randomly with time. A_p is an amplitude of the polarized component, while A_{ux} and A_{uy} are assumed to be slowly varying random functions of time. The Stokes parameters become as follows: $I = A_p^2 + \langle A_{ux}^2 \rangle + \langle A_{uy}^2 \rangle$, $Q = A_p^2 \cos 2\theta + \langle A_{ux}^2 \rangle - \langle A_{uy}^2 \rangle$, $U = A_p^2 \sin 2\theta \cos \phi$ and $V = -A_p^2 \sin 2\theta \sin \phi$. Since the relation $\langle A_{ux}^2 \rangle = \langle A_{uy}^2 \rangle \equiv (1/2)\langle A_u^2 \rangle$ generally holds for unpolarized light, the Stokes vector is simply expressed as

$$S = A_p^2 \begin{pmatrix} 1 \\ \cos 2\theta \\ \sin 2\theta \cos \phi \\ -\sin 2\theta \sin \phi \end{pmatrix} + \langle A_u^2 \rangle \begin{pmatrix} 1 \\ 0 \\ 0 \\ 0 \end{pmatrix}, \qquad (2.101)$$

where the first term in the right-hand side corresponds to the polarized component, while the second term to the unpolarized component. The degree of polarization is generally defined as

$$P = \frac{A_p^2}{A_p^2 + \langle A_u^2 \rangle} = \frac{\sqrt{Q^2 + U^2 + V^2}}{I}. \qquad (2.102)$$

We apply this relation to a case where unpolarized light is obliquely incident and reflected at an interface between two isotropic media with different refractive indices. Since the reflectivities for s- and p-polarizations are different for oblique incidence, the polarized component will be generated in spite of unpolarized light employed for the incidence. For this purpose, the reflectivities for s- and p-polarizations are separately calculated using the Fresnel's law of Eqs. (2.63) and (2.67). Further, we consider a general case where the polarization direction for s- or p-component is deviated from x- or y-coordinate by an angle of η, as shown in Fig. 2.20. Then,

$$A_{ux} = r_s E_{us} \cos \eta - r_p E_{up} \sin \eta,$$

$$A_{uy} = r_s E_{us} \sin \eta + r_p E_{up} \cos \eta,$$

where $E_{uj} = A_{uj} \exp[i(kz - \omega t + \phi_j(t))]$ with $j = s, p$. Thus the components of Stokes vector become

$$I = |r_s|^2 \langle |A_{us}|^2 \rangle + |r_p|^2 \langle |A_{up}|^2 \rangle,$$

$$Q = \left(|r_s|^2 \langle |A_{us}|^2 \rangle - |r_p|^2 \langle |A_{up}|^2 \rangle\right)(\cos^2 \eta - \sin^2 \eta),$$

$$U = \left(|r_s|^2 \langle |A_{us}|^2 \rangle - |r_p|^2 \langle |A_{up}|^2 \rangle\right) 2 \cos \eta \sin \eta,$$

$$V = 0.$$

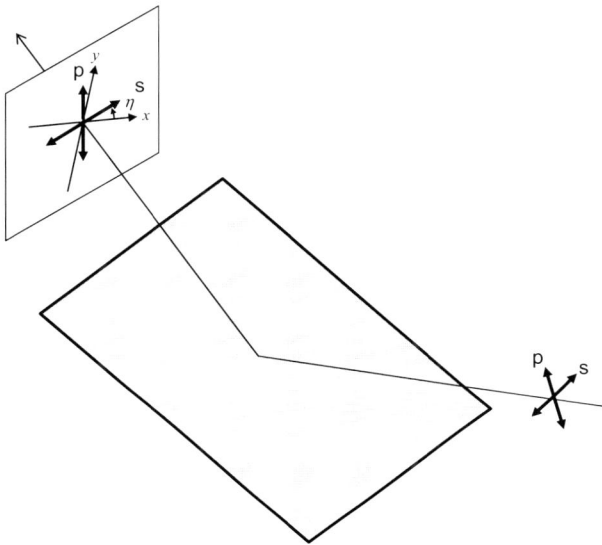

Figure 2.20   Geometry of the reflection of unpolarized light at a surface.

The degree of polarization for the reflected light is evaluated by the relation $P = \sqrt{Q^2 + U^2 + V^2}/I = (|r_s|^2 - |r_p|^2)/(|r_s|^2 + |r_p|^2)$, under the conditions of $\langle |A_{us}|^2 \rangle = \langle |A_{up}|^2 \rangle$ and $|r_s|^2 > |r_p|^2$. It is reasonable that P does not depend on η and hence on the direction of the coordinate system. The result is shown in Fig. 2.21. The degree of polarization thus obtained increases with increasing incidence angle, takes a maximum value of unity at a Brewster angle and then decreases again. Thus it is rather surprising that completely unpolarized light produces completely polarized light only through reflection.

The polarization of light is known to play an important role in various nanophotonic structures in nature to display the brilliant structural colors in animals. Recently it was found that anisotropic reflection for s- and p-polarizations significantly affects the structure-based color mixing in some butterflies and moths, and circularly polarized reflection displays brilliant colors in some kinds of beetles. It is also known that even eye perceptions in some kinds of animals are polarization sensitive.

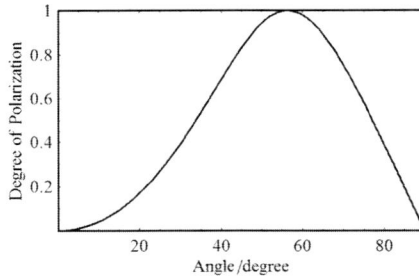

Figure 2.21 Degree of polarization calculated for the reflected light at an interface of two isotropic media when unpolarized light is incident from a medium with the refractive index of 1.0 to that of 1.5.

2.3 Vision and Color

2.3.1 *What Is Color?*

A word "color" is one of the most frequently used words in daily life. Since color and its expression bear various emotional and social meanings, we are able to add even human feelings to materials and phenomena by indicating a color. However, if one asks "what is color?", you may feel at a loss for an answer.

One could say that the scientific research on color was started by an English physicist, Isaac Newton. He is now famous for a man who discovered the law of gravitation, but he also performed a great deal of experiments on optics, which were summarized in his famous book *Opticks* [Newton (1704)]. Even now, we know the importance of his research from his name still remaining as Newton's ring and the Newtonian telescope.

Newton performed the most fundamental experiment on color using an experimental arrangement shown in Fig. 2.22. In this experiment, he introduced a thin light beam from the sunlight into the first prism, which made a white beam dispersed into various colors according to rainbow colors. Next, using a lens, he led the light beams thus dispersed into the second prism, which made them combined and generated a white beam again (Exercise (5)). Then, he led it to the third prism to disperse it again into rainbow colors.

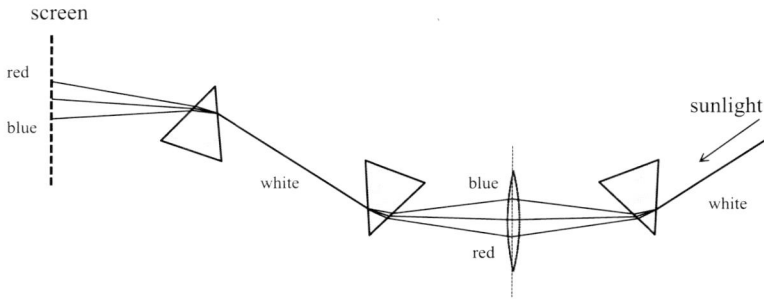

Figure 2.22   Newton's experiment on color.

This simple but extremely important experiment teaches us two essential physical bases concerning color. The first one is that a white color consists of various colors and the second one is that the difference of colors is determined by the angle of refraction at a prism apex. Thus, one can say that color is completely defined according to a purely physical law. Nowadays, we know that the variation of refraction angles depends on the dispersion relation of the refractive index of the matter. In an ordinary material, the refractive index is larger when the wavelength of light becomes shorter. Therefore, a light beam is largely deflected when it is blue, while it is small when it is red.

A hundred years later after Newton published *Opticks*, a German writer and scientist, Johann Wolfgang von Goethe published a book "Zur Farbenlehre (Theory of Colours)" [Goethe (1810)]. In this book, he discussed the color from a completely different standpoint. Above all, the most impressive expression is that when he was in an inn towards evening, he saw a "well-favoured girl with a brilliantly fair complexion, black hair, and a scarlet bodice" coming into the room. He looked attentively at her in half shadow. On the white wall after she turned away, he saw "a black face that was surrounded with a bright light, while the dress of the perfectly distinct figure appeared of a beautiful sea-green". He wrote the origin of this phenomenon in terms of a statement "the opposite colour producing each other successively on the retina". Namely, he insisted that color was a kind of physiological phenomena.

This experiment can be easily reproduced by fixing our eyes on a figure painted by red or blue for about 15 seconds, and abruptly changing our eyes to a piece of white paper (see Experiment IV). Then, according to the original color of the figure, we can see pale green or yellow colored pattern of nearly the same shape. Considering the Newton's discovery that a white color is made from a mixture of various colors, we notice that after gazing a colored figure, only the complementary color can be detected though light with various colors uniformly enters into our eyes. This phenomenon is called *color adaptation* and is considered as a kind of physiological functions of vision.

It is also known that a visual illusion called *chromatic induction* makes us feel a completely different color when it is adjacent to the other colors (see Experiment IV). Thus, it is clear that color cannot be discussed simply by a physical quantity such as the angle of refraction, but is deeply related to our visual functions including cognition process in brain. Hence, it is quite difficult to truly understand the meaning of color. However, it is at least necessary to understand the meaning of color from various standpoints, particularly when we consider the biological meaning of color created by nanophotonic structures.

2.3.2 *Vision*

Before describing a method to express the color quantitatively, we will briefly touch upon human vision in this section [Rodieck (1998); Nicholls *et al.* (2001); Kawamura (2010)]. The higher animals are generally equipped with an organization called *eye*, which is developed from nervous system. There are two kinds of eyes in the animal world: One is called *mosaic eye*, which consists of a lot of small eyes called *ommatidium*, and the other is called *camera eye*, which is equipped with a large lens in a round part of eye called *eyeball*. The former is an eye of insect, while the latter is one that is common in vertebrate.

In Fig. 2.23, we show a schematic view of human eyeball. A whole eyeball is surrounded by a membrane called *sclera*, a part of which is transparent and is called *cornea*. Inside the sclera, two membranes called *choroid* and *retina* exist. Light incident through

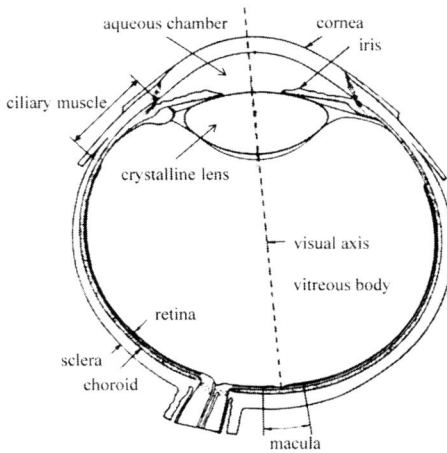

Figure 2.23   Structure of human eyeball [Hongo *et al.* (2000)].

the cornea is focused on a retina by a lens called *crystalline lens* after passing through an *aqueous chamber* containing *aqueous humor* located in front of the lens and also through a gelatin-like medium called *vitreous body* located behind it. The refractive index of cornea is known as 1.376, while those of aqueous humor and vitreous body are 1.336 and 1.33, respectively. The crystalline lens has a form of an asymmetric double-convex lens, whose refractive index is 1.386 at the surface, while it is 1.409 at the center. The crystalline lens is equipped with an *iris* in its front, which plays a role of iris diaphragm of a camera, and with *ciliary muscle* around it, which controls the curvature of the front surface of the lens by contraction and relaxation of the muscle and then changes the focal length of the lens.

The retina is developed from a part of brain during embryonic development process. At first, an eye appears as a tissue made from a double-layered membranes formed from a forebrain, which looks like a cup and thus called *optic cup*. A retina is then developed from the inner layer of the cup. The thickness of the retina is only 0.2 mm, which is usually divided into three cell layers, which are called *ganglion*, *bipolar* and *visual cell layer*, situating from the surface facing the cornea, as shown in Fig. 2.24. The visual cell layer is placed

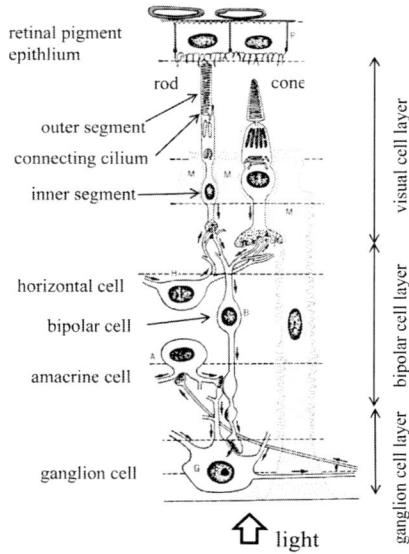

Figure 2.24 Cross section of retina. (Reproduced from [Yamada (1967)] with permission.)

at the most inner part of the membrane, within which two kinds of *visual cells* called *rod* and *cone* are distributed. The visual cells originate from *cilliated epithelial cells* covering the wall of *ventricle* of the brain.

The visual cell has a projection called *outer segment*, within which a lot of visual pigments are packed. This projection is resulted from the differentiation of cilium. The outer segment consists of a layered structure generated by folding a cellular membrane. As for an outer segment of a rod, the layered structure consists of disc-shaped layers, and several hundreds to a few thousands of discs are packed within an outer segment. The rod visual pigments given a name of *rhodopsin* are located in these discs, which contributes to light and darkness detection. On the other hand, in a cone, the outer segment is cone-shaped and the layered structure is made by folding the cell membrane, in which cone visual pigments contributing to the color detection are distributed. The outer segment is connected with the *inner segment*, in which a nucleus and a lot of mitochondria are located, with an aid of a structure called *connecting cilium*.

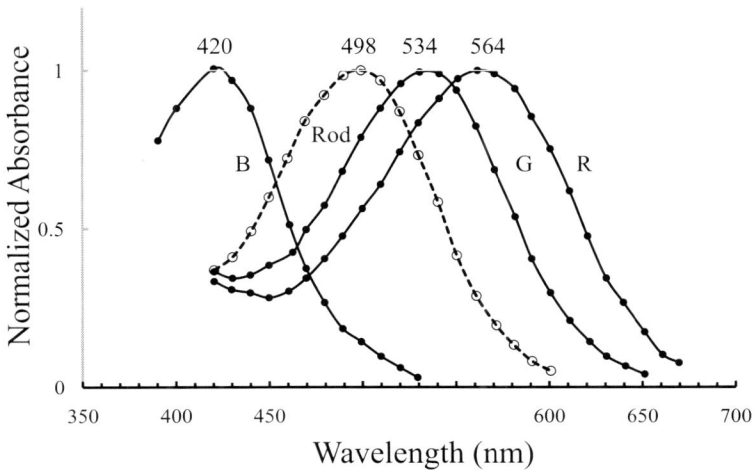

Figure 2.25 Spectral sensitivity curves for rod and three types of cones [Hongo *et al.* (2000)].

A rhodopsin molecule existing within the outer segment of a rod consists of an *11-cis retinal* chromophore combined with a protein called *opsin*, whose absorption spectrum has a peak around 500 nm (see Fig. 2.25). When light hits a rhodopsin molecule, 11-cis retinal is excited to an electronic excited state, which causes a cis-trans photoisomerization within a very short time. Following this isomerization of the chromophore, rhodopsin changes its conformation gradually by passing through the states known as *bleaching intermediates*, and finally, bleaches completely by dissociating into *all-trans retinal* and opsin. As a result of the conformational change, rhodopsin becomes "active" to trigger a series of reactions within a cytoplasm, which leads to the reduction of the inflow of Na^+ to the outer segment. Thus, the cell is hyperpolarized when it is stimulated by light[a].

---

[a]In general, the inside of a living cell is electrically negative (usually at $-30 \sim -60$ mV) compared with its outside. This level of negative voltage is mainly attained by the electrochemical equilibrium of K^+: its concentration is high in the inside of the cell, while it is low in the outside, and K^+ permeates the cell membrane through K^+ channel proteins present in the cell membrane. When a Na^+ channel becomes open as in the case of visual cells in the dark, Na^+ flows into the cell to neutralize the negativity in the cell to some extent. The electrical activity of a neuron is described

On the other hand, in cones, two to four types of cone pigment are present depending on the animal species, although 11-cis retinal chromophore commonly exists as in a rod pigment, rhodopsin. In human, there are three types of cone visual pigment. The absorption spectra of the resulting cone pigments have peaks at 420, 530 and 560 nm, corresponding to visual pigments sensing blue, green and red, respectively (see Fig. 2.25). Each type of cone pigment is expressed in a single cell-type of cone so that we have blue-, green- and red-sensitive cones. In cone pigments, as in rhodopsin, 11-cis retinal is isomerized to all-trans retinal on light absorption, which induces the conformational change in the pigment to trigger a series of reactions to reduce the influx of Na^+ into the outer segment.

Alll-trans retinal thus produced after light illumination is converted to all-trans *retinol* and then transported to *retinal pigment epithelium*, where 11-cis retinal is reproduced by enzymatic reactions. Then, it is sent back to visual cell and combined with opsin to produce rod or cone visual pigment again.

Rods and cones are both visual cells, and yet they have many different features. In a human eyeball, an optically central axis called *visual axis* is present. The retina around this region is highly sensitive and is called *macula*, where the density of cone cells is the highest, while rods are hardly distributed. On the contrary, rods are rather highly distributed in a surrounding region. Further, the sensitivity of a rod is generally higher than that of cone so that the rods mainly contribute to vision in dim environment, whereas the cones are responsible for color vision in light environment. Both rods and cones evoke sustained electrical signals during a period of light stimulus. However, the duration of a signal evoked by a brief light flash is by far shorter in cones than that in rods so that cones are superior to detect moving objects.

We can actually know the presence and distribution of rods through the following experiences: In the evening, we can hardly detect colors and thus only a monochromatic scenery is seen.

---

by the voltage change relative to that at the resting state. When the influx of Na^+ is reduced in the light in visual cells, the inside of the cell becomes more negative compared with the voltage in the dark. This negative-going voltage change is called *hyperpolarization*, or it is said that the cell is *hyperpolarized*. When a positive-going voltage change takes place, it is called *depolarization* or it is said that the cell is *depolarized*.

Further, when we observe stars using an astronomical telescope, we sometimes feel better to see the surrounding area instead of the center of visual field. These facts clearly demonstrate the presence of rods, because under dim environment, only rods detect light, which are distributed in a surrounding region of visual field.

Unlike usual neuronal cells, the electrical signal produced by a visual cell is not spike-like but a slow graded signal whose amplitude is somewhat logarithmically related to the illuminating light intensity. This signal is transported to the brain after data processing in the second cell layer through neuronal circuits formed by *horizontal*, *bipolar* and *amacrine cells*. Although the detailed mechanisms of the signal processing among these cells are not fully understood yet, here we will briefly describe the outline of the data processing.

A signal from a visual cell is first sent to horizontal and bipolar cells. On the bipolar cell level, two types of important signal processing, which are directly related to our contrast vision, are known to take place. One is to create *on* and *off* responses relating to the detection of a temporal change in the brightness, and the other is to produce a *center-* and-*surround* receptive field organization relating to the detection of a spatial change in the brightness.

Correspondingly, there are two types of bipolar cell called *on-* and *off-bipolar cells*. *On*-bipolar cells respond to light with a sustained depolarization and *off*-bipolar cells with a sustained hyperpolarization. In warm blooded animals, a cone cell connects to both these two types of bipolar cell, while a rod connects only to one bipolar cell called *rod bipolar cell*, as shown in Fig. 2.26. Rod bipolar cells respond to light with a sustained depolarization, and this signal is converged to the cone's *on-* and *off*-pathway through one of the sub-classes of amacrine cells called *AII amacrine*. On the other hand, the depolarization signal evoked at the beginning of the *on* signal in the *on*-bipolar cell and that at the end of the *off* signal in *off*-bipolar cell are mixed in *on-off amacrine cell* to produce an *on-off* signal, which is a paired depolarization signal evoked at the beginning and ending of a light stimulus (see Fig. 2.28a).

The center-surround receptive field organization is performed as follows: When a visual cell located just below a bipolar cell is illuminated by a spot of light of a sufficiently small size, the

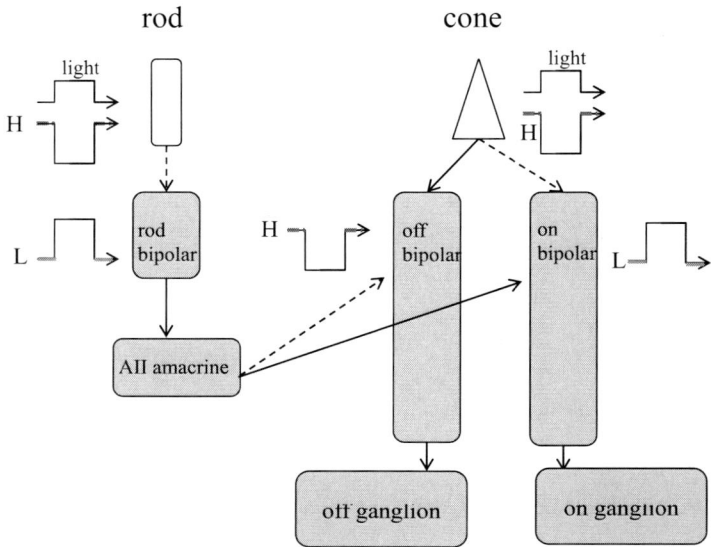

Figure 2.26 Cellular network generating the *on* and *off* responses, which is related to the detection of a temporal change in the brightness. H and L indicate the depolarized and hyperpolarized states in the resting state under dark, while solid and dashed arrows indicate the inductions of the non-inverted and inverted voltage changes to a cell.

bipolar cell will receive a signal from the visual cell: The bipolar cell will be depolarized, if it is the *on* type (Fig. 2.27a). When a surrounding region of the bipolar cell of interest is illuminated by a donut-shaped light, the same *on*-bipolar cell responds with a hyperpolarization which is brought about from surrounding visual cells not directly connected to this bipolar cell. This *on*-bipolar cell, therefore, responds with depolarization to a stimulus at the center and with hyperpolarization in the surround (Fig. 2.27b). When the cell is the *off*-type, the polarity of the response is reversed: the cell responds with the hyperpolarization to the center illumination and with the depolarization to the surround illumination. On the contrary, when uniform illumination is given to a retina, a visual cell at the center hyperpolarizes, while it receives a depolarization signal from the surround. These signals are summed in the visual cell. As a result, the depolarization induced in an *on*-bipolar cell becomes

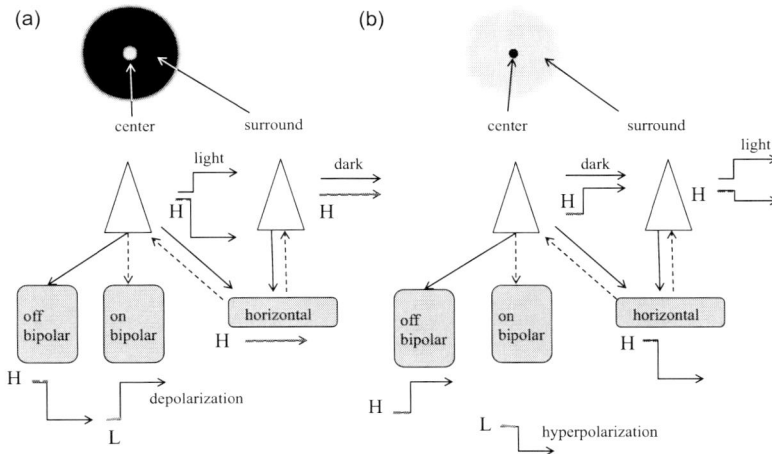

Figure 2.27 Schematic diagrams showing the center-surround receptive field organizations under (a) spot and (b) donut-shaped illumination. Under spot illumination, only a visual cell at the center is hyperpolarized, which leads to the depolarization (hyperpolarization) in the *on* (*off*)-bipolar cell located just above the visual cell. On the other hand, under donut-shaped illumination, visual cells located in the surround response to cause the hyperpolarization, which results in the depolarization in the visual cell at the center through horizontal cells and then causes the hyperpolarization (depolarization) in the *on* (*off*)-bipolar cell. Note that a visual cell in the surround also receives the depolarization signals from its surround. Thus, the concept of the center and surround is relative.

considerably smaller than that evoked during spot illumination at the center.

After all, the best stimulus to an *on*-bipolar cell becomes a white spot with black background for the depolarization, while a black spot with white background is preferable to the hyperpolarization. The production of the center-surround antagonistic electrical responses in bipolar cells owes to horizontal cells: these horizontal cells receive inputs from neighboring visual cells that do not directly connect to the bipolar cell of interest, and transmit negative feedback signals to a visual cell which connects to this bipolar cell. As a result, by illumination of the surroundings, a sign-inverted electrical response is evoked in this bipolar cell. This negative-

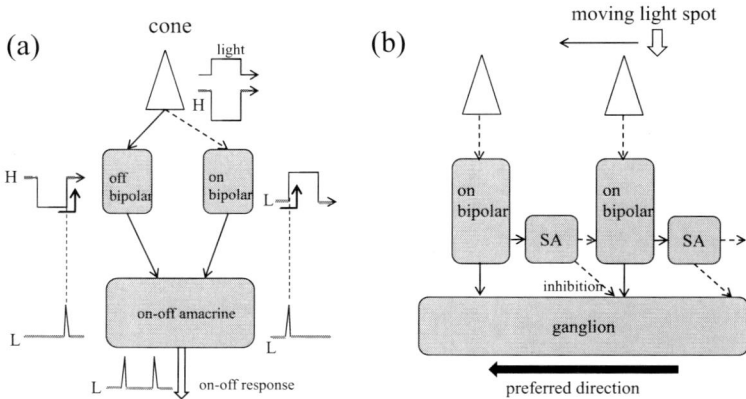

Figure 2.28 (a) Rising edges in *on-* and *off*-bipolar cells are detected in an *on-off* amacrine cell to bring an *on-off* response. (b) Mechanism of the motion detection on the level of ganglion cell. SA's are *starburst amacrine cells*, which align one-dimensionally and give the inhibitory signals to the neighboring *on*-bipolar cells and ganglion. This determines the inhibitory directionality, which brings a *preferred* direction that is reversed to it.

feedback circuit formed by horizontal cells and visual cells is one of the examples of the neuronal mechanism of lateral inhibition, a well-known circuit in sensory physiology. This mechanism is also thought to enhance the detection of the image where the light intensity is very different at the boundary and hence to contribute to edge emphasis.

The signal from a bipolar cell is further transmitted to amacrine and ganglion cells. Many sub-classes are known in amacrine cells, and their roles are believed to be diverse. One of their functions is to converge the signal from rod bipolar cell to the cone *on-* and *off*-pathway (Fig. 2.26). Another is related with the motion detection on the level of ganglion cells: a group of a class of amacrine cells aligning one-dimensionally connects to a ganglion cell with inhibitory inputs (Fig. 2.28). There is a delay of the inhibitory inputs to the ganglion cell if an object moves in a *preferred* direction so that the ganglion cell receives excitatory inputs from *on*-bipolar cells. When an object moves in a non-preferred direction, the inhibitory inputs reach to the ganglion cell in advance of the excitatory inputs from the *on-*

bipolar cells so that the ganglion cell does not evoke responses. As a result, this ganglion cell shows directional selectivity.

Ganglion cells are nerve cells that fire spike discharges to send the signals to the brain. Depending on the type of the visual information they send to the brain, the ganglion cells are further classified: one type called *on type*, in which the number of spikes is increased under light illumination, another one called *off type*, in which the number of spikes is reduced, and the other one called *on-off type* that shows spikes at the beginning and end of the light stimulus.

Color information is processed in a similar way. The color information is sensed by three types of cones as three primary colors. In the second cell layer, the similar negative feedback system seems to be operating and antagonistic color is selectively detected. For example, if an *on*-bipolar cell located at the center received red signal, the surrounding cells provides the *off* signal for green signal. Although many such color combinations are believed to be present in the second cell layer, its detail is not fully understood. Anyway, it is believed that the output signal is selectively generated where the color variation is large[a].

In view of a relation with the perception of colors, the followings points should be noted: (1) Two kinds of visual cells, rods and cones, are present; rods mediate night vision, while cones do daylight vision. The spatial distribution of rods and cones differs in a retina. The durations of an electrical signal and sensitivities considerably differ between rod and cone. (2) The sense of color in man is expressed as three primary colors, which correspond to three types of cones (Exercise (6)). (3) The color information is pre-processed within a retina before being sent to the brain. What we recognize as a color by our eyes is the result of our recognition that is processed within our retina and brain, which differs considerably from what we measure using a monochromator.

---

[a]For example, in gold fish, one type of a bipolar cell has a center response of depolarization to green light and a surround response of hyperpolarization to the same green light. The same cell is hyperpolarized with red light at the center while depolarized at the surround.

2.3.3 *How to Express a Color Quantitatively*

2.3.3.1 Munsell color system

Whenever we draw a painting, select clothes and gifts, and display a color on a computer, we have to express the color in a somewhat quantitative way. In order to express the color, it is a custom to refer to a color in a paintbox or of material, and sometimes to express by adding a modifying word to a color. For example, to express a red color, we often use red, scarlet, vermilion, carmine, crimson, . . . , or wine-red, blood-red, cherry-red, . . . , and also a blazing red, fiery red, dark red, ...

However, it is a practical problem that color felt by man differs considerably between individuals. One thinks it red or blue, while another thinks it orange or bluish green. Thus, it is absolutely necessary to standardize the color to quantitatively express it. This was a particularly serious problem among artists from the end of the 19th to the beginning of the 20th century. An American painter and professor Albert Henry Munsell was one of such artists, who established a method to standardize the color system [Munsell (1905)]. This is nowadays known as *Munsell color system*, which is still frequently used in various artistic and technological fields concerning vision.

He considered that color has the following three attributes: *hue*, *chroma* and *lightness*. Hue expresses a variation of tint, while chroma does a degree of saturation and lightness does brightness or darkness. He first considered two extremes for each of these three attributes and divided it equidistantly between these extremes according to visual perception. Then, he made a standard sample of color for each color, which is called *color chip*, and constructed a *color atlas* in which these color chips were arranged regularly.

At first, let us take up hue. He considered red (R), yellow (Y), green (G), blue (B) and purple (P) as fundamental hues, and arranged them in a ring shape to form a *color wheel*. Further, he put intermediate hues between the adjacent two standardized hues and denote them as symbols of YR, GY, BG, PB, and RP. For example, a hue between yellow and red is called YR, while that between green and yellow is called GY. Thus, a color wheel is divided into ten hues and then ten color chips are created. Occasionally, each of ten

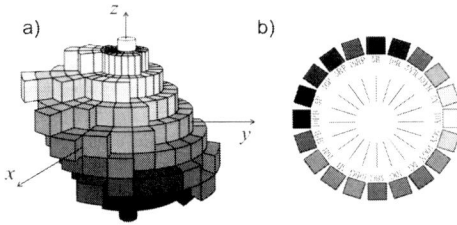

Figure 2.29 Munsell color system [Ikeda (2005)].

hues is further divided into 2, 4 or 10. We show an example of the color wheel in Fig. 2.29b, where each hue is divided into two and totally 20 hues are presented. The divided hues are expressed by numbers such that R is divided into $5R$ and $10R$, for example. The hue thus determined is expressed by a symbol H, which is usually called *Munsell hue*.

Next, we consider lightness. The lightness is expressed by a symbol of V and is called *Munsell value*. For ideal black, incident light of any color is completely absorbed and the lightness is defined as $V = 0$, while for ideal white, that of any color is completely reflected and it is defined as $V = 10$. Between these two extremes, the lightness is divided into ten equidistantly. Then, the lightness was standardized by making color chips. Finally, *Munsell chroma* is defined as C, which expresses the saturation of color. Munsell put $C = 0$ for achromatic color, and increased the number as $1, 2, 3 \ldots$, as the degree of saturation was increased.

In the Munsell color system, each color chip is expressed by a symbol HV/C for chromatic colors and NV for achromatic colors. For example, $5R4/10$ expresses that the hue is red as is expressed by $5R$ and the lightness is on the fourth stage among ten, while the chroma is on the tenth. On the other hand, $N8$ is an achromatic color with the lightness is on the eighth stage among ten. Further, these three attributes are sometimes expressed in a 3D space as *Munsell color solid* shown in Fig. 2.29a. In this figure, the vertical axis, z axis, expresses the lightness, while the color ring is expressed on the xy plane. The hue is arranged along the circumference, while the chroma is expressed by a radius. On a plane giving a constant value of H in the color solid, the difference of colors that are determined solely by chroma and lightness is called *color tone*.

Although, Munsell performed the above procedures entirely through visual perception of man, these attributes are now quantitatively connected with parameters used in XYZ color system, which will be described later. For example, the lightness is quantitatively expressed by the following formula:

$$Y_C = 1.2219V - 0.2311V^2 + 0.23951V^3$$
$$-0.021009V^4 + 0.0008408V^5, \qquad (2.103)$$

where Y_C is a parameter to express the lightness in XYZ color system[a].

2.3.3.2 RGB color system

In the Munsell color system, a color chip should be made for each color. Such a color system is usually called *color appearance system*. On the other hand, color system described from now on provides a method to create a color by mixing standard colors and is called *color mixing system*. Here, we will describe one such example of *RGB color system*, which was established in Commission Internationale de l'Eclairage (CIE) in 1931.

At first, we determine three light waves with three different colors called *reference color stimuli* and denote them as [R], [G] and [B], which corresponds to the wavelengths of monochromatic light of 700.0 nm, 546.1 nm, and 435.8 nm, respectively. It is also defined that additive color mixing of these light waves with amounts of 1.0000 lm for [R], 4.5907 lm for [G] and 0.0601 lm for [B] generates white color[b]. 546.1 and 435.8 nm correspond to emission lines of Hg

---

[a]C in a subscript indicates that a tungsten lamp with a correlated color temperature of 2855.6 K is used as a light source after inserting a filter so as to make a correlated color temperature to be 6744 K. Color temperature is originally determined by a spectrum of black body radiation. The correlated color temperature is determined by a color perceived visually so as to coincide with that of the black body radiation at that temperature.

[b]lm (lumen) is a unit known as one of psychophysical quantities. The psychophysical quantity is a quantity, to which a physical quantity is re-estimated by a psychological quantity based on human perception. 1 lm is a flux of light that is emitted by a point light source of 1 cd (candela) within a solid angle of 1 sr (steradian). 1 cd is the luminous intensity, in a given direction, of a source that emits monochromatic

lamp, whereas 700.0 nm is selected from a wavelength region where human perception does not change much with changing wavelength.

If an unknown color is given, one can determine the color by mixing the above three primary stimuli appropriately to find a coincidence with the given color by human perception, which is called *color matching experiment*. When monochromatic light is used as a test, the proportions of the three light intensities will be determined to that wavelength. By changing the wavelength while keeping the radiation power constant, three curves corresponding to the three primary stimuli will be obtained over a visible wavelength range of 380–780 nm. Namely, if monochromatic light with a wavelength of λ is given, the following coefficients are to be determined:

$$[F_\lambda] = R[R] + G[G] + B[B]. \qquad (2.104)$$

By changing the wavelength, three curves corresponding to R, G and B are obtained.

These curves are then normalized so that the integrated areas over a visible range will be equal to each other, which are called *rgb color matching functions* and are denoted as $\bar{r}(\lambda)$, $\bar{g}(\lambda)$ and $\bar{b}(\lambda)$. Since the above experiment was naturally dependent on individuals, CIE employed the result of the Wright-Guild experiment as a standard [Wright (1929); Guild (1932)]. The color matching functions thus determined are shown in Fig. 2.30a. In this figure, we notice that $\bar{r}(\lambda)$ and $\bar{g}(\lambda)$ are zero at 435.8 nm, while $\bar{r}(\lambda)$ and $\bar{b}(\lambda)$ are zero at 546.1 nm. These indicate the color matching in these wavelengths can be performed only by single colors.

Below 546.1 nm, however, it is found that the color matching cannot be accomplished by mixing [G] and [B] because of the deficiency of brilliance. To reduce the brilliance of the light source, simple mixing of red to the monochromatic light source is found to give a good result. Thus, below 546.1 nm, the relation $[F_\lambda] + R[R] = G[G] + B[B]$ will hold, which causes a negative sign in $\bar{r}(\lambda)$ below 546.1 nm such that $[F_\lambda] = -R[R] + G[G] + B[B]$.

---

radiation at a frequency of 540×10^{12} Hz and that has a radiant intensity in that direction of $1/683$ W·sr^{-1}. This frequency is equivalent to a wavelength of 555 nm, which corresponds to a peak wavelength of relative luminous efficiency curve. The wavelength other than 555 nm is obtained by dividing $1/683$ W·sr^{-1} by relative luminous efficiency curve.

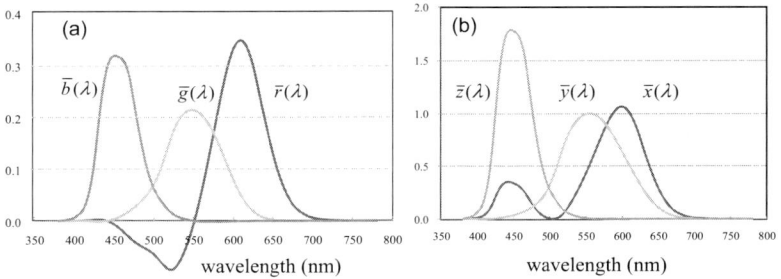

Figure 2.30 Color matching functions for (a) the RGB and (b) XYZ color systems.

2.3.3.3 XYZ color system

Since the addition of a negative amount of color did not sit well, CIE examined another color system, in which all the color matching functions would take positive values. In the same year of 1931, CIE proposed *CIE 1931 standard colorimetric system (XYZ)* or simply *XYZ color system*, where the color matching functions, $\bar{x}(\lambda)$, $\bar{y}(\lambda)$ and $\bar{z}(\lambda)$, were obtained by a linear transformation of the rgb color matching functions, which were called *CIE color matching functions* or now simply *color matching functions*.

The linear transformation is performed in the following way:

$$\begin{pmatrix} \bar{x}(\lambda) \\ \bar{y}(\lambda) \\ \bar{z}(\lambda) \end{pmatrix} = \begin{pmatrix} 2.7689 & 1.7517 & 1.1302 \\ 1.0000 & 4.5907 & 0.0601 \\ 0.0000 & 0.0565 & 5.5943 \end{pmatrix} \cdot \begin{pmatrix} \bar{r}(\lambda) \\ \bar{g}(\lambda) \\ \bar{b}(\lambda) \end{pmatrix}, \quad (2.105)$$

and its inverse transformation is given as

$$\begin{pmatrix} \bar{r}(\lambda) \\ \bar{g}(\lambda) \\ \bar{b}(\lambda) \end{pmatrix} = \begin{pmatrix} 0.41845 & -0.15865 & -0.082834 \\ -0.09116 & 0.25242 & 0.01571 \\ 0.00092 & -0.00255 & 0.17860 \end{pmatrix} \cdot \begin{pmatrix} \bar{x}(\lambda) \\ \bar{y}(\lambda) \\ \bar{z}(\lambda) \end{pmatrix}.$$

$$(2.106)$$

Since rgb color matching functions, $\bar{r}(\lambda)$, $\bar{g}(\lambda)$ and $\bar{b}(\lambda)$, are normalized within a wavelength range between 380 and 780 nm, and also additive color mixing of [R] 1.0000 lm, [G] 4.5907 lm and [B] 0.0601 lm gives achromatic white light with 5.6508 lm, the transformation is devised to make $\bar{y}$ agree with *spectral luminous efficiency* in *photopic vision*[a]. The obtained color matching functions are shown in Fig. 2.30b. Notice that $\bar{y}$ becomes unity at 555 nm, which is a maximum wavelength of spectral luminous efficiency.

By using these color matching functions, various quantities used in XYZ color system are derived. First, tristimulus values, X, Y and Z, for a light source are given as follows:

$$X = k \int_{380}^{780} S(\lambda)\bar{x}(\lambda)d\lambda,$$
$$Y = k \int_{380}^{780} S(\lambda)\bar{y}(\lambda)d\lambda, \qquad (2.107)$$
$$Z = k \int_{380}^{780} S(\lambda)\bar{z}(\lambda)d\lambda,$$

where $S(\lambda)$ is a spectral power distribution. k is a constant of proportionality, and is determined so as to agree with the photometric quantity. If $S(\lambda)$ is given as an absolute value at each wavelength, it is convenient to use a value of $k = 683$ $(\mathrm{lm \cdot W^{-1}})$.

In a similar way, tristimulus values for reflection are given as

$$X = K \int_{380}^{780} S(\lambda)\bar{x}(\lambda)R(\lambda)d\lambda,$$
$$Y = K \int_{380}^{780} S(\lambda)\bar{y}(\lambda)R(\lambda)d\lambda, \qquad (2.108)$$
$$Z = K \int_{380}^{780} S(\lambda)\bar{z}(\lambda)R(\lambda)d\lambda,$$

where $R(\lambda)$ is a spectral reflectance measured at a fixed direction within a fixed viewing angular range that is normally set at $2°$. Further K is determined as

$$K = 100/ \int_{380}^{780} S(\lambda)\bar{y}(\lambda)d\lambda, \qquad (2.109)$$

which means that the condition $Y = 100$ is obtained for a material with 100% reflectance at all the wavelengths. Tristimulus values for

---

[a]Spectral luminous efficiency is the spectral sensitivity of vision perceived by man, which differs by the brightness of the environment. Under bright condition (*photopic vision*), three kinds of cones work in eye, while under dim condition (*scotopic vision*), only rods work, whose spectral sensitivity differ from the sum of sensitivities in cones.

transmission are given in the same way as

$$X = K \int_{380}^{780} S(\lambda)\bar{x}(\lambda)T(\lambda)\mathrm{d}\lambda,$$

$$Y = K \int_{380}^{780} S(\lambda)\bar{y}(\lambda)T(\lambda)\mathrm{d}\lambda, \qquad (2.110)$$

$$Z = K \int_{380}^{780} S(\lambda)\bar{z}(\lambda)T(\lambda)\mathrm{d}\lambda,$$

for spectral transmittance $T(\lambda)$. According to the demands in later times, CIE proposed the utilization of a much wider viewing angular range of $10°$ in 1964 and the color system based on this viewing angle was called *CIE 1964 supplementary standard colorimetric system ($X_{10}Y_{10}Z_{10}$)* or simply $X_{10}Y_{10}Z_{10}$ *color system*.

In XYZ color system, the following *chromaticity coordinates* are defined:

$$x = \frac{X}{X + Y + Z},$$

$$y = \frac{Y}{X + Y + Z}, \qquad (2.111)$$

$$z = \frac{Z}{X + Y + Z},$$

where by using a relation of $x + y + z = 1$, independent variables are found to be only two. A color space spanned by x and y is called *xy chromaticity diagram*. If we employ a monochromatic light wave for a light source, the spectral power distribution is replaced by $S(\lambda') \propto \delta(\lambda' - \lambda)$ and then the equations in Eq. (2.107) are reduced to $X \propto \bar{x}(\lambda)$, $Y \propto \bar{y}(\lambda)$ and $Z \propto \bar{z}(\lambda)$. Thus, a pure color with a wavelength of λ is expressed by these coordinates as

$$x = \frac{\bar{x}(\lambda)}{\bar{x}(\lambda) + \bar{y}(\lambda) + \bar{z}(\lambda)},$$

$$y = \frac{\bar{y}(\lambda)}{\bar{x}(\lambda) + \bar{y}(\lambda) + \bar{z}(\lambda)}, \qquad (2.112)$$

which are called *spectral chromaticity coordinates*. By changing λ from 380 to 780 nm and drawing a spectrum locus, we obtain a chromaticity diagram shown in Fig. 2.31. The spectrum locus starts from the lower left, rises upward and descends to the lower right. Thus the diagram is essentially expressed by a shape of a pot placed

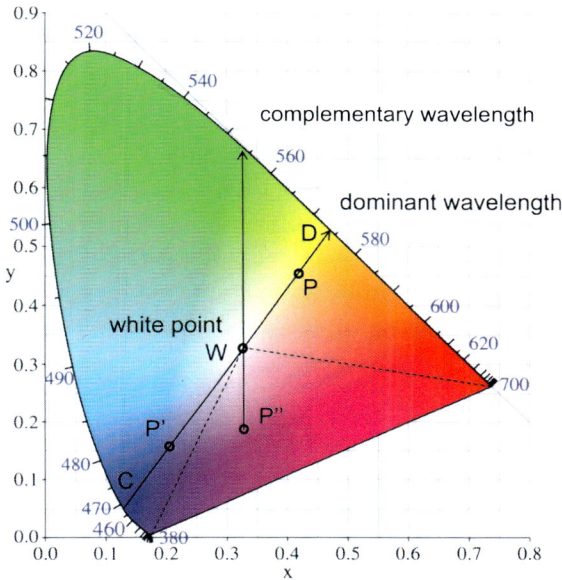

Figure 2.31   Chromaticity diagram in XYZ color system.

upside-down and slightly slanted. All the body colors are expressed as points within a figure obtained by combining the two ends.

A *white point*, W, is obtained by calculating Eq. (2.107) after putting $S(\lambda)$ to be constant. Since $\bar{r}(\lambda)$, $\bar{g}(\lambda)$ and $\bar{b}(\lambda)$ are normalized within this wavelength range, the relation $X = Y = Z$ is obtained from a set of equations of Eq. (2.111), which leads to $x_w = y_w = 1/3$, where x_w and y_w are coordinates of the white point, as shown in this figure. An intersection point D with a perimeter for a straight line passing through points of P and W is called *dominant wavelength* for the point P, while the opposite extension of the line gives another intersection point C called *complementary wavelength*, which corresponds to a dominant wavelength for P'. On the other hand, a point, P'', located within a triangle created by combining a white point W with two ends, has no apparent dominant wavelength. In this case, a straight line is extended to the opposite direction and the intersection point with a perimeter, corresponding to complementary wavelength, is usually employed.

The ratio of distances $\overline{PW}$ and $\overline{DW}$ is called *excitation purity* and becomes a measure of the color saturation.

Thus, in the chromaticity diagram, we can read the hue as a position along the perimeter, while the chroma is expressed by a distance from the white point. On the other hand, the lightness cannot be expressed in this diagram. If one will explicitly express the lightness, it is convenient to use a coordinate (x, y, Y) in a 3D coordinate system, where Y contains the information of lightness.

2.3.3.4 CIELAB and CIELUV color spaces

Although the XYZ color system thus created gave an intuitive way to express hue and chroma of color, MacAdam found that the color matching experiment around a point in a chromaticity diagram showed a certain areal range, within which one could not discriminate the color by the eye [MacAdam (1937)]. He found that the area thus determined differed largely according to the coordinates. He expressed the area by an ellipse, which is now called *MacAdam ellipse* as shown in Fig. 2.32a. Note that the size of ellipse is enlarged by ten times. It is clear that the ellipse is considerably elongated, whose size is radially distributed from the lower left and takes a maximum at the top of the diagram. A large ellipse means that it is hard to determine the color and also to define the *color difference* in the chromaticity diagram.

In 1976, CIE proposed two new color systems to attain *uniform color space*, which considerably improved the apparent disagreement between perceptible color difference and geometrical one in a diagram. These color systems are called *CIE 1976 L*a*b** color space* and *CIE 1976 L*u*v** color space*, or simply *CIELAB* and *CIELUV color spaces*, respectively. In the CIELAB color space, the lightness L^* is defined as

$$L^* = \begin{cases} 116(Y/Y_n)^{1/3} - 16 & \cdots\cdots Y/Y_n > 0.008856, \\ 903.29(Y/Y_n) & \cdots\cdots Y/Y_n \leq 0.008856, \end{cases} \quad (2.113)$$

and is called *CIE 1976 lightness*. On the other hand, the color coordinates a^* and b^* are defined as

$$\begin{cases} a^* = 500[f(X/X_n) - f(Y/Y_n)], \\ b^* = 200[f(Y/Y_n) - f(Z/Z_n)], \end{cases} \quad (2.114)$$

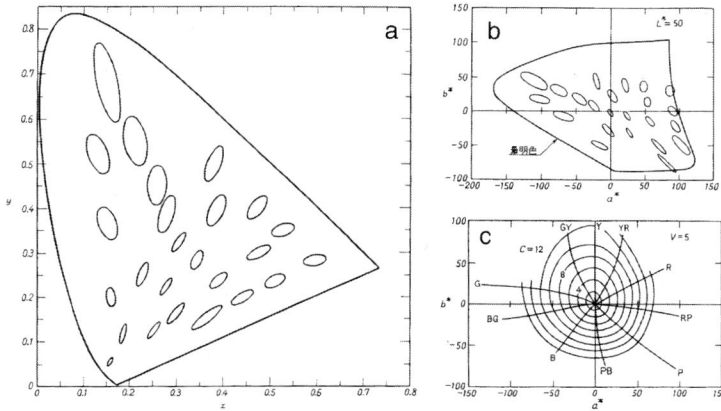

Figure 2.32 MacAdam ellipses in chromaticity diagrams for (a) the XYZ color system and (b) the CIELAB color space. (c) Metric chroma C^* and metric hue angle θ in the CIELAB chromaticity diagram are expressed nearly by concentric circles and lines radiated in all directions, respectively, in the a^*b^* coordinate. (Reproduced from JIS Z8729 with permission.)

where $f(s)$ is given as

$$f(s) = \begin{cases} s^{1/3} & \cdots\cdots s > 0.008856, \\ 7.78s + 16/116 & \cdots\cdots s \leq 0.008856. \end{cases} \quad (2.115)$$

Here, X_n, Y_n and Z_n are tristimulus values of X, Y and Z obtained by using a *perfect reflecting diffuser* under the illuminations of *CIE standard illuminant* or *supplementary standard illuminant*[a]. In Fig. 2.32b, we show a chromaticity diagram for CIELAB color space. It is clear that the size distribution of MacAdam ellipses is considerably improved.

Further, CIE proposed another color space called *CIELUV color space* in 1980. The transformation in this case was slightly simple and the color coordinates u^* and v^* are given as

$$\begin{cases} u^* = 13L^*(u' - u'_n), \\ v^* = 13L^*(v' - v'_n), \end{cases} \quad (2.116)$$

---

[a]Illuminant is a radiation whose relative spectral distribution is provided. CIE standard illuminant indicates illuminant A and *daylight illuminant* D_{65}, whose correlated color temperatures are 2856 and 6504 K, respectively. Supplementary standard illuminant indicates daylight illuminants D_{50}, D_{55} and D_{75} in addition to illuminant C. The daylight illuminant is that close to the daylight.

where

$$\begin{cases} u' = 4X/(X + 15Y + 3Z), \\ v' = 9Y/(X + 15Y + 3Z), \end{cases} \tag{2.117}$$

and u'_n and v'_n are chromaticity coordinates corresponds to a perfect reflecting diffuser under the illuminations of CIE standard illuminant or supplementary standard illuminant. On the other hand, as for L^*, the CIE 1976 lightness is employed.

The chromaticity diagram obtained by these perceptibly uniform color spaces is called *UCS (uniform-chromaticity scale) chromaticity diagram*. By using these chromaticity coordinates, it is possible to express the color difference that is close to perceptible one. If we include the lightness to the chromaticity coordinates and consider a color space in 3D, the color difference between points 1 and 2 is generally expressed in terms of the differences of lightness, chroma and hue. If we choose CIELAB as an example and follow CIE 1994, we can express the color difference ΔE_{21} as

$$\Delta E_{21} = \left(\Delta L_{21}^{*2} + \Delta C_{21}^{*2} + \Delta H_{21}^{*2}\right)^{1/2}, \tag{2.118}$$

where ΔL_{21}^*, ΔC_{21}^* and ΔH_{21}^* are *lightness*, *chroma* and *hue differences*. On the other hand, the lightness and chroma differences are separately defined as

$$\Delta L_{21}^* = L_2^* - L_1^*,$$
$$\Delta C_{21}^* = \left(a_2^{*2} + b_2^{*2}\right)^{1/2} - \left(a_1^{*2} + b_1^{*2}\right)^{1/2}.$$

Originally, as shown in Fig. 2.33, the distance between two spatially separated points in 3D space should be simply expressed as

$$\Delta E_{21}^2 = (L_2^* - L_1^*)^2 + (a_2^* - a_1^*)^2 + (b_2^* - b_1^*)^2. \tag{2.119}$$

However, since we have defined the lightness and chroma differences as above, slightly complicated expression is required for ΔH_{21}^{*2}. At first, we transform ΔC_{21}^{*2} into

$$\Delta C_{21}^{*2} = \left\{\left(a_2^{*2} + b_2^{*2}\right)^{1/2} - \left(a_1^{*2} + b_1^{*2}\right)^{1/2}\right\}^2$$
$$= a_2^{*2} + b_2^{*2} + a_1^{*2} + b_1^{*2} - 2\left(a_2^{*2} + b_2^{*2}\right)^{1/2}\left(a_1^{*2} + b_1^{*2}\right)^{1/2},$$

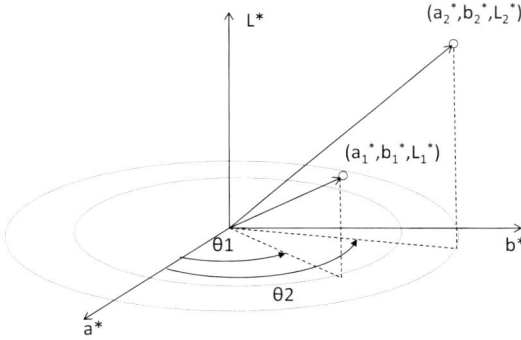

Figure 2.33   Color difference defined in the CIELAB color space.

which leads to the following expression for ΔH_{21}^{*2} using Eq. (2.119):

$$
\begin{aligned}
\Delta H_{21}^{*2} &= \Delta E_{21}^{2} - \Delta L_{21}^{*2} - \Delta C_{21}^{*2} \\
&= (a_2^* - a_1^*)^2 + (b_2^* - b_1^*)^2 - \left\{ (a_2^{*2} + b_2^{*2})^{1/2} - (a_1^{*2} + b_1^{*2})^{1/2} \right\}^2 \\
&= 2 \left(a_2^{*2} + b_2^{*2} \right)^{1/2} \left(a_1^{*2} + b_1^{*2} \right)^{1/2} - 2a_1^* a_2^* - 2b_1^* b_2^* \\
&= 2C_1^* C_2^* \left[1 - (a_1^* a_2^* + b_1^* b_2^*) / \left\{ (a_2^{*2} + b_2^{*2})^{1/2} (a_1^{*2} + b_1^{*2})^{1/2} \right\} \right] \\
&= 2C_1^* C_2^* [1 - \cos(\theta_1 - \theta_2)], \qquad\qquad (2.120)
\end{aligned}
$$

where C^* and θ are radial and angular parts in polar coordinate system and are called *metric chroma* and *metric hue angle*, respectively, which are defined as

$$
C^* \equiv \left\{ a^{*2} + b^{*2} \right\}^{1/2},
$$
$$
\theta \equiv \tan^{-1}(b^*/a^*).
$$

Here, it is defined that ΔH_{21} takes a value of $\Delta H_{21} > 0$ when $\theta_2 > \theta_1$. If we plot C^* and θ thus defined in a chromaticity diagram, it is found that the former and latter are expressed nearly by concentric circles and lines radiated in all directions, as shown in Fig. 2.32c.

Exercises

(1) Confirm the continuity conditions for E_z, H_x, and H_z at an interface shown in Fig. 2.9.

(2) Confirm the energy conservation expressed in Eq. (2.73) using Eqs. (2.63), (2.64), (2.67), (2.68), and (2.72).
(3) Design a Fresnel rhomb using a material having the refractive index of 1.5.
(4) Explain the polarization state after a linearly polarized light beam with the 45° inclined polarization direction is reflected by a flat ideal metal plate placed at 45° to the propagation direction.
(5) Explain how to place a lens to perform the Newton's experiment shown in Fig. 2.22.
(6) Explain the difference between the three primary colors of light and painting from a physiological point of view.

Experiments

I. Measurement of the Light Velocity

Materials: a pulsed laser, a PIN or avalanche photodiode, an oscilloscope, a mirror and a glass plate, a DC power source for the photodiode, cables for connecting the photodiode with the oscilloscope, a sheet of white paper and a measuring tape.

Procedure: Prepare a pulsed laser whose pulse width is less than 10 ns. Since 1 ns corresponds to 30 cm for light to propagate, the narrower, the better. The second harmonics of Q-switched YAG laser is one of such candidates, which usually possesses a pulse width of 6 ns with a green color. A part of the output pulse is split by a glass plate, while the other is reflected back by a mirror that is placed far from the laser (see Fig. 2.34). The pulse deflected by the glass plate and that reflected by the mirror are combined on a white paper, behind which a fast photodiode is placed. The PIN or avalanche type photodiode with a fast response corresponding to the pulse width is preferable. The output of the photodiode is led into an input of a fast oscilloscope with 50 Ω input impedance. You can see two peaks corresponding to the above two pulses. It is better to adjust the intensity of the two peaks by slightly changing the inclination angle of the mirror or glass plate. The differences in traveling time and distance directly give the light velocity. In our experience, placing

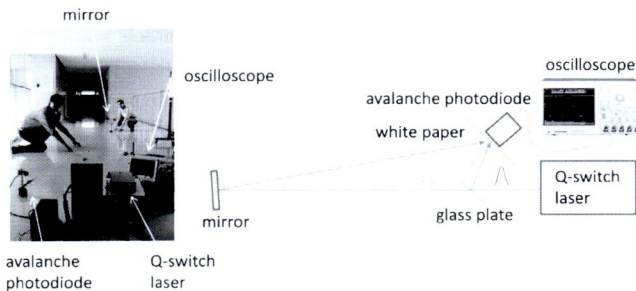

Figure 2.34 Experiment I. Experimental arrangement for the light-velocity measurement.

the mirror at 10–50 m distant from the laser gives an accuracy of two or three figures. The similar experiment can be performed by propagating an electric pulse within a coaxial cable instead of propagating a light pulse, which gives a velocity of electric pulse within a cable.

II. Polarization of Light

Materials: two sheets of film polarizer, a transparent plastic sheet or material such as a cellophane adhesive tape, a transparent set square and a CD cover case, a liquid crystal display, a laser whose output is linearly polarized, and a glass plate.

Procedure: (1) Look at the blue sky through a polarizer and you will find it is considerably polarized particularly at right angles to the direction of a ray of the sun. (2) Investigate the reflected light from various materials and you will be find the reflected light is also polarized due to the presence of Brewster's angle. (3) Look at the liquid crystal display, which is also polarized with the polarization direction of 45° to the horizontal plane (see Figs. 2.35c and d). (4) Superpose two sheets of film polarizers and rotate one of them. This experiment confirms that light does not pass through the sheet at cross polarization, while it passes at parallel polarization. (5) Put a transparent set square or a CD cover case between the polarizers with cross polarization, and you will find a part of light is passed with variously coloring (see Fig. 2.35a). Paste a cellophane adhesive

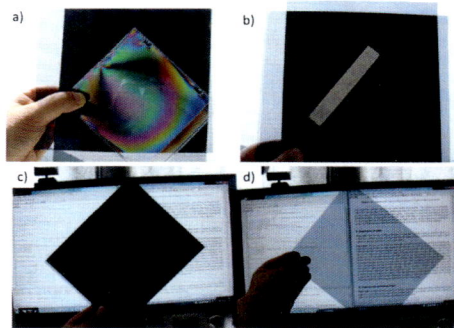

Figure 2.35 Experiment II. Two polarizers in cross polarization are observed with (a) a transparent CD cover and (b) a cellophane adhesive tape placed between them. (c,d) A liquid crystal display screen is seen through a polarizer in two different directions.

tape on one of the polarizers and superpose the other polarizer on it. Then you will see through it even when the polarizers are in cross polarization (see Fig. 2.35b). This phenomenon is only visible when a tape is pasted around $45°$ to the polarization direction. Superposing the adhesive tapes will give various colors depending on the number of the tapes. (6) Using a linearly polarized laser, determine the Brewster angle by minimizing the reflected light intensity from a glass plate.

III. Experiments on Human Vision 1

Materials: a sheet of paper, on which a pattern of black and white radial stripes are painted, and a rotating apparatus whose rotation rate is controllable and readable.

Procedure: Set a sheet of paper with a black-and-white pattern on a rotating apparatus. Look at the sheet while increasing rotation rate and record the rate when you cannot discriminate the stripes. Try this experiment under bright and dim conditions, and you will find the considerable difference between the limiting frequencies under these conditions. The differences reflect a slow response of rods, while under bright condition, cones give a fast response.

Figure 2.36   Experiment III. Rotating apparatus using an optical chopper.

IV. Experiments on Human Vision 2

Materials: a sheet of paper, on which two circles are painted by red and blue, and a display screen.

Procedure: (a) Fix your eyes at a center of two circles painted by red and blue on a sheet. Hold that for about 15 seconds under bright condition (see Fig. 2.37a), and abruptly turn your eyes to another sheet of white paper. You will find pale green or yellow pattern of nearly the same shape on the white paper. This phenomenon is called *color adaptation*. (b) Write letters with a light reddish color (for example, RGB = (253, 116, 3)) on a display screen.

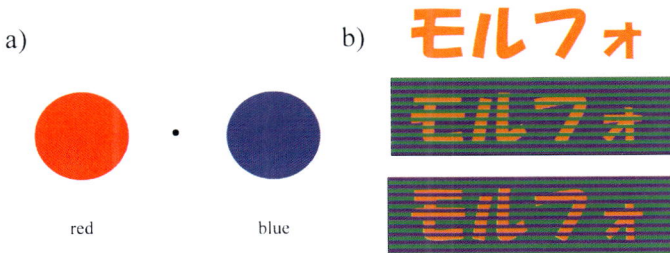

a)

red   blue

b) モルフォ

Figure 2.37   Experiment IV. (a) Paint red and blue circles on a sheet of white paper and fix your eyes on a point at the center. Abruptly turning your eyes to another sheet of white paper, you will find faint circles with complementary color on the sheet. (b) Write letters with a light reddish color on a display screen. Then, the background color is changed into violet or green. Adding stripes on it with green or violet will make you feel that the letters are painted differently.

Then, the background color is changed into violet (112, 48, 160) or green (0,134,0). Addition of stripes with green or violet will make you feel that the letters are painted with a different color (see Fig. 2.37b). This phenomenon is called *chromatic induction*. Investigate the change of the perceived color when the colors of letters, background and stripes, and also their sizes, are changed. It is interesting to compare the perceived color with the RGB colors generated on a display screen and to plot it on a chromaticity diagram. The trajectory on the diagram will give you a quantitative evaluation of this phenomenon.

Chapter 3

Thin-Layer Interference

3.1 Interference of Light

3.1.1 Interference

The interference of light is one of the most important optical phenomena in the field of bionanophotonics. At first, we will summarize its features for the convenience of the later discussion. The interference is a phenomenon, in which waves of more than two are superposed together to cause various patterns in temporal and spatial domains. Since it is based on the superposition principle described in the previous section, and most of interesting phenomena appearing in bionanophotonics originate from this simple principle, here we will consider it generally.

Consider two plane waves traveling toward a direction determined by a wave vector $\mathbf{k}_1$ with an angular frequency of ω_1 and toward that by $\mathbf{k}_2$ with ω_2, which are expressed as

$$\tilde{u}_1 = u_{10}e^{i(\mathbf{k}_1 \cdot \mathbf{r} - \omega_1 t)},$$
$$\tilde{u}_2 = u_{20}e^{i(\mathbf{k}_2 \cdot \mathbf{r} - \omega_2 t)},$$

where u_{10} and u_{20} denote the amplitudes of the two waves, and we have employed a complex representation. These two waves are

Bionanophotonics: An Introductory Textbook
Shuichi Kinoshita
Copyright © 2013 Pan Stanford Publishing Pte. Ltd.
ISBN 978-981-4364-71-3 (Hardcover), 978-981-4364-72-0 (eBook)
www.panstanford.com

assumed to be superposed with each other, which results in

$$\tilde{u}_1 + \tilde{u}_2 = u_{10}e^{i(\mathbf{k}_1\cdot\mathbf{r}-\omega_1 t)} + u_{20}e^{i(\mathbf{k}_2\cdot\mathbf{r}-\omega_2 t)}.$$

As is described later, we will put $u_{10} = u_{20} \equiv u_0$ to maximize the interference effect. Then, the above sum of the two wave functions will be transformed into

$$\tilde{u}_1 + \tilde{u}_2 = u_0 e^{i\{(\mathbf{k}_1+\mathbf{k}_2)\cdot\mathbf{r}-(\omega_1+\omega_2)t\}/2}$$
$$\times \left(e^{i\{(\mathbf{k}_1-\mathbf{k}_2)\cdot\mathbf{r}-(\omega_1-\omega_2)t\}/2} + c.c.\right)$$
$$= 2u_0 e^{i\{(\mathbf{k}_1+\mathbf{k}_2)\cdot\mathbf{r}-(\omega_1+\omega_2)t\}/2}$$
$$\times \cos \frac{(\mathbf{k}_1 - \mathbf{k}_2)\cdot\mathbf{r} - (\omega_1 - \omega_2)t}{2}. \tag{3.1}$$

Thus, the interference effect appears as a generation of an apparent new wave having an average wave vector and angular frequency, which is then modulated by the differences of the wave vectors and angular frequencies.

Generation of interference fringe ($\omega_1 = \omega_2$)   Consider first a simple case, where the angular frequencies of the two waves are assumed to be equal to each other, that is, $\omega_1 = \omega_2 \equiv \omega$. In this case, the sum of the two wave functions is reduced to

$$\tilde{u}_1 + \tilde{u}_2 = 2u_0 \cos(\mathbf{K}\cdot\mathbf{r}/2)e^{i\{(\mathbf{k}_1+\mathbf{k}_2)\cdot\mathbf{r}/2-\omega t\}}, \tag{3.2}$$

where we put $\mathbf{k}_1 - \mathbf{k}_2 \equiv \mathbf{K}$. This relation shows that a new apparent wave propagates toward an average direction of the two waves, while its amplitude is given as $2u_0 \cos(\mathbf{K}\cdot\mathbf{r}/2)$, which means the spatial modulation due to the presence of a cos factor. The direction of the apparent wave propagation and that of the spatial modulation are determined by vectors of $(\mathbf{k}_1 + \mathbf{k}_2)/2$ and $\mathbf{K}/2$, respectively, whose relationships are illustrated in Fig. 3.1a. Thus, the direction of the propagation and that of the spatial modulation are orthogonal to each other.

The pitch of the spatial modulation is generally determined by a relation $\Lambda = 2\pi/K$. Further, by using Fig. 3.1a, it is proved that $K = 2k\sin(\theta/2)$ with $k = 2\pi/\lambda$. Thus, the pitch of the modulation is expressed by $\Lambda = \lambda/\{2\sin(\theta/2)\}$. From this relation, it is noticed that the pitch is related to the crossing angle of θ and is inversely proportional to $\sin(\theta/2)$. Therefore, with increasing crossing angle, the pitch will decrease and finally it takes a minimum value of

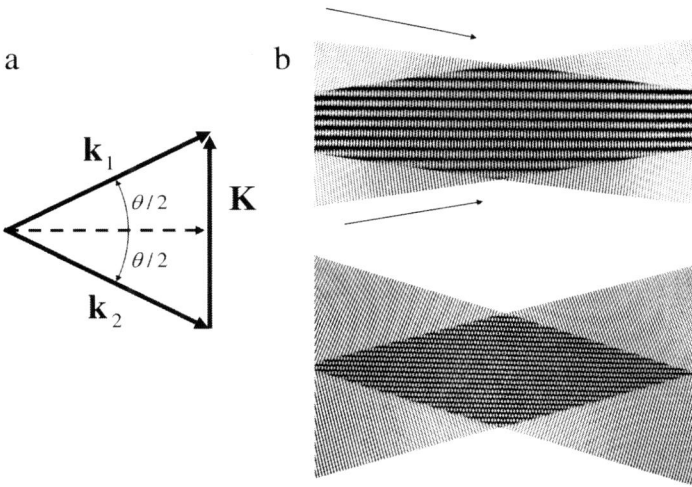

Figure 3.1 Principle of the interference of light. (a) Wave vector conservation rule in case of light interference. $\mathbf{k}_1$ and $\mathbf{k}_2$ are the wave vectors of two incident light waves, while the vector $\mathbf{K}$ is concerned with a fringe generated by the two crossing waves. (b) Demonstrations of the fringes visualized by overlapping two transparent sheets, on which many regular line segments are drawn.

$\lambda/2$, while with decreasing crossing angle, it becomes larger and finally it diverges to infinity in principle. Since the intensity of the wave is proportional to the square of its amplitude, the spatial modulation will be observed according to a factor $\cos^2(\mathbf{K} \cdot \mathbf{r}/2)$, which is essentially stationary and is called *interference fringe*.

The visual demonstration of interference fringe can be easily given by placing two transparent sheets one another, on which many line segments are drawn in parallel with a regular interval. Each line segment corresponds to a wave front of a plane wave. By making the directions of the two "waves" slightly different, apparent "interference fringe" becomes visible. With changing crossing angle, the change of the fringe pitch is easily seen. In reality, this phenomenon is not due to interference fringe, but to moiré fringe. However, since the periodic patterns thus created are similar to each other, it becomes a good demonstration to understand the principle of the interference due to traveling waves.

Enhancement or extinction of waves ($\omega_1 = \omega_2$ and $\mathbf{k}_1 = \mathbf{k}_2$) If the two waves with the same angular frequencies travel toward the same directions, the enhancement or extinction of the waves occur. To demonstrate it, we put $\omega_1 = \omega_2 \equiv \omega$ and $\mathbf{k}_1 = \mathbf{k}_2 \equiv \mathbf{k}$. Further, we consider a general case where the two waves have different initial phases. Then, their sum becomes

$$\tilde{u}_1 + \tilde{u}_2 = (u_0 e^{i\phi_1} + u_0 e^{i\phi_2}) e^{i(\mathbf{k}\cdot\mathbf{r}-\omega t)}$$

$$= 2u_0 \cos \frac{\phi_1 - \phi_2}{2} e^{i\{\mathbf{k}\cdot\mathbf{r}-\omega t+(\phi_1+\phi_2)/2\}},$$

where ϕ_1 and ϕ_2 are the initial phases for the two waves. An apparent wave newly generated by the superposition is that with the average initial phase and its amplitude is characterized by a factor $2u_0 \cos\{(\phi_1 - \phi_2)/2\}$, which is dependent only on the phase difference $\Delta\phi \equiv \phi_1 - \phi_2$ between the two waves. Thus, the amplitude of the superposed wave is directly connected with the phase difference of the original two waves. For example, in case $\Delta\phi = 2m\pi$ with m an integer, the amplitude takes its maximum value of $2u_0$, while in case $\Delta\phi = (2m + 1)\pi$, it becomes zero so that the two waves will completely disappear owing to the interference.

Standing wave ($\omega_1 = \omega_2$ and $\mathbf{k}_1 = -\mathbf{k}_2$) On the other hand, if the two waves travel toward opposite directions, the sum of the two waves become

$$\tilde{u}_1 + \tilde{u}_2 = 2u_0 e^{-i\omega t} \cos(\mathbf{k}\cdot\mathbf{r}), \tag{3.3}$$

where we put $\omega_1 = \omega_2 \equiv \omega$ and $\mathbf{k}_1 = -\mathbf{k}_2 \equiv \mathbf{k}$. Thus, the spatial and temporal parts of the superposed wave are completely separable, which means that the wave does not propagate and only oscillates at its place. Such a wave is called *standing wave* or *stationary wave*. The oscillation frequency is the same as the original wave and the amplitude becomes twice the original wave. The spatial intensity variation in this case is determined by a function of $\cos^2(\mathbf{k}\cdot\mathbf{r})$, whose period is obtained as $\Delta r = \lambda/2$. We show a typical example of the standing wave in Fig. 3.2.

General case ($\omega_1 \neq \omega_2$ and $\mathbf{k}_1 \neq \mathbf{k}_2$) In a general case of $\omega_1 \neq \omega_2$ and $\mathbf{k}_1 \neq \mathbf{k}_2$, the superposition of the two waves is simply expressed by a general relation, Eq. (3.1), of

$$\tilde{u}_1 + \tilde{u}_2 = 2u_0 \cos \frac{\mathbf{K}\cdot\mathbf{r} - \Delta\omega t}{2} e^{i\{(\mathbf{k}_1+\mathbf{k}_2)\cdot\mathbf{r}-(\omega_1+\omega_2)t\}/2}, \tag{3.4}$$

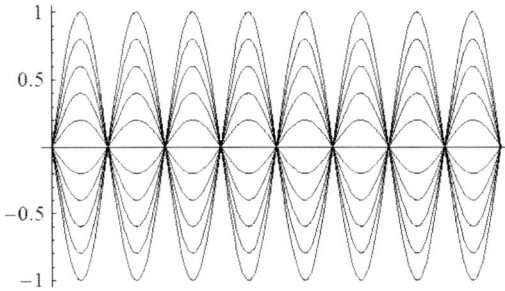

Figure 3.2 Example of standing wave.

where $\Delta\omega = \omega_1 - \omega_2$. This formula is understandable in terms that a new wave having an average wave vector and an average angular frequency is created through the superposition of the waves, while its amplitude is modulated spatially and temporally owing to the second wave propagating toward the direction of $\mathbf{K}$ with the angular frequency of $\Delta\omega/2$. Thus, a moving interference fringe is generated, whose velocity is determined by $\Delta\omega/|\mathbf{K}|$.

The magnitude of $\mathbf{K}$ is simply determined by the crossing angle and the wave vectors of the two incident waves so that it is roughly on the order of the inverse of the wavelength of light, while the angular frequency difference $\Delta\omega$ can be made smaller or larger, depending on the two waves employed. Thus, the velocity of the moving interference fringe can be freely made slower or faster.

As an special case of this moving fringe, we consider that the two waves have nearly the same angular frequencies and travel toward the same direction along the x axis. The superposition of the two wave functions is expressed as

$$\tilde{u}_1 + \tilde{u}_2 = 2u_0 \cos\{(Kx - \Delta\omega t)/2\} e^{i\{(k_1+k_2)x - (\omega_1+\omega_2)t\}/2}. \qquad (3.5)$$

When $\Delta\omega$ is not so large as compared with ω_1 or ω_2, the amplitude of the superposed wave oscillates slowly as compared with that of the original waves, which generates a *beat*, as shown in Fig. 3.3.

The application of the above moving fringe is not so popular. In nonlinear spectroscopy, however, there is a method to utilize this phenomenon, which is known as four wave mixing spectroscopy, where the third light wave is incident to adjust its angle to generate Bragg diffraction concerned with this moving fringe. This causes the

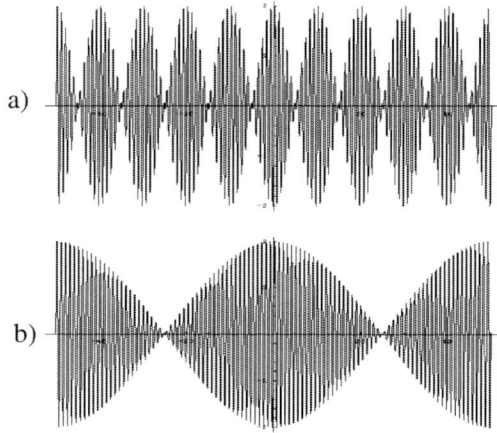

Figure 3.3 Example of light beat. The differences of the angular frequencies of two waves are set at (a) 10 % and (b) 2 % of the original frequency.

fourth light to be emitted, but with an angular frequency different from incident one owing to the Doppler effect caused by the moving fringe. This spectroscopy is often used to probe how and what extent the moving fringe is generated within the material through light-matter interaction.

Finally, we will extend our discussion to include a case of $u_{10} \neq u_{20}$. When $\omega_1 = \omega_2$, the cycle-averaged intensity of the superposed waves is expressed as

$$
\begin{aligned}
I &= \frac{1}{2}|\tilde{u}_1 + \tilde{u}_2|^2 \\
&= \frac{1}{2}u_{10}^2 + \frac{1}{2}u_{20}^2 + u_{10}u_{20}\cos \mathbf{K} \cdot \mathbf{r},
\end{aligned}
\tag{3.6}
$$

where the first and second terms in the right-hand side indicate the intensities of the individual waves, while the last term indicates the interference between them. Since the cos term changes from -1 to $+1$, the intensity changes from $(1/2)(u_{10}-u_{20})^2$ to $(1/2)(u_{10}+u_{20})^2$. If we define visibility as

$$
\begin{aligned}
V &= \frac{(1/2)(u_{10} + u_{20})^2 - (1/2)(u_{10} - u_{20})^2}{(1/2)(u_{10} + u_{20})^2 + (1/2)(u_{10} - u_{20})^2} \\
&= \frac{2u_{10}u_{20}}{u_{10}^2 + u_{20}^2},
\end{aligned}
\tag{3.7}
$$

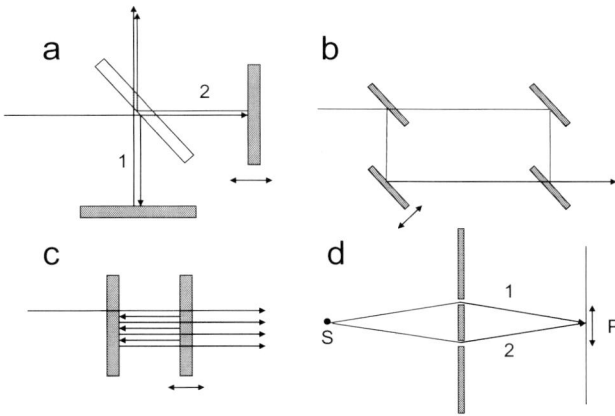

Figure 3.4 Examples of interferometers and interference experiment. (a) Michelson, (b) Mach-Zehnder, (c) Fabry-Perot, and (d) Young's interference experiment.

the maximum visibility is obtained when $u_{10} = u_{20}$. Thus, as we have assumed at the beginning, various interference phenomena described above are clearly visible only when the two waves have nearly the same amplitudes.

3.1.2 *Coherence*

Next, we consider a case where a light wave interferes with itself. To realize this situation, the light wave should be divided into two and they should be superposed again. Thus, special devices are anyway necessary. In Fig. 3.4, we show examples of such devices: (a) Michelson's interferometer, (b) Mach-Zehnder interferometer, and (c) Fabry-Perot interferometer, while (d) shows Young's interference experiment, which is a typical example of interference experiments.

Michelson's Interferometer Here, we will focus our attention on the Michelson's interferometer and the Young's interference experiment. First, we will show the essential point of the first example. The Michelson's interferometer consists of one half-mirror and two flat mirrors (Fig. 3.5). A light wave incident from the left of the interferometer first hits a half-mirror and is divided into

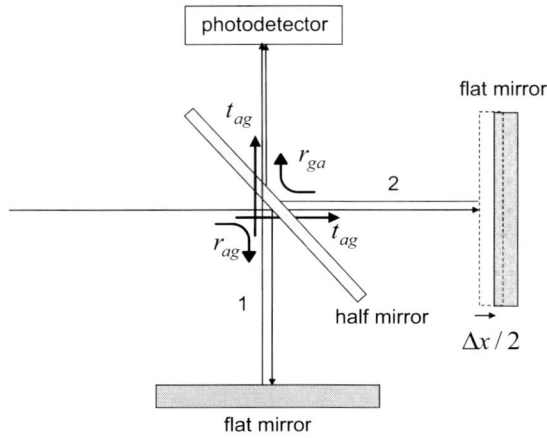

Figure 3.5 Principle of the Michelson's interferometer.

two: One is reflected (pathway 1) and the other is transmitted (pathway 2). The reflected wave is further reflected by a flat mirror and then passes through the half-mirror, while the transmitted wave is reflected by a mirror and then reflected by the half-mirror. These two light waves are superposed and then detected by a photodetector. It is possible to move one of the flat mirrors in parallel to the incident wave so that the relative path length, and hence relative phase, between two pathways can be made variable.

Let us denote the amplitudes of the waves taking these two pathways as $\tilde{u}_1$ and $\tilde{u}_2$, while that of the incident wave as $\tilde{u}_0$. The amplitude reflectivity and transmittance from air to glass are denoted as r_{ag} and t_{ag} with those in the opposite case as r_{ga} and t_{ga}. Then, we obtain

$$\tilde{u} = \tilde{u}_1 + \tilde{u}_2$$
$$= \left(r_{ag} t_{ag} + t_{ag} r_{ga} e^{ik\Delta x} \right) \tilde{u}_0.$$
$$= r_{ag} t_{ag} \left(1 - e^{ik\Delta x} \right) \tilde{u}_0.$$

where we have used the relation $r_{ag} = -r_{ga}$ and have neglected the reflection at the back side of the half mirror. We have also neglected the phase delay based on the path length traveling to the photodetector. We have further put the relative phase difference between these two pathways as $k\Delta x$. Then, the cycle-averaged intensity at the photodetector becomes

$$I = \frac{1}{2}|\tilde{u}|^2$$
$$= u_0^2 \left|r_{ag}t_{ag}\right|^2 (1 - \cos k\Delta x), \tag{3.8}$$

where we put $|\tilde{u}_0|^2 = u_0^2$. This relation shows that the detected signal in the Michelson's interferometer will sinusoidally oscillate with varying Δx, while at $\Delta x = 0$, it gives zero.

It is interesting to raise a simple question: Where does the light wave go away, when the two waves completely disappear owing to interference? The answer of this question is obtained if we calculate the light wave traveling backward to the incident direction. If we denote the amplitudes of these waves as $\tilde{u}_{1b}$ and $\tilde{u}_{2b}$, their sum becomes

$$\tilde{u}_b = \tilde{u}_{1b} + \tilde{u}_{2b}$$
$$= \left(r_{ag}^2 + t_{ag}t_{ga}e^{ik\Delta x}\right)\tilde{u}_0.$$

When the Michelson's interferometer gives null signal, that is, in case of $\cos k\Delta x = 1$, the right-hand side of the second relation becomes $(r_{ag}^2 + t_{ag}t_{ga})\tilde{u}_0$. Since the Fresnel's law states that $r_{ag}^2 + t_{ag}t_{ga} = 1$, it turns out to be $\tilde{u}_0$, which means all the energies of the light waves go backward to the incident direction. Thus, the energy of light will conserve also in case of light interference (Exercise (1)).

Young's interference experiment   Next, we consider the Young's interference experiment. As shown in Fig. 3.4d, this experiment is essentially based on a double-slit experiment, where a point light source designated as S emits light that hits two narrow slits. The light wave passing through the slit will be diffracted, that is, the direction of the propagation is extended. Thus, the two waves diffracted at the two slits interfere on a screen, which gives an interference pattern.

We will designate the two pathways as 1 and 2, where the first one passes through the upper slit and the second one through the lower one. The light waves interfering at a point P on the screen is expressed as

$$\tilde{u}(P) = \tilde{u}_1(P) + \tilde{u}_2(P)$$
$$= \left(a_1 e^{ikr_1} + a_2 e^{ikr_2}\right)\tilde{u}_0,$$

where we take the amplitudes and optical path lengths along the pathways 1 and 2 as a_1 and a_2, and r_1 and r_2, respectively. Here, we take a_1 and a_2 as real, because we consider that the phase difference between the two pathways will be solely determined by the optical-path-length difference.

The cycle-averaged intensity at P is given as

$$I(P) = \frac{1}{2}|\tilde{u}(P)|^2$$
$$= \left(\frac{1}{2}a_1^2 + \frac{1}{2}a_2^2 + a_1 a_2 \cos k\Delta x\right) u_0^2,$$

where we put $\Delta x = r_1 - r_2$. Putting $a_1 = a_2 \equiv a$, we obtain

$$I(P) = a^2 u_0^2 (1 + \cos k\Delta x). \tag{3.9}$$

Thus, apart from a sign in front of the cos term, exactly the same expression is obtained.

These interference experiments can be more generally described as follows: Consider two light waves with the same angular frequency such as

$$\tilde{u}_1 = u_{10} e^{i(\mathbf{k}_1 \cdot \mathbf{r} - \omega t + \phi_1)},$$
$$\tilde{u}_2 = u_{20} e^{i(\mathbf{k}_2 \cdot \mathbf{r} - \omega t + \phi_2)},$$

where u_{10} and u_{20} are assumed to be real. Then, the intensity of the light waves detected at a point $\mathbf{r}$ is described as

$$I = \frac{1}{2}|\tilde{u}_1 + \tilde{u}_2|^2$$
$$= \frac{1}{2}|\tilde{u}_1|^2 + \frac{1}{2}|\tilde{u}_2|^2 + \text{Re}\left\{\tilde{u}_1^* \tilde{u}_2\right\}$$
$$= \frac{1}{2}u_{10}^2 + \frac{1}{2}u_{20}^2 + u_{10}u_{20} \cos\{(\mathbf{k}_1 - \mathbf{k}_2)\cdot\mathbf{r} + (\phi_1 - \phi_2)\}$$
$$= I_1 + I_2 + 2\sqrt{I_1 I_2} \cos\{(\mathbf{k}_1 - \mathbf{k}_2)\cdot\mathbf{r} + (\phi_1 - \phi_2)\}, \tag{3.10}$$

where $I_j = (1/2)u_{j0}^2$ with $j = 1, 2$. This is exactly the same result as has been derived in the preceding subsection (see Eq. (3.6)). Thus, the intensity oscillates between $(\sqrt{I_1} + \sqrt{I_2})^2$ and $(\sqrt{I_1} - \sqrt{I_2})^2$ according to the phase difference of $\phi_1 - \phi_2$.

Coherence Function   The term "coherence" is now commonly used in scientific and non-scientific fields, but its original meaning is rather restricted in wave optics and is associated with the following question: To what extent light interferes with itself? In general, a light wave is called *coherent* when it interferes with itself however the phase difference may increase. On the other hand, a wave is called *incoherent* when it does not interfere however the phase difference may decrease.

Since any kind of wave can, in principle, interfere with itself, thus, both terms of "coherent" and "incoherent" are not absolute but should be relative. It is generally understood that the concept of interference is deeply connected with a duration needed for the detection. Namely, if the maximum time difference, within which the interference is observable, is much smaller than a detection duration, we call such a light wave as incoherent, while we call it coherent if the interference is always observable beyond the detection duration.

In this sense, the interference discussed so far is confined to a case of ideally coherent light wave, because we consider only a monochromatic light wave. On the other hand, ordinary light is an ensemble of waves with various angular frequencies and thus more or less assumes incoherent character. In Fig. 3.6, we show an example of six light waves, each having a different angular frequency by an amount of 4%. The upper figure shows the waves simply superposed with each other, while the lower two figures are the square of the sum of their amplitudes. It is clear that initially the oscillation is clearly seen, while it fades out owing to the difference of the angular frequencies at later times.

There are generally two ways to describe such a partially coherent state. One is to consider such light waves as an ensemble of waves with various angular frequencies, while the other is to consider that a light wave is subject to random modulations in amplitude and phase. Since it is mathematically easy to consider the latter case, we will employ it at first.

Let a general form of a wave function as
$$\tilde{u} = \tilde{A}(t - \mathbf{k} \cdot \mathbf{r}/\omega)e^{i(\mathbf{k} \cdot \mathbf{r} - \omega t)}, \tag{3.11}$$
which states that its temporally and spatially modulated amplitude also propagates with the velocity of the wave. The modulation of the

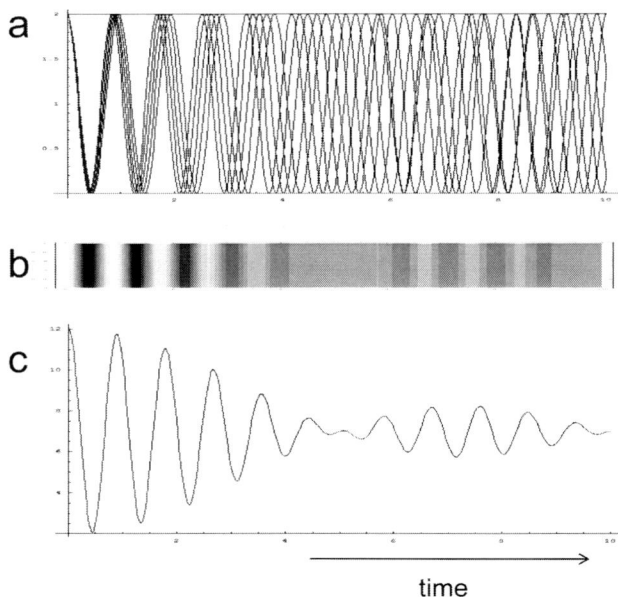

Figure 3.6 Interference of six waves with slightly different angular frequencies. In (a), the waves are drawn one over another, (b) and (c) are the square of the sum of their amplitudes. The angular frequency of each wave differs by 4% from wave to wave.

amplitude can be divided into a pure amplitude part and a phase part, which will be described as

$$\tilde{A}(t - \mathbf{k} \cdot \mathbf{r}/\omega) = \left|\tilde{A}(t - \mathbf{k} \cdot \mathbf{r}/\omega)\right| e^{\mathrm{i}\phi(t - \mathbf{k} \cdot \mathbf{r}/\omega)}. \tag{3.12}$$

The modulations appearing in the amplitude and phase are commonly called *AM* (*amplitude modulation*) and *FM* (*frequency modulation*).

For simplicity, we will consider only a case of phase modulation. Thus, a light wave subject to random phase modulation and its phase-delayed wave are described as

$$\tilde{u}_1 = A_1 e^{\mathrm{i}\{kr_1 - \omega t + \phi(t - kr_1/\omega)\}},$$
$$\tilde{u}_2 = A_2 e^{\mathrm{i}\{kr_2 - \omega t + \phi(t - kr_2/\omega)\}},$$

where the phase delay is incorporated as a path-length difference in terms of r_1 and r_2. Further, putting $kr_j/\omega = \tau_j$ with $j = 1, 2$, we obtain the superposition of these two waves as

$$I = \frac{1}{2}\left\langle |\tilde{u}_1 + \tilde{u}_2|^2 \right\rangle$$
$$= I_1 + I_2 + 2\sqrt{I_1 I_2}\left\langle \cos\left[\omega(\tau_1 - \tau_2) + \phi(t - \tau_1) - \phi(t - \tau_2)\right]\right\rangle$$
$$= I_1 + I_2 + 2\sqrt{I_1 I_2}\left\langle \cos\left[\omega\tau + \Delta\phi\right]\right\rangle,$$

where $\Delta\phi \equiv \phi(t - \tau_1) - \phi(t - \tau_2)$ and $\tau \equiv \tau_1 - \tau_2$, while $\langle\cdots\rangle$ indicates the averaging with respect to ensemble or time. The function $\langle\cos\left[\omega\tau + \Delta\phi\right]\rangle$ becomes zero when $\Delta\phi$ is completely random, while it becomes $\cos\omega\tau$ when the phase does not change randomly, i.e. $\Delta\phi = 0$. Thus, it is reasonable to put the term as

$$\langle\cos\left[\omega\tau + \Delta\phi\right]\rangle = \gamma(\tau)\cos\omega\tau, \tag{3.13}$$

where we call $\gamma(\tau)$ *coherence function*.

For ordinary light, the relation $\gamma(\tau) \rightarrow 1$ holds naturally when $\tau \rightarrow 0$, because $\Delta\phi = 0$ always holds for $\tau = 0$, while the relation $\gamma(\tau) \rightarrow 0$ generally holds when $\tau \rightarrow \infty$. On the other hand, for ideally coherent light, $\gamma(\tau) = 1$ holds irrespectively of τ. We illustrate general trends of $\gamma(\tau)$ in Fig. 3.7. A time characterizing the coherent character of light is called *coherence time*, which we denote as τ_c in the figure. On the other hand, a length during which light shows coherence is called *coherence length*, which is defined as $l_c = c\tau_c$, where c is the light velocity. Typical values of coherence lengths are known to be 30 cm for a 546 nm line of Hg lamp, 300 m for a single mode laser, while only 3 μm is expected for light from an electric bulb. Thus, under a usual detection condition, light from electric bulb is always incoherent, while laser light is usually coherent.

General treatment of coherence function   Since absolute coherence is obtained only from a completely monochromatic wave, the concept of coherence is deeply connected with the monochromaticity of light. Next, we will investigate the relation between these two apparently different concepts. According to Eq. (3.11), let us describe general formulas for two ordinary light waves as

$$\tilde{u}_1 = \tilde{A}_1(t - \tau_1)e^{-i\omega(t-\tau_1)},$$
$$\tilde{u}_2 = \tilde{A}_2(t - \tau_2)e^{-i\omega(t-\tau_2)},$$

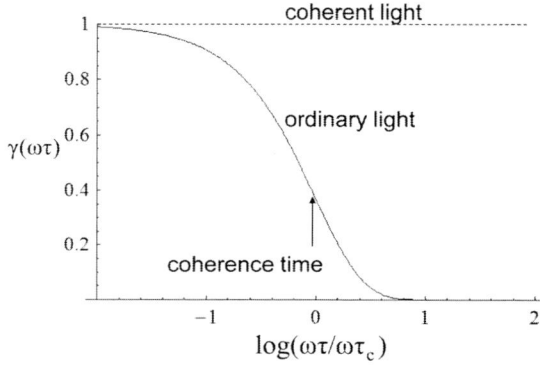

Figure 3.7 Typical examples of the coherence functions for coherent and ordinary light.

where we put $\tau_j \equiv \mathbf{k}_j \cdot \mathbf{r}/\omega$ with $j = 1, 2$. Then, the cycle-averaged intensity of the light waves interfering with each other is expressed as

$$
\begin{aligned}
I &= \frac{1}{2}\left\langle |\tilde{u}_1 + \tilde{u}_2|^2 \right\rangle \\
&= \frac{1}{2}\left\langle \left| \tilde{A}_1(t - \tau_1)e^{i\omega\tau_1} + \tilde{A}_2(t - \tau_2)e^{i\omega\tau_2} \right|^2 \right\rangle \\
&= \frac{1}{2}\left\langle \left| \tilde{A}_1(t - \tau_1) \right|^2 \right\rangle + \frac{1}{2}\left\langle \left| \tilde{A}_2(t - \tau_1) \right|^2 \right\rangle \\
&\quad + \mathrm{Re}\left\{ \left\langle \tilde{A}_1^*(t - \tau_1)\tilde{A}_2(t - \tau_2) \right\rangle e^{-i\omega\tau} \right\} \\
&= \langle I_1 \rangle + \langle I_2 \rangle + \mathrm{Re}\left\{ \left\langle \tilde{A}_1^*(t - \tau_1)\tilde{A}_2(t - \tau_2) \right\rangle e^{-i\omega\tau} \right\},
\end{aligned} \tag{3.14}
$$

where we put $\tau = \tau_1 - \tau_2$ and $\langle I_j \rangle = (1/2)\langle |\tilde{A}_j(t - \tau_j)|^2 \rangle$. Further, we have assumed that the amplitude, $\tilde{A}_j(t)$, does not change during an oscillation period of the light wave.[a]

If we put $t - \tau_1 \equiv t_1$, then $t - \tau_2 = t_1 + \tau_1 - \tau_2 = t_1 + \tau$ and hence

$$
\begin{aligned}
\langle \tilde{A}_1^*(t - \tau_1)\tilde{A}_2(t - \tau_2) \rangle &= \langle \tilde{A}_1^*(t_1)\tilde{A}_2(t_1 + \tau) \rangle \\
&= \frac{1}{T_d}\int_{-T_d/2}^{T_d/2} dt_1\, \tilde{A}_1^*(t_1)\tilde{A}_2(t_1 + \tau),
\end{aligned}
$$

where T_d corresponds to a time to take the time average of data, which usually corresponds to a detection duration. If the correlation

---

[a]This assumption is necessary when the cycle average is calculated.

time of the light wave is much shorter[a] than T_d, the upper and lower limits of the integration can be extended to infinity such that

$$\langle \tilde{A}_1^*(t - \tau_1)\tilde{A}_2(t - \tau_2)\rangle \approx \lim_{T_d \to \infty} \frac{1}{T_d} \int_{-T_d/2}^{T_d/2} dt_1 \, \tilde{A}_1^*(t_1)\tilde{A}_2(t_1 + \tau)$$

$$\equiv \Gamma_{12}(\tau), \tag{3.15}$$

where $\Gamma_{ij}(\tau)$ is generally called *mutual coherence function* or *the first-order mutual correlation function*. Particularly, for $\Gamma_{jj}(\tau)$, it is simply called *coherence function* or *the first-order autocorrelation function*. Using the coherence function thus defined, we have derived the cycle-averaged intensity of the interfering light waves as

$$I = \langle I_1 \rangle + \langle I_2 \rangle + 2\sqrt{\langle I_1 \rangle \langle I_2 \rangle} \, \text{Re}\left\{ \frac{\Gamma_{12}(\tau)}{\sqrt{\Gamma_{11}(0)\Gamma_{22}(0)}} e^{-i\omega\tau} \right\},$$

where we put $\Gamma_{jj}(0) = \langle A_j^*(t_1)A_j(t_1)\rangle = 2\langle I_j \rangle$. Using this formula, we define a *normalized first-order mutual coherence function* as

$$\gamma_{12}^{(1)}(\tau) = \frac{\Gamma_{12}(\tau)}{\sqrt{\Gamma_{11}(0)\Gamma_{22}(0)}}, \tag{3.16}$$

and thus we obtain

$$I = \langle I_1 \rangle + \langle I_2 \rangle + 2\sqrt{\langle I_1 \rangle \langle I_2 \rangle} \, \text{Re}\{\gamma_{12}^{(1)}(\tau)e^{-i\omega\tau}\}. \tag{3.17}$$

If a light wave is divided into two and a time delay is given to one of them, then the intensity of the superposed waves should be expressed by a normalized coherence function of $\gamma_{11}^{(1)}(\tau)$. Thus, when $\gamma_{11}^{(1)}(\tau) = 1$, $I = \langle I_1 \rangle + \langle I_2 \rangle + 2\sqrt{\langle I_1 \rangle \langle I_2 \rangle} \cos\omega\tau$, which expresses a case of completely coherent light wave, while when $\gamma_{11}^{(1)}(\tau) = 0$, the intensity becomes $I = \langle I_1 \rangle + \langle I_2 \rangle$, which expresses that of completely incoherent light wave.

Wiener-Khinchin theorem   Next, we will investigate the relation between the autocorrelation function and the spectrum of light. For this purpose, we put

$$\tilde{u}(t) = \tilde{A}(t - \tau_0)e^{-i\omega_0(t-\tau_0)},$$

---

[a]The condition that the correlation time, τ_c, should be shorter than the time for averaging, T_d, is inevitable to obtain the averaged quantity. This is because τ_c gives a measure for a light wave to maintain its constant phase. Hence, when T_d is shorter than τ_c, nonzero value will be obtained for any averaging process even if the delay time τ is set to be much larger than τ_c and thus the averaging cannot be performed.

and consider the autocorrelation function of the wave function,

$$\langle \tilde{u}^*(t)\tilde{u}(t+\tau)\rangle = \langle \tilde{A}^*(t-\tau_0)\tilde{A}(t+\tau-\tau_0)\rangle e^{-i\omega_0\tau}$$
$$= \Gamma(\tau)e^{-i\omega_0\tau}, \qquad (3.18)$$

where we put

$$\Gamma(\tau) = \langle \tilde{A}^*(t-\tau_0)\tilde{A}(t+\tau-\tau_0)\rangle = \langle \tilde{A}^*(t)\tilde{A}(t+\tau)\rangle, \qquad (3.19)$$

which holds in a stationary case. We further consider that randomly varying complex amplitude of the wave is described by the sum of waves with various angular frequencies such that

$$\tilde{A}(t) = \int d\omega\, \tilde{a}(\omega)e^{-i\omega t}, \qquad (3.20)$$

where $\tilde{a}(\omega)$ is a complex amplitude, which expresses to what extent a wave with an angular frequency ω is included in the original wave. Inserting this relation into Eq. (3.19), we obtain

$$\Gamma(\tau) = \int d\omega \int d\omega'\, \langle \tilde{a}^*(\omega)\tilde{a}(\omega')\rangle e^{i\omega t}e^{-i\omega'(t+\tau)}.$$

In the case of stationary light, the autocorrelation function should not depend on time t. Thus, in order to cancel t, $\omega = \omega'$ in the integrand should be automatically satisfied. Hence, the integrand should have a form $\langle \tilde{a}^*(\omega)\tilde{a}(\omega')\rangle = G(\omega)\delta(\omega-\omega')$, where $\delta(\omega-\omega')$ is a delta function. Inserting this into the above relation, we obtain

$$\Gamma(\tau) = \int d\omega\, G(\omega)e^{-i\omega\tau}, \qquad (3.21)$$

or inversely,

$$G(\omega) = \frac{1}{2\pi}\int d\tau\, \Gamma(\tau)e^{i\omega\tau}, \qquad (3.22)$$

where $G(\omega) \equiv \langle |\tilde{a}(\omega)|^2 \rangle$ is the square of the amplitude for an angular frequency of ω and corresponds to *spectral broadening*. These two relations, Eqs. (3.21) and (3.22), are called *Wiener-Khinchin theorem* and become quite important to evaluate the characteristics of light, because they directly combine correlation function with spectral broadening. A fact that these two quantities are combined through Fourier transform states that a light wave giving a short correlation time will show a broad spectrum, while that giving long one will show a narrow spectrum.

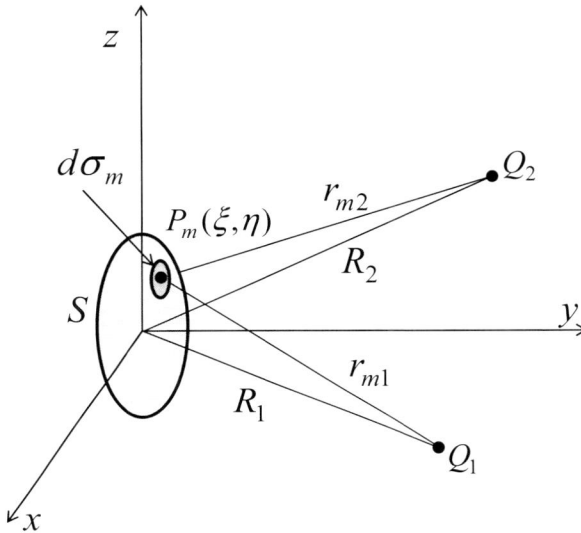

Figure 3.8 Geometry to explain the spatial coherence.

By using this relation and Eq. (3.18), the Fourier transform of the autocorrelation of the wave function can be calculated as

$$S(\omega) = \frac{1}{2\pi} \int d\tau \; \langle \tilde{u}^*(t)\tilde{u}(t+\tau) \rangle e^{i\omega\tau}$$

$$= \frac{1}{2\pi} \int d\tau \; \Gamma(\tau) e^{i(\omega-\omega_0)\tau} = G(\omega - \omega_0), \qquad (3.23)$$

where $S(\omega)$ corresponds to a so-called *spectrum of light*. Thus, the Fourier transform of the autocorrelation function of a wave function directly gives the spectrum of light.

3.1.3 *Spatial Coherence*

So far, we have only considered the coherence in time domain, which is often called *temporal coherence*. However, since a wave function has a spatial part in addition to a temporal part, it is possible to consider *spatial coherence* in contrast to temporal coherence.

Consider a disk-shaped light source designated as S shown in Fig. 3.8, which is assumed to emit light in all directions uniformly.

The light is to be detected at two different points, Q_1 and Q_2. Further, we divide the light source into small areas, $d\sigma_m$, around a point P_m, where m is an integer. The distances between this point and the detecting points are denoted as r_{m1} and r_{m2}, respectively, while those from the coordinate origin are denoted as R_1 and R_2.

The wave function of light emitted from $d\sigma_m$, which will be observed at each point of Q_j, can be described as

$$\tilde{u}_{mj}(t) = \tilde{A}_m(t - \tau_{mj}) \frac{e^{-i\omega(t-\tau_{mj})}}{r_{mj}}, \qquad (3.24)$$

where we put $\tau_{mj} \equiv k r_{mj}/\omega$ and assume that the amplitude of wave decreases in inverse proportion to a distance from the light source. The wave function of this type is also a solution of wave equation and is called *spherical wave*, which will be described later in detail (Sec. 6.3.1). Then, a mutual correlation function of the amplitude of the wave functions evaluated at Q_1 and Q_2 becomes

$$\Gamma_{12}(\tau) = \left\langle \sum_{m,n} \tilde{u}_{m1}^*(t)\tilde{u}_{n2}(t) \right\rangle$$

$$= \left\langle \sum_m \tilde{A}_m^*(t - \tau_{m1})\tilde{A}_m(t - \tau_{m2}) \right\rangle \frac{e^{-i\omega(\tau_{m1}-\tau_{m2})}}{r_{m1}r_{m2}},$$

where τ in the left-hand side is defined as $\tau \equiv k(R_1 - R_2)/\omega$ and we have assumed that light emitted from different areas of the light source have no correlation with each other[a]. We further assume that the light source has a sufficiently long coherence length in comparison with the difference in the distances from the light source and the two detecting points. This assumption is not realistic but is based on our consideration that only spatial coherence is taken into account. Then,

$$\Gamma_{12}(\tau) \approx \Gamma_{12}(0) = \sum_m \left\langle \tilde{A}_m^*(t)\tilde{A}_m(t) \right\rangle \frac{e^{-i\omega(\tau_{m1}-\tau_{m2})}}{r_{m1}r_{m2}}$$

$$= \sum_m I_m \frac{e^{-i\omega(\tau_{m1}-\tau_{m2})}}{r_{m1}r_{m2}} \rightarrow \int_S d\sigma \, I(\xi,\eta) \frac{e^{ik(r_2-r_1)}}{r_1 r_2}, \qquad (3.25)$$

where we put $I_m \equiv \left\langle \tilde{A}_m^*(t)\tilde{A}_m(t) \right\rangle$ under the assumption of stationary light. Further, instead of calculating the sum over all the

---

[a]Namely, we have assumed $\left\langle \tilde{A}_m^*(t - \tau_{m1})\tilde{A}_n(t - \tau_{n2}) \right\rangle = 0$ for $m \neq n$.

space points on a plane of the light source, we have integrated over an area of S after replacing the coordinate of a point P_m by (ξ, η) and r_{mj} by $r_j(\xi, \eta)$ with $j = 1, 2$. The final expression is known as *van Cittert-Zernike theorem.*

Detecting at a distant place   If the distance between the light source and detecting point is sufficiently long compared with the size of the light source, the following approximation is applicable:

$$r_j = \sqrt{(x_j - \xi)^2 + (y_j - \eta)^2 + (R_j^2 - x_j^2 - y_j^2)}$$

$$\approx R_j \left\{ 1 - \frac{x_j\xi + y_j\eta}{R_j^2} \right\},$$

where (x_j, y_j) is the (x, y)-coordinate of a detecting point Q_j. By inserting this approximation into Eq. (3.25), the phase factor in the exponent becomes

$$k(r_2 - r_1) \approx k(R_2 - R_1) - k \left(\frac{x_2\xi + y_2\eta}{R_2} - \frac{x_1\xi + y_1\eta}{R_1} \right)$$

$$= k(R_2 - R_1) - k \left\{ \left(\frac{x_2}{R_2} - \frac{x_1}{R_1} \right) \xi + \left(\frac{y_2}{R_2} - \frac{y_1}{R_1} \right) \eta \right\}$$

$$= k(R_2 - R_1) - (\Delta k_x \xi + \Delta k_y \eta),$$

where $\Delta k_x \equiv k(x_2/R_2 - x_1/R_1)$ and $\Delta k_y \equiv k(y_2/R_2 - y_1/R_1)$. Thus, Eq. (3.25) becomes

$$\Gamma_{12}(0) \approx \frac{1}{R_1 R_2} e^{ik(R_2 - R_1)} \int_S d\xi\, d\eta\; I(\xi, \eta)\; e^{-i(\Delta k_x \xi + \Delta k_y \eta)}, \qquad (3.26)$$

or

$$\Gamma_{12}(0) \approx \frac{1}{R_1 R_2} e^{ik(R_2 - R_1)} \int_S d\mathbf{r}\; I(\mathbf{r})\; e^{-i\Delta \mathbf{k} \cdot \mathbf{r}}. \qquad (3.27)$$

Therefore, a spatial correlation function is essentially equivalent to the Fourier transform of the intensity distribution of the light source. Consider a simple case where the detecting points are located on a plane almost perpendicular to the incident direction. In this case, the above relation is describable as in the following: if the size of the light source becomes larger, the distribution of $\Delta \mathbf{k}$ becomes smaller and hence the detections at two separated points show spatial correlation only when the detecting points are placed sufficiently near to each other. On the other hand, if the size of the light source

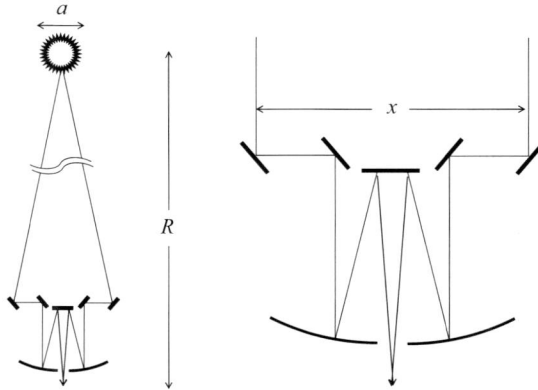

Figure 3.9   Michelson stellar interferometer.

is smaller, long spatial coherence is obtainable. In this sense, a point light source has, in principle, an infinite length of spatial correlation.

The characteristic length of spatial correlation l_c will be defined as the spatial separation of two detecting points corresponding to $|\Delta \mathbf{k}|$, which we call *spatial coherence length*, hereafter. In one dimension, l_c can be roughly estimated by approximating Eq. (3.26) as $\Gamma_{12}(0) \propto \int_S d\xi\ I(\xi) \exp[i\Delta k\xi]$. This relation gives a rough estimation of Δk under an assumption that the light source emits light uniformly. That is, $\Delta k \approx 2\pi/a$ with a the size of the light source. On the other hand, Δk can be combined with l_c through the relation $\Delta k = kl_c/R$ (see the right figure of Fig. 3.9 with $x \approx l_c$), where R is a distance between the light source and detecting point. Thus, l_c is obtained as

$$l_c \approx 2\pi R/(ka) = \lambda R/a. \tag{3.28}$$

Experiments on spatial coherence   This relation has been often used in astronomy to estimate the diameters of stars from interference experiments. If we observe the sun light on the earth, a spatial correlation length l_c is estimated at 5×10^{-5} m when we use the parameters, $a \sim 1.4 \times 10^9$ (m), $R \sim 1.5 \times 10^{11}$ (m) and $\lambda \sim 5 \times 10^{-7}$ (m). On the other hand, when we observe light from a fix star, say Sirius, which is located much far from the earth, l_c amounts to 40 m for $a \sim 10^9$ (m) and $R \sim 8 \times 10^{16}$ (m). Michelson proposed this method to obtain the diameters of stars and realized it by constructing a so-called Michelson stellar interferometer, as shown in Fig. 3.9 [Michelson and Pease (1921)]. On the other hand, Hanbury Brown and Twiss performed an intensity correlation experiment instead of directly interfering light arriving at distant positions, and succeeded to largely expand the detector distance, and hence to reduce the detectable diameter of stars [Hanbury Brown and Twiss (1956)] (Exercise (2)).

Even in a laboratory, spatial coherence sometimes becomes the topics of a talk when a light source other than a laser, or that consisting of scatterers illuminated by a laser, is used for the interference experiment. Consider a pinhole of the diameter of 1 mm, which is illuminated by an ordinary light source, say a low-pressure mercury lamp with the wavelength of 546 nm for example, and perform a Young's interference experiment using a double slit placed at a distance of 30 cm separated from the pinhole. The spatial coherence length in this case becomes $x = 0.546 \times 3 \times 10^4 / 10^3 = 16$ (μm). Thus, a separation of two slits should be sufficiently smaller than 16 μm, otherwise no interference fringe will be observed.

3.2 Thin-Film Interference

Thin-film interference is one of the simplest mechanisms to produce structural colors, yet it plays an important role in bionanophotonics and has a wide variety. Typical examples of thin-film interference can be found in soap bubble, oil film and antireflection coating on glasses, as shown in Fig. 3.10, which is anyway characterized by pale coloration accompanying color change with viewing angle.

The principle of this mechanism is explained as follows: Consider a case where a plane wave of light is incident on a thin film of thickness d and refractive index n_b with an angle of incidence θ_a. Then, the light waves reflected at two surfaces will interfere with each other as shown in Fig. 3.11. Thus, the thin-film interference can be regarded as one of the interferometers described in the previous section. In fact, if the reflectivities at the surfaces increase sufficiently, it just corresponds to a Fabry-Perot interferometer shown in Fig. 3.4c. The difference of the optical path lengths, ΔL,

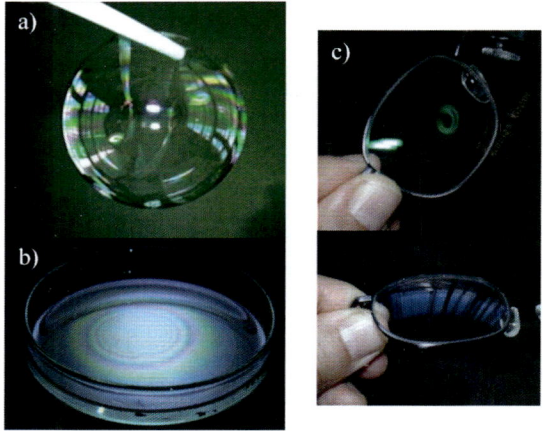

Figure 3.10 Examples of thin-film interference: (a) soap bubble, (b) oil film and (c) antireflection coating on glasses.

evaluated at AA' for the two reflected light waves is obtained as $\Delta L = n_b(\overline{OB} + \overline{BA'}) - n_a\overline{OA}$, which results in

$$\Delta L = 2n_b d / \cos\theta_b - 2n_a d \tan\theta_b \sin\theta_a$$
$$= 2n_b d / \cos\theta_b - 2n_b d \sin^2\theta_b / \cos\theta_b$$
$$= 2n_b d \cos\theta_b,$$

where θ_b and n_a are the angle of refraction and the refractive index of the medium on the incident side, respectively. Further, we have used the Snell's law of $n_a \sin\theta_a = n_b \sin\theta_b$.

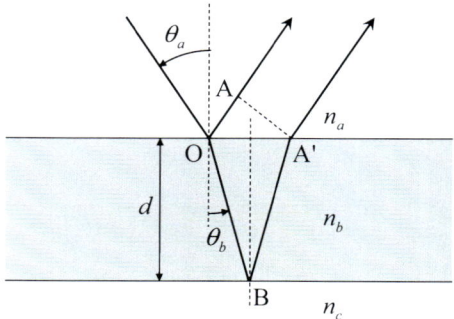

Figure 3.11 Simple mechanism for thin-film interference.

In general, the interference condition differs whether the thin film is attached to a material having a higher refractive index or not. The former is a case for monolayer antireflection coating on glasses and binoculars. A typical example of the latter is a soap bubble. According to the Fresnel's law (Eqs. (2.63), (2.64), (2.67), and (2.68)), the difference is expressed in terms of the phase change during the reflection at a surface. When light is incident from a material with a smaller refractive index to that with higher one, the phase of the wave changes by π, while it does not in the opposite case. For example, when light is incident on a soap bubble, light passes through air ($n = 1.0$) $\rightarrow$ water ($n \sim 1.33$) $\rightarrow$ air ($n = 1.0$). Thus the reflection at the first surface changes its phase, while that at the second one does not. In case of monolayer antireflection coating on glasses, light passes through air ($n = 1.0$) $\rightarrow$ coating material ($1.0 < n < 1.5$) $\rightarrow$ glass ($n \sim 1.5$), in which the phase changes occur at both surfaces. Since the amount of each phase change is exactly π, they do not eventually affect the interference condition.

If we take only the reflections at two surfaces into account, a simple expression for the interference condition is obtained by considering both the amplitude reflectivity at each surface and the difference of the optical path lengths given above. The amplitude of the reflected light wave is then given by the sum of two components:

$$\tilde{u} \propto r_{ab} + r_{bc}e^{2ikn_b d \cos\theta_b},$$

where k is a wavenumber in vacuum, and r_{ab} and r_{bc} are the amplitude reflectivities for light waves propagating from a to b and from b to c, respectively, with c denoting a medium on the transmission side. In case of antireflection coating, assuming $r_{ab} = r_{bc}$, we obtain

$$\tilde{u} \propto r_{ab} \left(e^{-ikn_b d \cos\theta_b} + e^{ikn_b d \cos\theta_b} \right) e^{ikn_b d \cos\theta_b}$$
$$= 2r_{ab} \cos\left(kn_b d \cos\theta_b\right) e^{ikn_b d \cos\theta_b}.$$

On the other hand, in case of soap bubble, a relation $r_{ab} = -r_{bc}$ holds and hence the following relation is obtained:

$$\tilde{u} \propto r_{ab} \left(e^{-ikn_b d \cos\theta_b} - e^{ikn_b d \cos\theta_b} \right) e^{ik_0 n_b d \cos\theta_b}$$
$$= -2ir_{ab} \sin\left(kn_b d \cos\theta_b\right) e^{ikn_b d \cos\theta_b}.$$

Since the power reflectivity is proportional to a square of the absolute value of the amplitude, the interference condition in antireflection coating is derived from the relation

$$|\tilde{u}|^2 \propto \cos^2\left(kn_b d \cos\theta_b\right).$$

Thus, $|\tilde{u}|^2$ takes a maximum when a relation $kn_0 d \cos\theta_b = m\pi$ is satisfied, which leads to the well-known condition for constructive interference:

$$2n_b d \cos\theta_b = m\lambda, \tag{3.29}$$

where we put $k = 2\pi/\lambda$ with λ a wavelength giving the maximum reflectivity and m is an integer expressing the order of interference.

On the other hand, that for soap bubble, the condition is obtained from the relation

$$|\tilde{u}|^2 \propto \sin^2\left(kn_b d \cos\theta_b\right).$$

Since $|\tilde{u}|^2$ takes a maximum when $kn_0 d \cos\theta_b = (2m-1)\pi/2$ in this case, the condition for constructive interference is given as

$$2n_b d \cos\theta_b = \left(m - \frac{1}{2}\right)\lambda. \tag{3.30}$$

The conditions for destructive interference are obtained in a similar way and are given as reverse relations to each other. Namely, Eq. (3.30) becomes the condition for destructive interference for antireflection coating, while Eq. (3.29) becomes that for soap bubble.

In the above treatment, we have considered only one-time reflection at each surface and have neglected the transmission at each surface. However, as shown in Fig. 3.12, when a material has high reflectivity at the surface, multiple reflection more than two times will considerably affect the interference in addition to the decrease in transmittance. An exact calculation considering such effects gives the amplitude reflectivity and transmittance expressed as

$$r = r_{ab} + t_{ab}r_{bc}t_{ba}e^{i\phi} + t_{ab}r_{bc}r_{ba}r_{bc}t_{ba}e^{2i\phi} + \cdots$$
$$= r_{ab} + t_{ab}r_{bc}t_{ba}e^{i\phi}\kappa, \tag{3.31}$$

and

$$t = t_{ab}t_{bc}e^{i\phi/2} + t_{ab}r_{bc}r_{ba}t_{bc}e^{3i\phi/2} + \cdots$$
$$= t_{ab}t_{bc}e^{i\phi/2}\kappa, \tag{3.32}$$

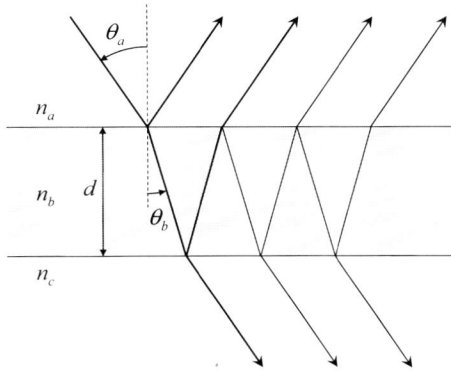

Figure 3.12 Thin-film interference considering multiple reflection.

where

$$\kappa = 1/\left(1 - r_{bc}r_{ba}e^{i\phi}\right),\qquad(3.33)$$

and

$$\phi = 4\pi n_b d \cos\theta_b / \lambda.\qquad(3.34)$$

r_{ab} and t_{ab} are the amplitude reflectivity and transmittance at a surface for the light propagating from a to b, respectively, and are obtained from the Fresnel's law (Eqs. (2.63), (2.64), (2.67) and (2.68)). Then, the (power) reflectivity and transmittance are given as $R = |r|^2$ and $T = \{n_c \cos\theta_c/(n_a \cos\theta_a)\}|t|^2$ according to Eq. (2.72), where we assume $\mu_1 = \mu_2 = \mu_3 = \mu_0$. It is easily verified that the interference conditions given by Eqs. (3.29) and (3.30) correspond to $\kappa = 1$ with $t_{ab}t_{ba} = 1$ and $|r_{ab}| = |r_{bc}|$.

The examples calculated in cases of soap bubble and antireflection coating are shown in Fig. 3.13, where we set a wavelength giving maximum or minimum reflectivity at $\lambda = 500$ nm for $m = 1$. At first glance, one will notice that a reflection spectrum seems to be upside down to the other. Further, the reflectivity is generally low in each case and smoothly changes with wavelength. Thus, simple thin-film interference gives only weak dependence on the wavelength with low reflectivity. It is also noticed that the approximate relations considering only the reflections at the two surfaces can reproduce the reflection spectra considerably well as long as the present parameter set is employed.

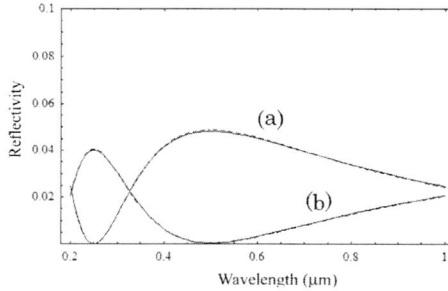

Figure 3.13 Reflection spectra of a film under normal incidence, where (a) soap-bubble (Eq. (3.30)) and (b) antireflection (Eq. (3.29)) conditions are satisfied at the wavelength of 0.5 μm with $m = 1$. The refractive index and thickness of the film are set at 1.25 and 0.1 μm, respectively. The film is assumed to contact with (a) air ($n = 1.0$) and (b) a material ($n = 1.5$), respectively. Dashed lines are the approximate results calculated by putting $\kappa = 1$, $t_{ab}t_{ba} = 1$ and $|r_{ab}| = |r_{bc}|$ in Eqs. (3.31) and (3.32).

Next, we show the incidence-angle dependence in Fig. 3.14. First, we notice that the results are strongly polarization-dependent. For s-polarization, the reflectivity rapidly increases as the incident angle is increased, which is mainly due to the increase of the amplitude reflectivity at the surface due to the Fresnel's law. It is also noticed that slight shift of the peak or dip is observed. This comes from the decrease of the optical path length difference with increasing incidence angle. Thus, the reflection due to thin-film interference will be more vividly perceptible at large angles. On the other hand,

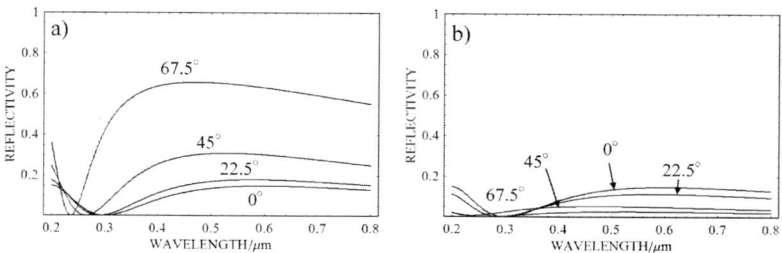

Figure 3.14 Reflection spectra of a film in air for various angles of incidence with (a) s- and (b) p-polarizations. The thicknesses and refractive indices of the film are set at 0.1 μm and 1.5, respectively.

the reflectivity rather decreases for *p*-polarization with increasing incidence angle, though the peak shift is also observed. The reason of this phenomenon comes from a fact that the reflection at the surface is strongly suppressed by the presence of Brewster's angle for *p*-polarization. Thus, for oblique incidence, the contribution of *s*-polarization becomes overwhelming. However, since the reflection spectrum is broad and is only weakly dependent on the incidence angle, the color of the reflected light will be only weakly perceptible.

In order to enhance the reflectivity of thin-film interference even under normal incidence, it is necessary to increase the reflectivity at each surface. To realize this, a material having a higher refractive index will be preferable. TiO_2 ($n = 2.76$ for rutile type and 2.52 for anatase type) film is a good candidate for that purpose and is actually used as *interference pearl luster pigment*. This pigment involves a thin silica flake with both surfaces coated with TiO_2. The thickness of the coating is so adjusted that thin-film interference occurs efficiently in a visible region.

We have performed calculations on the reflectivity due to a thin film with varying refractive index, which are shown in Fig. 3.15. In this figure, the thickness of the film is so chosen that the optical path length of the film under normal incidence always becomes 0.6 μm. It is clear that with increasing refractive index, the reflectivity increases as well. However, when the refractive index increases much more, the overall reflectivity tends to be saturated. Furthermore, the presence of dips, which are originally based on the condition of destructive interference, becomes more remarkable with their widths being gradually narrower. This is because at large difference of refractive indices between the film and the surrounding medium, the reflectivity at the surface becomes considerably larger so that thin-film interference can be regarded as a kind of Fabry-Perot (FP) interferometer (see Fig. 3.4c), since multiple reflections at both surfaces occur effectively. Thus, the dips become narrow wavelength windows for transmitted light, which is the most important function of the FP interferometer. Thus, to make thin-film interference work efficiently as a coloring method, it is necessary to select appropriate reflectivity at the surfaces.

When a thin film is attached to a metal surface, the similar coloration is attained if the kinds of metals are properly selected.

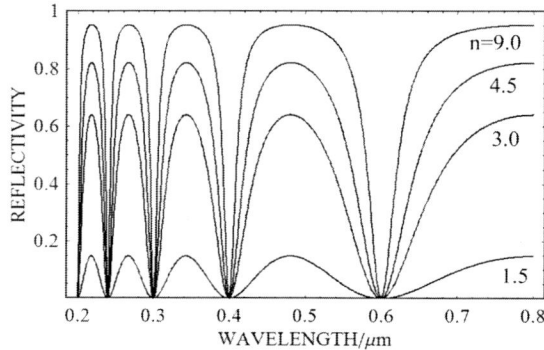

Figure 3.15 Reflection spectra of a film with various refractive indices in air. The thickness of the film is so chosen that the optical path length always agrees with 0.6 μm.

In this case, a thin film is often manufactured by oxidizing the surface of a metal substrate. Typical examples calculated for Ti and Al metals are shown in Fig. 3.16. Although the dispersion of refractive index should be incorporated into the calculation, here we simplify the calculations by employing a constant refractive index of 2.76 for a TiO_2 thin film with a Ti substrate with the refractive

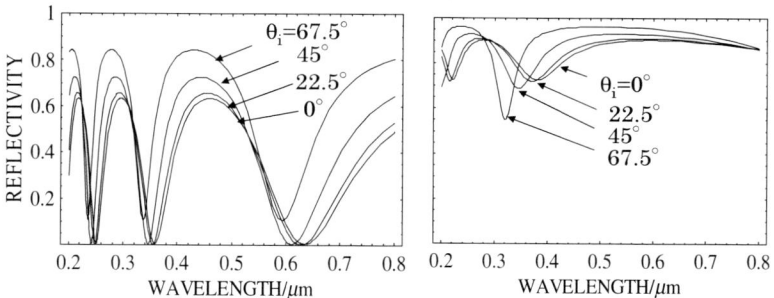

Figure 3.16 Reflection spectra of bulk metal coated with a thin oxide film, calculated for various angles of incidence. The complex refractive indices of the metals employed are (a) 2.16 + 2.93i and (b) 0.68 + 4.80i, which correspond to Ti and Al, respectively. The thicknesses and refractive indices of their oxide films are set at (a) 0.15 μm and 2.76, and (b) 0.15 μm and 1.761, respectively,

index of $2.16 + 2.93i$ and that of 1.761 for a Al_2O_3 film with an Al substrate of $0.68 + 4.80i$. When the reflectivity of metal is high enough as in a case of Al_2O_3-Al system, the total reflectivity is almost saturated and no particular coloration is expected. However, under appropriate reflectivity as in TiO_2-Ti system, the specific coloration and iridescence are clearly observed. The similar effect is obtainable when both surfaces of a transparent film are coated by metal layers with appropriate thicknesses. The pigment containing Al flakes, both sides of which were coated first by MgF_2 and then by Cr or Ti, were first manufactured by Flex Products and were called *optically variable pigment*® (see Sec. 3.6.8), which are now used widely for industrial materials.

When the thickness of the film is increased, regular multiple peak appears in the reflection spectrum owing to the higher-order interference of $m > 1$. These examples are shown in Fig. 3.17. In this figure, we show thin-film interference due to films with the thicknesses of 0.3 and 0.5 μm. It is clear that the number of peaks increases with increasing thickness. If the thickness and refractive index are properly selected, multiple peak appearing in a visible region causes a special effect called *non-spectral color*[a] to human vision. This is in distinct contrast to ordinary *spectral color*, where a single wavelength of light is perceived in eye.

The non-spectral color is deeply connected to the spectral sensitivities of vision and is explained in terms that more than two color-receptors among three are sensed simultaneously. Violet and red are spectral colors, whereas purple and magenta are non-spectral ones. White is also a non-spectral color, where all the three color-receptors are sensed. The multiple reflection peak due to thin-film interference is possible to cause the non-spectral color, if its peak separation fits that of the response of the color receptors. In addition, the angular dependence is more pronounced in the higher-order interference than that for the first order owing to the narrow spectral widths (compare Fig. 3.17 with Fig. 3.14), which causes the color change due to the non-spectral mechanism more

---

[a]The perceived color not determined by a single wavelength of light is called non-spectral color. Cyan and magenta are their examples: Cyan is the mixture of green and blue, and is a complementary color to red, while magenta is that of red and blue, and is a complementary to green.

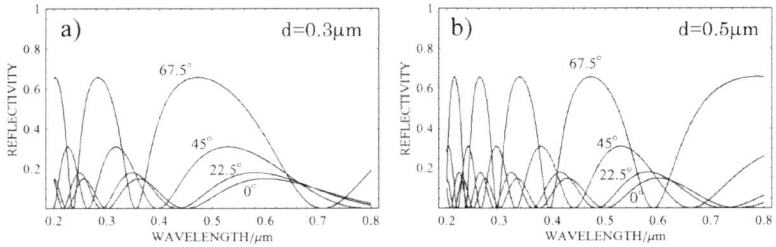

Figure 3.17 Reflection spectra of a film with the thickness of (a) 0.3 and (b) 0.5 μm for various angles of incidence under *s*-polarization. The refractive index of the film is set at 1.5 with that of the surrounding medium of 1.0.

specifically. A typical example has been recently discovered in the neck feather of rock dove, which shows peculiar green/purple two-color iridescence (see Sec. 3.6.5).

3.3 Multilayer Interference

Multilayer interference is also very familiar to us because it has been frequently employed to produce colors in nature and in artificial products. Its examples are shown in Fig. 3.18. Multilayer interference is qualitatively understood in terms that a pair of thin layers pile periodically. Consider two layers designated as A and B with the thickness d_A and d_B, and the refractive indices n_A and n_B,

Figure 3.18 Example of multilayer interference: laser mirror with high reflectance.

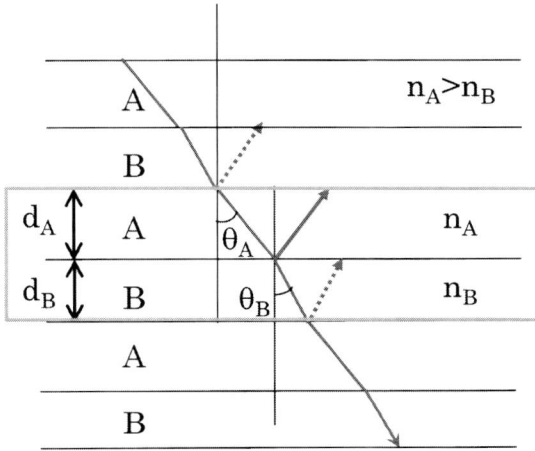

Figure 3.19 Schematic illustration of multilayer interference. Complete constructive interference is obtained when the sum of A and B layers is regarded as a film satisfying the thin-film interference condition for antireflection coating, while a single A or B layer satisfies the soap-bubble condition.

respectively, as shown in Fig. 3.19. We assume $n_A > n_B$ for the present. If we consider a certain pair of AB layers, the phases of the reflected light at the upper and lower B-A interfaces both change by π. Thus, the relation similar to the antireflection coating in thin-film interference of Eq. (3.29) is applicable. Considering that the layer consists of two layers, we obtain

$$2(n_A d_A \cos\theta_A + n_B d_B \cos\theta_B) = m\lambda, \qquad (3.35)$$

for constructive interference with the angles of refraction in A and B layers as θ_A and θ_B. On the other hand, if we consider only A layer within AB layer, the phase of the reflected light does not change at an A-B interface, while it changes at an B-A interface. Thus, if in addition to Eq. (3.35), a soap-bubble relation of Eq. (3.30),

$$2n_A d_A \cos\theta_A = \left(m' - \frac{1}{2}\right)\lambda, \qquad (3.36)$$

is further satisfied, the reflected light from the A-B interface adds to that from the B-A interfaces so that the multilayer will give the maximum reflectivity. Here, $m' \le m$ should be satisfied because of the restriction of the thickness. In particular, $m = 1$ for Eqs. (3.35) and $m' = 1$ for Eq. (3.36) correspond to the lowest-order case where the optical path lengths in the A and B layers are equal to each other. Land called this case *ideal multilayer* [Land (1972)]. On the other hand, if the thickness of A layer does not satisfy Eq. (3.36), while the sum of A and B layers satisfies Eq. (3.35), the reflection at the $A - B$ interface works destructively so that the peak reflectivity will decrease. This case is called *non-ideal multilayer*. Ideal multilayer is effective to increase the reflectivity, while its reflection band width is increased considerably. On the other hand, non-ideal multilayer has a feature to suppress the reflection band width, which is sometimes efficient to show the color more distinctly.

The above interpretation is, however, too simple to understand the mechanism of the multilayer interference. In fact, it is only applicable when the difference in refractive indices of the two layers is small enough. Otherwise, multiple reflection modifies the interference condition to a large extent. Thus, quantitative evaluation of wavelength-dependent reflectivity is generally complex, but is important to understand the functions of bionanophotonics, because most of the bionanophotonic structures in nature originate from multilayer interference or its analogue.

The method to calculate the reflectivity for multilayered structure has been frequently described since Lord Rayleigh presented the paper [Rayleigh (1917)]. Nowadays, the reflectivity and transmittance are often calculated through several conventional methods including transfer matrix, iterative, and Huxley's method (see [Kinoshita (2008)]). Some of these methods are suitable for the evaluation of natural bionanophotonic structures, because a multilayer with arbitrary refractive indices and thicknesses even without any periodicity is treatable.

3.3.1 *Transfer Matrix Method*

3.3.1.1 Derivation of transfer matrix (normal incidence)

Here, we will show two typical methods to calculate the reflectivity and transmittance of multilayer. First, we show a transfer matrix method [Born and Wolf (1959)]. This method is applicable to a

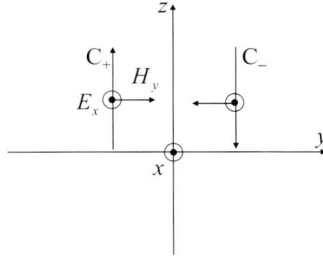

Figure 3.20 Coordinates employed in transfer matrix method.

system that does not have any periodicity of layers without any restriction on thickness, refractive index and number of layers.

Put an electric field of x-polarized light propagating toward the z direction as $E_x(z)$, as shown in Fig. 3.20. Then, the electric field of light within a layer is generally expressed as the sum of waves traveling toward the positive and negative directions along the z axis as

$$E_x(z)e^{-i\omega t} = \left(C_+e^{ikz} + C_-e^{-ikz}\right)e^{-i\omega t}, \qquad (3.37)$$

where $k = n\omega/c$ is a wave number of light in a medium with ω, c and n the angular frequency and velocity of light in vacuum, and the refractive index of the medium, respectively. C_+ and C_- are constants to be determined by a boundary condition. From the Maxwell equation of $\nabla \times \mathbf{E} = i\omega\mu_0\mathbf{H}$, the magnetic field has only a y-component and is expressed as

$$
\begin{aligned}
H_y(z)e^{-i\omega t} &= \frac{1}{i\omega\mu_0}\frac{\partial E_x(z)}{\partial z}e^{-i\omega t}\\
&= \frac{1}{i\omega\mu_0}\left\{C_+(ik)e^{ikz} + C_-(-ik)e^{-ikz}\right\}e^{-i\omega t}\\
&= \frac{k}{\omega\mu_0}\left(C_+e^{ikz} - C_-e^{-ikz}\right)e^{-i\omega t}, \qquad (3.38)
\end{aligned}
$$

where we have employed the permeability of vacuum, μ_0, instead of μ.

At $z = 0$, Eqs. (3.37) and (3.38) become

$$E_x(0) = C_+ + C_-,$$

$$H_y(0) = \frac{k}{\omega\mu_0}(C_+ - C_-).$$

Solving these equations with respect to C_+ and C_-, we obtain

$$C_+ = \frac{1}{2}\left\{E_x(0) + \frac{\omega\mu_0}{k}H_y(0)\right\},$$

$$C_- = \frac{1}{2}\left\{E_x(0) - \frac{\omega\mu_0}{k}H_y(0)\right\}.$$

Inserting these solutions into Eqs. (3.37) and (3.38), we obtain

$$E_x(z) = \frac{1}{2}\left\{E_x(0) + \frac{\omega\mu_0}{k}H_y(0)\right\}e^{ikz} + \frac{1}{2}\left\{E_x(0) - \frac{\omega\mu_0}{k}H_y(0)\right\}e^{-ikz}$$

$$= E_x(0)\cos kz + \frac{i\omega\mu_0}{k}H_y(0)\sin kz, \tag{3.39}$$

$$H_y(z) = \frac{k}{\omega\mu_0}\left[\frac{1}{2}\left\{E_x(0) + \frac{\omega\mu_0}{k}H_y(0)\right\}e^{ikz}\right.$$

$$\left. -\frac{1}{2}\left\{E_x(0) - \frac{\omega\mu_0}{k}H_y(0)\right\}e^{-ikz}\right]$$

$$= \frac{ik}{\omega\mu_0}E_x(0)\sin kz + H_y(0)\cos kz, \tag{3.40}$$

which can be expressed in a matrix form:

$$\begin{pmatrix}E_x(z)\\H_y(z)\end{pmatrix} = \begin{pmatrix}\cos kz & i(\omega\mu_0/k)\sin kz\\i(k/(\omega\mu_0))\sin kz & \cos kz\end{pmatrix}\begin{pmatrix}E_x(0)\\H_y(0)\end{pmatrix}.$$

or

$$\begin{pmatrix}E_x(0)\\H_y(0)\end{pmatrix} = \begin{pmatrix}\cos kz & -i(\omega\mu_0/k)\sin kz\\-i(k/(\omega\mu_0))\sin kz & \cos kz\end{pmatrix}\begin{pmatrix}E_x(z)\\H_y(z)\end{pmatrix}$$

$$= \begin{pmatrix}\cos kz & -i(\eta_0/n)\sin kz\\-i(n/\eta_0)\sin kz & \cos kz\end{pmatrix}\begin{pmatrix}E_x(z)\\H_y(z)\end{pmatrix}$$

$$\equiv M(z)\begin{pmatrix}E_x(z)\\H_y(z)\end{pmatrix}, \tag{3.41}$$

where a matrix $M(z)$ defined here is called *transfer matrix*[a]. $\eta_0 \equiv \sqrt{\mu_0/\epsilon_0}$ is the radiation impedance of vacuum with ϵ_0 permittivity of vacuum

In a multilayer consisting of N layers, the above relation holds in each layer and at a boundary between adjacent two layers, the

[a]When the reflection and transmission properties of a multilayer are calculated, incident and reflected light waves should be considered in an incident space, while only a transmitted light wave is considered in an output space. Thus, it is sometimes convenient to consider that the transmitted light is transformed from the incident and reflected light. According to this custom, we inversely define the transfer matrix.

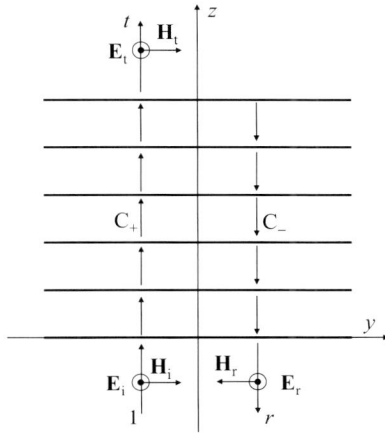

Figure 3.21 Transfer matrix method applied to the calculation of multilayer reflection and transmission.

tangential components of the electric and magnetic fields in one layer are connected with those in the other, as shown in Fig. 3.21. Then, the electric and magnetic fields at the final surface of the last layer are connected with the initial surface of the first layer simply as

$$\begin{pmatrix} E_{x1}(0) \\ H_{y1}(0) \end{pmatrix} = M_1(d_1)M_2(d_2)\cdots M_N(d_N)\begin{pmatrix} E_{xN}(d_N) \\ H_{yN}(d_N) \end{pmatrix}$$

$$\equiv \begin{pmatrix} A & B \\ C & D \end{pmatrix}\begin{pmatrix} E_{xN}(d_N) \\ H_{yN}(d_N) \end{pmatrix}, \tag{3.42}$$

where E_{xj}, H_{yj} and d_j are the x-component of the electric field and the y-component of the magnetic field, and the thickness with respect to the jth layer, respectively.

If we put the amplitude reflectivity and transmittance of a multilayer as r and t, respectively, as shown in Fig. 3.21, and define the transfer matrix as

$$\begin{pmatrix} 1+r \\ (k_i/(\omega\mu_0))(1-r) \end{pmatrix} = \begin{pmatrix} A & B \\ C & D \end{pmatrix}\begin{pmatrix} t \\ (k_t/(\omega\mu_0))t \end{pmatrix}, \tag{3.43}$$

then the calculation of this matrix gives

$$1+r = t\{A + B(k_t/(\omega\mu_0))\},$$
$$1-r = t\{C + D(k_t/(\omega\mu_0))\}/(k_i/(\omega\mu_0)),$$

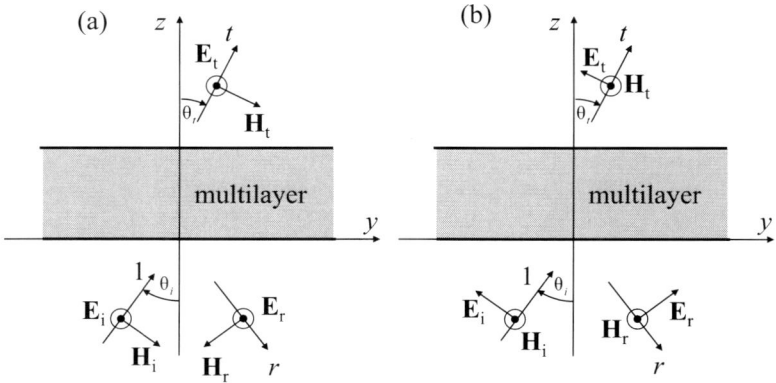

Figure 3.22 Transfer matrix method applied to multilayer reflection and transmission calculations under oblique incidence. (a) *s*- and (b) *p*-polarizations.

which yields

$$
\begin{aligned}
r &= \frac{k_i A + (k_i k_t/(\omega\mu_0))B - \omega\mu_0 C - k_t D}{k_i A + (k_i k_t/(\omega\mu_0))B + \omega\mu_0 C + k_t D} \\
&= \frac{n_i A + (n_i n_t/\eta_0)B - \eta_0 C - n_t D}{n_i A + (n_i n_t/\eta_0)B + \eta_0 C + n_t D},
\end{aligned} \tag{3.44}
$$

and

$$
\begin{aligned}
t &= \frac{2k_i}{k_i A + (k_i k_t/(\omega\mu_0))B + \omega\mu_0 C + k_t D} \\
&= \frac{2n_i}{n_i A + (n_i n_t/\eta_0)B + \eta_0 C + n_t D},
\end{aligned} \tag{3.45}
$$

where $k_i \equiv n_i \omega/c$ and $k_t \equiv n_t \omega/c$ are wave numbers of light in incident and output spaces with n_i and n_t their refractive indices. By using these expressions, the (power) reflectivity and transmittance are described as $R = |r|^2$ and $T = (n_t/n_i)|t|^2$, respectively.

3.3.1.2 Derivation of transfer matrix (oblique incidence)

s-polarization For oblique incidence, the result is dependent on the polarization of incident light and the above expression will be slightly modified. For the polarization perpendicular to the *yz* incident plane (*s*-polarization), the continuity of the electric field

E_x and the magnetic field H_y should be guaranteed at the boundary. Thus, we put[a]

$$E_x(z) = \left(C_+ e^{ikz\cos\theta} + C_- e^{-ikz\cos\theta}\right) e^{iky\sin\theta},$$

$$H_y(z) = \frac{k\cos\theta}{\omega\mu_0}\left(C_+ e^{ikz\cos\theta} - C_- e^{-ikz\cos\theta}\right) e^{iky\sin\theta}.$$

At $z = 0$, these expressions become

$$E_x(0) = (C_+ + C_-)e^{iky\sin\theta},$$

$$H_y(0) = \frac{k\cos\theta}{\omega\mu_0}(C_+ - C_-)e^{iky\sin\theta}.$$

Solving these equations with respect to C_+ and C_- gives

$$C_+ = \frac{1}{2}\left\{E_x(0) + \frac{\omega\mu_0}{k\cos\theta}H_y(0)\right\} e^{-iky\sin\theta},$$

$$C_- = \frac{1}{2}\left\{E_x(0) - \frac{\omega\mu_0}{k\cos\theta}H_y(0)\right\} e^{-iky\sin\theta},$$

which lead to

$$E_x(z) = E_x(0)\cos(kz\cos\theta) + \frac{i\omega\mu_0}{k\cos\theta}H_y(0)\sin(kz\cos\theta),$$

$$H_y(z) = \frac{ik\cos\theta}{\omega\mu_0}E_x(0)\sin(kz\cos\theta) + H_y(0)\cos(kz\cos\theta).$$

The transfer matrix becomes

$$M(z) = \begin{pmatrix} \cos(kz\cos\theta) & -i(\omega\mu_0/(k\cos\theta))\sin(kz\cos\theta) \\ -i(k\cos\theta/(\omega\mu_0))\sin(kz\cos\theta) & \cos(kz\cos\theta) \end{pmatrix}.$$

$$= \begin{pmatrix} \cos(kz\cos\theta) & -i(\eta_0/(n\cos\theta))\sin(kz\cos\theta) \\ -i(n\cos\theta/\eta_0)\sin(kz\cos\theta) & \cos(kz\cos\theta) \end{pmatrix},$$

$$(3.46)$$

where we put $\omega\mu_0/k = \eta_0/n$. Further, the reflectivity and transmittance are calculated from the relation

$$\begin{pmatrix} 1+r \\ (k_i\cos\theta_i/(\omega\mu_0))(1-r) \end{pmatrix} = \begin{pmatrix} A & B \\ C & D \end{pmatrix}\begin{pmatrix} t \\ (k_t\cos\theta_t/(\omega\mu_0))t \end{pmatrix},$$

$$(3.47)$$

where θ_i and θ_t are azimuth angles in incident and output spaces. These results are essentially equivalent to those of normal incidence,

---

[a] In these formulas, a variable y formally appears on their right-hand sides, which will, however, vanish as a common factor owing to the translational invariance within the xy plane.

if we simply replace k as $k \to k \cos \theta$. Therefore, if we put $k \cos \theta \to k^{(s)}$, then the amplitude reflectivity and transmittance are expressed as

$$
\begin{aligned}
r^{(s)} &= \frac{k_i^{(s)} A + \left(k_i^{(s)} k_t^{(s)} / (\omega \mu_0)\right) B - \omega \mu_0 C - k_t^{(s)} D}{k_i^{(s)} A + \left(k_i^{(s)} k_t^{(s)} / (\omega \mu_0)\right) B + \omega \mu_0 C + k_t^{(s)} D} \\
&= \frac{n_i A \cos \theta_i + (n_i n_t \cos \theta_i \cos \theta_t / \eta_0) B - \eta_0 C - n_t D \cos \theta_t}{n_i A \cos \theta_i + (n_i n_t \cos \theta_i \cos \theta_t / \eta_0) B + \eta_0 C + n_t D \cos \theta_t},
\end{aligned}
$$

$$ \tag{3.48} $$

and

$$
\begin{aligned}
t^{(s)} &= \frac{2 k_i^{(s)}}{k_i^{(s)} A + \left(k_i^{(s)} k_t^{(s)} / (\omega \mu_0)\right) B + \omega \mu_0 C + k_t^{(s)} D} \\
&= \frac{2 n_i \cos \theta_i}{n_i A \cos \theta_i + (n_i n_t \cos \theta_i \cos \theta_t / \eta_0) B + \eta_0 C + n_t D \cos \theta_t}.
\end{aligned}
$$

$$ \tag{3.49} $$

***p*-polarization**   On the other hand, for the polarization parallel to the incident plane (*p*-polarization), the continuity of the electric field E_y and the magnetic field H_x should hold at the boundary. We put in this case as

$$
\begin{aligned}
H_x(z) &= \left(C_+ e^{ikz \cos \theta} + C_- e^{-ikz \cos \theta}\right) e^{iky \sin \theta}, \\
E_y(z) &= -\frac{1}{i\omega\epsilon} \frac{\partial H_x(z)}{\partial z} \\
&= -\frac{k \cos \theta}{\omega\epsilon} \left(C_+ e^{ikz \cos \theta} - C_- e^{-ikz \cos \theta}\right) e^{iky \sin \theta},
\end{aligned}
$$

where ϵ is the permittivity of the medium. At $z = 0$, these expressions become

$$
\begin{aligned}
H_x(0) &= (C_+ + C_-) e^{iky \sin \theta}, \\
E_y(0) &= -\frac{k \cos \theta}{\omega\epsilon} (C_+ - C_-) e^{iky \sin \theta},
\end{aligned}
$$

which lead to the solutions

$$
\begin{aligned}
C_+ &= \frac{1}{2} \left\{ -\frac{\omega\epsilon}{k \cos \theta} E_y(0) + H_x(0) \right\} e^{-iky \sin \theta}, \\
C_- &= \frac{1}{2} \left\{ \frac{\omega\epsilon}{k \cos \theta} E_y(0) + H_x(0) \right\} e^{-iky \sin \theta}.
\end{aligned}
$$

Then, $H_x(z)$ and $E_y(z)$ are obtained as

$$H_x(z) = -\frac{i\omega\epsilon}{k\cos\theta} E_y(0)\sin(kz\cos\theta) + H_x(0)\cos(kz\cos\theta),$$

$$E_y(z) = E_y(0)\cos(kz\cos\theta) - \frac{ik\cos\theta}{\omega\epsilon} H_x(0)\sin(kz\cos\theta).$$

The transfer matrix in this case is defined in a similar way as

$$\begin{pmatrix} E_y(0) \\ H_x(0) \end{pmatrix} \equiv M(z) \begin{pmatrix} E_y(z) \\ H_x(z) \end{pmatrix}, \tag{3.50}$$

and is expressed as

$$M(z) = \begin{pmatrix} \cos(kz\cos\theta) & i(k\cos\theta/(\omega\epsilon))\sin(kz\cos\theta) \\ i(\omega\epsilon/(k\cos\theta))\sin(kz\cos\theta) & \cos(kz\cos\theta) \end{pmatrix}$$

$$= \begin{pmatrix} \cos(kz\cos\theta) & i(\eta_0\cos\theta/n)\sin(kz\cos\theta) \\ i(n/(\eta_0\cos\theta))\sin(kz\cos\theta) & \cos(kz\cos\theta) \end{pmatrix}, \tag{3.51}$$

where we put $\omega\epsilon/k = \omega n^2\epsilon_0/(n\omega/c) = nc\epsilon_0 = n\sqrt{\epsilon_0/\mu_0} = n/\eta_0$ with $\epsilon = n^2\epsilon_0$. Further, the reflectivity and transmittance are calculated from the relation

$$\begin{pmatrix} (-1+r)\cos\theta_i \\ k_i(1+r)/(\omega\mu_0) \end{pmatrix} = \begin{pmatrix} A & B \\ C & D \end{pmatrix} \begin{pmatrix} -t\cos\theta_t \\ k_t t/(\omega\mu_0) \end{pmatrix}, \tag{3.52}$$

which leads to

$$-1+r = -t\{A\cos\theta_t/\cos\theta_i - B(k_t/(\omega\mu_0\cos\theta_i))\},$$

$$1+r = -t\{C\cos\theta_t - D(k_t/(\omega\mu_0))\}/(k_i/(\omega\mu_0)).$$

The amplitude reflectivity and transmittance for p-polarization are then expressed as

$$r^{(p)} = \frac{-k_i^{(p)}A + \left(k_i^{(p)}k_t^{(p)}/(\omega\mu_0)\right)B - \omega\mu_0 C + k_t^{(p)}D}{k_i^{(p)}A - \left(k_i^{(p)}k_t^{(p)}/(\omega\mu_0)\right)B - \omega\mu_0 C + k_t^{(p)}D}$$

$$= \frac{-n_i A/\cos\theta_i + n_i n_t B/(\eta_0\cos\theta_i\cos\theta_t) - \eta_0 C + n_t D/\cos\theta_t}{n_i A/\cos\theta_i - n_i n_t B/(\eta_0\cos\theta_i\cos\theta_t) - \eta_0 C + n_t D/\cos\theta_t}, \tag{3.53}$$

and

$$t^{(p)} = \frac{2k_i^{(p)}(\cos\theta_i/\cos\theta_t)}{k_i^{(p)}A - \left(k_i^{(p)}k_t^{(p)}/(\omega\mu_0)\right)B - \omega\mu_0 C + k_t^{(p)}D}$$

$$= \frac{2n_i/\cos\theta_t}{n_i A/\cos\theta_i - n_i n_t B/(\eta_0\cos\theta_i\cos\theta_t) - \eta_0 C + n_t D/\cos\theta_t}, \tag{3.54}$$

where we put $k_j^{(p)} = k_j/\cos\theta_j$ with $j = i, t$.

The features of transfer matrix method are summarized as follows: (1) The reflectivity and transmittance for an arbitrary multilayer are obtained without any periodicity in layer thickness and refractive index, and (2) when the optical thickness of the layer satisfies a simple relation, the reflectivity can be expressed as a simple analytical form as described in the following sections, whereas (3) complicated matrix calculations are sometimes required.

3.3.1.3 Simple relations derived from transfer matrix

We will derive simple relations from the transfer matrix method described above. Consider a transfer matrix for the jth layer under normal incidence, which is expressed as

$$M_j(d_j) = \begin{pmatrix} \cos n_j k_0 d_j & -i(\eta_0/n_j)\sin n_j k_0 d_j \\ -i(n_j/\eta_0)\sin n_j k_0 d_j & \cos n_j k_0 d_j \end{pmatrix}, \quad (3.55)$$

where d_j, n_j and k_0 are the thickness and refractive index of the jth layer, and the wavenumber of light in vacuum. The interference condition for the multilayer consisting of periodic two layers designated as A and B is generally expressed as $2(n_A d_A + n_B d_B) = m\lambda$ under normal incidence, where $d_{A,B}$ and $n_{A,B}$ are the thickness and refractive index of A or B layer. Further, for ideal multilayer, a relation of $n_A d_A = n_B d_B$ is satisfied.

At first, we consider the lowest-order interference of $m = 1$ for ideal multilayer. In this case, $n_j k_0 d_j = \pi/2$ should hold, from the consideration of the above relation. Hence, the relation

$$M_j(d_j) = \begin{pmatrix} 0 & -i(\eta_0/n_j) \\ -i(n_j/\eta_0) & 0 \end{pmatrix}$$

is obtained. Assume that the number of layers constituting the multilayer is totally N, which consists of alternately arranged A and B layers. If the number N is even, the total transfer matrix can be transformed into

$$\begin{pmatrix} A & B \\ C & D \end{pmatrix} = \left[\begin{pmatrix} 0 & -i(\eta_0/n_A) \\ -i(n_A/\eta_0) & 0 \end{pmatrix} \begin{pmatrix} 0 & -i(\eta_0/n_B) \\ -i(n_B/\eta_0) & 0 \end{pmatrix} \right]^{N/2}$$

$$= \begin{pmatrix} -n_B/n_A & 0 \\ 0 & -n_A/n_B \end{pmatrix}^{N/2}$$

$$= (-1)^{N/2} \begin{pmatrix} (n_B/n_A)^{N/2} & 0 \\ 0 & (n_A/n_B)^{N/2} \end{pmatrix}.$$

Inserting this relation into Eq. (3.44), we obtain the amplitude reflectivity r as

$$r = \frac{n_i \, (n_B/n_A)^{N/2} - n_t \, (n_A/n_B)^{N/2}}{n_i \, (n_B/n_A)^{N/2} + n_t \, (n_A/n_B)^{N/2}} = \frac{1-\gamma}{1+\gamma}, \qquad (3.56)$$

where we put $\gamma \equiv (n_t/n_i)(n_A/n_B)^N$. This relation states the amplitude reflectivity takes a value of ± 1 when $\gamma \to 0$ or $\gamma \to \infty$, which indicate that the large difference of refractive index increases the reflectivity. Completely the same relation holds for odd N, although γ should be put as $\gamma = n_A^2/(n_i n_t) \cdot (n_A/n_B)^{N-1}$ (Exercise (3)).

Next, we consider the higher-order interference for an ideal multilayer. In this case, a factor $n_j k_0 d_j$ is expressed as $n_j k_0 d_j = m\pi/2$ with $m > 1$. Thus, for odd orders, the relations $\cos n_j k_0 d_j = 0$ and $\sin n_j k_0 d_j = \pm 1$ are found to hold so that the same result as in a case of $m = 1$ will be obtained. On the other hand, for even orders, $\cos n_j k_0 d_j = (-1)^{m/2}$ and $\sin n_j k_0 d_j = 0$ hold and the transfer matrix becomes

$$\begin{pmatrix} A & B \\ C & D \end{pmatrix} = \begin{pmatrix} (-1)^{m/2} & 0 \\ 0 & (-1)^{m/2} \end{pmatrix}^N = (-1)^{mN/2} \begin{pmatrix} 1 & 0 \\ 0 & 1 \end{pmatrix}. \quad (3.57)$$

Hence, the amplitude reflectivity simply becomes

$$r = \frac{n_i - n_t}{n_i + n_t}, \qquad (3.58)$$

which means that if the refractive indices of incident and output spaces are the same, no reflection occurs irrespective of the number of layers. Such an alteration rule is clearly demonstrated as the appearance/disappearance of the reflection peaks, as shown in Fig. 3.26b.

3.3.2 Huxley's Method

Huxley derived a completely different expression for a periodic multilayer [Huxley (1968)]. He considered that p layers (designated as b) of a higher refractive index were embedded in a medium (designated as a) of a lower refractive index, as shown in Fig. 3.23. Then, the recurrence equations for the field amplitudes evaluated at the center of the a layers become

$$a_{j-1}^- = \rho a_{j-1}^+ + \tau a_j^- \quad \text{and} \quad a_j^+ = \tau a_{j-1}^+ + \rho a_j^-, \qquad (3.59)$$

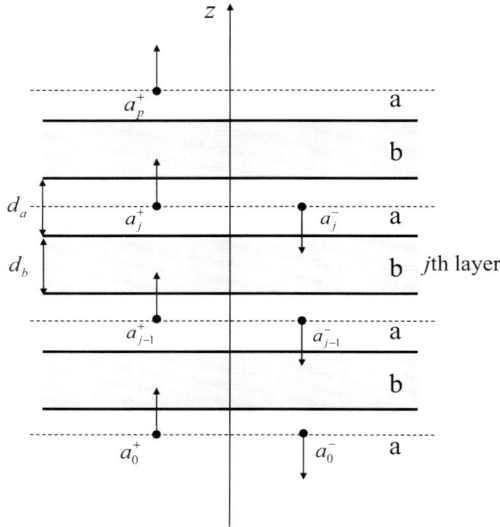

Figure 3.23 Coordinates employed in Huxley's method.

where a_j^+ and a_j^- are the field amplitudes for $+z$ and $-z$ propagating waves in the jth layer. ρ and τ are the amplitude reflectivity and transmittance of a single layer of b evaluated at the center of the a layer, as shown in Fig. 3.24. Actually, ρ and τ are obtained as follows: At first, we use the relations of thin-film interference (Eqs. (3.31) and (3.32)) described as

$$r = r_{ab} + \frac{t_{ab} r_{bc} t_{ba} e^{i\phi}}{1 - r_{bc} r_{ba} e^{i\phi}} \quad \text{and} \quad t = \frac{t_{ab} t_{bc} e^{i\phi/2}}{1 - r_{bc} r_{ba} e^{i\phi}}.$$

Then, these relations are rewritten after putting $c \rightarrow a$ and replacing the notations as $r_{ba} \rightarrow r$ and $\phi \rightarrow -2\phi_b$ with $\phi_b = -2\pi n_b d_b / \lambda$. Further, considering the phase shift within the a layers, we transform r and t as $r \rightarrow \rho \exp[i\phi_a]$ and $t \rightarrow \tau \exp[i\phi_a]$, where a phase factor $\exp[i\phi_a]$ is introduced because the amplitude reflectance and transmittance are evaluated at the points of A and B, respectively, with $\phi_a = -2\pi n_a d_a / \lambda$. Thus, we obtain

$$\rho = \left\{ -r + \frac{r(1 - r^2)e^{-2i\phi_b}}{1 - r^2 e^{-2i\phi_b}} \right\} e^{-i\phi_a} = -r \frac{1 - e^{-2i\phi_b}}{1 - r^2 e^{-2i\phi_b}} e^{-i\phi_a}, \quad (3.60)$$

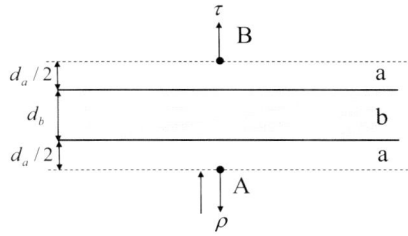

Figure 3.24 Thin-film interference evaluated at points of A and B.

and

$$\tau = \frac{(1 - r^2)e^{-i\phi_b}}{1 - r^2 e^{-2i\phi_b}} e^{-i\phi_a}, \tag{3.61}$$

where we have used a relation of $r_{ab}^2 + t_{ab}t_{ba} = 1$.

Huxley considered the following trial functions for the solutions of the recurrence equations, Eq. (3.59):

$$a_j^+ = a_0^+ \mu^j \quad \text{and} \quad a_j^- = h a_0^+ \mu^j, \tag{3.62}$$

where μ and h are unknown constants. Inserting these functions into Eq. (3.59), we obtain

$$h = \rho + \tau h \mu, \tag{3.63}$$

$$\mu = \tau + \rho h \mu. \tag{3.64}$$

Further, eliminating either μ or h from the above equations, we obtain the following two quadratic equations:

$$\mu^2 - \mu(1 + \tau^2 - \rho^2)/\tau + 1 = 0, \tag{3.65}$$

$$h^2 - h(1 - \tau^2 + \rho^2)/\rho + 1 = 0. \tag{3.66}$$

By eliminating the product μh from Eqs. (3.63) and (3.64), we confirm that μ and h are connected with each other through a linear relation of

$$h = (\tau\mu - \tau^2 + \rho^2)/\rho. \tag{3.67}$$

Since Eq. (3.59) is linear with respect to a_j^+ or a_j^-, the superposition of the two independent solutions, $h_{1,2}$ or $\mu_{1,2}$, gives general solutions[a]:

$$a_j^+ = \alpha a_0^+ \mu_1^j + (1 - \alpha) a_0^+ \mu_2^j, \tag{3.68}$$

$$a_j^- = \alpha h_1 a_0^+ \mu_1^j + (1 - \alpha) h_2 a_0^+ \mu_2^j, \tag{3.69}$$

---

[a] The general solution for the first one should be given as $a_j^+ = \alpha a_0^+ \mu_1^j + \beta a_0^+ \mu_2^j$ with arbitrary constants α and β. However, when $j = 0$, a relation $\alpha + \beta = 1$ should be satisfied, which leads to $\beta = 1 - \alpha$. The second solution is derived directly from a condition that $a_j^- = h a_j^+$ is satisfied for each set of 1 and 2.

where α is a constant. The suffices 1 and 2 to μ and h correspond to negative and positive signs in the solutions of Eqs. (3.65) and (3.66), respectively. The following relations are generally derived from Eqs. (3.65) and (3.66) by inserting ρ of Eq. (3.60) and τ of Eq. (3.61):

$$\mu_1 + \mu_2 = (1 + \tau^2 - \rho^2)/\tau = 2\frac{\cos 2\phi - r^2}{1 - r^2}, \tag{3.70}$$

$$\mu_1 \mu_2 = 1, \tag{3.71}$$

and

$$h_1 + h_2 = (1 - \tau^2 + \rho^2)/\rho = -\frac{2\cos\phi}{r}, \tag{3.72}$$

$$h_1 h_2 = 1, \tag{3.73}$$

The last equalities in Eqs. (3.70) and (3.72) are obtained as follows: By using Eqs. (3.60) and (3.61), the left-hand sides of the equalities are further transformed into

$(1 + \tau^2 - \rho^2)/\tau$

$$= \frac{\left\{ \left(1 - r^2 e^{-2i\phi}\right)^2 + (1 - r^2)^2 e^{-4i\phi} - r^2 \left(1 - e^{-2i\phi}\right)^2 e^{-2i\phi} \right\} / \left(1 - r^2 e^{-2i\phi}\right)^2}{(1 - r^2)e^{-2i\phi}/\left(1 - r^2 e^{-2i\phi}\right)}$$

$$= \frac{\left(1 - r^2 e^{-2i\phi}\right)\left(1 - 2r^2 e^{-2i\phi} + e^{-4i\phi}\right) / \left(1 - r^2 e^{-2i\phi}\right)^2}{(1 - r^2)e^{-2i\phi}/\left(1 - r^2 e^{-2i\phi}\right)}$$

$$= \frac{e^{2i\phi} - 2r^2 + e^{-2i\phi}}{1 - r^2} = 2\frac{\cos 2\phi - r^2}{1 - r^2}, \tag{3.74}$$

and

$(1 - \tau^2 + \rho^2)/\rho$

$$= \frac{\left\{ \left(1 - r^2 e^{-2i\phi}\right)^2 - (1 - r^2)^2 e^{-4i\phi} + r^2 \left(1 - e^{-2i\phi}\right)^2 e^{-2i\phi} \right\} / \left(1 - r^2 e^{-2i\phi}\right)^2}{-r\left(1 - e^{-2i\phi}\right) e^{-i\phi}/\left(1 - r^2 e^{-2i\phi}\right)}$$

$$= \frac{(1 - e^{-4i\phi})(1 - r^2 e^{-2i\phi})/(1 - r^2 e^{-2i\phi})^2}{-r\left(1 - e^{-2i\phi}\right) e^{-i\phi}/\left(1 - r^2 e^{-2i\phi}\right)}$$

$$= \frac{1 - e^{-4i\phi}}{-r(1 - e^{-2i\phi})e^{-i\phi}} = -\frac{2\cos\phi}{r}, \tag{3.75}$$

where we assume an ideal multilayer with $\phi_a = \phi_b \equiv \phi$ for simplicity.

The boundary condition $a_p^- = 0$ in Eq. (3.69) gives $\alpha h_1 a_0^+ \mu_1^p + (1 - \alpha)h_2 a_0^+ \mu_2^p = 0$, which results in

$$\alpha = h_2/(h_2 - m^2 h_1), \tag{3.76}$$

with $m^2 = (\mu_1/\mu_2)^p$. Thus, the power reflectivity and transmittance are obtained as

$$R = \left|\frac{a_0^-}{a_0^+}\right|^2 = |\alpha h_1 + (1-\alpha)h_2|^2$$

$$= \left|\frac{h_1 h_2 + h_2^2 - m^2 h_1 h_2 - h_2^2}{h_2 - m^2 h_1}\right|^2 = \left|\frac{1 - m^2}{h_2 - m^2 h_1}\right|^2, \quad (3.77)$$

and

$$T = \left|\frac{a_p^+}{a_0^+}\right|^2 = |\alpha \mu_1^p + (1-\alpha)\mu_2^p|^2$$

$$= \left|\frac{h_2 \mu_1^p + h_2 \mu_2^p - m^2 h_1 \mu_2^p - h_2 \mu_2^p}{h_2 - m^2 h_1}\right|^2$$

$$= \left|\frac{(h_2 - h_1)\mu_1^p}{h_2 - m^2 h_1}\right|^2 = \left|\frac{(h_2 - h_1)m}{h_2 - m^2 h_1}\right|^2, \quad (3.78)$$

where we have used $h_1 h_2 = 1$ in the first relation. The last equality in the second relation is derived by using a relation of $\mu_1^p = (\mu_1/\mu_2)^{p/2} = m$ with $\mu_1 \mu_2 = 1$.

The physical meaning of Eq. (3.77) is explained as follows: First, using the relations Eqs. (3.70)–(3.73), we have converted Eqs. (3.65) and (3.66) into

$$\mu^2 - 2\frac{\cos 2\phi - r^2}{1 - r^2}\mu + 1 = 0, \quad (3.79)$$

$$h^2 + 2\frac{\cos \phi}{r}h + 1 = 0. \quad (3.80)$$

Then, the discriminants of the above two quadratic equations, D_μ and D_h, become

$$D_\mu/4 \equiv \left(\frac{\cos 2\phi - r^2}{1 - r^2}\right)^2 - 1 = \frac{(\cos 2\phi - 1)(\cos 2\phi + 1 - 2r^2)}{(1 - r^2)^2}$$

$$= \frac{4(1 - \cos^2 \phi)(r^2 - \cos^2 \phi)}{(1 - r^2)^2},$$

$$D_h/4 \equiv \left(\frac{\cos^2 \phi}{r}\right)^2 - 1 = \frac{\cos^2 \phi - r^2}{r^2}.$$

Thus, it is clear that D_μ and D_h are complementary with respect to the term $\cos^2 \phi - r^2$. Hence, when $\mu_{1,2}$ are real, $h_{1,2}$ become complex, and *vice versa*.

(1) Case for $\cos^2 \phi - r^2 < 0$

The solutions $\mu_{1,2}$ are real, while $h_{1,2}$ are complex having the absolute values of unity through the relation Eq. (3.73). The reflectivity in this case is further calculated as

$$R = \left| \frac{1 - m^2}{(h_1 + h_2)(1 - m^2)/2 - (h_1 - h_2)(1 + m^2)/2} \right|^2 ,$$

where the first and second terms in the denominator are real and pure imaginary, respectively. Hence, it is transformed into

$$R = \frac{(1 - m^2)^2}{(h_1 + h_2)^2(1 - m^2)^2/4 - (h_1 - h_2)^2(1 + m^2)^2/4}$$

$$= \frac{1}{\cos^2 \phi/r^2 - \{(1 + m^2)/(1 - m^2)\}^2 (\cos^2 \phi - r^2)/r^2}$$

$$= \left\{ 1 + \frac{4m^2}{(1 - m^2)^2} \left(1 - \frac{\cos^2 \phi}{r^2} \right) \right\}^{-1} , \qquad (3.81)$$

where we have used the relation $(h_1 - h_2)^2/4 = (h_1 + h_2)^2/4 - 1 = (\cos^2 \phi - r^2)/r^2$. Since we have defined $\mu_{1,2}$ as $\mu_1 < \mu_2$, the relation $m^2 = (\mu_1/\mu_2)^p < 1$ holds naturally so that m^2 tends to zero with increasing number of layers, p, and hence the reflectivity becomes unity. Thus, the relation $\cos^2 \phi - r^2 < 0$ gives the range of reflection band, which just corresponds to one-dimensional photonic band gap, which will be described later (see Sec. 5.1.4).

(2) Case for $\cos^2 \phi - r^2 > 0$

The solutions $h_{1,2}$ are real, while $\mu_{1,2}$ are complex having the absolute values of unity. Putting $\mu_{1,2} \equiv \exp[\mp i\theta]$, we calculate the reflectivity as

$$R = \left| (h_1 + h_2)/2 - \frac{1 + m^2}{1 - m^2}(h_1 - h_2)/2 \right|^{-2}$$

$$= \left\{ \frac{(h_1 + h_2)^2}{4} + \cot^2 p\theta \cdot \frac{(h_1 - h_2)^2}{4} \right\}^{-1} = \left\{ 1 + \frac{\cos^2 \phi - r^2}{r^2 \sin^2 p\theta} \right\}^{-1} .$$

$$(3.82)$$

where we have used a relation $(1 + m^2)/(1 - m^2) = -i \cot p\theta$. The term including $\sin^2 p\theta$ expresses the oscillation appearing in the tails on both sides of the reflection band (Exercise (4)). When $\sin^2 p\theta = 0$, $R \to 0$, while in case of $\sin^2 p\theta = 1$, $R \neq 0$, which gives an envelope function for the tails of the reflection band:

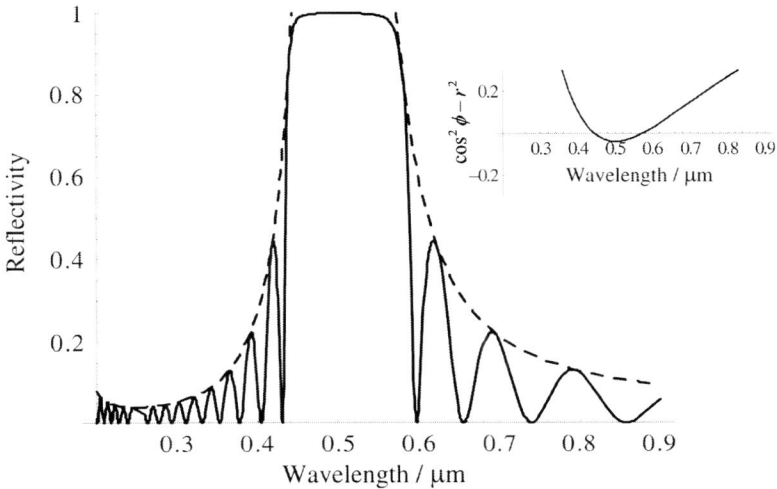

Figure 3.25 Reflection spectra for an ideal multilayer consisting of 10 layers of 0.0833 μm thickness with the refractive index of 1.5, which are embedded in a medium with the refractive index of 1.0. Solid line is calculated using Eq. (3.77), while dashed line is calculated using Eq. (3.83). The parameter, $\cos^2\theta - r^2$, characterizing the reflection properties is also plotted in the inset.

$$R_{\text{env}} = \frac{1}{1 + (\cos^2\phi - r^2)/r^2}. \quad (3.83)$$

In Fig. 3.25, we show an example of the reflectivity calculated using Eq. (3.77) and an envelope function R_{env} of Eq. (3.83).

Huxley's expression for the reflectivity of multilayer has the following features: (1) The reflectivity is obtained through arithmetic computation without iteration or matrix calculation, and (2) reflection bandwidth and oscillation appearing in the tails of the reflection band are obtained analytically, whereas (3) applicable multilayers are considerably limited.

3.3.3 *Features of Multilayer Reflection*

In this subsection, we will investigate the features of multilayer reflection by calculating the reflection spectra of various types of multilayers using the above calculation methods. First, we show,

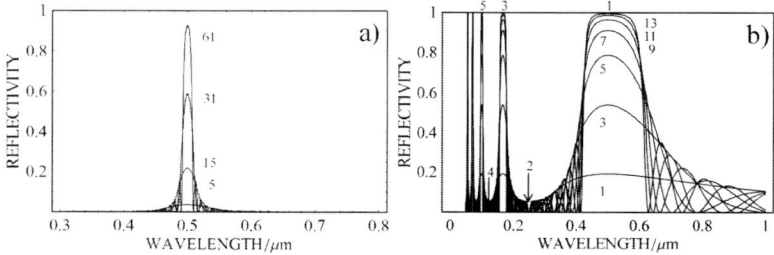

Figure 3.26 Reflection spectra for ideal multilayers consisting of alternate layers with the refractive indices of (a) 1.6 and 1.55, and (b) 1.6 and 1.0. The thicknesses of the layers are set at (a) 78.1 and 80.6 nm, and (b) 78.1 nm and 125 nm, respectively. The numbers of layers are indicated in the figure. In (b), interference orders are also shown. The refractive indices of the incident and output media are assumed to be the same, which are set at (a) 1.55 and (b) 1.0.

in Fig. 3.26, the reflection spectra of ideal multilayer with varying number of layers under normal incidence. In (a), we show a case where the difference of the refractive indices between the two layers is very small ($\Delta n = 0.05$), which is a typical case of the multilayer composed of polymer materials. Since the optical path lengths of both layers are set at 0.125 μm, the peak wavelength of the reflectivity is nearly constant at 500 nm. With increasing number of layers, the peak reflectivity increases rapidly, while the bandwidth decreases gradually.

When the difference of the refractive indices increases, the maximum reflectivity almost reaches unity for only several layers. In Fig. 3.26b, we show the reflection spectra of multilayers for various numbers of layers with the difference of the refractive indices of 0.6. It is found that only 13 layers are enough to make the maximum reflectivity reach almost unity. It is clear that the bandwidth in this case is considerably broader compared with that in the small difference of refractive indices.

The bandwidth for small refractive-index difference can be easily estimated from the difference of the wavelengths giving the minimum reflectivity in the nearest oscillations on both sides. These wavelengths are evaluated from the condition that the reflected light from $(N + 1)$ interfaces interferes destructively and is expressed

N layers

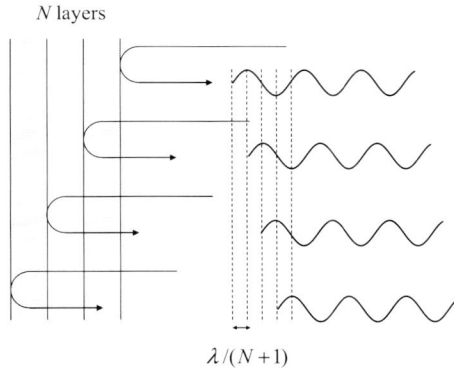

$\lambda/(N+1)$

Figure 3.27 Schematic diagram to explain the destructive interference observed in the tails of a reflection band.

as $2\hat{d} = \{m \pm 1/(N+1)\}\lambda_{\mp}$, where $\hat{d}$ is the optical path length of one layer and m is an integer. This relation is obtained under the assumption of only one-time reflection at each interface, as schematically shown in Fig. 3.27. Using this relation, the bandwidth $\Delta\lambda$ is estimated as a half of the wavelength difference between two adjacent null points:

$$\Delta\lambda \approx (\lambda_+ - \lambda_-)/2 = \left[2\hat{d}/\{m-1/(N+1)\} - 2\hat{d}/\{m+1/(N+1)\}\right]/2$$
$$= 4(N+1)\hat{d}/[2\{m^2(N+1)^2 - 1\}] \approx 2\hat{d}/(m^2 N), \qquad (3.84)$$

where the last relation is obtained under an assumption of large N. Thus, the bandwidth is inversely proportional to the number of layers.

On the other hand, the maximum reflectivity under normal incidence is attained when the relations $2n_A d_A = 2n_B d_B = (m' - 1/2)\lambda$ are satisfied for ideal multilayer. The amplitude reflectivity at this wavelength is obtained using transfer matrix method (Eqs. (3.43) and (3.44)) as $r = (1 - \gamma)/(1 + \gamma)$, where γ satisfies the following relation: $\gamma = (n_t/n_i) \cdot (n_A/n_B)^N$ for even N and $\gamma = n_A^2/(n_i n_t) \cdot (n_A/n_B)^{N-1}$ for odd N. Thus, the absolute value of the reflectivity comes close to unity when the number of layers is increased, irrespective of the magnitudes of the refractive indices of n_A and n_B. Therefore, in order to obtain the high reflectivity for the multilayer having small refractive-index difference, it is necessary to pile up many layers, which inevitably reduces the bandwidth.

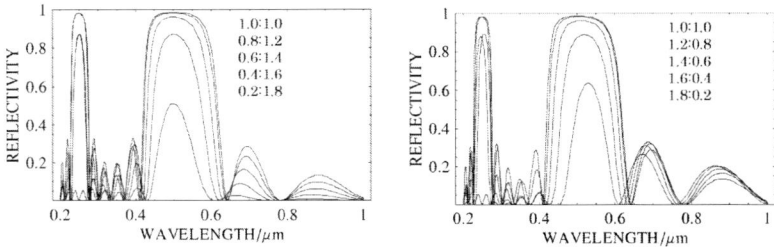

Figure 3.28 Reflection spectra of non-ideal multilayers of 11 alternate layers having the refractive indices of 1.6 and 1.0, respectively. The optical path length of the sum of these two layers is set constant at 0.25 μm, while the ratio of the optical path lengths for the higher refractive index to that for the lower one is changed as is indicated in the figure.

The above relation with respect to r also indicates that the reflectivity easily reaches unity even for small N when the difference of the refractive indices is increased. However, it is difficult to estimate its bandwidth using the approximation employed above, because multiple reflection occurs overwhelmingly. In such a case, it is rather easy to consider the multilayer as a kind of one-dimensional photonic crystal, which will be described in detail in Sec. 5.1.4.3.

On the contrary, the wavelength at which $2n_A d_A = 2n_B d_B = m'\lambda$ is satisfied, the reflection from A-B and B-A interfaces destructively interferes. The reflectivity at this wavelength becomes $r = (n_i - n_t)/(n_i + n_t)$ and becomes null when the refractive indices of incident and output regions are the same. We can easily find this effect in Fig. 3.26b, where the reflection peaks at $1/2$, $1/4$, $1/6, \ldots$ of the peak wavelength corresponding to $m = 1$ completely disappear, whereas those of $1/3$, $1/5$, $1/7, \ldots$ remain unchanged.

Next, we will show a case of non-ideal multilayer. In Fig. 3.28, we show the calculated result corresponding to Fig. 3.26b with 11 layers. In this figure, the ratio of the optical path length of A layer with higher refractive index increases or decreases, while the sum of the optical path lengths of A and B layers is kept constant. When the deviation from the ideal multilayer becomes remarkable, the bandwidth reduces prominently in addition to the decrease in the peak reflectivity. At the same time, the peak position shifts toward longer or shorter wavelength as the width of the A layer

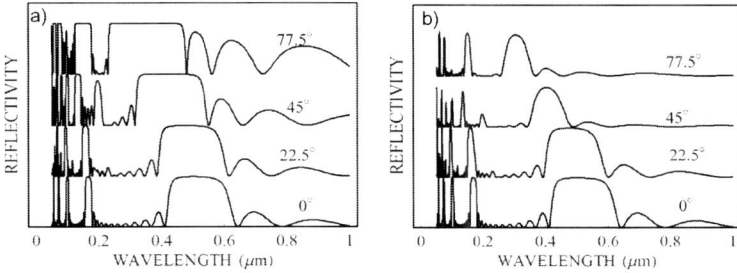

Figure 3.29 Incidence-angle dependence of the reflection spectra from an ideal multilayer of 11 alternate layers having the refractive indices of 1.6 and 1.0, respectively, under the illumination of (a) *s*- and (b) *p*-polarized light. The thicknesses of these layers are set at 78.1 and 125 nm, respectively.

decreases or increases. It is also noticed that the reflectivity at the wavelengths of $1/2, 1/4, 1/6, \ldots$ of the peak wavelength at $m = 1$ restores quickly. It is clear that a simple relation of Eq. (3.35) on the analogy of thin-film interference no longer holds, since the peak wavelength shifts definitely. The non-ideal multilayers are frequently observed in the natural products, which will be intended to exhibit a reduced bandwidth at the sacrifice of the extremely high reflectivity or to display the high-order reflection peaks that are hidden in case of ideal multilayer. The former effect is possibly beneficial to display distinct coloring while maintaining the considerably high reflectivity.

In Fig. 3.29, we show the angular dependence of reflection spectrum of a multilayer for large difference of refractive indices. The calculations for *s*- and *p*-polarization show the prominent peak shift with changing angle of incidence, as is expected. It is rather surprising that the change in the bandwidth seems to be quite different for these two polarizations. Namely, the reflection bandwidth under *s*-polarization remarkably increases, while that under *p*-polarization rather decreases. This is due to the variation of the reflectivity at the interfaces owing to the presence of Brewster's angle for *p*-polarization and also to the deviation from the ideal multilayer with changing angle of incidence.

In Figs. 3.30a and b, we show an example for the reflectivity and transmittance for the multilayer, where one of the two layers

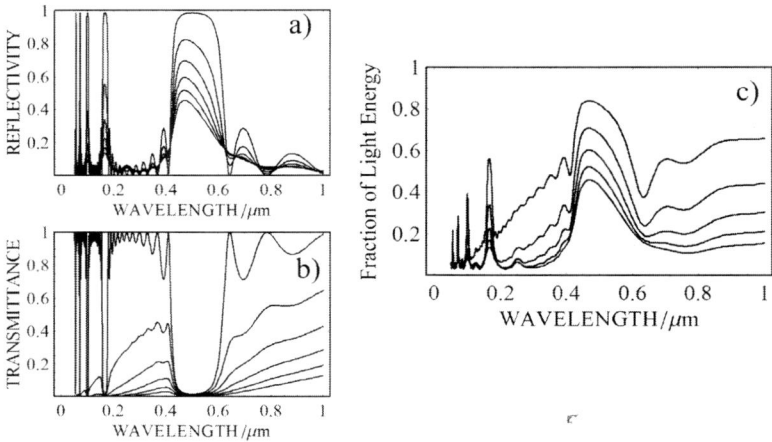

Figure 3.30 Effect of absorption on the multilayer reflection under normal incidence, in which 11 alternate layers of the thicknesses of 78.1 nm and of 125 nm are assumed to constitute the multilayer. The refractive index of the former is varied as 1.6, 1.6+0.05i, 1.6+0.1i, 1.6+0.15i, 1.6+0.2i and 1.6+0.25i (from top to bottom for (a) and (b)), while that of the latter is kept constant at 1.0. (a) Reflectivity and (b) transmittance. (c) Light energy calculated from the sum of the reflectivity and transmittance. Note that it is exactly 1.0 in (c) when the absorption is absent.

shows the absorptive behavior. With increasing imaginary part of the refractive index, the reflectivity naturally decreases with narrowing bandwidth and diffuses the tail structures without much changing peak wavelength. However, the most remarkable is that the transmittance of the background region decreases quickly. In addition, the remarkable restoring of the higher-order reflection bands is observed. This behavior is more easily understood when we calculate the total light energy from the sum of the reflected and transmitted light intensity, as shown in Fig. 3.30c. With increasing imaginary part, the fraction of light energy naturally decreases from unity. Remarkable is that only light energy that belongs to the highly reflective band tends to remain. Thus, the introduction of absorptive material contributes mainly to the background reduction in multilayer reflection. Since the transmitted light more or less contributes to the backscattering of light due to underlying

(a)   (b)   (c)

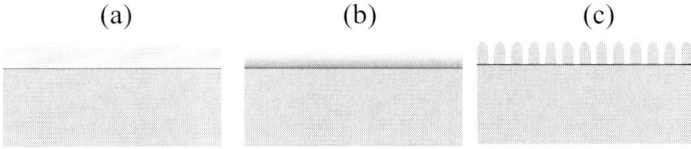

Figure 3.31 Various types of antireflective surface modulation. (a) Mono-layer or multilayer, (b) continuously changing refractive index, and (c) moth-eye structures.

structures, it will be inevitable for iridescent animals to visually emphasize the reflection color by reducing the background.

3.4 Antireflection Effect

3.4.1 *Monolayer and Multilayer Antireflection Coatings*

Antireflection at a surface is basically accomplished by eliminating the reflected light by means of the interference of light. As shown in Fig. 3.31, the following three methods have been frequently employed or considered so far: One is to coat a discrete film of monolayer or multilayer on a surface, the second is to prepare a surface having continuously changing refractive index, and the last one is a moth-eye type, which is composed of discrete nano-elements arranged regularly on a surface.

Consider first a case of monolayer or multilayer coating. To investigate the optical properties of monolayer or multilayer coating on a substrate, a transfer matrix introduced in Sec. 3.3.1 is very useful. The transfer matrix for a monolayer on a substrate with the refractive index of n_t is expressed as

$$\begin{pmatrix} A & B \\ C & D \end{pmatrix} = \begin{pmatrix} \cos n_1 k_0 d_1 & -i(\eta_0/n_1)\sin n_1 k_0 d_1 \\ -i(n_1/\eta_0)\sin n_1 k_0 d_1 & \cos n_1 k_0 d_1 \end{pmatrix}, \quad (3.85)$$

where d_1 and n_1 are the thickness and refractive index of a coating material, and the other parameters are defined as before. If the thickness of the coating layer satisfies the relation $n_1 k_0 d_1 = \pi/2$, that is, $n_1 d_1 = \lambda_{min}/4$, where λ_{min} is the wavelength giving the minimum reflectivity, then the transfer matrix and the corresponding amplitude reflectivity become

$$\begin{pmatrix} A & B \\ C & D \end{pmatrix} = \begin{pmatrix} 0 & -i(\eta_0/n_1) \\ -i(n_1/\eta_0) & 0 \end{pmatrix}, \tag{3.86}$$

and

$$r = \frac{n_i A + (n_i n_t/\eta_0) B - \eta_0 C - n_t D}{n_i A + (n_i n_t/\eta_0) B + \eta_0 C + n_t D}$$

$$= \frac{-(n_i n_t/\eta_0) i(\eta_0/n_1) + \eta_0 i(n_1/\eta_0)}{-(n_i n_t/\eta_0) i(\eta_0/n_1) - \eta_0 i(n_1/\eta_0)}$$

$$= \frac{n_i n_t/n_1 - n_1}{n_i n_t/n_1 + n_1} = \frac{n_i n_t - n_1^2}{n_i n_t + n_1^2}. \tag{3.87}$$

Thus, the antireflection is attainable under the condition that

$$r = \frac{n_i n_t - n_1^2}{n_i n_t + n_1^2} = 0 \rightarrow n_1 = \sqrt{n_i n_t}. \tag{3.88}$$

In case of two-layer coating, if the following two conditions, $n_1 d_1 = \lambda_{min}/4$ and $n_2 d_2 = \lambda_{min}/4$, are satisfied, the transfer matrix becomes

$$\begin{pmatrix} A & B \\ C & D \end{pmatrix} = \begin{pmatrix} 0 & -i(\eta_0/n_1) \\ -i(n_1/\eta_0) & 0 \end{pmatrix} \begin{pmatrix} 0 & -i(\eta_0/n_2) \\ -i(n_2/\eta_0) & 0 \end{pmatrix}$$

$$= \begin{pmatrix} -n_2/n_1 & 0 \\ 0 & -n_1/n_2 \end{pmatrix}, \tag{3.89}$$

which leads to

$$r = \frac{n_i n_2^2 - n_t n_1^2}{n_i n_2^2 + n_t n_1^2} = 0 \rightarrow n_2 = n_1 \sqrt{n_t/n_i}. \tag{3.90}$$

In three-layer coating with $n_1 d_1 = \lambda_{min}/4$, $n_2 d_2 = \lambda_{min}/2$ and $n_3 d_3 = \lambda_{min}/4$, the transfer matrix becomes

$$\begin{pmatrix} A & B \\ C & D \end{pmatrix} = \begin{pmatrix} 0 & -i(\eta_0/n_1) \\ -i(n_1/\eta_0) & 0 \end{pmatrix} \begin{pmatrix} -1 & 0 \\ 0 & -1 \end{pmatrix}$$

$$\times \begin{pmatrix} 0 & -i(\eta_0/n_3) \\ -i(n_3/\eta_0) & 0 \end{pmatrix}$$

$$= \begin{pmatrix} n_3/n_1 & 0 \\ 0 & n_1/n_3 \end{pmatrix}, \tag{3.91}$$

and the same relation as Eq. (3.90), $n_3 = n_1\sqrt{n_t/n_i}$, is obtained, though the reflection spectrum will be changed due to the presence

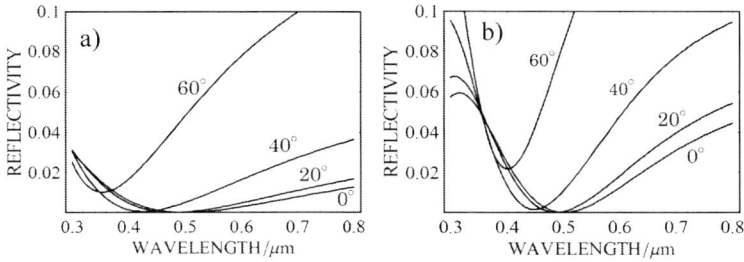

Figure 3.32 (a) Incidence-angle dependence of reflection spectra for monolayer antireflection coating under s-polarized light illumination with the refractive indices of the layer and substrate as 1.225 and 1.50. The thickness of the layer is set to satisfy the condition that their optical path lengths are equal to $\lambda_{min}/4$, where $\lambda_{min} = 500$ (nm). (b) Incidence-angle dependence of two-layer anti-reflector under s-polarized light illumination with the refractive indices of the first and second layers, and substrate as 1.38, 1.69, and 1.50.

of the second layer, which increases the degree of freedom to adjust the parameters.

We calculate the reflection spectra for various antireflection coatings according to a transfer matrix method. The typical results for monolayer and two-layer cases are shown in Fig. 3.32. In these figures, a wavelength giving the minimum reflectivity in each case is set at 500 nm under normal incidence. Further, the refractive index of a substrate is set at 1.50, while that of the coating material is so adjusted to satisfy a condition, Eq. (3.88) or (3.90). It is found that the reflectivity is strictly zero at 500 nm and maintains a low level around this wavelength under normal incidence, while with increasing incident angle, the low-reflectivity region shifts to shorter wavelengths and the minimum reflectivity increases with making low-reflectivity region narrower. This tendency is more remarkable when two-layer coating is employed. Thus, the monolayer coating gives a rather good result with respect to both a wavelength range and to an incidence-angle variation.

However, in order to satisfy the relation of Eq. (3.88), it is necessary to prepare an unrealistic material with the refractive index of 1.225 in the present case of the substrate of 1.50. On the other hand, in case of two- or three-layer coating, the restriction

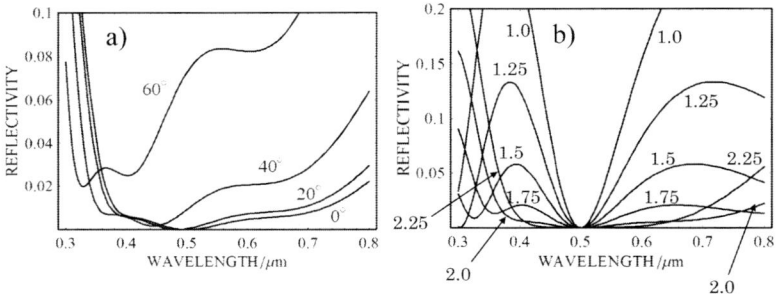

Figure 3.33 (a) Incidence-angle dependence of reflection spectra for three-layer antireflection coating under s-polarized light illumination, where the refractive indices of the first, second and third layers, and substrate are set at 1.38, 2.00, 1.69 and 1.50, respectively. The thicknesses of the layers are set to satisfy the condition that their optical path lengths are equal to $\lambda_{min}/4$, $\lambda_{min}/2$ and $\lambda_{min}/4$, where $\lambda_{min} = 500$ (nm). (b) Reflection spectrum for three-layer multilayer under normal incidence, where the refractive indices of the first and third layers, and substrate are set constant at 1.38, 1.69 and 1.50, respectively. The index of the second layer is varied as 1.0, 1.25, 1.5, 1.75, 2.0 and 2.25.

on the refractive index is greatly relaxed, because the ratio of the refractive indices of the two layers satisfies the relation, $n_2/n_1 = \sqrt{n_t/n_i}$ or $n_3/n_1 = \sqrt{n_t/n_i}$. Actually, this is easily attainable even when we employ $n_t = 1.5$ and $n_i = 1.0$, if only we prepare materials satisfying $n_2/n_1 = 1.225$ for two-layer coating or $n_3/n_1 = 1.225$ for three-layer one.

In this respect, the two- or three-layer antireflection will be superior. However, as is shown above, owing to the lack of degree of freedom, two-layer coating gives a rather poor result. If we compare two- and three-layer antireflection, the latter is clearly superior to the former especially when the appropriate value of refractive index is selected for the middle layer. In Fig. 3.33, we show a case of three-layer coating. In the left figure, we set the refractive indices of the three layers and that of the substrate at 1.38, 2.00, 1.69 and 1.5, respectively, and calculate the incidence-angle dependence. The wavelength dependence of the reflectivity is generally gentle which reminds us of that of monolayer coating.

In the right figure, we show the results with changing refractive index of the middle layer, while satisfying the condition of $n_2 d_2 =$

$\lambda_{min}/2$. It is clear that the wavelength dependence is largely dependent on the refractive index. In general, with increasing refractive index, a range of low-reflectivity region tends to increase. Thus, in three-layer coating, it is promising that a proper choice of refractive indices will offer excellent antireflection coating without employing materials having unusual refractive indices.

3.4.2 Continuously Changing Refractive Index and Moth-Eye Type Antireflection

When a refractive index abruptly changes at an interface between two materials, the power reflectivity at the interface is simply calculated according to the Fresnel's law such that $R = |(n_1 - n_2)/(n_1 + n_2)|^2$, where n_1 and n_2 are the refractive indices of the materials. For example, if $n_1 = 1.0$ and $n_2 = 1.5$, R amounts to 0.04. Thus, 4% of incident light is automatically reflected. However, if the refractive index at the interface gradually changes, what will happen? It easily comes to mind that the reflection at the interface is regarded as due to an ensemble of thin slices whose refractive indices gradually change with a distance from the surface. The situation is illustrated in Fig. 3.34. In such a case, only a small amount of light will be reflected at the interfaces between the adjacent thin slices and also the reflected light at many interfaces will interfere with each other. Therefore, if the condition of destructive interference will be satisfied, the antireflection effect is expected to occur.

Before making a quantitative evaluation of this effect, we will briefly estimate its effect qualitatively. Consider a case where a refractive index varies linearly within a region with the thickness of d and let the average refractive index in this varying region be $\bar{n}$, as shown in Fig. 3.34. Then, the phase difference between the light waves reflected at a frontal and real plane of the region will be roughly estimated at $2k_0\bar{n}d$, where k_0 is a wavenumber of light in vacuum and is expressed as $k_0 = 2\pi/\lambda$ with λ wavelength of light in vacuum. When this phase difference is equal to $2m\pi$ with m an integer, the destructive interference will occur. Since the complete destruction of interference will not occur otherwise, the reflectivity

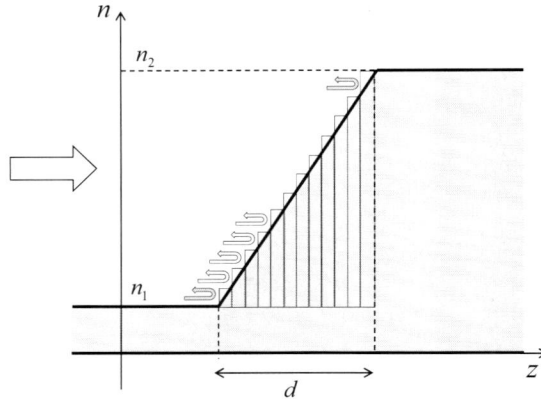

Figure 3.34 Principle of the reflection at a surface with a gradually changing refractive index, in which the reflectivity is determined by a sum of waves reflected at all the slices. Thus, the destructive interference will occur when $2\bar{n}d = m\lambda$, where $\bar{n}$ and λ are an average refractive index and the wavelength of light with m an integer.

will show an oscillation against the change in parameters such as $\bar{n}$, d and λ.

The reflectivity of light from a surface with gradually changing refractive index was first treated mathematically by Lord Rayleigh [Rayleigh (1880)], who considered a case where the refractive index changed in a quadratic manner as a function of depth. Here, we will investigate a case where the refractive index is varied linearly or with a Gaussian function from 1.0 to 1.5 within a given thickness. For this calculation, we have divided the region into 100 slices and have calculated the reflectivity using an iterative method.

The thickness-dependent reflectivity thus obtained is shown in Fig. 3.35, which indicates a clear decrease of reflectivity is observed irrespective of the interfacial shape. In case of linear dependence, the reflectivity abruptly decreases around $d = 90$ (nm) and shows an oscillation toward larger d's. The oscillation shows minima around 200, 400 and 600 nm. Since the average refractive index is simply obtained as $(n_2 - n_1)/2$ in a linear case, the reflection minima obtained from the above estimation will be simply expressed by $d_{\min} = m\lambda/(2\bar{n})$, which leads to 200, 400 and 600 nm for $m = 1, 2$

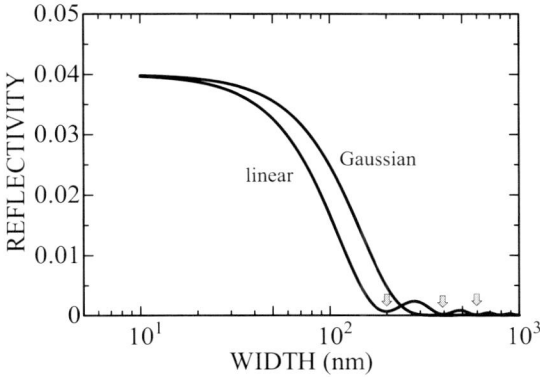

Figure 3.35 Thickness dependence of the reflectivity at a surface with continuously changing refractive indices. The changes of the refractive indices are considered to obey a Gaussian and linear function. The refractive index change following a Gaussian function is calculated using a function $n_j = (n_2 - n_1) \exp[-\{(jd/N)/(d/2)\}^2] + n_1$, where n_j is the refractive index of the jth slice with the total number of slices N. Arrows indicate the positions of destructive interference estimated qualitatively in a case of linearly changing refractive index (see text).

and 3, respectively, with $\lambda = 500$ (nm) and $\bar{n} = 1.25$. Thus, the above estimation is quite in good agreement with the calculation. On the other hand, in the Gaussian case, the reflectivity rapidly decreases around $d = 110$ (nm), shows a smooth decrease with increasing thickness, and above $d = 300$ (nm), eventually no reflection seems to occur.

In Fig. 3.36, we show the reflection spectrum for a Gaussian interfacial function with the width of 104 nm for various incidence angles. It is soon noticed that the reflection spectra differ considerably from those of monolayer or multilayer coating. Namely, the reflectivity monotonically decreases toward shorter wavelengths and seems to have a cutoff wavelength, below which almost complete antireflection is attained. With increasing incidence angle, the cutoff wavelength shifts gradually to shorter wavelengths, but the absolute reflectivity is by far smaller than that of a multilayer case. Considering that the optical path length within a slice decreases as the incidence angle is increased, the cutoff wavelength λ_{cut} is roughly given as $\lambda_{cut} = 2\bar{n}d\cos\theta$, where θ is

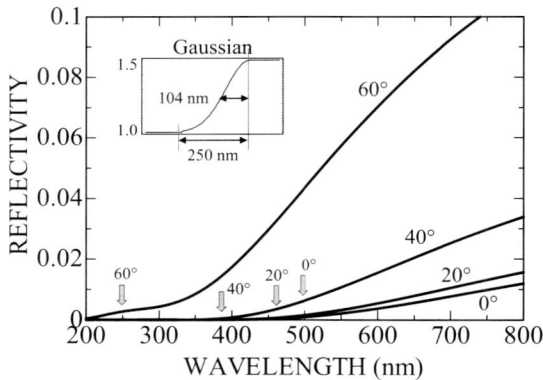

Figure 3.36 Incidence-angle dependence of reflection at an interface between a substrate of $n = 1.5$ and air with continuously varying refractive index, which obeys a Gaussian function with the half width of 104 nm (shown as inset). The region of 250 nm in thickness is divided into 100 layers and the reflectivity is calculated using an iterative method. Arrows indicate the approximate positions of cutoff wavelength estimated qualitatively for 0, 20, 40 and 60° incidence (see text).

an angle of incidence. We have marked their positions as arrows in Fig. 3.36, in which we have employed a value of $\bar{n}d \approx 250$ (nm). The estimated cutoff wavelength is qualitatively in good agreement with the calcualtion. Thus, if the thickness d is sufficiently large, this type of coating will give quite efficient antireflection capability with respect to both ranges of wavelength and incidence angle.

Thus, a gradual change of refractive index at an interface is ideal to reduce the interfacial reflection particularly at shorter wavelengths. However, to produce a gradation around the surface is quite difficult, because the refraction index should be down to 1.0 for use in air. Although the effort to fabricate such material has been in progress [Xi *et al.* (2007)], moth-eye type structure attracts attention as an alternative candidate. In insect world, small projections of several tens or a hundred nanometer sometimes cover a whole surface of cornea and are called *corneal nipple*. The first and most remarkable observations were reported by Bernhard *et al.* on the surface of moth's eye so that this type of structure has been often called *moth-eye structure* [Bernhard and Miller (1962)].

It is generally believed that these projections have a function of antireflection. Since the regular projections are arranged in 2D with their spacing considerably smaller than the wavelength of light, such a structure is regarded as the zeroth-order 2D diffraction grating described later. That is, the spacing is so small that any higher-order diffraction spots cannot be generated except for the zeroth-order one (see Sec. 4.5). In such a case, simple reflection and transmission of light will occur as if such a structured film is regarded as a simple film. Thus, if we consider a layer containing projections as a film having an apparent refractive index that depends on the depth of the film, the above results for the continuously changing refractive index can be directly applicable. In fact, recent calculations using RCWA (rigorous coupled wave analysis) method [Moharam and Gaylord (1981)] proved that the calculation due to sliced multilayer model reproduced the reflection spectra considerably well [Sun *et al.* (2008)]. According to the shape of the projection, the apparent refractive index will change gradually from the top to base of the projection and will give a similar profile to continuously changing type. Moreover, at a tip of projection, the effective refractive index eventually becomes 1.0. Thus, a complete structure for continuously changing refractive index is realized by moth-eye type.

3.5 Cholesteric Liquid Crystal

3.5.1 *Light Propagation in Cholesteric Liquid Crystal*

Cholesteric liquid crystal takes a layered structure, within which molecules take a preferred direction, on an average, that gradually changes from layer to layer. Therefore, it takes a form of helix as shown in Fig. 3.37. A periodicity of the helix is generally expressed by its pitch, P. However, in cholesteric liquid crystal, molecules can take two opposite directions randomly, while their molecular axes direct to the preferred direction. As a result, a half of the helical pitch $P/2$ determines the periodicity. Thus, it is regarded as a kind of multilayer that is specially designed for circular polarization, because this helical structure reflects light selectively with respect to wavelength and circular polarization. In the present subsection,

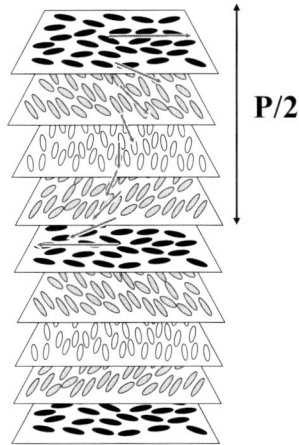

Figure 3.37 Schematic view of cholesteric liquid crystal (illustrated by N. Okamoto). The orientation of molecules is unidirectional on an average within a plane so that a half of the helical pitch $P/2$ determines the periodicity.

we will explain the optical properties of cholesteric liquid crystal in detail. Although a lot of review articles and books are now available, here we will follow excellent books written by de Gennes [de Genne (1974)] and by Orihara [Orihara (2004)].

Let us first consider that anisotropic molecules are aligned unidirectionally, on an average, within the xy plane. The average orientation of the molecules in a plane is designated as a vector **n** that is dependent on the height z, as shown in Fig. 3.38. We assume that the molecules are randomly oriented along **n** so that **n** and $-$**n** are essentially equivalent, which is the usual case for nematic liquid crystal. The azimuth of **n** is assumed to be linearly dependent on z, which guarantees a helical structure along the z axis. We thus put $\mathbf{n}(z) = (\cos q_0 z, \sin q_0 z, 0)$, where $q_0 = \pm 2\pi/P$ gives a wave vector concerning the periodicity of the helical structure. Hereafter, we take $q_0 > 0$, for simplicity, indicating a clockwise helix.

If the relative permittivities parallel and perpendicular to **n** within a plane are denoted as $\epsilon_{\parallel}$ and $\epsilon_{\perp}$, respectively, a relative dielectric tensor evaluated in a plane of $z = z$ is expressed by using

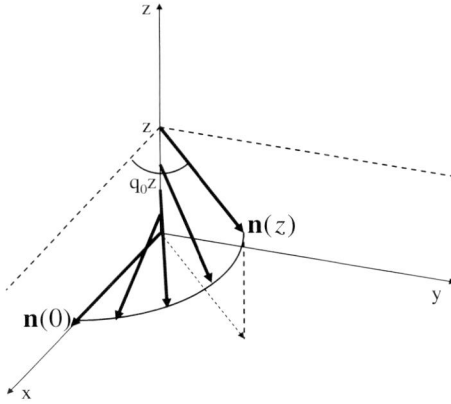

Figure 3.38 The average orientation of molecules in a plane is designated as **n** that is dependent on the height z. The molecules are assumed to be randomly oriented along **n**, and thus **n** and $-$**n** are equivalent. The azimuth angle of **n** is assumed to be linearly dependent on z, which guarantees the helical structure along the z axis.

a rotation around the z axis as

$$\epsilon(z) = R_z^{-1} \begin{pmatrix} \epsilon_{\parallel} & 0 \\ 0 & \epsilon_{\perp} \end{pmatrix} R_z \tag{3.92}$$

where R_z is a rotation matrix that makes a point rotate clockwise by an angle of $q_0 z$ around the z axis within the xy plane, and is expressed as

$$R_z = \begin{pmatrix} \cos q_0 z & \sin q_0 z \\ -\sin q_0 z & \cos q_0 z \end{pmatrix}. \tag{3.93}$$

Performing the matrix calculation actually, we obtain

$$
\begin{aligned}
\epsilon(z) &= \begin{pmatrix} \cos q_0 z & -\sin q_0 z \\ \sin q_0 z & \cos q_0 z \end{pmatrix} \begin{pmatrix} \epsilon_{\parallel} \cos q_0 z & \epsilon_{\parallel} \sin q_0 z \\ -\epsilon_{\perp} \sin q_0 z & \epsilon_{\perp} \cos q_0 z \end{pmatrix} \\
&= \begin{pmatrix} \epsilon_{\parallel} \cos^2 q_0 z + \epsilon_{\perp} \sin^2 q_0 z & (\epsilon_{\parallel} - \epsilon_{\perp}) \sin q_0 z \cos q_0 z \\ (\epsilon_{\parallel} - \epsilon_{\perp}) \sin q_0 z \cos q_0 z & \epsilon_{\parallel} \sin^2 q_0 z + \epsilon_{\perp} \cos^2 q_0 z \end{pmatrix} \\
&= \bar{\epsilon} \begin{pmatrix} 1 & 0 \\ 0 & 1 \end{pmatrix} + \frac{\epsilon_a}{2} \begin{pmatrix} \cos 2q_0 z & \sin 2q_0 z \\ \sin 2q_0 z & -\cos 2q_0 z \end{pmatrix}, \tag{3.94}
\end{aligned}
$$

where the last relation is obtained by inserting $\epsilon_{\parallel} = \bar{\epsilon} + \epsilon_a/2$ and $\epsilon_{\perp} = \bar{\epsilon} - \epsilon_a/2$. Here, $\bar{\epsilon} \equiv (\epsilon_{\parallel} + \epsilon_{\perp})/2$ is an average permittivity, while $\epsilon_a \equiv \epsilon_{\parallel} - \epsilon_{\perp}$ expresses the anisotropy (we assume $\epsilon_a > 0$, hereafter).

Consider a z-propagating light wave satisfying the wave equation

$$d^2\mathbf{E}(z)/dz^2 = -(\omega/c)^2\epsilon(z)\mathbf{E}(z), \tag{3.95}$$

where the electric field is assumed to have a form of $\mathbf{E}(z, t) = \mathbf{E}(z)\exp[i(kz - \omega t)]$ with k, ω and c the wave number, angular frequency and light velocity in vacuum, respectively. In a helical system, it is convenient to take right- (+ sign) and left-handed circularly (− sign) polarized light as its bases[a]:

$$E^{\pm}(z) = E_x(z) \pm iE_y(z). \tag{3.96}$$

Using the relation

$$\begin{pmatrix} E^+(z) \\ E^-(z) \end{pmatrix} = \begin{pmatrix} 1 & i \\ 1 & -i \end{pmatrix} \begin{pmatrix} E_x(z) \\ E_y(z) \end{pmatrix},$$

we multiply the same transformation matrix from the left of Eq. (3.95) and obtain

$$-\frac{d^2}{dz^2}\begin{pmatrix} E^+(z) \\ E^-(z) \end{pmatrix} = \left(\frac{\omega}{c}\right)^2 \begin{pmatrix} 1 & i \\ 1 & -i \end{pmatrix} \epsilon(z) \begin{pmatrix} E_x(z) \\ E_y(z) \end{pmatrix}.$$

Further, using the relation

$$\begin{pmatrix} 1 & i \\ 1 & -i \end{pmatrix} \epsilon(z) = \bar{\epsilon} \begin{pmatrix} 1 & i \\ 1 & -i \end{pmatrix} + \frac{\epsilon_a}{2} \begin{pmatrix} e^{2iq_0 z} & -ie^{2iq_0 z} \\ e^{-2iq_0 z} & ie^{-2iq_0 z} \end{pmatrix}$$

$$= \begin{pmatrix} \bar{\epsilon} + (\epsilon_a/2)e^{2iq_0 z} & i\bar{\epsilon} - i(\epsilon_a/2)e^{2iq_0 z} \\ \bar{\epsilon} + (\epsilon_a/2)e^{-2iq_0 z} & -i\bar{\epsilon} + i(\epsilon_a/2)e^{2iq_0 z} \end{pmatrix}$$

$$= \begin{pmatrix} \bar{\epsilon} & (\epsilon_a/2)e^{2iq_0 z} \\ (\epsilon_a/2)e^{-2iq_0 z} & \bar{\epsilon} \end{pmatrix} \begin{pmatrix} 1 & i \\ 1 & -i \end{pmatrix},$$

we finally obtain

$$-\frac{d^2}{dz^2}\begin{pmatrix} E^+(z) \\ E^-(z) \end{pmatrix} = \begin{pmatrix} k_0^2 & k_1^2 e^{2iq_0 z} \\ k_1^2 e^{-2iq_0 z} & k_0^2 \end{pmatrix} \begin{pmatrix} E^+(z) \\ E^-(z) \end{pmatrix}, \tag{3.97}$$

where we put $k_0^2 = (\omega/c)^2\bar{\epsilon}$ and $k_1^2 = (\omega/c)^2\epsilon_a/2$.

The above wave equation states that light propagates in a material whose permittivity varies spatially along the z axis with the

---

[a]Since the electric fields for right- and left-handed circular polarization states are expressed as $E_y = -iE_x$ and $E_y = iE_x$, respectively, according to Eq. (2.90), inserting these expressions into Eq. (3.96) gives $E^+ = 2E_x$ and $E^- = 0$ are obtained for the right-handed circular polarization, while $E^+ = 0$ and $E^- = 2E_x$ for the left-handed circular polarization. Thus, $E^{\pm}$ become the bases for the right- and left-handed circularly polarized light.

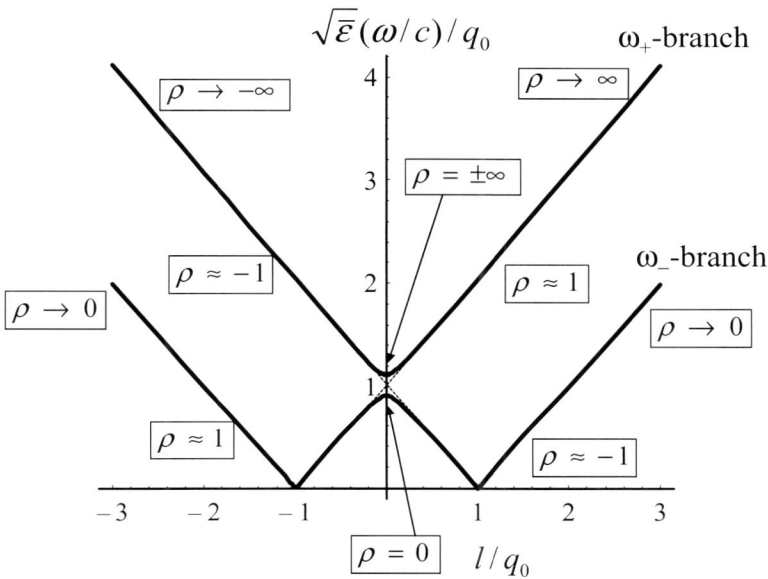

Figure 3.39 Dispersion relation of light in cholesteric liquid crystal, calculated using Eq. (3.102) with $\epsilon_a/2\bar{\epsilon} = 0.2$. Thick lines indicate two branches, $\omega/c = \sqrt{\bar{\epsilon}}|l-q_0|$ and $\omega/c = \sqrt{\bar{\epsilon}}|l+q_0|$. ρ indicates the polarization state for elliptical polarization.

period of $2\pi/q_0$. In such a case, it is a custom to express the material properties in terms of modes with the periodicity of q_0 as in a case of photonic crystal (see Sec. 5.1.3). Thus, light will, in principle, couple with various modes when it propagates in the material. Here, however, we confine ourselves within only two branches among such modes, $\omega/c = \sqrt{\bar{\epsilon}}|l - q_0|$ and $\omega/c = \sqrt{\bar{\epsilon}}|l + q_0|$, which are to cross at $l = 0$, where l is a wavenumber of the light mode within cholesteric liquid crystal (see Fig. 3.39).

By using these two modes, we will investigate the light propagation around the crossing point. For this purpose, we assume the electric fields are expressed by the following forms:

$$E^+(z) = a e^{i(l+q_0)z},$$
$$E^-(z) = b e^{i(l-q_0)z}. \tag{3.98}$$

Inserting Eq. (3.98) into Eq. (3.97), we obtain a set of homogeneous equations with respect to a and b:

$$(l + q_0)^2 a = k_0^2 a + k_1^2 b,$$
$$(l - q_0)^2 b = k_1^2 a + k_0^2 b,$$

which can be rewritten in a matrix form as

$$\begin{pmatrix} (l + q_0)^2 - k_0^2 & -k_1^2 \\ -k_1^2 & (l - q_0)^2 - k_0^2 \end{pmatrix} \begin{pmatrix} a \\ b \end{pmatrix} = 0. \qquad (3.99)$$

Nonzero solutions are obtained by putting the determinant of the coefficient matrix to be zero, which results in

$$\{(l + q_0)^2 - k_0^2\} \{(l - q_0)^2 - k_0^2\} - k_1^4 = 0. \qquad (3.100)$$

Rewriting this as

$$(l^2 + q_o^2 - k_0^2)^2 - 4q_0^2 l^2 - k_1^4 = 0, \qquad (3.101)$$

and inserting $k_0^2 = (\omega/c)^2 \bar{\epsilon}$ and $k_1^2 = (\omega/c)^2 \epsilon_a/2$ into this equation, we obtain

$$\left\{ \bar{\epsilon}^2 - \left(\frac{\epsilon_a}{2}\right)^2 \right\} \left(\frac{\omega}{c}\right)^4 - 2(l^2 + q_0^2)\bar{\epsilon} \left(\frac{\omega}{c}\right)^2 + (l^2 - q_0^2)^2 = 0,$$

which is a quadratic equation with respect to $(\omega/c)^2$. Solving it, we obtain

$$\left(\frac{\omega}{c}\right)^2 = \frac{1}{\bar{\epsilon}} \cdot \frac{(l^2 + q_0^2) \pm \sqrt{4q_0^2 l^2 + K^2 (l^2 - q_0^2)^2}}{1 - K^2}, \qquad (3.102)$$

which gives the dispersion relation of ω against l. Here, we put $K \equiv \epsilon_a/(2\bar{\epsilon}) = (n_\parallel^2 - n_\perp^2)/(n_\parallel^2 + n_\perp^2)$ with $\epsilon_\parallel = n_\parallel^2$ and $\epsilon_\perp = n_\perp^2$.

In Fig. 3.39, we plot an example of the dispersion relations calculated using Eq. (3.102). Since $K^2 = \{(n_\parallel^2 - n_\perp^2)/(n_\perp^2 + n_\parallel^2)\}^2 < 1$, the right-hand side of Eq. (3.102) is always positive irrespective of l and hence ω can have two positive values. Also since l appears as a form of l^2, the curve is symmetric with respect to $l = 0$. It is noticed that a band gap is present at $l = 0$, which results in the generation of two separate branches. We call them $\omega_+(l)$ and $\omega_-(l)$ branch, which correspond to the $\pm$ sign in Eq. (3.102), respectively. It is noticed that within the band gap, only $\omega_-(l)$ mode exists as a propagation mode. This means that for a clockwise helix, a band gap exists only for right-handed circularly polarized light, while no such gap is present for left-handed circularly polarized light. The positions of the band gap edges are easily obtained as $\omega_+(0) = cq_0/n_\perp$ and $\omega_-(0) = cq_0/n_\parallel$ from the relation

$$\left(\frac{\omega}{c}\right)^2 = \frac{q_0^2}{\bar{\epsilon}} \frac{1 \pm \{\epsilon_a/(2\bar{\epsilon})\}}{1 - \{\epsilon_a/(2\bar{\epsilon})\}^2} = \frac{q_0^2}{\bar{\epsilon}} \frac{1}{1 \mp \{\epsilon_a/(2\bar{\epsilon})\}}$$

$$= q_0^2 \frac{2}{(\epsilon_\parallel + \epsilon_\perp) \mp (\epsilon_\parallel - \epsilon_\perp)} = \frac{q_0^2}{\epsilon_\perp}, \frac{q_0^2}{\epsilon_\parallel}$$

which is obtained by inserting $l = 0$ into Eq. (3.102).

On the contrary, Eq. (3.101) can be solved with respect to l. For this purpose, Eq. (3.101) is transformed into

$$l^4 - 2(q_0^2 + k_0^2)l^2 + (q_0^2 - k_0^2)^2 - k_1^4 = 0,$$

and is solved with repect to l^2, which results in

$$l = \pm\sqrt{q_0^2 + k_0^2 \pm \sqrt{4q_0^2 k_0^2 + k_1^4}}. \tag{3.103}$$

This relation generally gives four real values except for a case

$$q_0^2 + k_0^2 - \sqrt{4q_0^2 k_0^2 + k_1^4} < 0, \tag{3.104}$$

which is rewritten as $(q_0^2 + k_0^2)^2 < 4q_0^2 k_0^2 + k_1^4$. Then, this relation is further transformed into $|q_0^2 - k_0^2| < k_1^2$. When $q_0^2 > k_0^2$ is satisfied, we obtain $q_0^2 < (\omega/c)^2(\bar{\epsilon} + \epsilon_a/2) = (\omega/c)^2\epsilon_\parallel$, which leads to $\omega > cq_0/n_\parallel = \omega_-(0)$. On the other hand, when $q_0^2 < k_0^2$, this relation is reduced to $\omega < cq_0/n_\perp = \omega_+(0)$. Thus, the condition of Eq. (3.104) is simply reduced to

$$\omega_-(0) < \omega < \omega_+(0). \tag{3.105}$$

This is nothing but a band gap, and hence within this region, two of the four solutions become imaginary, indicating that only light waves with particular states can selectively propagate, while those with the other states cannot propagate and will decay at a surface.

When Eq. (3.104) holds, the wave vector l takes a pure imaginary

$$l = \pm i\sqrt{\sqrt{4q_0^2 k_0^2 + k_1^4} - (q_0^2 + k_0^2)},$$

which can be expanded under the assumption of $k_0^2 \approx q_0^2 \gg k_1^2$ as

$$l = \pm i\sqrt{2k_0 q_0 \{1 + (1/2)k_1^4/(4k_0^2 q_0^2) + \cdots\} - (q_0^2 + k_0^2)}$$

$$\approx \pm i\sqrt{k_1^4/(4k_0 q_0) - (k_0 - q_0)^2}. \tag{3.106}$$

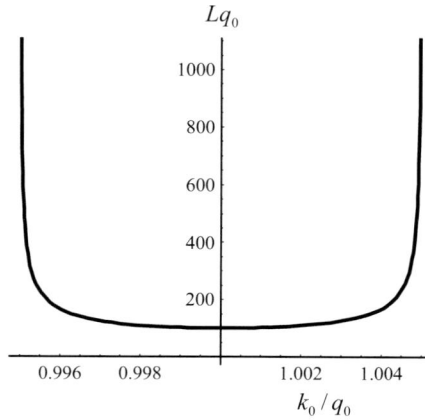

Figure 3.40 Penetration length into cholesteric liquid crystal calculated using Eq. (3.106) for $k_1/q_0 = 0.1$.

If we employ the positive sign, the above relation expresses a penetration length into a liquid crystal, which is defined as $L \equiv 1/(2|l|)$. We plot L in Fig. 3.40. It is found that under the condition of $k_0 = q_0$, L takes a minimum value of $L_{\min} = q_0/k_1^2 = k_0^2/(q_0 k_1^2) = (P/2\pi)(n_\parallel^2 + n_\perp^2)/(n_\parallel^2 - n_\perp^2)$ and becomes divergent at band edges. Hence, the penetration length becomes smaller when the anisotropy of refractive index is larger.

3.5.2 *Polarization States*

Next, we consider a polarization state within a cholesteric liquid crystal. A light wave propagating toward the z direction is generally expressed as

$$E_x(z, t) = \frac{1}{2}\left(E_x(z)e^{-i\omega t} + E_x^*(z)e^{i\omega t}\right),$$

$$E_y(z, t) = \frac{1}{2}\left(E_y(z)e^{-i\omega t} + E_y^*(z)e^{i\omega t}\right).$$

These are the expressions in a real representation, taking linear polarization states as their bases. If we consider a function

$$\breve{E}(z, t) \equiv E_x(z, t) + iE_y(z, t), \tag{3.107}$$

it is proved that it becomes an expression in a complex representation, taking circular polarization states as bases:

$$\check{E}(z,t) = \frac{1}{2}\left\{E_x(z) + iE_y(z)\right\} e^{-i\omega t} + \frac{1}{2}\left\{E_x(z) - iE_y(z)\right\}^* e^{i\omega t}$$

$$= \frac{1}{2}\left(E^+(z)e^{-i\omega t} + E^{-*}(z)e^{i\omega t}\right). \tag{3.108}$$

In a complex plane, a locus of a point corresponding to $\check{E}(z,t)$ expresses a polarization state of light propagating along the z axis. For example, if we consider a light wave propagating to the $+z$ direction with a clockwise rotation of polarization as

$$E_x(z,t) = \frac{1}{2}\left(r_1 e^{i(kz-\omega t)} + r_1 e^{-i(kz-\omega t)}\right)$$
$$= r_1 \cos\theta,$$
$$E_y(z,t) = -\frac{i}{2}\left(r_2 e^{i(kz-\omega t)} - r_2 e^{-i(kz-\omega t)}\right)$$
$$= r_2 \sin\theta,$$

where we put $kz - \omega t = \theta$, and set $E_x(z) = r_1 \exp[ikz]$ and $E_y(z) = r_2 \exp[i(kz - \pi/2)] = -ir_2 \exp[ikz]$ with real numbers of r_1 and r_2. In case of $r_1 = r_2$, these relations express a state of right-handed circular polarization. Thus, we generally obtain

$$\check{E}(z,t) = r_1 \cos\theta + ir_2 \sin\theta.$$

The locus of a point expressed as $\check{E}(z,t) \equiv X + iY$ in a complex plane is obtained by a relation

$$\frac{X^2}{r_1^2} + \frac{Y^2}{r_2^2} = 1,$$

which is an ellipse.

We define a polarization state for elliptical polarization as

$$\rho = \frac{r_2}{r_1}. \tag{3.109}$$

Then, we should consider the following four cases:

$$\begin{cases} \text{(i)} \;\; r_1 > 0, \;\; r_2 > 0 \;\; \cdots\cdots \;\; \rho > 0, \\ \text{(ii)} \;\; r_1 > 0, \;\; r_2 < 0 \;\; \cdots\cdots \;\; \rho < 0, \\ \text{(iii)} \;\; r_1 < 0, \;\; r_2 < 0 \;\; \cdots\cdots \;\; \rho > 0, \\ \text{(iv)} \;\; r_1 < 0, \;\; r_2 > 0 \;\; \cdots\cdots \;\; \rho < 0, \end{cases} \tag{3.110}$$

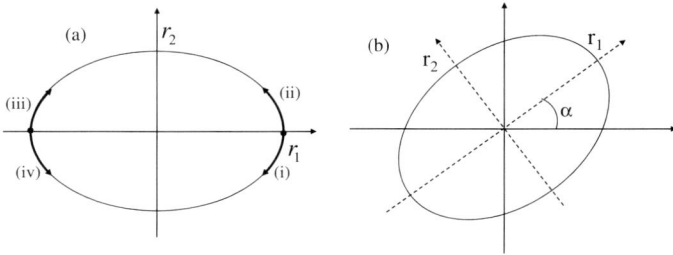

Figure 3.41 (a) Polarization state expressing elliptical polarization. (i)~(iv) correspond to Eq. (3.110). Arrows indicate the directions of loci evaluated at $z = 0$ when t is increased, that is, θ is decreased. (b) Polarization state where an ellipse is inclined by an angle of α.

which are illustrated in Fig. 3.41a, where $\rho > 0$ and $\rho < 0$ mean the electric field obeys clockwise and counterclockwise rotation along an elliptical orbit with time when we see the orbit from the $+z$ direction. Particularly, $\rho = 1$ and -1 indicate right- and left-handed circular polarizations, while $\rho = 0$ and $\pm\infty$ indicate the linear polarizations.

Using this expression and Eq. (3.108), we will investigate the polarization states for propagation modes expressed by Eq. (3.98):

$$\breve{E}(z,t) = \frac{1}{2}\left(a e^{i\{(l+q_0)z-\omega t\}} + b e^{-i\{(l-q_0)z-\omega t\}}\right)$$

$$= \frac{1}{2}e^{iq_0 z}\left\{\left(\frac{a+b}{2} + \frac{a-b}{2}\right)e^{i(lz-\omega t)}\right.$$

$$\left. + \left(\frac{a+b}{2} - \frac{a-b}{2}\right)e^{-i(lz-\omega t)}\right\}$$

$$\equiv e^{i\alpha}(r_1 \cos\theta + i r_2 \sin\theta), \tag{3.111}$$

where we put $\alpha = q_0 z$, $\theta = lz - \omega t$, $r_1 = (a+b)/2$ and $r_2 = (a-b)/2$. The last expression indicates that the electric field draws an elliptical orbit that is inclined by an angle of α, as shown in Fig. 3.41b. The polarization state in this case is expressed as

$$\rho = \frac{r_2}{r_1} = \frac{a-b}{a+b}. \tag{3.112}$$

On the other hand, Eq. (3.99) gives the ratio of a/b or b/a for a given set of ω and l, and hence

$$a/b = k_1^2/\{(l+q_0)^2 - k_0^2\}, \tag{3.113}$$

or

$$b/a = k_1^2/\{(l - q_0)^2 - k_0^2\}. \tag{3.114}$$

For ω_- branch, employing Eq. (3.113), we can calculate ρ as

$$\rho = \frac{k_1^2 + k_0^2 - (l + q_0)^2}{k_1^2 - k_0^2 + (l + q_0)^2} = \frac{(\omega/c)^2 \epsilon_\parallel - (l + q_0)^2}{(\omega/c)^2 \epsilon_\perp + (l + q_0)^2}. \tag{3.115}$$

From this relation, various polarization states are predicted for sets of ω and l. For example, when $l = 0$, $\omega \to \omega_-(0) = cq_0/n_\parallel$ so that ρ becomes $\rho = (q_0^2 - q_0^2)/\{q_0^2(\epsilon_\perp/\epsilon_\parallel) - q_0^2\} = 0$. When $l \approx q_0$, $\omega \to 0$, then $\rho \to -4q_0^2/(4q_0^2) = -1$. On the other hand, when $l \to \pm\infty$, ω takes an asymptotic value[a] of $\omega \to c|l|/n_\parallel$ so that $\rho \to (l^2 \epsilon_\parallel/n_\parallel^2 - l^2)/(l^2 \epsilon_\perp/n_\parallel^2 + l^2) = 0$. These results are summarized in Fig. 3.39.

Essentially, $+z$-propagating light consists of the mixture of light waves with right- and left-handed circular polarizations. However, in the frequency region lower than a band gap, both light waves originate from ω_- branch. On the other hand, in the frequency region higher than the band gap, a light wave with right-handed circular polarization originates from ω_+ branch, while that with left-handed circular one is from ω_- branch. Therefore, it is expected that a kind of crossover occurs around a band gap. It is also noticed that at large frequencies, both branches tend to give linear polarizations that are perpendicular to each other.

On the other hand, at the band edge, one of the light waves tends to have a linear polarization, but does not propagate eventually, indicating that a stationary wave is generated through the interference of $+z$ and $-z$-propagating light waves. Thus, cholesteric liquid crystal offer various interesting optical phenomena and will benefit us as a lot of visual applications. In natural world, on the contrary, this type of structure is commonly distributed as a typical material of the outer layer of the skin and has been utilized to display peculiar visual effect.

---

[a]This relation is obtained from Eq. (3.102). For $l \to \pm\infty$, $(\omega/c)^2 \to (1/\bar{\epsilon})(l^2 \pm Kl^2)/(1 - K^2) = l^2/\{\bar{\epsilon}(1 \mp K)\}$. In case of ω_- mode, it reduces to $\omega_- \to \sqrt{c^2 l^2/\{\bar{\epsilon}(1 + K)\}} = c|l|\sqrt{1/\epsilon_\parallel} = c|l|/n_\parallel$.

3.6 Examples of Thin-Layer Interference

3.6.1 *Beetles*

3.6.1.1 Insect cuticle

Before showing multilayered structure, we will first explain the surface structure of insect (see [Neville (1975)]). The surface of insect is generally covered with *cuticle*, which becomes a barrier for various inner organisms to the outer world and has a function to maintain homeostasis against the change of environment (see Fig. 3.44b). The cuticle is composed of various complex chemicals including proteins, lipids, polyphenols, chitins, and so on, which are complicatedly intertwined with each other to form a solid material. In general, the outer part of insect body is divided into three parts: *epicuticle*, *procuticle* and *epidermis*, from outer to inner part.

Epicuticle is defined as a layer lack of chitin. It constitutes an outermost layer of insect body, which is generally composed of four layers: *cement layer* (*tectocuticle*), *wax layer* (*lipid cuticle*), *outer epicuticle* (sometimes called *cuticlin*) and *inner epicuticle* (*dense layer* or *protein epicuticle*). It is common that these four layers are not fully furnished. The latter two are secreted in sequence directly from epidermal cells before ecdysis. In particular, outer epicuticle is known as the first layer to be secreted, while the former two are secreted just before or after ecdysis.

Cement layer is secreted through a pipe called *dermal gland* that perpendicularly breaks through the surface layers, and covers an underlying wax layer to protect from the outer space. It consists of various materials, dependent on the species, such as protein with polyphenol, tanned protein with lipid, wax stabilized with shellac, and so on.

Wax layer consists of lipids of a labile nature and has a function of water control by covering a whole body. It is secreted through thin pipes called *wax canal* of typically 6 nm in diameter. A large number of wax canals fuse into a small number of thicker pipes called *pore canal*, which is originally a projection of an epidermal cell, and takes a form of twisted ribbon within the procuticle. Within a wax canal, wax filaments are present, which bundle into a large size at their roots and are connected with those in a pore canal. The role of these filaments is believed to be a sort of wick to smoothly transport labile lipid on the outer epicuticle. The filament is believed to have esterase activity, and thus the synthesis of wax is performed within the canals.

The outer epicuticle consists of trilaminar membrane with the thickness up to 18 nm. At first, a partial layer appears as patches, which fuse into a full layer. The function of the outer epicuticle is to resist enzymes in moulting fluid. The inner epicuticle of typically 0.5–2.0 μm thickness consists of tanned protein with polyphenol or lipid, which usually shows laminations. The optical properties of this layer are isotropic, while those of the outer epicuticle are anisotropic.

Procuticle is secreted after the epicuticle is secreted, and is defined as a layer that involves chitin. This layer is further divided into *exocuticle* and *endocuticle*. The exocuticle is secreted before ecdysis, and becomes tanned and harder after ecdysis. Thus, the color of exocuticle is usually brown or black, which is a result of tanning or melanin formation. The exocuticle contains cuticle microfibrils of ~3 nm diameter, which are composed of unidirectionally grown chitin crystallites embedded into protein matrix. These microfibrils are arranged uniformly in a plane parallel to the surface and their directions gradually change with a depth from a surface, thus causing anticlockwise helical structure throughout the exocuticle. The pitch of the helix lies in a visible wavelength range in scarab beetles, which causes peculiar structural colors in these insects. The endocuticle appears as growth and digestive layer, which also contains bundles of chitin microfibrils. Between endocuticle and epidermis, cuticle deposition zone of 1 μm thickness is located.

Chitin consists of polysaccharide chains of poly-*N*-acetylglucos-amine with varying lengths [Neville *et al.* (1976)]. The molecular chains of chitin are associated together in a highly ordered manner to form a crystallite of 2.8 nm diameter. Chitin usually exists in three polymorphic forms, α-, β- and ψ-chitin, depending on the arrangement of the molecular chains in crystallites. In arthropod cuticle, only α-chitin exists, where the molecular chains are arranged in an anti-parallel form. The crystallites pack in a hexagonal or pseudo-hexagonal array but the extent and degree of perfection vary considerably. Although chitin has never been found free of proteins, the information concerning the bonding with protein has not been clarified.

The cuticle of newly emerged insect is soft and pale pink, but soon it darkens and becomes harder, which is known as *tanning* and *sclerotization*. However, these two processes should be considered to be different concerning chemical reactions for cross-linking, because even albino can form hard cuticle. In spite of that, they are closely related with each other. Various mechanisms have been proposed so far, which assume small molecular tanning agent such as derivatives of tyrosine produced during ecdysis, biphenyl linking, and also protein groups. Anyway these small molecular groups containing movable electrons cross-link protein molecules causing tanned protein network. Chitin crystallites also become a part of the network.

Tanning pigment is sometimes called sclerotin, but its molecular structure is not unique. It may be defined as a pigment formed during cuticular hardening, and involving quinonoid and phenolic character. On the other hand, melanin is another pigment not having unique chemical structure, which is defined as a dark and insoluble pigment produced by oxidation of polyhydric phenols. Melanins are sometimes classified into several classes such as eumelanins (black, indole-type), phaeomelanins (yellow, red, or brown) and allomelanins (black, catechol-type). Melanin is often bound to protein to form melanoproteins. Although the melanin formation is sometimes included into sclerotization, these two processes should be also treated as two independent processes and actually, they sometimes take place in different time courses.

3.6.1.2 Jewel beetles, leaf beetles, and tiger beetles

Elytra of beetles are conspicuous in their brilliant and lustrous reflections, and have attracted people's eyes for a long period. In East Asia, more than a thousand years ago, wings of jewel beetles were actually used for the decoration of craft work. One of such examples is the Tamamushi Shrine (Tamamushi no Zushi), which was presumably made in the 7th century and is one of the most famous architectures in Japan.

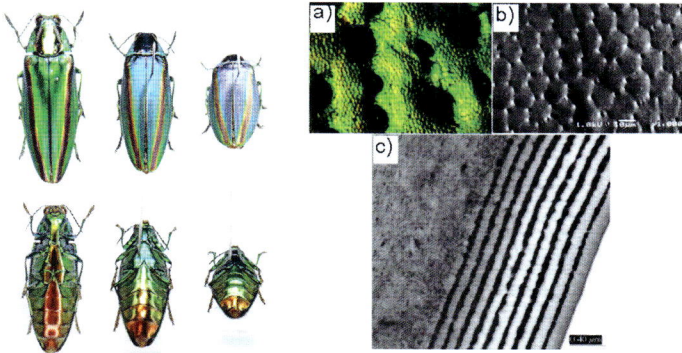

Figure 3.42 Left: Color changes on the dorsal and ventral sides of a jewel beetle, *Chrysochroa fulguidissima*. Right: (a) Optical and (b) scanning electron microscopic views, and (c) the cross section of the elytron. Scale bars in (b) and (c) are 10 μm and 0.4 μm, respectively. (Courtesy of Prof. T. Hariyama for TEM image, and the others are reproduced from [S. Kinoshita, *Cell Technol.* **22**, 1113 (2003)].)

We will first show the characteristics of iridescent colors of a jewel beetle, *Chrysochroa fulguidissima*. In Fig. 3.42, we show the dorsal and ventral sides of the jewel beetle. Both sides are lustrous and brilliant, and are glaring when one sees under specular reflection. The overall view shows brilliant green with copper brown stripes in the dorsal side, while in the ventral side, it is lustrous copper brown. When we change a viewing angle from normal to an elytron to oblique to it, it is easily noticed that the dorsal color gradually changes from green to dark blue, while in the ventral side, it changes from copper to green. It is further noticed the color change is rather continuous and isotropic, which is quite in contrast to that of the *Morpho* butterfly, where the rather abrupt color change from blue to violet is observed and the reflection of light is strongly anisotropic. Under a right-handed circular polarization analyzer, however, we find no change in appearance, suggesting that a helical structure possibly existing below the surface layer does not contribute to this color.

When we observe the elytron under an optical microscope, many depressions of ∼50 μm in diameter are found, which are randomly distributed and cover the whole elytron (Fig. 3.42a). Thus, the surface of the elytron is more or less modulated. Under a high magnification, it is found that the surface is completely covered with polygonal patterns of irregular pentagons or hexagons with a typical size of 10 μm (Fig. 3.42b). No sign of microstructures giving optical interference is observed. Thus, for the further studies, we need an electron microscope.

The first electron microscopic observation of this type of beetle was performed by Durrer and Villinger, who examined a buprestid beetle, *Euchroma gigantea* [Durrer and Villiger (1972)]. Under an electron microscope, they found 5 layers of electron-dense regions that were regularly arranged near the surface at the transverse cross section of an elytron. They deduced that these layers were composed of melanin. The uppermost electron-lucent layer had the thickness of 0.2–0.4 μm, which was followed by the electron-dense layers of 0.09-μm thickness for a copper brown region and 0.06 μm for a green region, while the electron-lucent layers between the dense layers were rather constant with 0.06-μm thickness. They also noticed that an arch-shaped modulation at the surface changed the angle of incidence to the multilayered system and contributed to various colorings of this beetle.

Schultz and Rankin investigated several species of tiger beetles using optical measurement, chemical treatment, and electron microscopic observation [Schultz and Rankin (1985a,b)]. The results are visualized by their illustration shown in Fig. 3.43, in which the surface of elytral cuticle is divided into three regions:

a   b

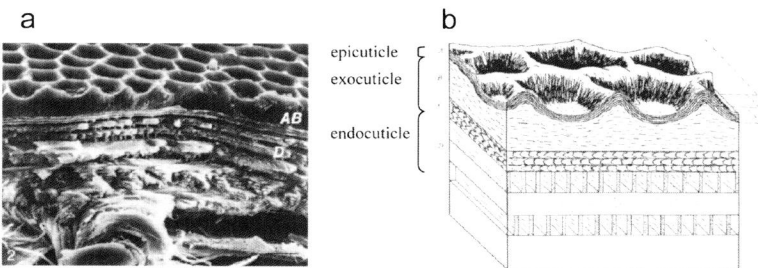

epicuticle
exocuticle

endocuticle

Figure 3.43 (a) Cross section of cuticular layer of the elytron of tiger beetle, *Cicindela splendida*, and (b) its schematic illustration. (Reproduced from [Schultz and Rankin (1985a)] with permission.)

epicuticle, exocuticle and endocuticle. The epicuticle region shows the thickness of 1–2 μm, and is alveolate with polygonal ridges and hollows with the depth of 3.5 μm. Just below the alveoli, 4–9 electron-dense layers are located with the thickness of 0.03–0.1 μm, which are composed of melanin granules. Between the layers, electron-lucent regions with the thickness of 0.06–0.125 μm are present. The uppermost region of the epicuticle is covered with electron-lucent layer of 20–30 nm thickness. Below the epicuticle, outer exocuticle is present, which is composed of electron-dense 25–35 lamellae of 0.055–0.1 μm interval, and may constitute imperfect helical structure. Both epicuticle and outer exocuticle show dark brown and are thought to be melanized. Below it, the inner exocuticle of peculiar helical pattern and the endocuticle of plywood structures are seen.

Kurachi *et al.* investigated a leaf beetle, *Plateumaris sericea*, which showed regional color variations from blue to red (450–750 nm) [Kurachi *et al.* (2002)]. Through electron microscopic observation, they found that a layered structure consisting of only 3 electron-dense layers, contributed to their specific colorings. The thickness of the dense layer varied from 60–100 nm for blue specimen, to 70–160 nm for bronze one, and to 100–140 nm for red one. They simulated the reflection spectrum using the refractive indices of 1.73 and 1.40 for electron-dense and lucent layers, respectively, and obtained good agreement with the measured spectrum for each color group.

Hariyama *et al.*, on the other hand, reported on the structural color of the jewel beetle, *Chrysochroa fulguidissima*, and showed that the peaks of the reflection spectra for green and copper brown regions of the elytra were located at 550 nm and 750–950 nm, respectively [Hariyama *et al.* (2005)]. He performed an electron microscopic observation and found that the epicuticle was covered with several alternate layers of electron-dense and lucent regions, as shown in Fig. 3.42c. The layer interval varied with the color of the elytron and ranged over 100–200 nm. Further, they found that the electron-dense layer was somewhat irregular in a strict sense. They used a densitometer to convert the change in electron density to that in refractive index, and found the interface of multilayer was not optically discontinuous but rather gently modulated like a sinusoidal curve. They calculated the reflectivity under this model and obtained fairly good agreement.

Thus, it has been experimentally clarified that the elytra of these lustrous beetles are covered with multilayer systems, which appear as alternate stripes of electron-dense and lucent regions in a transmission electron microscopic image. The color seems to be directly connected with the interval of these layers. Therefore, there is no doubt that multilayer interference is a cause of the structural colors in these beetles.

Nevertheless, it is sometimes difficult to reproduce the reflection spectrum and the angle-dependent reflectivity using a formalism described in the previous section. The difficulty mainly comes from how to determine the refractive indices for electron-dense and lucent regions. If the electron-dense layers are composed of melanin granules, the complex refractive index should be determined correctly. However, index determination is not an easy task: The elytron surface is not flat, but is curved and usually contains many irregular depressions, which may cause scattering loss, even when the exact reflectivity has to be measured. These structural characteristics prevent us from directly applying conventional optical techniques for teh index determination, such as ellipsometry, since these techniques require a perfectly specular surface.

Recently, we have reported a new experimental procedure that can directly determine the refractive indices of individual layers in the natural multilayer systems [Yoshioka and Kinoshita (2011)]. This procedure is based on preparing a semi-frontal thin section of the sample and measuring the reflection and transmission spectra under a microscope, which are then analyzed as thin-film interference with real and imaginary components of the refractive index as parameters. The thickness of a sample is separately determined in an epoxy resin region using a known refractive indices for the wavelength range from 400 to 700 nm, which are determined beforehand for a prism-shaped epoxy resin by means of a minimum-deviation-angle method[a].

---

[a]The refractive index of a prism-shaped material is obtained by an angle between the incident and emerging ray when it is minimized by rotating the prism. If we denote the angle of an apex of the prism and the minimum deflection angle as α

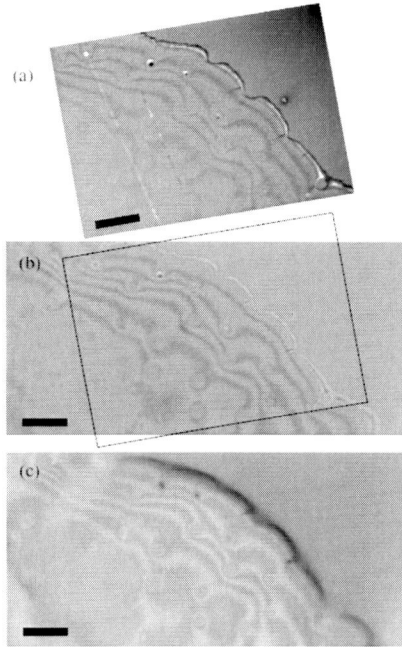

Figure 3.44 Semi-frontal thin section of the elytron of jewel beetle. (a) TEM image, (b) optical image under transmission illumination, and (c) that under epi-illumination. A rectangular region in (b) is tilted in order to compare the image of (a). Note that the upper right region is the epoxy resin. Scale bar: 10 μm. (Reproduced from [Yoshioka and Kinoshita (2011)] with permission.)

In Fig. 3.44a, we show a TEM image of a semi-frontal section of the elytron. Since the section is cut almost in parallel to the surface, the multilayer appears as repetitive stripes with the widths more than 1 μm. The stripes seem to be very wavy, because the layers are not perfectly flat. In Figs. 3.44b and c, we show the optical images measured under an optical microscope observed at exactly the same region. It is clear that dark stripes obtained under a transmission microscope completely agree with those obtained in TEM image, suggesting the electron-dense region is actually light absorbent. This assumption is further confirmed by comparing the same region with

and β, respectively, the refractive index is obtained through a relation $n = \sin\{(\alpha + \beta)/2\}/\sin(\alpha/2)$.

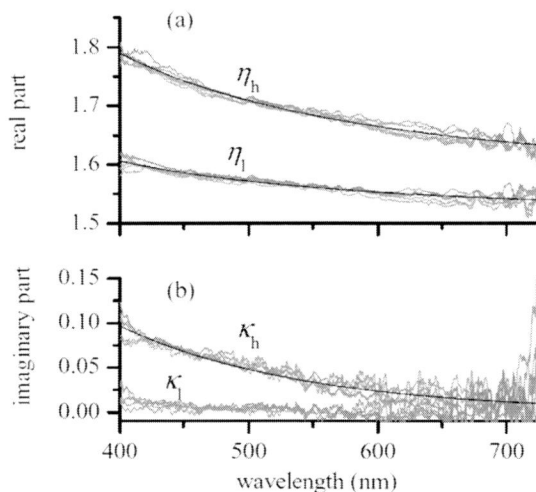

Figure 3.45 (a) Real and (b) imaginary parts of the wavelength-dependent complex refractive indices for two types of materials that comprise jewel beetle's multilayer structure, which are denoted as $\eta_h + i\kappa_h$ and $\eta_l + i\kappa_l$ for high- and low-index layers, respectively. Six gray curves are obtained from six different pairs of reflectance and transmittance spectra. Black curves are drawn using the approximate expressions shown in the text. (Reproduced from [Yoshioka and Kinoshita (2011)] with permission.)

the observation under epi-illumination: It is soon noticed that the contrast in the stripe pattern in the latter image appears opposite to that in the above image.

These images make us possible to directly determine the refractive indices for both layers by quantitatively performing transmission and reflection spectral measurements under a microscope and by analyzing them using the relations of thin-film interference given in Eqs. (3.31) and (3.32). The results are shown in Fig. 3.45, where $\eta_{h,l}$ and $\kappa_{h,l}$ are real and imaginary components of the complex refractive indices for the electron-dense and lucent layers, respectively. It is found that all the refractive indices are smooth functions of wavelength and increase with decreasing wavelength. Particularly, it is noticed that both real and imaginary components for electron-dense layer are higher than those for electron-lucent layer. It is found that the wavelength dependence for electron-dense and lucent layers is well approximated by the following empirical formula that will be applicable within a range of 400–700 nm:

$$\eta_l(\lambda) = 1.51 + \frac{1.53 \times 10^{-2}}{\lambda^2}, \qquad \kappa_l(\lambda) \approx 0$$

$$\eta_h(\lambda) = 1.56 + \frac{3.60 \times 10^{-2}}{\lambda^2}, \qquad \kappa_h(\lambda) = 1.62 \exp(-\lambda/0.142),$$

where λ should be given in a unit of μm.

However, it should be considered that even if the accurate values are employed as the refractive indices, the complete understanding of the optical properties of multilayer in natural world need further considerations. One is how to include the effect of the surface modulation like alveoli and another is the irregularity of the multilayer, which is inevitable in the natural products. These effects on the optical responses should be more thoroughly investigated, since they are essential to express true appearances of the animals.

3.6.2 *Butterflies and Moths*

3.6.2.1 Butterfly scales

We will first explain the most important structure of butterflies and moths to exhibit their colors. Since there are excellent works on butterfly scales by Ghiradella [Ghiradella (1991, 1998)] and also good books have been already published [Nijhout (1991)], readers will be better to refer to these works for a deeper understanding. Here, we will briefly review these works for a clear understanding of structure-based colors of butterflies and moths.

Butterflies and moths are characterized by their wings covered with scales. The scales are of a thin plate-like form, whose typical dimensions are 0.2 mm in length, 0.1 mm in width and 0.003 mm in thickness, which cover a whole wing like tiles on a roof or a dense tapestry. The butterfly wing is typically covered with two types of scales (sometimes three types are present): one is a *cover scale* (*glass scale*) and the other a *ground scale* (*basal scale*). The cover scales are variously specialized in shape and play a variety of roles on the wing, whereas the ground scale usually has a rectangular form and seems to be less specialized.

A scale is considered to be a flattened form of a *bristle* that has a long, cylindrical shape with a hollow inside. Both a scale and bristle have a complicated interior structure and are called *macrochaete*, which is in contrast to a simple hair called *microchaete*. Bristles are distributed in all orders of insects, whereas scales are restricted in several orders: Coleoptera (beetles), Diptera (flies, horseflies), Psocoptera (booklices), Trichoptera (caddisflies) and Lepidoptera (butterflies and moths).

The scales of butterflies and moths originate from a single-layered epithelial cell in a developing wing [Kühn (1955)]. Some of the epithelial cells have scale primordia and become scale mother cells that are scattered irregularly on the wing. They will soon line up in rows with regular spacing. A scale mother cell divides twice: the first one is oriented perpendicular to a wing plane and a resulting daughter cell located at the inner side degenerates. On the other hand, the second one takes place at an angle of 45° to a surface, which results in a scale-forming cell at the proximal part and a socket-forming cell at the distal part. A scale-forming cell pushes a narrow cylindrical projection above a plane of the wing, while a socket-forming cell extends a projection around a neck of the scale projection. The scale projection grows and then flattens to form a scale. When it is fully grown, cuticle secretion begins along the scale.

A fully developed scale is of a thin, plate-like shape and is clearly asymmetric with respect to its surface: a surface faced to a wing membrane is flat and featureless, while the obverse surface is decorated with elaborate structures. In Fig. 3.46, we show the morphology of a typical butterfly scale [Ghiradella (1998)]. In the frontal surface, parallel running *ridges*, which are decorated by an oblique *lamellar structure*, partly overlapping with each other. At the side of a ridge, regular pleated structures, called *microrib*, are running nearly perpendicular to the lamellar structure. Between ridges, *crossribs* connect the neighboring ridges together. There is a window between the crossribs, which connects an interior to an open space. Below a ridge, pillars, called *trabeculae*, support the architecture of a whole scale. The structural colors in the scales of butterfly wings is eventually expressed as elaborations in various parts of a scale.

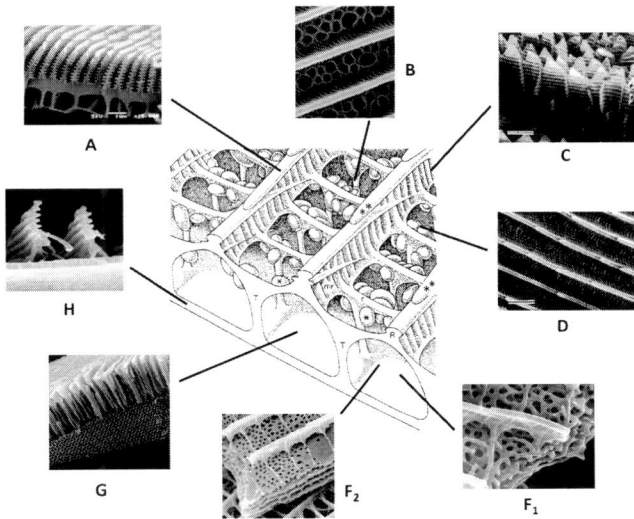

Figure 3.46 Elaborate structures created at various sites in a butterfly scale: (1) ridge-iridescence type, A and C, (2) flat-iridescence type, B and D, and (3) lumen-iridescence type, F and G. We have further added H type into base-iridescence type, and have divided F type into two, F_1 and F_2. In the photograph of H, ridge-lamellae (A type) is also seen in addition to a thin base film (H type). (Reproduced from [Ghiradella (1991)] with permission.)

3.6.2.2 Overview of structural colors in butterflies and moths

Next, we will overview butterflies and moths from a viewpoint of elaborate structures decorated on their scales. We will again refer to excellent works performed by Ghiradella, who extensively investigated the morphological characteristics of scales in conjunction with their functions, and further proposed rational classification for scales specialized to structural colors [Ghiradella (1991, 1998)]. The result is summarized in her well-known illustration shown in Fig. 3.46. She classified the specialized scales into three categories, according to the location: (1) ridge, (2) flat between ridges and (3) lumen at the lower side of a scale.

First, we show a *ridge-iridescence type*. In this type, the structure of ridge is sometimes specialized by increasing its height and broadening the width of lamellae, which constitutes a partially stacking multilayer. This specialized structure produces quasi-multilayer interference as will be described in the *Morpho* blue (type A; see Sec. 3.6.3). In this type, microribs run perpendicularly to the lamellae and seem to little contribute to the optical response. However, when the gradient of lamellae becomes steep, both lamellae and microribs will contribute to the optical responses as two independent periodic structures. In an extreme case, lamellae stand nearly perpendicularly to a base plane, and periodic microribs now contribute to optical interference. Ghiradella exemplified a scale of double-spotted owl, *Eryphanis aesacus*, for this type (type C).

Next, we show a *flat-iridescence type*. Ghiradella showed three types as examples of this type. One is a case where the flat is elaborated into randomly distributed alveoli (type B), and giant blue swallowtail, *Papilio zalmoxis*, is exemplified. The second one is that the flat is covered with rather regularly running microribs that are perpendicular to the ridges, which make the flat look like a wide plane with pleats (type D). The example is seen in green-patched looper, *Diachrysia balluca*. The third one is the modification of type B, whose flat is filled with plates-and-pores pattern (type E), and is characteristic of pierid androconical scale. The extreme case of this type is that obverse and reverse planes of the scale are fused together to make a thin film, on which ridges are regularly running (type H, which I dare to add into the Ghiradella's illustration). This type of scale sometimes produces thin-film interference, which was reported to have a function of an optical diffuser in case of a cover scale in *Morpho didius* [Yoshioka and Kinoshita (2004)]. The stiffness of the scale in the last case is clearly weak so that even under a scanning electron microscope, a scale is sometimes turned up by a function of electron beam.

The last one is *lumen-iridescence type*, in which a lumen located below the ridges is filled with elaborated structures. She exemplified two cases of this type. One is that a lumen is filled with body-lamellar structure, which will produce multilayer interference (type F). The body-lamella sometimes takes a form of true or curved multilayer, that bearing randomly distributed holes, or sometimes it takes a form of multilayered network of fibrous materials. Thus, simple multilayer interference no longer holds and a sophisticated treatment that takes into account both regular multilayer and irregular structure is anyway necessary. The second type is that a lumen is filled with three-dimensional lattice of a photonic crystal (type G; see Sec. 5.3.3). This type is widely distributed in butterfly species. The photonic crystal in butterflies is not a complete crystal, but always accompanies the domains of a small size with their orientations changing from domain to domain. Anyway, the trabeculae in the lumen of a scale disappear in lumen-iridescence type and the elaborate structures itself maintain mechanically the shape of a scale.

In addition to these microscopic structures, macroscopic structure often gives a special visual effect on the butterfly or moth wing. One of such effects occurs even on the order of a scale size: butterfly/moth scales are sometimes curved in a longitudinal direction, which produces the special angular effect and also polarization effect, as is seen in a scale of Madagascan sunset moth, *Chrysiridia ripheus* (see Sec. 3.6.2.3) [Yoshioka and Kinoshita (2007); Yoshioka *et al.* (2008)]. The optical interaction between cover and ground scales will also produce an important effect. For example, in *Morpho didius*, overlapping of light-diffusive cover scales on iridescent ground scales gives a matting wing, where the size of a cover scale compared with an iridescent ground scale plays a decisive role [Yoshioka and Kinoshita (2004)]. It is also known that the cover-ground interaction enhances the efficiency of light scattering, which makes a wing whitish as in the pierid butterfly [Stavenga *et al.* (2006a)].

Finally, the distribution of pigment definitely affects a visual effect of a wing. In case of *Morpho sulkowskyi*, the lack of melanin pigment clearly shows a pearl-like whitish wing, while the presence of melanin pigment at the lower part of ground scale enhances blue coloring of *Morpho rhetenor*, by absorbing light of complementary colors. In this case, the pigment distribution in wing membrane and ventral scales should be also taken into account in addition to iridescent scales on a dorsal side. In fact, in *Morpho cypris*, these components are proved to contribute evenly to producing white spots owing to light scattering [Yoshioka and Kinoshita (2006)]. In the following, we will show the structural colors in various butterfly/moth wings and investigate their origins from various viewpoints such as structural elaborations, macroscopic structure and chromatic effect.

3.6.2.3 Thin-layer interference in butterflies and moths

Although many butterfly species are expected to utilize the multilayer interference for producing structural colors, the systematic investigations on the microscopic structures are very few. Here, we will review a surveying work performed by Tilley and Eliot on scales of 54 species of lycaenids (Lycaenidae) together with 7 species of metalmarks (Riodininae) [Tilley and Eliot (2002)].

They classified the iridescent scales from a structural aspect and divided them into three types except for species with undifferentiated structures. One is the *Morpho* type scale, to which some of lycaenids and metalmarks belong. This type is characterized by well-developing ridges with periodic lateral modulation, which stand on a hollow interior held by trabeculae. The ridges are standing closely enough, between which deep dips are present.

The characteristics of this type in lycaenids are peculiar, because microribs mainly contribute to structural colors, which are normally arranged vertically to a scale plane. This is in contrast to a case of *Morpho* butterflies, in which lamellar structure mainly contributes to the colors. The microribs in these species incline by 20–40° from the vertical direction so that the iridescent color is observed only when a wing is tilted. On the other hand, the structure found in metalmarks is the same as that of *Morpho* butterflies and the lamellar structure is mainly responsible for the colors. Thus, the *Morpho* type in their classifications corresponds to a type A or C in Ghiradella's classification.

The second type is the *Urania* type, which mainly belongs to a type F or G in Ghiradella's classification. The name of *Urania* type comes after Süffert's observation [Süffert (1924)] that a scale had a layered structure on the underside of a scale as in Madagascan sunset moth, *Urania riphenus* (now called *Chrysiridia ripheus*). This type gives a major cause of the structural colors in lycaenids, and is widely distributed throughout the whole subfamilies. The *Urania*-type bears either multilayer or photonic-crystal structures in the interior of the scale.

In Fig. 3.47, we show typical examples of this type. In scales of Japanese oakblue, *Arhopala japonica japonica*, a multilayer with 4–5 layers is found below sparsely distributed ridges, which has many

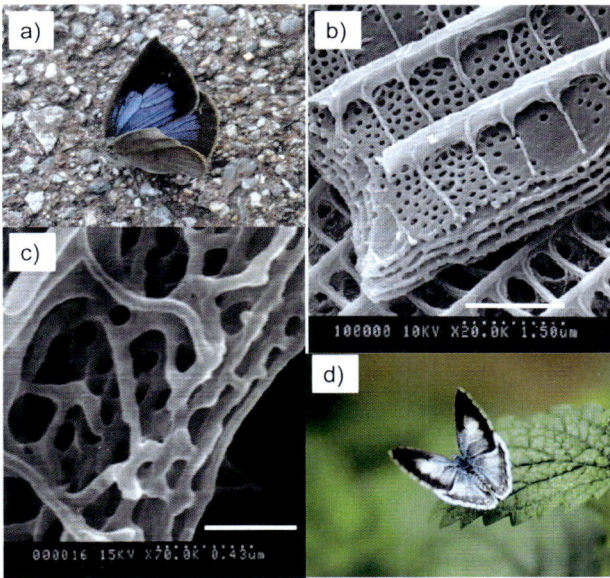

Figure 3.47 Scanning electron microscopic images of iridescent scales of licaenid butterflies [Kinoshita and Yoshioka (2005b)]. (a,b) Japanese oakblue, *Arhopala japonica japonica*, and (c,d) holly blue, *Celastrina argiolus ladonides*.

pepper-pot perforations particularly on the top of the layers. On the other hand, in scales of holly blue, *Celastrina argiolus ladonides*, a multilayer is constituted of fibrous network. Since these types are considered to be the examples of the modifications of type F, we name them as F_1 and F_2, and add them to Fig. 3.46. It is expected that when these modifications are introduced, the layer-to-layer interference is definitely weakened, and hence both the wavelength selectivity and reflectivity are reduced considerably. This is particularly prominent in F_2 type, which results in pale blue coloration, while the diffusivity of light is remarkably increased.

In addition to these examples, many variations of this type were reported. In a scale of orchid tit, *Chliaria othona*, only a single layer with pepper-pot perforation is present. However, this butterfly develops microribs on the ridge. Furthermore, apical scales do not possess even a pepper-pot structure and only the *Morpho*-type structure develops. Thus, this species is located at an intermediate position between the *Morpho*- and *Urania*-type. In blue iridescent scales of pointed ciliate blue, *Anthene lycaenina*, the upper surface of the interior is almost completely closed by a membrane to make a "bag", which shows holes here and there, into which we can see pepper-pot structure inside. The scales of cycad hairstreak, *Eumaeus minijas*, show the "satin" type surface between ridges, say type D. The scales of green hairstreak, *Callophrys rubi*, and Chapman's green haistreak, *C. avis*, have not a layered structure but possess crystallites, thus belong to type G.

The third type is *piliform* and *plume* scales. The piliform scale is observed in provence hairstreak, *Tomares ballus*, with pepper-pot structures inside. The rest species possess undifferentiated colored scales and have no significant internal structure, yet they show the iridescence.

From the structural and phylogenetic considerations, Tilley and Eliot speculated that an old type of iridescent scale was of the *Urania*-type and the transmutation to the *Morpho*-type took place in the past. During the evolutional process, the loss of the multilayer structure and the recreation of trabeculae might be attained. *Chliaria othona* made it possible to imagine how the transformation actually progressed during the evolutional step.

Color mixing within a scale   The peculiar pit structure found in green-banded peacock swallowtail (emerald swallowtail), *Papilio palinurus*, was analyzed by Vukusic *et al.* (Fig. 3.48) [Vukusic *et al.* (2000, 2001)]. They investigated a scale of this butterfly and found that many concavities were distributed on the scale, which were of a hemispherical shape having a diameter 4–6 µm with the depth of 0.5–3 µm. Below a round concavity, a curved multilayer consisting of about 20 alternate layers of cuticle and air was present.

They deduced that optical mixing was a main reason for this green-color production. Namely, when white light was incident on this structure, yellow light hitting directly on the bottom of a concavity was selectively reflected, while blue light hitting on the side of a concavity was reflected, hit again on the opposite side of the concavity wall and was then reflected back to a reverse direction. Thus, only yellow and blue colors were selectively retro-reflected,

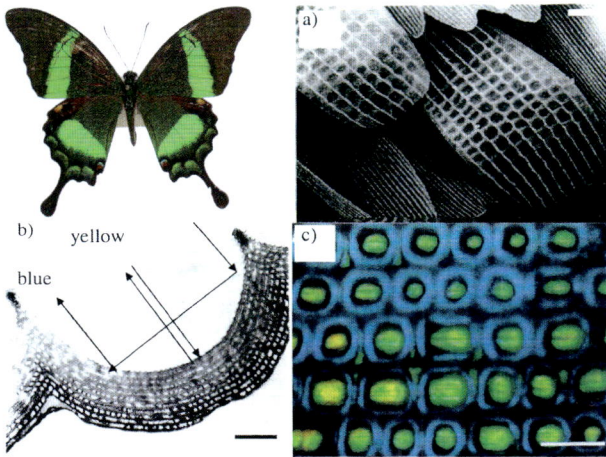

Figure 3.48 Color mixing mechanism due to concavities on a scale of green-banded peacock swallowtail, *Papilio palinurus*. Scale bars are (a) 10 µm, (b) 1 µm and (c) 10 µm. (Reproduced from [Vukusic *et al.* (2001)] with permission. Specimen of the butterfly is in the possession of The Nature and Human Activities, Hyogo, Japan and the photograph was taken by Dr. M. Kambe.) .

which produced green color in our eyes. Since the oblique reflection of blue color at the wall of the concavity was strongly polarization-dependent, the color change with the polarization direction of incident light was also observed. They analyzed the reflection spectrum using a multilayer interference model and found good agreement with the experiments. They compared this result with the similar *Papilio* butterflies, *P. ulysses* (blue mountain swallowtail), and *P. blumei* (green swallowtail), where the concavities were found to be shallow and the usual multilayer reflection was expected, while the diffusivity of light in these butterflies was realized by their concaved layers[a].

Color mixing on an inter-scale level Beautiful moths belonging to a genus *Chrysiridia* inhabiting in Madagascar and East Africa, and *Urania* in neotropical attracted great attention of scientists in 1920s,

---

[a]By convention, we will abbreviate a scientific name such as *P. ulysses* for *Papilio ulysses* when it appears repeatedly.

Figure 3.49 Madagascan sunset moth, *Chrysiridia ripheus*, and the enlarged views of the scales at variously colored positions [Yoshioka and Kinoshita (2007)]. Notice narrow reflective bands are running in parallel with each other.

who performed surveying works on structurally colored animals. Süffert conducted a detailed work on *C. croesus* (East African sunset moth), and placed this moth as a typical example of structure-based coloring, and called the *Urania*-type [Süffert (1924)]. Mason also investigated a scale of *C. ripheus* (Madagascan sunset moth) and concluded that the color was produced by multiple thin films in a scale with the superposition of ridges and crossribs on the outer surface [Mason (1927a)]. The essential structures of scales of the genera of *Urania* and *Chrysiridia* were later confirmed by an electron microscope [Lippert and Gentil (1959)]. Thus, it is now believed that simple multilayer interference explains the structural colors of these moths.

Recently, we have investigated the wing of *C. ripheus* in detail using an optical microscope and have found a peculiar pattern under epi-illumination, as shown in Fig. 3.49 [Yoshioka and Kinoshita (2007); Yoshioka *et al.* (2008)]. Namely, neighboring scales constitute narrow reflective band running laterally, which form separate parallel bands over a wing. It is also noticed that a pale color with different hue covers an area between these bands. For example, in the red-purplish area of a ventral hind wing, the narrow bands are seen purplish, while broad yellow band covers the rest of the area. Since the sizes of these areas are too small to

a)

300 μm

Incident light

2 μm

b)

reflectance / arb. units

s

p

s-p

400 500 600 700 800

wavelength / nm

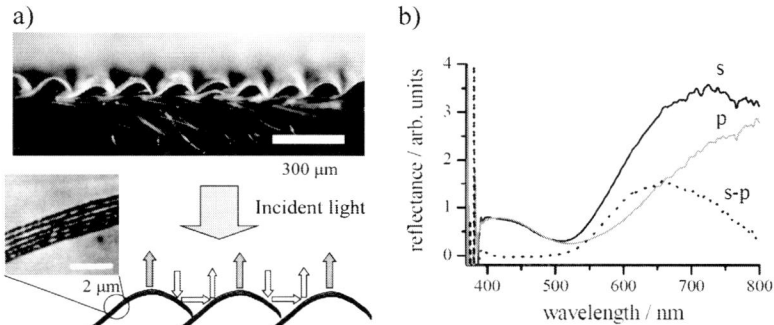

Figure 3.50 (a) SEM image (upper) of the longitudinal section of strongly curved scales of Madagascan sunset moth, *Chrysiridia ripheus*, and the schematic illustration (lower) showing the principle of inter-scale color mixing. TEM of the longitudinal cross section is also shown in the inset [Yoshioka and Kinoshita (2007)]. (b) Reflection spectra of the wing measured under unpolarized light with a polarization analyzer directed to *s*- and *p*-polarizations [Yoshioka and Kinoshita (2007)]. The broken line is the difference of the two, which corresponds to retro-reflection component due to the reflection at the curved part of the scale.

be distinguishable to the naked eye, a color mixing on a scale level is realized.

The optical and electron microscopic observations of the longitudinal cross section show that the distal part of a scale is strongly curved to make a hemisphere (Fig. 3.50), while in the proximal part, they are rather flat, where the hemisphere part of an adjacent scale overlaps. Owing to this structure, the two scales form a valley-like deep groove between them. Below these specialized scales (cover scales), flat ground scales are arranged. Under a scanning electron microscope, the surface of a cover scale is smooth with the very thin ridges of 0.4 μm width running sparsely separated by 3.5 μm.

Under the surface of a curved scale, a multilayered structure is clearly seen using a transmission electron microscope (Fig. 3.50a). The multilayer is also curved along the surface of the scale, and the number of the layers are 6 at the top of the scale with decreasing number to 2 toward the proximal part. The thickness of the dense layer is 170±20 nm, while the lucent layer is somewhat distributed from 100 to 150 nm. It is clear that the narrow purplish band observed above comes solely from a specular reflection at the top of the curved scale, while the broad yellow band from the retro-reflection at the curved part (see the illustration in Fig. 3.50a).

In order to confirm this expectation, the reflection spectra are measured under the illumination of unpolarized light with a polarization analyzer directed to s- and p-polarizations. The reflection spectra thus obtained consists of a small peak around 400 nm, which is essentially independent of polarization, and a broad peak around 700 nm, which is largely dependent on the polarization. We subtract the refection spectrum obtained for p-polarization from that for s-polarization and obtain a broad peak around 650 nm.

Considering the condition of multilayer interference, we have concluded that the peak at 400 nm comes from the second-order ($m = 2$) peak of 394 nm under normal incidence on the top of the curved scale, while the latter broad peak is the sum of the first-order ($m = 1$) peak under normal incidence (787 nm) and the first-order ($m = 1$) peak under oblique incidence on the curved part (650 nm). These results shows the color mixing is of an inter-scale phenomenon due to a specialized form of the overlapping curved scales, and also realized by using the sum of the first- and second-order multilayer reflections. Thus, this color mixing is strongly polarization-sensitive even under the illumination of unpolarized light.

3.6.3 *The* Morpho *Blue*

3.6.3.1 Mechanisms of the *Morpho* blue

Since *Morpho* butterflies have played a central role in the history of bionanophotonic research, we will describe their characteristics in a separate section. *Morpho* butterflies exclusively inhabit in the Central and South America, and belong to a tribe Morphini, a subfamily Morphinae, a family Nymphalidae, according to a recent classification. At present, about 80 species are recorded in this tribe and most of their males assume brilliantly blue (see Fig. 3.51), although there are some species that do not show apparent structural colors such as *M. catenarius* (see Fig. 3.51d) and *M. perseus*.

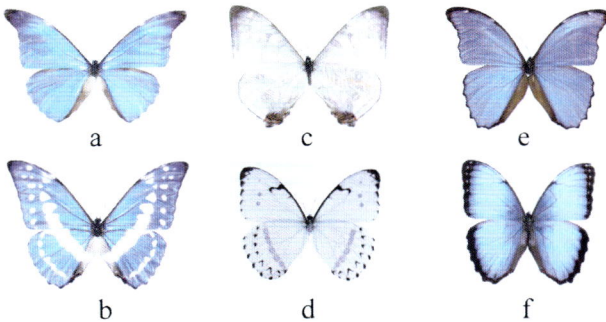

Figure 3.51 Dorsal sides of the male (a) *Morpho rhetenor*, (b) *Morpho cypris*, (c) *Morpho sulkowskyi*, (d) *Morpho catenarius*, (e) *Morpho didius*, and (f) *Morpho peleides*. (The photographs were taken by Ms. N. Okamoto.)

The first scientific description on the "*Morpho* blue" probably appeared in Walter's book [Walter (1895)]. Although it is sometimes difficult to obtain this book, we can know his description partly through a paper written by Lord Rayleigh (John William Strutt, 3rd Baron Rayleigh) [Rayleigh (1919)]. According to this paper, Walter performed an immersion experiment to clarify the characteristics of the *Morpho* wing. He soaked a wing into liquids having various refractive indices and found that the wing color changed from blue to green with increasing refractive indices and finally it lost the brilliancy. From this experiment, he insisted that the mechanism of the *Morpho* blue was due to "surface color" that was often observed in colored materials.

Lord Rayleigh opposed to this theory and proposed a new theory based on interference of light due to multilayer interference. Many scientists in the fields of physics, chemistry and biology took part in this problem, such as Michelson, Onslow, Süffert, Merritt, Mason, Lord Rayleigh (Robert John Strutt, 4th Baron Rayleigh) and so on [Michelson (1911); Onslow (1923); Süffert (1924); Merritt (1925); Mason (1927a); Rayleigh (1930)]. These researchers performed various experiments on the *Morpho* wing such as the reflection measurement, microscopic observation, immersion test in liquids, application of pressure, and so on. With analyzing experimental results, it was gradually believed that the interference of light was a major cause of the *Morpho* blue.

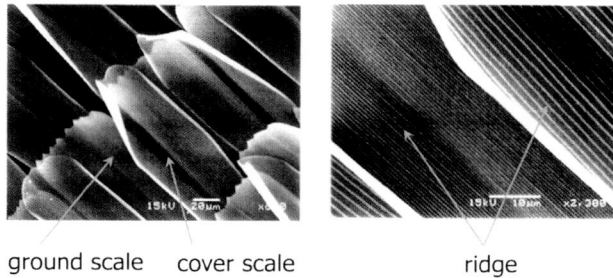

ground scale   cover scale   ridge

Figure 3.52   Scanning electron microscopic images of a wing of the male *Morpho didius*, showing alternately arranged cover and ground scales [Kinoshita and Yoshioka (2005b)]. Scale bars are 20 μm and 10 μm, respectively.

In 1942, Anderson and Richards [Anderson and Richards (1942)], and Gentil [Gentil (1942)] performed independently historic observations on the *Morpho* wing using a newly developed electron microscope, and found a periodic structure on a wing scale. These observations seemed to confirm the validity of the light interference, and it was generally believed that the *Morpho* blue was due to multilayer interference. Thereafter, researchers such as Lippert and Gentil [Lippert and Gentil (1959)], Hirata and Ohsako [Hirata and Ohsako (1966)], and Ghiradella [Ghiradella (1974, 1984)] extensively investigated the periodic structure using an electron microscope.

In general, the wing of the *Morpho* butterfly is covered with two kinds of scales called cover and ground scale, as shown in Fig. 3.52. On each scale, straight lines called ridge are running almost equidistantly. The ridge is generally well developed with the height of 2–3 μm, whose cross section is beautifully decorated by elaborate lamellar structure, as shown in Fig. 3.54. Since the ridges are densely distributed in the ground scale, it is considered that the ground scale is mainly responsible for the *Morpho* blue. Ghiradella reported that the ground scale of the *Morpho* butterfly was a classic example of ridge-iridescent scales, and the lamellae were prominently stacked alternately on either side of a ridge to form the interference mirror [Ghiradella (1984)]. Further, the lamellae was found to be attached obliquely to a base plane of a scale.

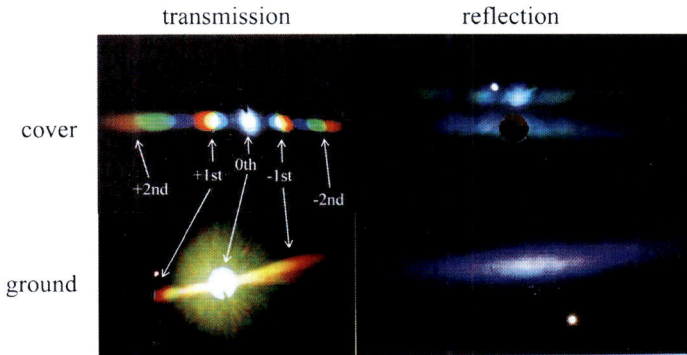

Figure 3.53 Transmission and reflection patterns of a single cover or ground scale of *Morpho didius* illuminated by white light [Yoshioka and Kinoshita (2004)]. Several spots seen on the transmission side correspond to diffraction spots due to spacings between regularly arranged ridges.

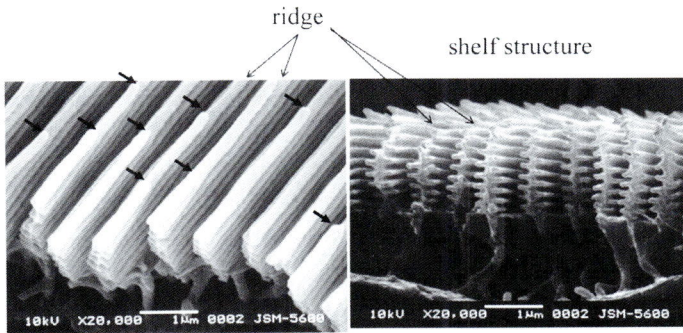

Figure 3.54 Scanning electron microscopic images of a ground scale of *Morpho didius*, seen from an oblique direction and at the cross section [Kinoshita *et al.* (2002a)]. Arrows in the left figure shows the positions of the upper ends of obliquely running shelves.

After the 1990s, various industrial demands to apply the *Morpho* blue to painting, textile, decoration and display increased, which promoted research studies on the *Morpho* butterfly rapidly. Tabata *et al.* reported a detailed measurement on the angular dependence of the reflection from the *Morpho* wing and analyzed the reflection spectrum using a model of multilayer interference [Tabata *et al.*

(1996)]. Later, their work achieved success as a form of new iridescent fiber called *Morphotex* [Iohara *et al.* (2000)]. Vukusic *et al.* reported a detailed measurement of the angle-dependent reflection on a single-scale level and showed that clear diffraction spots were found on a transmission side, while only a slender pattern was found on a reflection side with a dip at the center (see Fig. 3.53) [Vukusic *et al.* (1999)]. They also reported that the maximum reflectivity of a scale reached as high as 70%.

We reported the detailed experiments on the angular dependence of the reflection from the *Morpho didius* (see Fig. 3.51e) and proposed a new model to explain both a reflection spectrum and its angular dependence by taking diffraction effect and irregularity into account [Kinoshita *et al.* (2002a,b); Kinoshita and Yoshioka (2005a); Kinoshita (2008)]. This model was based on a fact that the width of shelves stacking on a ridge was so narrow that light interference was mainly achieved through the diffraction effect due to each shelf rather than multilayer interference that inevitably accompanied multiple reflection among shelves (instead of calling lamellar structure, let us call the periodic structure found in a ridge "shelf structure" as compared to a book shelf in a library).

Further, in order to avoid creating diffraction spots, which necessarily appeared when microstructures were arranged regularly, it was considered that the irregularity was introduced into the heights of ridges. The presence of the irregularity in height was easily observed from the random distribution of the upper ends of obliquely equipped shelves, as shown in Fig. 3.54. The irregularity thus introduced disturbed the constructive interference among ridges, and as a result the reflection pattern originated from a simple sum of the light diffraction from each ridge. In addition, owing to a slender shape of a shelf, the diffraction of light occurred only perpendicularly to a ridge, while almost specular reflection occurred in a direction parallel to it. Thus, our model explained that regularly arranged shelves produced the *Morpho* blue, while it became diffusive only in one direction owing to the irregularity and anisotropic shape of the microstructure. This speculation was confirmed for the most part by the experiments and calculations by analytical and numerical methods.

Research on *Morpho* butterflies rapidly developed after that. Particularly, numerical calculations using a structural model resembling the actual shape of microstructures were performed one after another [Plattner (2004)], and the angular dependence of the reflection for various *Morpho* species was investigated by Berthier *et al.* [Berthier *et al.* (2003, 2006)] and Kambe [Kambe (2008)]. Our group further progressed research on the *Morpho* wing from a view point how they were seen in human vision. One research was concerned with the role of cover scales covering over ground scales in the *M. didius* wing [Yoshioka and Kinoshita (2004)]. This was because species with a small size of cover scale showed glossy blue wings, while those with a large size showed matte blue ones. Detailed experiments clarified that a cover scale played a role of the wavelength-selective optical diffuser based on the coexistence of the ridge reflection and thin-film interference.

Another one was related with the origin of white spots in *M. cypris* (see Fig. 3.51b) [Yoshioka and Kinoshita (2006)]. The white spots in this species were peculiar because the microstructures found in blue and white parts showed completely the same structures. We investigated the origin of white spots and found that melanin pigment played an important role to show the wing bluer. On the other hand, to show whitish, the butterfly removed the melanin pigment completely from both ventral and dorsal scales and even from a wing membrane, which enhanced the isotropic light scattering and then caused the whitish appearance in our eye.

Thus, even a species of the *Morpho* butterfly possesses a lot of fascinating devices and functions in herself, which have been produced during a long evolutional process and thus will become an excellent model system for bionanophotonics.[a]

3.6.3.2 *Morpho* mimicry

Another stream of the *Morpho* research has been quickly developing in this decade. That is to mimic the *Morpho* blue to fabricate the elaborate structure using recently developed nanotechnology. This

---

[a]Our recent research using non-standard FDTD calculations will be presented in Sec. 7.3

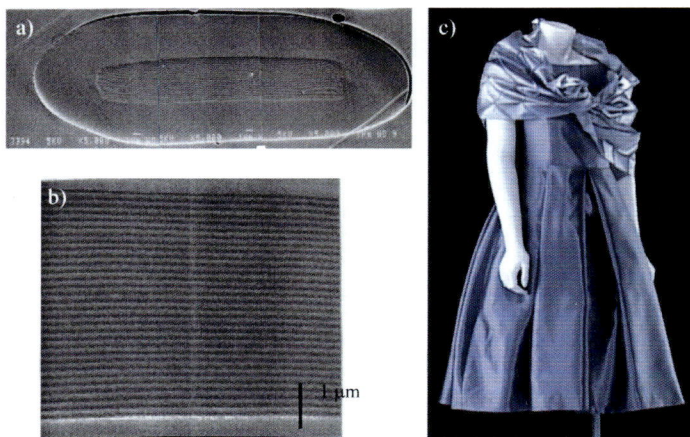

Figure 3.55 (a) SEM image of the cross section of a flattened polyester fiber mimicking *Morpho* butterflies. (b) The fiber includes 61 alternate layers of nylon 6 and polyester with the thickness of 70–90 nm. A dress woven using this fiber is shown in the right side [Courtesy of Teijin Fibers Limited].

work is classified roughly into two categories: one is to mimic the *Morpho* blue irrespectively of mimicking the microstructure, and the other is to mimic its microstructure itself. The former work is connected with the industrial demands related to vision such as painting, ink, display, fiber, cosmetics and so on, while the latter to the photonic technology to search for new photonic materials and their fabrication methods.

In a textile world, a quite sophisticated fiber was invented by mimicking the scale structure of the *Morpho* butterfly [Iohara *et al.* (2000)]. This fiber was made of polyester and had a flattened shape of 15–17 μm thickness, within which 61 layers of nylon 6 and polyester with the thickness of 70–90 nm were incorporated as shown in Fig. 3.55. Because of the multi-layered structure, the wavelength-selective reflection and the change of color with viewing angle were achieved. Moreover, the flat shape of the fiber made it possible to align the direction of the multilayer, which increased the effective reflectivity. They demonstrated it by weaving a wedding dress using this fiber. However, since the polymer materials employed had similar refractive indices ($n = 1.60$ for

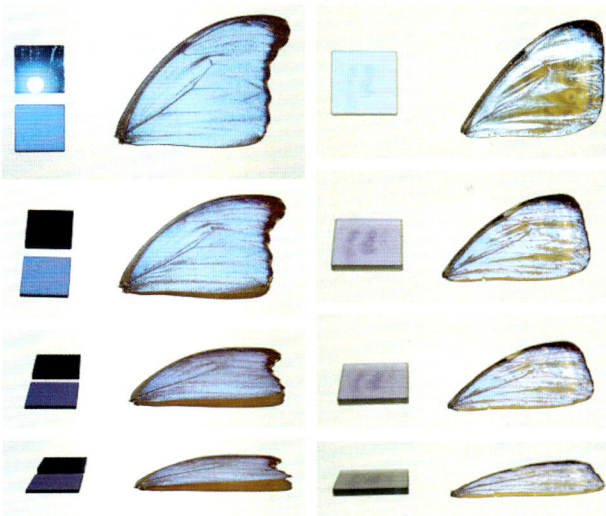

Figure 3.56 Left: Viewing angle dependence of *M. didius* wing and a multilayer-coated black substrate with (lower) and without (upper) irregular surface, obtained under the illumination of nearly normal to the wing/substrate. Right: that of *M. sulkowskyi* wing and a multilayer-coated transparent substrate with the irregular surface [S. Kinoshita and S. Yoshioka, Chemistry Today 377, 25 (2002)].

nylon 6 and $n = 1.55$ for polyester), the high reflectance was attained only within a small wavelength range. In fact, the wedding dress seemed to assume pale blue and was really in harmony with the ceremony, although it differed considerably from that of the *Morpho* butterfly concerning the color impact and luster.

We also reproduced the *Morpho* blue by mimicking its coloration mechanism [Yoshioka *et al.* (2004); Saito *et al.* (2004)]. First, we prepared a transparent or black glass plate. One surface was polished roughly to produce the irregular surface and then multilayer consisting of alternate layers of SiO_2 ($n = 1.47$) and Ta_2O_5 ($n = 1.97$) was coated on this rough surface. Thus, the coexistence of regularity and irregularity was actually introduced, in addition to the cooperation with the pigment. The result is shown in Fig. 3.56. The reproduced substrate looked quite similar to the *Morpho* wing, because the strong blue reflection was accompanied by diffusivity and the color change into violet was clearly observed in the oblique view. This was in contrast with the case when the roughness was not introduced, where the blue color was only visible under specular reflection, otherwise the substrate looked black. In this case, the roughness was a key point to fabricate this type of substrate, because if it was too rough, multiple scattering made the substrate whitish, whereas if it was too smooth, specular reflection would occur. Furthermore, the anisotropic reflection could not be introduced in the present method. Thus, controlled randomness was anyway necessary.

Therefore, we attempted to introduce the controlled randomness using dry etching method. First, electron-beam resist was placed on the substrate, and then electron-beam lithography followed by ion-beam etching was performed. Thereafter, multilayer coating using SiO_2 and Ta_2O_5 layers was laid. In order to introduce the regularity, we considered a unit rectangle, whose size was ~ 2 μm in length and 0.3 μm in width. These units were randomly distributed on a quartz substrate with their heights set at 0.11 μm, which were designed to cancel specular reflection at 440 nm under normal incidence. The result is shown in Fig. 3.57. The upper photograph shows a

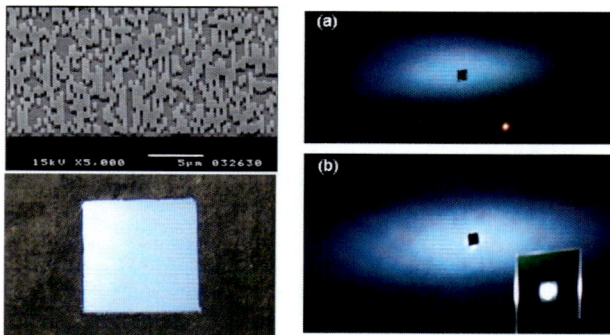

Figure 3.57 Left: (upper) SEM image of a glass substrate, whose surface is processed with randomly distributed tips of a size $0.3 \times \sim 2$ μm^2 with a two-valued random height of 0 or 0.11 μm. (lower) Multilayer coating consisting of SiO_2 and TiO_2 on the substrate of 6×6 mm^2 reveals the brilliancy similar to that of *Morpho* butterflies. Right: The reflection patterns obtained from (a) the *M. didius* wing and (b) the fabricated substrate [Reproduced from [Yoshioka *et al.* (2004)] with permission].

SEM image of the substrate before coating and the lower one the photograph of the substrate whose size is 6×6 mm^2. The reflection pattern, shown in the right side of the figure, almost completely agrees with that of a *Morpho* wing. It is thus demonstrated in order to reproduce the *Morpho* blue, we need not mimic the structure itself but only does its mechanism.

Saito *et al.* extended further our method of fabricating the *Morpho* substrate to be useful for the mass production [Saito *et al.* (2006)]. He employed nanocasting lithography (NCL) to obtain the fine pattern replication using various polymer materials. They prepared a quartz master plate according to the method described above, and UV curable resin was spin coated. UV light then illuminated after the coated surface was covered with a glass plate. The UV exposure made the resin to adhere the glass plate, while the surface of the master quartz plate was covered with antisticking surfactant beforehand. After removing the glass plate with the replicated microstructure from the master plate, TiO$_2$ and SiO$_2$ layers were alternately deposited. The fabricated replica looked very similar to the master plate under a scanning microscopic observation. Thus a low cost reproduction is attainable, which promises the cost down and mass production of the *Morpho* substrate.

Completely different approach to mimic the microstructure of the *Morpho* butterfly itself was progressed by Matsui *et al.*, who utilized a focused-ion-beam chemical-vapor-deposition (FIB-CVD) technique to reproduce the complete structure of a ridge on the *Morpho* scale [Watanabe *et al.* (2005a,b)]. The method is explained as follows: Ga$^+$ ion beam is focused on a sample surface with a small focal size of typically 10 nm, where aromatic hydrocarbon precursor gas is injected through a nozzle, which makes the decomposed carbon atoms deposited on the surface. They used phenanthrene (C$_{14}$H$_{10}$) as a precursor and it was found that the deposited material was mostly amorphous diamond-like carbon, which promised hard and transparent qualities. They demonstrated to make a ridge by this method, in which the lamellar distance was so adjusted to agree with that of the *Morpho* butterfly and even to reproduce the alternate structure of the lamella (see Fig. 3.58). The time to fabricate a structure with 20 μm in length was reported to be 55 min.

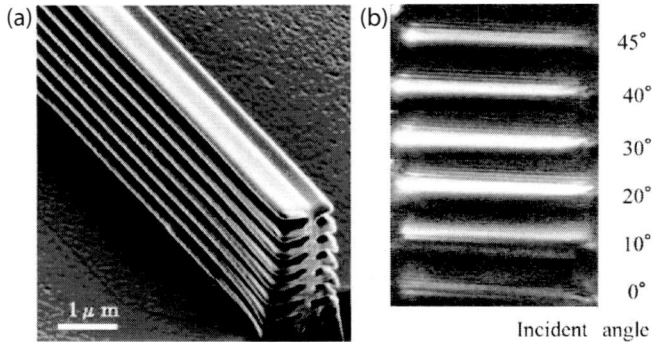

Figure 3.58 Microstructure of the *Morpho* butterfly fabricated by focused-ion-beam chemical-vapor-deposition method. (a) SEM image of the structure and (b) optical microscopic image observed under 0–45° incidence angle of white light. (Reproduced from [Watanabe *et al.* (2005b)] with permission.)

The replica illuminated by white light was found to actually glitter blue to violet under microscopic observation.

The replication of the *Morpho* scale was also reported by Huang *et al.*, who aimed to mimic the photonic structure of butterfly scales to find out new photonic devices [Huang *et al.* (2006)]. They employed atomic layer deposition (ALD) technique to replicate the wing structure of *M. peleides* (see Fig. 3.51f) using Al_2O_3 under the temperature of 100°C. They found the wing color was changed from blue to green, yellow and pink according to the thickness of the Al_2O_3 coating of 10–40 nm. After the coating, the sample was annealing at 800°C for 3 hours to crystallize the coating and to burn out the organic compounds in the wing. Then, the completely hollow replica was obtained, which consisted of robust polycrystalline Al_2O_3 hollow structure with a crystal grain size smaller than 3 nm. They made transmission electron microscopic observation and also performed optical measurement. The reflection spectrum thus obtained had a clear peak at 420 nm, which was slightly shorter than that of the original wing itself.

Thus recent rapidly developing nanotechnology has proved that even the most elaborate structures in nature can be reproduced. Therefore, the future of the study on structural colors is extremely promising, because the optical and physical investigations on their mechanisms will be immediately confirmed by fabricating the microstructures themselves.

3.6.4 *Motile Iridophore of Neon Tetra*

3.6.4.1 Iridophore in fish

The colors of fish skins are generally controlled by a variety of specialized cells called *chromatophores* [Fujii (1993a)]. Their beautiful colors and patterns, silvery flank and spectacular color changes are all achieved by the combined action of these cells. Generally, the chromatophores are classified into several types according to pigmentary granules and microstructures involved in cells. Above all, the following five chromatophores are widely known and are universally distributed in fish: *melanophore, xanthophore, erythrophore, leucophore* and *iridophore*. Among these, the last two are related to structural colors, while the first three contribute to pigmentary colors. In addition, the shapes of cells belonging to the first four are known to be dendritic, while the iridophores are usually round.

Iridophores play a central role in the structural colors in fish skin. They usually possess definite stacks of purine crystals, i.e. guanine, hypoxanthine or uric acid, inside the cells, which constitute a multilayer to reflect light of a specific color and/or a silvery color. Usually the iridophores can be further divided into *static* and *motile* ones according to the responses to physiological changes. The most of the iridophores distributed in skin belong to static ones, which display silvery whitish flanks, brilliant blue skins and so on. This type of iridophore was extensively studied since the 1900s. On the contrary, the presence of motile iridophores was noticed in the 1980s and up to now, not so many species are known to exhibit the motilities against neurotransmitters, hormones and light stimulations. Neon tetra, damselfish and goby are such well-known examples.

3.6.4.2 Motile iridophore in neon tetra

The stacking structures of motile iridophores in neon tetra (see Fig. 3.59a), *Paracheirodon innesi*, inhabiting in the Amazon River, were

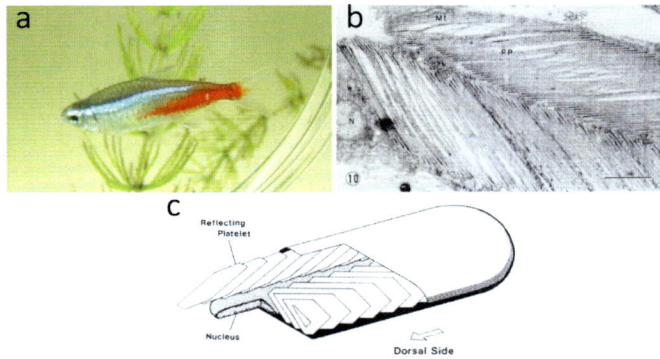

Figure 3.59 (a) Neon tetra, (b) TEM image of a pair of stacks of platelets in an iridophore and (c) their schematic illustration. (Reproduced from Figs. 3.10 and 3.11 of [Nagaishi and Oshima (1992)] with permission. The photograph of neon tetra was taken by Dr. S. Yoshioka).

first studied by Lythgoe and Shand [Lythgoe and Shand (1982)]. They noticed a pronounced color change was observed in neon tetra between day and night. In daytime, it showed a brilliant blue-green stripe, while it did dull violet-blue color at night. They performed various experiments and found that the color change occurred under light stimulation and the blue color changed into violet after placing it for 25–35 min in dark. The color changes were also observed under various physiological stimulations such as the osmotic change and addition of adrenaline. Later, using an immunofluorescence technique, Lythgoe *et al.* clarified the response to light was related to the presence of rhodopsin or porphyropsin in an iridophore [Lythgoe *et al.* (1984)].

They noticed the presence of two types of light-reflecting crystal in the dermis: One was a wide crystal distributed solely in the stripe zone with the thickness of 5–10 nm, and the other was a slender crystal of 43–108 nm thickness, which was distributed in silvery dermis, ventral region, and even in lateral stripe and iris. The former was clearly an active platelet, while the latter was inactive. They found a clear correlation between the spacing of the former platelets and the apparent colors, and considered that light stimulation caused sodium channels to open, resulting in an increase of the interval of platelets osmotically.

Nagaishi and Oshima reported the detailed observation on the iridophores in a blue stripe of neon tetra, in which they described that an iridophore had a rectangular flat shape with the dimension of 60–70 μm length, 15–20 μm width and 1–2 μm height [Nagaishi and Oshima (1989); Nagaishi *et al.* (1990); Nagaishi and Oshima (1992)]. Within the iridophore, a pair of stacks were located above a thin nucleus. In each stack, many guanine platelets in a hexagonal shape were enclosed by membranous structure and arranged regularly inclining at a constant angle to the surface, as shown in Figs. 3.59b and c. The inclination of the platelets would be useful to reflect light to a horizontal direction when illuminated from the dorsal direction. They performed a lot of experiments and concluded that the major cause for the color change was due to the change in inclination angle of crystal platelets. They called it *Venetian blind mechanism* and speculated that the motility of crystal inclination was controlled by tublin-dynein system.

In spite of these energetic experiments, two significant questions still arise: One is which mechanism of color change is actually valid in living fish, tilting or swelling, and the other is how the broad reflection spectrum is attained by so thin crystal platelets. In order to answer these questions, we performed the following experiments [Yoshioka *et al.* (2011)]: First, in order to evaluate the first question quantitatively, we performed a detailed optical study on a single stack of platelets inside an iridophore. For this purpose, an experimental system was constructed, which enabled to observe almost simultaneously the real image, reflection direction and spectrum from a single stack of platelets. Briefly, white light from a Xe lamp was first focused on a plane of an aperture stop, where a small pinhole was placed, and then the image of this pinhole was focused on a back focal plane of an objective so that the illuminating light became nearly collimated at a sample position. When a sample had a specular surface, reflected light focused by the objective lens would form a small spot again on the back focal plane. The intensity pattern of this plane was observed by a CCD camera. On the other hand, the real image of a sample was observed with another CCD camera and further using an optical fiber, the reflected light from a small part was analyzed by a spectrometer.

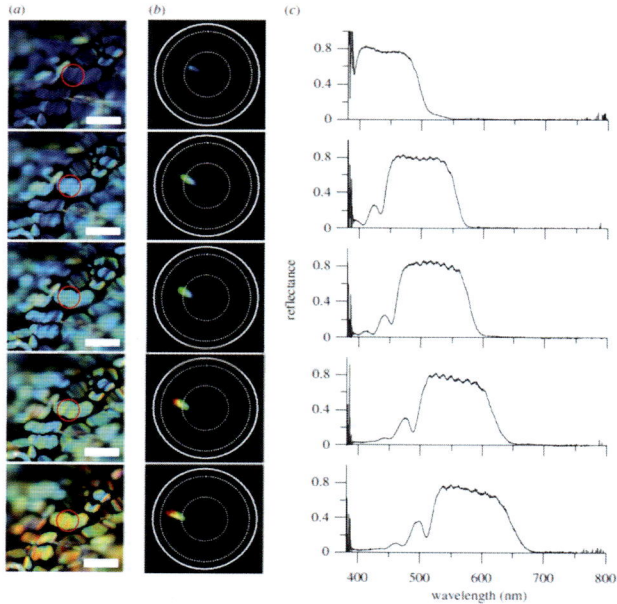

Figure 3.60 Change in the optical properties of a stack of platelets within an iridophore of neon tetra during the color change from blue to yellow [Yoshioka *et al.* (2011)]. (a) Real image of the iridophores during the color change. The red circle indicates the illuminated area for the measurements shown in (b) and (c) (scale bar is 20 μm). (b) The far-field patterns obtained by illuminating the area corresponding to the real images in (a). The two inner dotted circles indicate the angles of 15° and 30° with respect to the optical axis, and the solid circle indicates the maximum detectable angular range of 35.7°. (c) The reflectance spectra corresponding to the images in (a) and (b).

We show a typical example of the measurements in Fig. 3.60, where real images of platelet stacks, the image of the back focal plane corresponding to the reflection direction, and the reflection spectrum are shown during the color change from blue to yellow. The color change was induced by replacing the Ringer solution in a sample chamber by a solution having a potassium-ion concentration 50 times higher than the normal. It is clear that together with the color change from blue to yellow, the change of the reflection spectrum to the longer wavelengths is observed. In addition, the direction of reflection is found to change remarkably. From these experiment, we have shown that the relation between the wavelength of reflection and the tilt angle during color change is well explained by the Venetian blind model. This result directly proved that the swelling of the iridophore does not take part in the color change, because the swelling causes only the shift of reflection spectrum, but not the tilt angle to change.

The second question could be answered by measuring a platelet by an interference microscope. From this measurement, although the thicknesses of platelets were considerably distributed from platelet to platelet, it was found that the thickness was generally not uniform and considerably tapered with the thickest part of 60±5 nm. Although this value considerably differed from those reported previously, the reflection spectrum for oblique multilayer calculated using the thicknesses of 60 and 155 nm and the refractive indices of 1.83 and 1.37 for platelet and cytoplasm, respectively, could reproduce the measured spectrum fairly well.

Thus, the above questions are now all answered and the mechanism to move the inclination angle of the platelets will shift to the next question to be disputed how these platelets move simultaneously. Since these platelets are maintained gently by membranous structure, the measurement of fluctuations that will appear in the reflection intensity and direction seems to be quite interesting, because the information on the fluctuations is directly connected with the mechanism of the motility.

3.6.5 *Two-Color Iridescence in Rock Dove*

Next, we will show an example, in which even the simplest structure of a thin film produces an ingenious color effect to human and possibly animal's eyes [Yin *et al.* (2006); Yoshioka *et al.* (2007); Nakamura *et al.* (2008)]. The neck feather of commonly spotted rock dove, *Columba livia*, shows an unusual optical characteristic: green (purple) feathers located at the neck suddenly change their color into purple (green) only by slightly shifting a viewing angle, which is quite in contrast to a gradual color change observed in usual iridescence. The color of the rock dove's feather is mainly due to numerously sticking barbules, whose length, width and thickness

Figure 3.61 Commonly spotted rock dove, *Columba livia*, and the cross sections of burbules of its neck feather [Yoshioka *et al.* (2007); Nakamura *et al.* (2008)].

are typically 350, 40 and 3 μm, respectively. The cross section of a barbule is crescent-shaped (Fig. 3.61a), which produces a very broad reflection pattern even under normal incidence. It is remarkable that the reflection pattern from a single barbule of a green (purple) feather shows a sudden change from green (purple) to purple (green) with a rather distinctive gray boundary.

The electron microscopic image of the cross section of a barbule shows the presence of many melanin granules, which are enclosed by an outer cortex layer (Fig. 3.62b). The diameters of melanin granules are randomly distributed from 500 to 750 nm and their arrangement is also irregular. Hence, they will not contribute directly to the coloration. Thus, the cortex layer with a rather uniform thickness becomes an only candidate for the coloration, whose thickness ranges in 600–700 nm for a green feather and in 480–580 nm for a purple feather. In fact, the reflection spectra from each barbule displaying green, gray, or purple color show multiple sinusoidal reflection spectra (Fig. 3.62a), which are found to shift gradually according to the perceptible color. These spectra are found in very good agreement with the thin-film interference, if we assume, for example, a layer thickness and refractive index of 650 nm and 1.5 for a green feather.

Thus, the neck feather of rock dove is actually caused by the simplest mechanism of thin-film interference. However, the optical

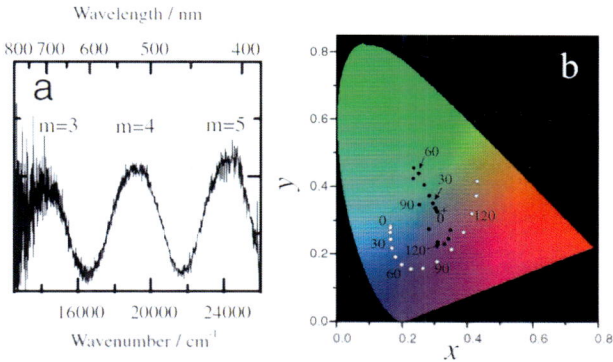

Figure 3.62 (a) Reflection spectrum from a green neck feather and (b) chromaticity diagram showing the loci of reflection due to thin-film interference at various angles of incidence [Yoshioka *et al.* (2007)]. The thicknesses of the films are set at 650 nm (closed circle) and 400 nm (open circle) with a refractive index of 1.5.

interference occurs in higher orders, which produce multiple peaks in a visible wavelength range: a green feather shows $m = 3$, 4 and 5 interference peaks at 700, 520 and 420 nm under an incidence angle of 30°. The most important point to produce the *two-color iridescence* described above is based on a fact that the separation of two adjacent peaks in wavenumber unit coincides with that of color matching functions of human vision for blue and red. When this condition is actually satisfied, one will perceive purple, while when the viewing angle is changed and hence the reflection peak shifts from red to green and also from blue to ultraviolet, one will perceive only green color. Between these colors, one feels gray because all the colors are perceived almost equally.

This situation is visually understood in terms of chromaticity diagram, where the two-color iridescence is expressed by a linear movement of a locus crossing a white (gray) zone ($x = 1/3$ and $y = 1/3$) as shown in Fig. 3.62b. It is clear that the multiple peaks due to thin-layer interference is suitable to enhance the two-color iridescence because the largest peak shift is obtained when the locus crosses a gray zone. However, to satisfy these conditions, it is necessary to strictly control the thickness of the cortex layer, because if the layer is thicker, the locus only moves near the gray

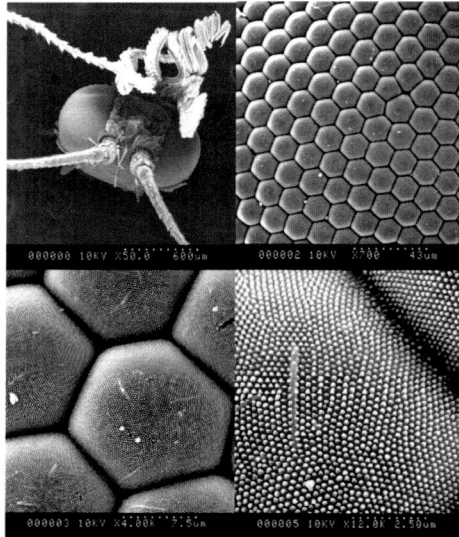

Figure 3.63 Moth-eye structure observed in an enlarged view of a compound eye of a moth, *Agrotera nemoralis*. The magnifications are 50, 700, 4000, and 12000 from the upper left to the lower right. (Courtesy of Dr. S. Yoshioka.)

zone and then no particular color sense is stimulated, while if it is thinner, the locus draws a circle like ordinary iridescent thin film. The two-color iridescence will be more effective to avian tetrachromatic vision, because the color matching functions in avian vision are much more well defined owing to the presence of color filter called oil drop.

3.6.6 *Moth-Eye Structure*

3.6.6.1 Natural moth-eye structures

Bernhard and Miller found that corneal surfaces of facets of some lepidopteran compound eyes were covered with congruent cone-shaped protuberances (see Figs. 3.63 and 3.64) [Bernhard and Miller (1962); Bernhard *et al.* (1965, 1968); Bernard and Miller (1968)]. They called them *corneal nipples*, which were present only on corneal chitin and covered its surface completely. The nipples

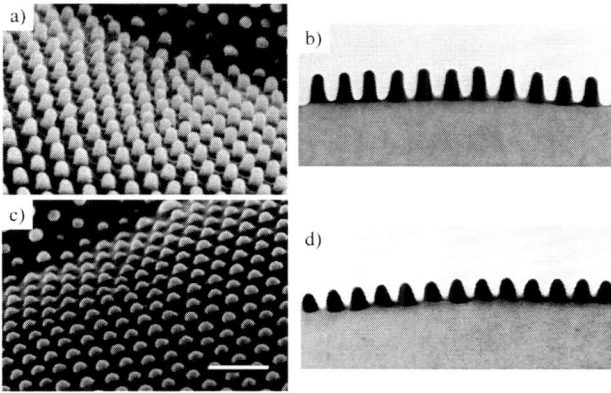

Figure 3.64 Corneal nipple arrays in butterflies. (a,b) The nymphalid *Polygonia c-aureum* and (c,d) the lycaenid *Pseudozizeeria maha*, showing the difference in nipple height and shape. Scale bar is 500 nm. (Reproduced from [Stavenga *et al.* (2006b)] with permission of The Royal Society.)

were around 200 nm in amplitude and were arranged closely in a hexagonal closest-packing with the separation of the adjacent nipples of 200 nm. The shape of the nipple somewhat resembled a hanging bell. The size and shape were, however, found to depend on species.

They thought such a regular and minute structure would give a function of antireflection effect to the corneal surface. This was because the interface between the corneal substance with a high refractive index of typically 1.57 and that of the outer space of 1.0, would give the reflection loss at the surface and moreover cause the glare that would be visible even from far away. Thus, the nipple structure anyway contributed to the impedance matching by gradually changing the refractive indices at the surface.

The distribution of corneal nipples was surveyed extensively by Bernhard *et al.*, who investigated 361 species that belonged to most of insect orders [Bernhard *et al.* (1970)]. Since the sizes and shapes of nipples were variously distributed, they classified them into groups I - III according to the amplitude of the nipples. Group I included the species without nipples or with those having the amplitude lower than 50 nm, group II with those of 50–200 nm, and group III with those higher than 200 nm. Thus, the medium-

and well developed nipples anyway belonged to groups II and III. They found species assuming corneal nipples were restricted within only four orders, Thysanura (silverfish), Diptera (flies), Trichoptera (caddisflies) and Lepidoptera (moths and butterflies). It was interesting that silverfish, quite distantly related to the other three, assumed corneal nipples, which stimulated the scientist's imagination in later years. For example, Bernhart *et al.* speculated that the distribution in silverfish implied the regression, rather than progression, from the well developed nipples with the common ancestors among these four species. The regression was incomplete for these orders, while most of the other orders did completely.

The hypothesis that corneal nipples appeared as a phylogenetic product to accommodate nocturnal activity in moths to increasing transmission of light, would be puzzled, because only Papilionidae (swallowtail) among various families of diurnal butterflies lack the nipples. In Lepidoptera, 95% of the species belonging to the families Pieridae (white butterfly), Lycaenidae (hairstreak) and Nympharidae (brush-footed butterfly) show the nipples, while the nipples are present in only 30% of Hesperiidae (skipper) and none of Papilionidae. In moth species, this ratio becomes 75%. The distribution in butterflies in the three families was later investigated by Stavenga *et al.* on 19 diurnal butterflies, which showed similar tendency that only small arrays of about 30 nm amplitude were found in Papilionidae, while in Pieridae and Nymphalidae, they ranged from 90 to 230 nm (see Fig. 3.64) [Stavenga *et al.* (2006b)].

Similar moth-eye structures seem to be widely distributed among the insect world. Yoshida *et al.* reported that the similar structure was found in the wing of a hawkmoth, *Cephonodes hylas* (hawkmoth), which lost almost all the wing scales at the time of emergence [Yoshida *et al.* (1996, 1997)]. The resultant transparent wing assumed nipples with the height of 250 nm with the separation of 200 nm in a hexagonal packing. They compared the reflectivity before and after crushing the protuberances, and found 70% reduction of reflectivity was observed.

Watson and Watson performed atomic force microscopic (AFM) investigation on wings of cicada (*Pflatoda claripennis*) and termite (Rhinotermitidae sp.), and found nipples with the height of 225–250 nm and spacing of 200 nm for the former, while somewhat rough protuberances spaced 700–1000 nm with the height of 150–250 nm for the latter [Watson and Watson (2004)]. They speculated the structure for the former would presumably offer the protection against the predators due to the antireflection effect, and the self-cleaning effect against small particles containing in water drop due to the structurally hydrophobic nature, in addition to the mechanical rigidity. On the other hand, that for the latter would only contribute to the mechanical rigidity.

Parker *et al.* investigated the compound eyes of Eocene fly in Baltic amber (45 Ma) [Parker *et al.* (1998)]. They found the corneal surface comprised of a series of parallel ridges with approximate sine-wave profile with the periodicity of 240 nm and the depth of 146 nm. They also found the similar structures in two extant Diptera species. They measured the reflectivity of a model substrate having a similar surface fabricated by a photoresist method,[a] and found that considerable reduction of the reflectivity was obtained for both *s*- and *p*-polarizations.

3.6.6.2 Fabrications of moth-eye structures

Probably the first attempt to fabricate the moth-eye structure was carried out by Bernhart *et al.* themselves [Bernhard *et al.* (1965, 1968)]. They scaled up the structure by 10^4 and performed a model experiment using a microwave as an incident source. The nipples fabricated were cone-shaped with the height of 1.5–6 cm and made from a mixture of 25% beeswax and 75% paraffin to accommodate to the permittivity of cuticle. They found considerable increase in the transmittance in a broad range of wavelength from 1.5 to 5 times the height of the nipples.

The direct fabrication of the moth-eye structure appeared immediately after the above optical and physiological investigations were reported. Clapham and Hutley attempted to fabricate a moth-eye structure using a photoresist method [Clapham and Hutley (1973)]. They utilized two collimated Kr^+ laser beams of 351 nm to intersect each other to generate interference fringe consisting of

---

[a]A photosensitive material called *resist* is a material that loses the resistance or susceptibility to an etching solvent when it is exposed to light, which are called *positive* and *negative photoresist*, respectively.

parallel lines with the spacing of 210 nm, then rotated the substrate by 90° to impose two such patterns at right angles. This procedure made it possible to create a sinusoidally modulated pattern inside a photoresist coated on a glass substrate. They observed the considerable reduction of reflection below a wavelength of 550 nm under normal incidence.

Recently, the moth-eye type antireflection has attracted considerable attention and has been practically applied as a new optical tool in a rapidly growing field of nanophotonics. Particularly, it is strongly expected as an efficient antireflection device for silicon solar panel because of its superior characteristics with respect to the acceptability for a wide range of wavelength and angle of incidence. This is because silicon has a high refractive index of 3.617 at 1000 nm, which causes 30% reflection loss at its bare surface.

In general, antireflection of a substrate having a high refractive index has been attained using the following two methods: One is a conventional monolayer or multilayer coating method, which is now an industrial standard for silicon solar panel to lay SiN_x coating through plasma-enhanced chemical vapor deposition so that the reflectivity can be minimized at around 600 nm under normal incidence. However, the antireflection efficiency thus obtained is strongly wavelength-dependent and highly affected by the angle of incidence. Furthermore, the instrumental burden is not so light.

Another method to attain the antireflection is to modify a surface anyway. This is further categorized into three according to the kinds of microstructures formed on a surface: (a) moth-eye, (b) nanoporous and (c) surface texturized structures. The moth-eye type is particularly superior because it is insensitive to wavelength and angular variations. Further, it has a merit that the microstructure can be made from a substrate itself.

In order to fabricate it, top-down or bottom-up method is normally applied. The former is to fabricate according to a basic design, while the latter to follow the self-assembly process. Several methods are known for the former such as interference, electron beam, and nanoimprint lithography. The method employed by Clapham and Hutley belongs to the first method. Although they used the photoresist layer as it was, it is now common to transfer

a)

b)

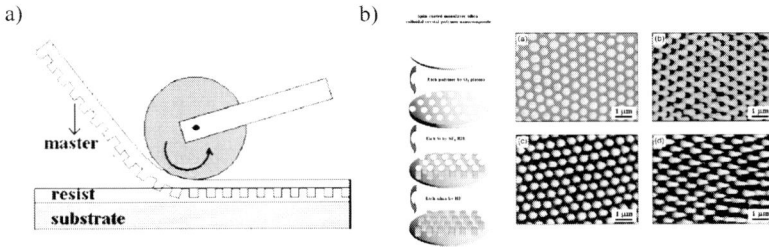

Figure 3.65 (a) Nanoimprint lithography using a soft roller technique. (b) Self-assembly lithography using colloidal template and SEM images of moth-eye structure thus fabricated. (Reproduced from (a) [Chen *et al.* (2009)] and (b) [Sun *et al.* (2008)] with permissions.)

the pattern obtained after developing to a substrate by ion etching [Lalanne and Morris (1997); Aydin *et al.* (2002)].

The electron beam lithography is carried out to lay an electron-beam resist layer on a substrate, on which electron beam is scanned according to a basic design [Kanamori *et al.* (1999, 2005); Boden and Bagnall (2008)]. Then, the pattern is transferred to a substrate by ion etching. This method gives a precise fabrication within a scanning plane but depends somewhat on experience in the depth direction. Further, it is eventually difficult to extend an area of lithograph. For example, Kanamori *et al.* reported to take 10.8 hours for an area of 1.2×1.2 mm^2 [Kanamori *et al.* (2005)].

The last one is nanoimprint lithography, which transfers a master pattern to a resist layer [Yu *et al.* (2003); Chen *et al.* (2009)]. There are many types such as thermoplastic and photo nanoimprint lithography. Chen *et al.* introduced soft roller nanoimprint technology, in which a flexible master transfered from a rigid master created by interference lithography was used to transfer a pattern to a resist layer on a rigid substrate (see Fig. 3.65a) [Chen *et al.* (2009)]. All these methods proved a good antireflection capability as the moth-eye structure.

On the other hand, not so many methods have been reported for a bottom-up method. Jiang *et al.* reported a method of colloidal templating, in which colloidal particles are arranged during a spin coating to form a 2D pseudo colloidal crystal on a silica substrate (see Fig. 3.65b). Subsequent etching by O$_2$ plasma and SF$_6$ reactive ion etching can transfer crystal arrangement to a silica substrate [Sun *et al.* (2007); Linn *et al.* (2007); Sun *et al.* (2008)]. They actually reported good broadband antireflection properties down to 2.5% within 350–850 nm under normal incidence.

The antireflection due to nanoporous material is directly based on a principle of gradually changing refractive index at an interface. This method was first proposed by Lord Rayleigh [Rayleigh (1880)], who mathematically connected the refractive index of a material with that of outer space continuously. Since a material showing the refractive index close to 1.0 could not be obtained actually, it was thought impossible to realize it. The moth-eye structure was one of the solutions for this problem, and nanoporous is another solution to make a refractive index to be close to unity by introducing porous material.

Although the ideas of these two methods are essentially the same, moth-eye structure is based on the regularity so that the size of each microstructure needs not be so small because the regularity reduces the scattering loss at a surface considerably. Hence, it is only to satisfy the condition of the zeroth-order diffraction for given angular and wavelength ranges. On the contrary, antireflection due to nanoporous structure is essentially based on irregular structure so that it is inevitable for reducing the scattering loss that the characteristic size should be small enough for light to regard it as continuous.

As an example, we show a report written by Xi *et al.*, who showed that oblique-angle deposition of SiO_2 or TiO_2 on a silicon substrate yields an array of nanorods of SiO_2 or TiO_2 with a tilt angle of $45°$ [Xi *et al.* (2007)]. Using this method, they achieved a minimum refractive index of 1.05 for SiO_2 layer. They fabricated a graded-index layer consisting of 3 TiO_2 and 2 SiO_2 layers so that the refractive indices follow a quintic-index profile and confirmed that the reflectivity reduced to around 0.3% for incident angles between 0 to $55°$ and to 0.5% at wavelengths from 574 to 1010 nm.

The surface texturized modification is to fabricate the depressions of a macroscopic size on a surface to make incident light to be reflected many times at their walls. Parretta *et al.* made many inverted pyramids of 10-µm size on a silicon substrate and obtained the reflectivity below 10% for both wide angular and wavelength

Figure 3.66 Scarab beetle, *Protaetia pryeri*, photographed (left) without and (right) with a right-handed circular polarization analyzer [Kinoshita and Yoshioka (2005b)].

ranges [Parretta *et al.* (1999)]. This method reminds us of a petal of pansy, on which a lot of protuberances of a macroscopic size are known to increase the absorption of light through multiple reflection and to enhance the colors of the petals [Bernhard *et al.* (1968)].

3.6.7 *Cholesteric Liquid Crystal: Scarab Beetles*

In Fig. 3.66, we first show a typical example of scarab beetles, *Protaetia pryeri*, inhabiting in Okinawa in Japan. We took these photographs under the illumination of natural light with and without a right-handed circular polarization analyzer placed in front of a camera lens. It is clear that it looks brass-colored without an analyzer, while it immediately changes into black when seen under the analyzer. These characteristics are generally observed in a fairly large number of scarab beetles, as we have shown the examples in Fig. 3.67. In most of these cases, beetles turn black or dark under a right-handed circular polarization analyzer, while they look more brilliant under a left-handed circular one. The exception is *Plusiotis resplendens*, called golden scarab beetle, which shows light-green under a left-handed circular polarization analyzer, while it turns orange-yellow under a right-handed circular one.

As described in Sec. 3.5, these features are characteristic of cholesteric liquid crystal. Cholesteric liquid crystal consists of a layered structure, within which molecules take a preferred direction, on an average, that gradually changes from layer to layer with

left   right   left   right

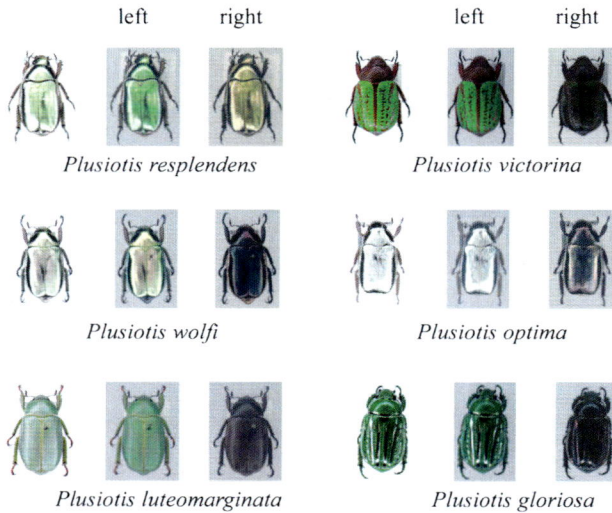

Plusiotis resplendens   *Plusiotis victorina*

Plusiotis wolfi   *Plusiotis optima*

Plusiotis luteomarginata   *Plusiotis gloriosa*

Figure 3.67 Various iridescent beetles viewed (left) without and with a (center) left- and (right) right-handed circular polarization analyzer, photographed by E. Nakamura and N. Arakawa, in the possession of The Nature and Human Activities, Hyogo, Japan.

increasing distance from a surface, and shows a helical structure, whose axis is normally perpendicular to the surface. The periodicity of the helical structure is determined by a pitch of the helix, P (see Fig. 3.37), and if the molecules within a plane is isotropic with respect to their orientations like a nematic liquid crystal, a half of the helical pitch, $P/2$, determines the periodicity. Thus, the selective reflection of light occurs when the wavelength of light, λ, becomes close to the helical pitch. In this case, a peak wavelength of the first reflection band is determined as $\lambda_p = (n_\parallel + n_\perp)P/2 = nP$, while its band width becomes $\Delta\lambda = |n_\parallel - n_\perp|P = \delta nP$, where $n_\parallel$ and $n_\perp$ are refractive indices corresponding to two optic axes in a uniaxial system, and are taken as those parallel and perpendicular to the surface. n and δn are the average refractive index and the birefringence, respectively.

The helical structure is often seen in optically active substances and is thought to be due to a fact that an energy of a system consisting of optically active molecules will be decreased if the molecules are piled with changing their directions little by little. Since the bodies of living things are made of optically active molecules, helical structures are commonly present in their bodies.

When the molecules are aligned in parallel to the surface and change their directions according to a distance from the surface, the section obliquely cut with respect to the surface will sometimes show a parabolic pattern, which was already reported at the beginning of the 20th century. The reason for the appearance of such a parabolic pattern was first analyzed by Booligand [Bouligand (1965)] and the relationship between this pattern and a helical structure was clarified. Just at the same time, Neville *et al.* found that helical and uniaxial structures were alternately arranged within the outer layer of grasshopper [Neville and Luke (1969a,b)]. The thicknesses of these two layers were found to be controlled by a circadian rhythm. If the pitch of the helical structure becomes on the order of wavelength of light, selective reflection will occur. In addition, in biological system, only left-handed helices exist so that light with left-handed circular polarization is selectively reflected. This is a main reason why a beetle turns black or dark when we see it through a right-handed circular polarization analyzer.

The first observation on anomalous reflection from the beetle's elytron was reported by Michelson, who investigated the reflection of light from the elytron of a golden scarab beetle, *Plusiotis resplendens* [Michelson (1911)]. He found that reflected light was circularly polarized even under normal incidence. The degree of the circular polarization was maximum at blue, decreased gradually toward a yellow region, completely depolarized in orange-yellow, and appeared again with the opposite direction in the red end. He considered the origin of this circular polarization as due to a screw structure of a molecular dimension, and assigned that the circular polarization came from the selective absorption for only one of the two polarizations. He also considered that the depolarization in the orange-yellow region, where the two circularly polarized components equally contributed to the reflection, was due to the inhomogeneity within the elytron.

Sixty years later, Neville *et al.* performed the experiments to understand the essential properties of this phenomenon [Neville and Caveney (1969); Neville and Luke (1971); Neville (1975, 1977,

1984)]. Neville and Caveney reported the detailed descriptions on the structural and optical properties in various species of scarab beetles [Neville and Caveney (1969)], in which they described: (1) The color of elytra of these species changed with changing viewing angle, (2) a combination of circular and linear polarizers capable of passing light with a different circular polarization made the elytra completely black, and (3) transmitted light through an excised surface layer showed an optical rotation.

They scraped the cuticle surface minutely and found that the unusual optical properties originated from the outer region of the exocuticle, where a regular lamellar pattern of 4–15 μm thickness with the apparent pitch of 153–198 nm was observed. On the contrary, the inner region of the exocuticle was sclerotized and showed a dark color with irregular lamellar structure. The lamellar spacing in the outer exocuticle normally changed systematically to cause the reflection in a wide wavelength range (see Fig. 3.68c). The excised cuticle layer showed a clear optical rotation for the transmitted light, the direction of which changed at a particular wavelength, indicating the selective reflection occurred around this wavelength. They found that the overall features were very similar to those of a well-known cholesteric liquid crystal.

Caveney compared two species of scarab beetles, *Plusiotis optima* and *P. resplendens* (see Fig. 3.67) [Caveney (1971)]. The former showed silvery color, while the latter did golden luster like brass. The thicknesses of the optically active layers for these beetles were found to be 16 and 22 μm, respectively. Remarkable was the layer of the latter species that showed a sandwiched structure as shown in Fig. 3.68a, within which 5 μm upper and 15 μm lower layer showed clear parabolic patterns, while a sandwiched layer of 1.81 μm did not show the similar pattern but showed unidirectional molecular architecture. The reflection spectra for circular polarizations were very peculiar because it showed a peak around a green region of 560 nm under left-handed circular polarization, while it shows a peak in an orange region around 575–624 nm under a right-handed circular one (Fig. 3.68b). The reflectivity was of a rather high value of 0.32–0.35 in each case, which amounted to the total reflectivity of 0.6–0.7. On the other hand, in *P. optima*, only light of a left-handed circular

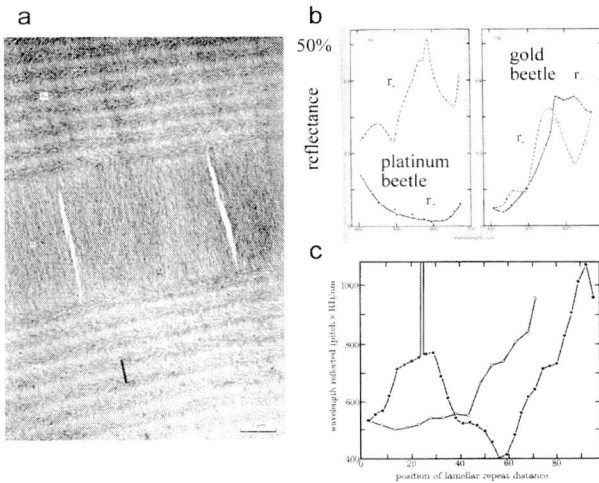

Figure 3.68 (a) Oblique section of the reflecting layer of golden scarab beetle, *Plusiotis resplendens*, (b) reflectivities under two circular polarizations for the elytra of platinum beetle, *P. optima*, and *P. resplendens*, and (c) spacing of the lamella against the lamellar position for *P. optima* (open circles) and *P. reslendens* (solid circles). The position of unidirectional layer is indicated as vertical lines. r_+ indicates the reflectivity under right-handed circular polarization, while r_- does the reflectivity under left-handed circular polarization. (Reproduced from [Caveney (1971)] with permission of The Royal Society.)

polarization was reflected with the maximum reflectivity of 0.5 (Fig. 3.68b).

The origin of this peculiar phenomenon comes from the presence of the sandwiched layer, which shows the optical anisotropy. Caveney calculated the phase change between different axes of the anisotropy using the measured refractive indices, and found this layer had a function of a perfect $\lambda/2$ retardation plate around 590 nm, which worked fairly well in a region of 550–650 nm.

What happens when a $\lambda/2$ retardation plate is inserted between two cholesteric liquid crystal of anti-clockwise helical system? Figure 3.69 shows the phenomenon schematically. When light with left-handed circular polarization is incident normally to the material, it is completely reflected by the upper parabolic layer, while that with right-handed circular one penetrates the first

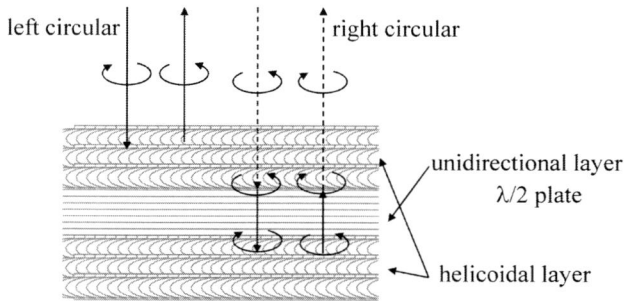

Figure 3.69 Schematic representation of the reflection from the elytron of *Plusiotis resplendens*. Light with left-handed circular polarization is reflected directly at the first helical layer, while that with right-handed circular one penetrates the first layer, which is converted to left-handed circular one, reflected at the second helical layer and then converted to right-handed circular one to pass the first layer.

layer without loss. The $\lambda/2$ retardation plate changes right-handed circular polarization into left-handed one when the light transmits. Then, the converted left-handed circular polarized light is selectively reflected by the second parabolic layer. The $\lambda/2$ retardation plate again changes the left-handed circular polarization of the reflected light into a right-handed one, which penetrates the first layer from the back side without loss and emerges out. Thus, both the left-handed circularly polarized light reflected at the first layer and right-handed circularly polarized light converted in this way contribute to the reflected light.

Caveney measured the anisotropic refractive indices of these layers and found that $n_{\parallel} = 1.603$ and $n_{\perp} = 1.700$ with $\delta n = 0.097$ for a helical layer and $n_{\parallel} = 1.535$ and $n_{\perp} = 1.701$ with $\delta n = 0.166$ for a unidirectional layer, where $n_{\parallel}$ and $n_{\perp}$ were the refractive indices parallel and perpendicular to the optic axis of the layer. These values were quite in contrast to those of ordinary scarab beetles, e.g. $n_{\parallel} = 1.580$ and $n_{\perp} = 1.598$ with $\delta n = 0.018$ known for *Potosia speciosissima*. He chemically extracted the exocuticle layer using NH_4OH solution and found high content of uric acid, volume fraction of 0.6–0.7, was distributed in these layers. Uric acid crystal is known to have a high refractive index of $n = 1.89$ with high birefringence. In order to make uric acid incorporation to generate large anisotropy, highly ordered orientation of crystals is anyway needed, which has not been clarified yet. Thus, quite interesting works concerning the beetle's elytra were done during the 1960s-1970s by these authors. However, it is rather strange that continuing studies have not appeared in these 30 years.

3.6.8 *Metallic Pigments*

In these twenty years, the industrial applications of structure-based colorations have been rapidly growing. This is mainly because they owe much to the recently developing nanotechnology, since structural colorations lay their bases mostly on structures in a nanometer range. The structurally colored materials thus manufactured can be usually classified into two: The first one is that structurally colored powders are mixed into the other materials such as paint and ink so that their abilities basically display after painting or writing. The other is that the materials themselves are used to display structural colors. The former is called metallic pigment, while we can see good examples for the latter in light-interference fiber, multilayer film, and artificial opal.

Here, we briefly review a recent trend in the metallic pigments. The metallic pigments are usually mixed with paint as a form of small flakes, which will be generally classified into several categories according to the supporting materials and the ways of interference employed: (1) those employing metal flakes as supporting materials, (2) those employing transparent flakes, (3) holographic pigments, and (4) liquid crystalline pigments.

In the first type, metallic flakes are used without any treatment or they are used after being coated by various inorganic materials. The former called *metallic luster pigment* uses various kinds of metal flakes such as Al, Cu, Zn, Sn, Al bronze and Au bronze, which shows metallic appearance with the colors characteristic of metals or alloys employed. Such metal flakes are further coated with colored materials such as Fe_2O_3 or SiO_2 containing organic pigments, which give strongly colored metallic appearance. On the other hand, metal flakes are sometimes coated by a transparent oxide film such as Al_2O_3 or SiO_2, whose thickness is adjusted to show interference effect in a visible region.

Figure 3.70 (a) Pearl luster pigment, (b) interference pearl luster pigment, and (c) interference metallic pigments. An example of metallic pigments painted on a car (Courtesy of Kansai Paint Co. Ltd.)

Much more intense interference effect is obtained by coating both sides of an Al flake with a combination of a layer of MgF_2 and a very thin metal layer of Cr or Ti on the outside (see Fig. 3.70c). The thickness to the MgF_2 layer is adjusted so as to show the interference of light. The interference effect is considerably enhanced by the presence of Al flake with high reflectivity and the outer metallic layer with moderate reflectivity. This type of pigment was first manufactured by Flex Products and was called *optically variable pigment* (OVP®), which has been extensively employed for a security of currency. The prominent change of color with viewing angle is generally called *flip-flop character*, and is one of the major characters of this type of metallic pigment. This character is now further improved in order to use for the painting of various products such as automobile and mobile phone. One such example is shown in Fig. 3.70.

The second one is generally called *pearl pigment* and corresponds to pigments using transparent supporting materials (see Fig. 3.70a). As transparent materials, mica, Al_2O_3, and SiO_2 are often used, which are further coated by TiO_2 layer to enhance the reflection at the interface, and by Fe_2O_3 to enhance the colors. Sometimes, the thicknesses of these coating layers are so adjusted to display the interference of light, which is called *interference pearl pigment* (see Fig. 3.70b). The coating is sometimes performed by a double layer of SnO_2 and TiO_2. Anyway, owing to the relatively low reflectivity of these pigments, light can easily penetrate into the paint layer in the deep, being partially reflected by various flakes, which gives pearl-like appearance and is favorable for the cosmetic use.

In the above pigments, thin-layer interference is a fundamental optical process for displaying their structural colors, which are further enhanced by the presence of metal or a material layer with the high refractive index. However, holographic pigment lays its basis on diffraction grating. This pigment is characterized by a metal layer with regularly arranged grooves of a submicron size, which is further sandwiched by inorganic or organic layers. Such a material is manufactured, for example, by successive evaporations of MgF_2, Al, and MgF_2 layers on a plastic film with regular grooves that have been embossed beforehand. The three-layer film thus fabricated is removed from the plastic film and is broken into fragments. Contrary to the metallic pigments shown above, this pigment displays various colors simultaneously owing to a function of diffraction grating.

The last one is based on the coloration due to cholesteric liquid crystal. As an industrial use, silicone polymer is often used, whose orientations are slightly changed from the surface to form an helix with a pitch of around 10 layers. Such a cholesteric liquid crystal film is cut into pieces and mix with paint, ink or plastic to show its unique color change.

Since the industrial world to use these pigments has been rapidly expanding with the progress of nanotechnology, the pigments described here are only a few examples of a massive amount of newly developed pigments. However, at the present stage, their fundamental coloration mechanisms are restricted within simple optical processes such as thin-layer interference and diffraction grating. Furthermore, they usually have only a single function, that is, to display. On the contrary, in the natural world, even a tiny butterfly utilizes various optical mechanisms with adding considerable modifications from a nanometer to millimeter scale. Further, they condenses their multi-functional mechanisms within their tiny wings such as tools for flying, displaying, and repelling of water. Thus, it seems quite necessary to make much more careful observations of the natural products so that we can elucidate, at least partly, their multi-functional mechanisms to bring out their true appearances.

Exercises

(1) We have solved the mystery of missing light, when the complete destructive interference occurs in the Michelson's interferometer. The Mach-Zehnder's interferometer directly explains its presence as a light wave propagating in a different direction. Confirm that the total light intensity is actually conserved during the interference experiment.
(2) Hanbury Brown and Twiss demonstrated that the diameter of star was obtainable from the intensity correlation experiment not from the direct interference experiment. Explain the principle of their measurement.
(3) In ideal multilayer, it is shown that Eq. (3.56) holds for even N with $\gamma = (n_t/n_i)(n_A/n_B)^N$. Confirm that the same relation holds for odd N when we put $\gamma = n_A^2/(n_i n_t) \cdot (n_A/n_B)^{N-1}$.
(4) Explain the physical origin of the oscillation observed on the tails of the reflection band due to a multilayer (see Fig. 3.25).

Experiments

I. Observation of Scarab Beetles under Circular Polarization Analyzers

Materials: right- and left-handed circular polarization analyzers[a] and several scarab beetles.

Procedure: Look at scarab beetles through right- and left-handed circular polarization analyzers, and you will find a drastic change of color as shown in Figs. 3.66 and 3.67.

II. Liquid Immersion Test on Butterfly Wings and Bird Feathers

Materials: butterfly wings (e.g. *Morpho* butterfly) and bird feathers (e.g. peacock), liquids with various refractive indices, and a laboratory dish.

---

[a]Circular polarization analyzer consists of a combination of a circular polarizer and a linear polarizer.

Procedure: Put a butterfly wing or bird feather into a laboratory dish and sprinkle a liquid over it. Marked color change will be observed if the wing colors are of a structural origin. With increasing refractive index, the color changes, for example, from blue to green or from ultraviolet to blue with decreasing brilliancy. When the refractive index approaches 1.5–1.6, the structural color will completely disappear. With increasing refractive index much more, the color will reappear again. The color change of this sort mainly comes from the change in interference condition within a butterfly scale or a feather barbule. If one employs a beetle's elytron, no such change will be observed though slight color change is usually observed due to the change of the refraction angle. This is because a liquid cannot soak into the beetle's elytron.

Chapter 4

Diffraction of Light and Diffraction Grating

4.1 Diffraction in 1D

Light propagates in a straight line through space. This is one of the well-known features of light. However, this description is true only when the space is uniform and there is no obstacle to hinder its propagation. If there is an obstacle, light will go round it and will diffusely propagate with extending propagation direction. This phenomenon is known as *diffraction*, which is one of the typical wave natures of light. The diffraction of light is familiar even in our daily life: When the sunlight goes though a narrow crack of a wall, one will notice that the size of a light spot casting on a floor is not the same as that of the crack, but is much larger than it. If one holds two pencils in contact up to the light, a few dark lines will be noticed between them. These phenomena are typical indications of the diffraction of light.

In the field of nanophotonics where microstructures have a size comparable with the wavelength of light or less, the diffraction of light often plays a key role. The typical and the most familiar example is the coloring of optical disk, on which narrow tracks carrying an

Bionanophotonics: An Introductory Textbook
Shuichi Kinoshita
Copyright © 2013 Pan Stanford Publishing Pte. Ltd.
ISBN 978-981-4364-71-3 (Hardcover), 978-981-4364-72-0 (eBook)
www.panstanford.com

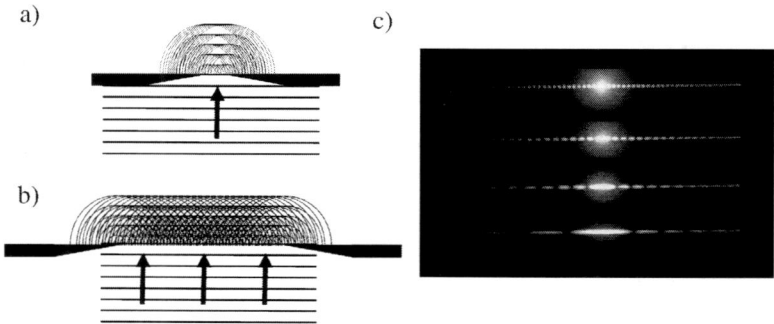

Figure 4.1 Diffraction phenomena explained by Huygens' principle in case where a slit width is (a) narrow and (b) wide. (c) The diffraction experiment using a slit with varying width. The slit is illuminated by a He-Ne laser with its width being narrower from top to bottom.

enormous amount of data are equipped with a constant interval. The diffraction makes the illuminating light to be diffusely reflected at each track and to be interfered with each other, which makes the optical disk rainbow-colored.

The mechanism of diffraction is often explained using Huygens' principle. Huygens' principle states that when a wave propagates in space, all the space points on the wave front generate secondary spherical waves and an envelope of these waves gives a new wave front. According to this principle, let us consider the diffraction phenomenon intuitively. In Fig. 4.1, we show the schematic illustration of this principle in case where a plane wave of light illuminates a narrow slit and goes through it. In this figure, the plane wave propagates from the bottom and its wave fronts are expressed as repeated straight lines. On the other hand, the secondary waves are expressed as repeated semicircular lines, the intervals of which are given by the wavelength of light.

It is soon understood that the envelope of the secondary waves has a flat region around a center of the slit, while it gives curved parts at both ends of the slit. If the slit width is narrower, the flat region becomes limited, while it becomes wider when the slit width is wider. Since a new wave propagates to a direction normal to this envelope, it is clear that the diffraction effect becomes more significant when a slit width, and hence a structural size, becomes smaller.

Slit with an infinite length   At first, we will explain this phenomenon using a simple 1D (one-dimensional) model as shown in Fig. 4.2a. Consider a slit with the width of a, which is illuminated by a plane wave that is normally incident to the slit. Let us put a center of the slit as the origin of the coordinates and take the x axis parallel to the slit width, while the z axis is to agree with the propagation direction of the incident light. If each point within a small portion dx emits secondary light after the illumination of the incident light, the amplitude of the emitted light toward θ direction, which is observed at a distance sufficiently far from the slit (far field), should be proportional to the following factor:

$$e^{i(\mathbf{k}\cdot\mathbf{r}-\omega t)}e^{-ikx\sin\theta}\,dx,$$

where $\mathbf{r}$ is a position vector of a point of observation and $\mathbf{k}$ is a wave vector of light propagating along the θ direction.

Performing the integration over the whole slit width to sum up all the contributions, we obtain

$$\tilde{u}_\theta \propto \int_{-a/2}^{a/2} dx\; e^{i(\mathbf{k}\cdot\mathbf{r}-\omega t)}e^{-ikx\sin\theta}$$

$$= e^{i(\mathbf{k}\cdot\mathbf{r}-\omega t)}\cdot \frac{e^{-ik(a/2)\sin\theta} - e^{ik(a/2)\sin\theta}}{-ik\sin\theta}$$

$$= e^{i(\mathbf{k}\cdot\mathbf{r}-\omega t)}\cdot \frac{a\sin\{k(a/2)\sin\theta\}}{k(a/2)\sin\theta}, \qquad (4.1)$$

where $\sin x/x$-type function appearing in the right-hand side is called *sinc function*, whose functional form is shown in Fig. 4.2b. Thus, the cycle-averaged intensity I is obtained as

$$I \propto \frac{1}{2}|\tilde{u}_\theta|^2 = \frac{a^2}{2}\left(\frac{\sin k_x a/2}{k_x a/2}\right)^2, \qquad (4.2)$$

where we put $k_x \equiv k\sin\theta$.

The calculated result for the cycle-averaged intensity is shown in Fig. 4.2c, which shows several small peaks on both sides in addition to a strong central peak. If we place a screen at a distance of l that is sufficiently far from the slit, a bright spot will appear on the screen when θ coincides with one of these peaks. Further, the spot will

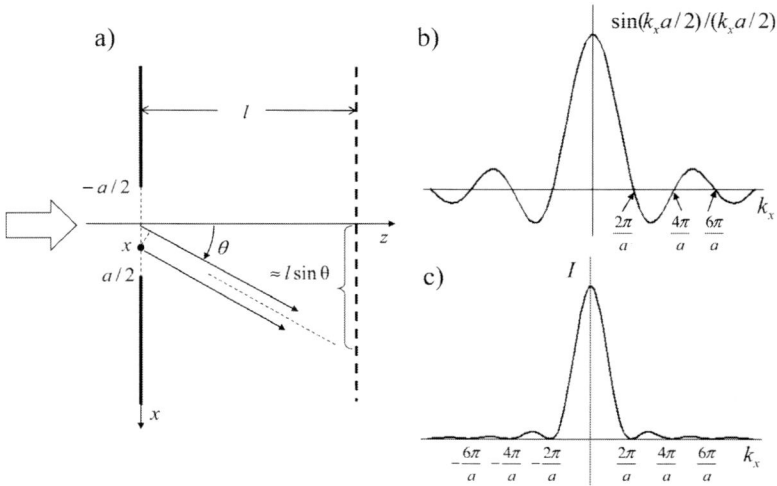

Figure 4.2 (a) Principle of the diffraction from a 1D slit, (b) a sinc function, and (c) far-field diffraction pattern calculated by squaring the sinc function.

be located at a point $l \tan \theta \approx l \sin \theta$ measured from the center. Since the horizontal axis k_x is also proportional to $\sin \theta$ and also the far-field pattern is expressed by this formula, the calculated result shown in Fig. 4.2c is directly comparable with the diffraction pattern on a screen.

As shown in Fig. 4.1c, the experimental result shows successive spots appearing almost equidistantly on both sides, whose interval tends to increase with reducing slit width. Further, a spot appearing at a center is found to be broader than the others. These features are well understandable by inspecting the functional form of Eq. (4.2). Namely, the function shown in Fig. 4.2c takes null values at $k_x = \pm 2m\pi / a$ with m an integer. Hence the interval of dark points appearing on both sides will be $\Delta k_x = 2\pi / a$, while around a center, it will become $4\pi / a$. Namely, it is expected that the interval of the diffraction spots becomes constant[a] and is inversely proportional to the slit width, while the length of the spot around a center is twice

---

[a]Strictly speaking, the interval of the brightest point in each spot is not equidistant because a denominator $(k_x a / 2)^2$ will slightly shift the peak position determined by a numerator $\sin^2 k_x a / 2$, whereas the intervals of dark points are strictly constant.

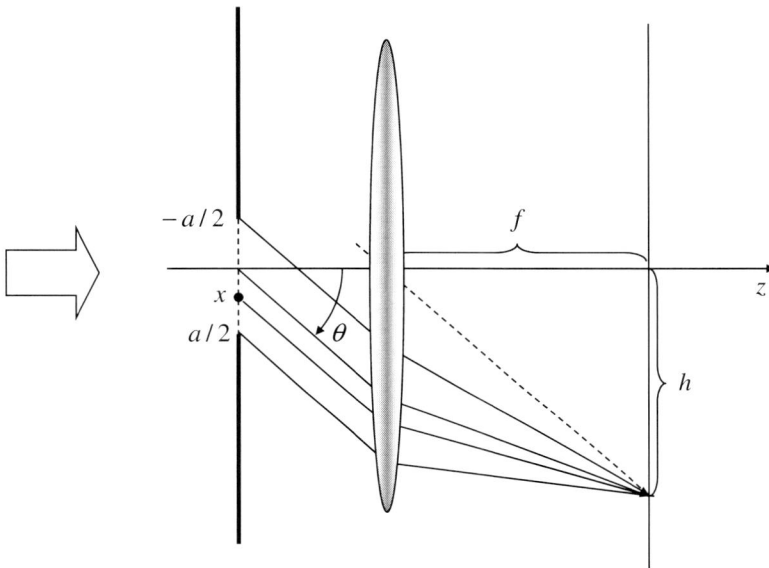

Figure 4.3 Fourier transform lens.

those on both sides. The experimental results shown in Fig. 4.1c just support this expectation.

Since the major component of angular dependence of diffraction appears as a central peak whose width is expressed roughly by a half of $4\pi/a$. Thus, $2\pi/a$ is a measure for the angular broadening due to the diffraction of light. From this inspection, the angular broadening is derived as

$$\theta \sim (2\pi/a)/k = \lambda/a, \tag{4.3}$$

using the relation $k_x \equiv k \sin\theta \sim k\theta$ for small θ.

Fourier transform lens   When we place a convex lens behind a slit, the light diffracted toward a direction of θ will converges into a single point on a focal plane. If we put its distance measured from the center as h, h is calculated as $h = f \tan\theta \approx f \sin\theta = fk_x/k$. Thus, the far-field pattern proportional to k_x can be obtained directly on a focal plane, which is essentially equivalent to that obtained by placing a screen sufficiently far from the slit. Since the function of the convex lens is to convert a propagation direction into a spatial position, such a lens is often called *Fourier transform lens*.

4.2 Diffraction in 2D

In a 2D (two-dimensional) case, a similar procedure can be applied as in a 1D case. However, analytical approach is limited only in a few cases. A general expression for 2D diffraction under normal incidence is expressed as

$$\tilde{u}(\mathbf{k}) = \iint dx dy \, A(x,y) e^{-i(k_x x + k_y y)} = \iint d\mathbf{r}' \, A(\mathbf{r}') e^{-i\mathbf{k} \cdot \mathbf{r}'}, \quad (4.4)$$

where an area contributing to the diffraction of light is expressed by a function $A(x,y)$ or $A(\mathbf{r}')$ with $\mathbf{r}'$ a position vector on a plane including the opening. Further, we have neglected a factor $\exp[i(\mathbf{k} \cdot \mathbf{r} - \omega t)]$ for simplicity. $A(\mathbf{r}')$ should be expressed as $A(\mathbf{r}') = 1$ within an opening, while it is expressed as $A(\mathbf{r}') = 0$ otherwise. If the opening is covered by a transparent material that shifts a phase of light, it is expressed as $A(\mathbf{r}') = \exp[i\phi]$, where ϕ is a phase shift.

Rectangular opening   First, we consider a rectangular opening with a width of a and a height of b. The amplitude of diffracted light in this case is expressed by

$$\tilde{u}(\mathbf{k}) = \int_{-a/2}^{a/2} dx \int_{-b/2}^{b/2} dy \, e^{-i(k_x x + k_y y)} = \int_{-a/2}^{a/2} dx \, e^{-ik_x x} \cdot \int_{-b/2}^{b/2} dy \, e^{-ik_y y}$$

$$= \frac{e^{ik_x a/2} - e^{-ik_x a/2}}{ik_x} \cdot \frac{e^{ik_y b/2} - e^{-ik_y b/2}}{ik_y}$$

$$= ab \cdot \frac{\sin(k_x a/2)}{k_x a/2} \cdot \frac{\sin(k_y b/2)}{k_y b/2}.$$

Hence, in case of a rectangular opening, the diffraction property is simply expressed by a product of two sinc functions corresponding to two sides of the rectangle.

Circular opening   In case of a circular opening, it is convenient to use polar coordinates instead of the Cartesian coordinates. We take the coordinates as are indicated in Fig. 4.4a, where we define $x = r' \cos\theta$ and $y = r' \sin\theta$. In addition, we consider angles α and β as

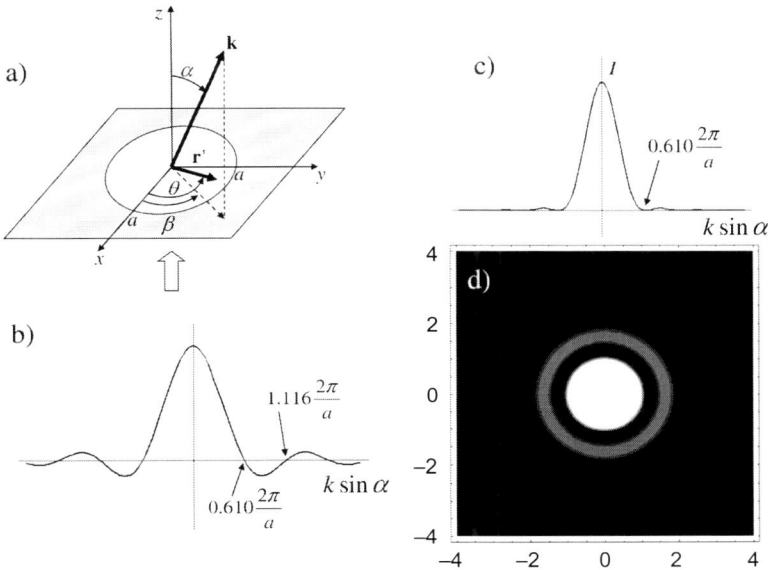

Figure 4.4 Diffraction of light due to a circular opening. (a) Coordinates, (b) amplitude and (c) intensity distribution of the diffracted light, and (d) diffraction pattern.

shown in the figure and put $k_x = k\sin\alpha\cos\beta$ and $k_y = k\sin\alpha\sin\beta$. Then, the amplitude of the diffracted light due to an circular opening with a radius a becomes

$$\tilde{u}(\mathbf{k}) = \int_0^a r'\mathrm{d}r' \int_0^{2\pi} \mathrm{d}\theta\, e^{-ikr'\sin\alpha(\cos\beta\cos\theta+\sin\beta\sin\theta)}$$

$$= \int_0^a r'\mathrm{d}r' \int_0^{2\pi} \mathrm{d}\theta\, e^{-ikr'\sin\alpha\cos(\beta-\theta)}, \tag{4.5}$$

where we have converted an areal element in the Cartesian coordinates of $\mathrm{d}x\mathrm{d}y$ into that in polar one of $r'\mathrm{d}r'\mathrm{d}\theta$.

Further, by putting $s \equiv kr'\sin\alpha$ and using a relation $\cos(\beta - \theta) = \sin(\theta - \beta + \pi/2) \equiv \sin\theta'$ with $\theta' = \theta - \beta + \pi/2$, the above expression is rewritten as

$$\tilde{u}(\mathbf{k}) = \int_0^a r'\mathrm{d}r' \int_{\pi/2-\beta}^{2\pi+\pi/2-\beta} \mathrm{d}\theta'\, e^{-is\sin\theta'}. \tag{4.6}$$

Then, we use the integral representation of a Bessel function of

$$J_n(s) = \frac{1}{2\pi} \int_{\alpha_0}^{2\pi+\alpha_0} e^{i(n\theta-s\sin\theta)}\,\mathrm{d}\theta, \tag{4.7}$$

and put $\alpha_0 = \pi/2 - \beta$ and $n = 0$, which yields

$$\tilde{u}(\mathbf{k}) = \int_0^a r' dr' \, 2\pi J_0(s) = 2\pi \int_0^a J_0(kr' \sin \alpha) r' dr'. \qquad (4.8)$$

Further, by using a relation known for Bessel functions, $\int s J_0(\gamma s) \, ds = (s/\gamma) J_1(\gamma s)$, the following expression is obtained:

$$\tilde{u}(\mathbf{k}) = \frac{2\pi a^2 J_1(ka \sin \alpha)}{ka \sin \alpha}. \qquad (4.9)$$

Hence, the cycle-averaged intensity becomes

$$I = \frac{1}{2} |\tilde{u}(\mathbf{k})|^2 = 2\pi^2 a^4 \left(\frac{J_1(ka \sin \alpha)}{ka \sin \alpha} \right)^2. \qquad (4.10)$$

In Figs. 4.4b and c, we plot an amplitude, $\tilde{u}(\mathbf{k})$, and its square against $k \sin \alpha$. The results are very similar to those obtained for 1D slit, but it is found that the zero-crossing points differ slightly. Namely, in the case of 1D slit, the zero-crossing points are expressed as $2\pi/a$, $4\pi/a$, $\cdots$ with a slit width of a, while in a circular opening, they are $0.610 \times 2\pi/a$, $1.116 \times 2\pi/a$, $\cdots$ for a radius of a. The diffraction due to a circular opening is thus characterized by a circular stripe pattern, as shown in Fig. 4.4d.

Resolution of optical instrument The above result is used to estimate the spatial resolutions of various optical instruments. Consider a case where a plane wave is incident to a convex lens and is focused into a spot on a focal plane. In this case, the convex lens functions as in the following two ways: One is a circular opening having a finite radius, and the other is a Fourier transform lens. That is, the first function diffracts light as in a circular opening expressed by Eq. (4.9), while the second one projects a far-field pattern onto a focal plane, which is schematically illustrated in Fig. 4.5.

If we denote a spatial resolution of a convex lens as Δy, which is defined by a radius at the first zero-crossing point on the focal plane, it becomes $\Delta y \approx 0.610 f \lambda / a$, since $k \sin \alpha_1 = 0.611 \times 2\pi/a$ and $\Delta y = f \tan \alpha_1 \approx f \sin \alpha_1$, where we put a diffraction angle to give the first zero-crossing point as α_1 with the focal length of the lens f. Thus, the minimum spot size obtainable by a convex lens under the illumination of a monochromatic plane wave is proportional to the wavelength of incident light and the focal length of the lens, and is inversely proportional to the radius of the lens.

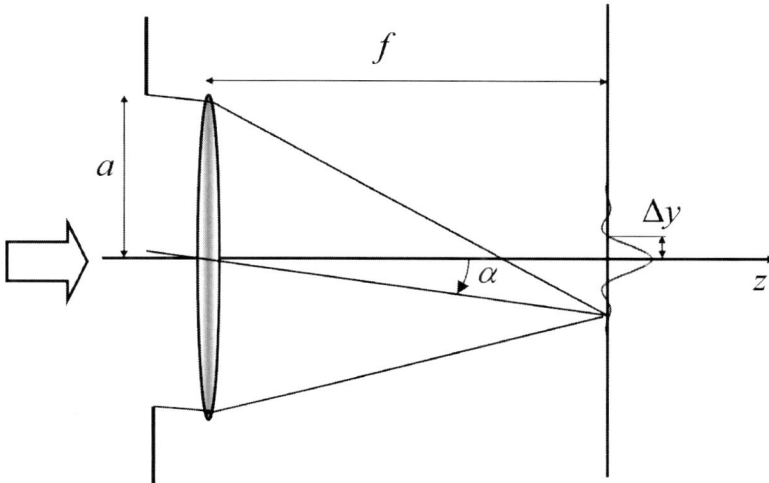

Figure 4.5 Function of a convex lens, which can be divided into two: that of a circular opening and of a Fourier transform lens. From this analysis, we can estimate the spatial resolution of a lens.

Usually, in optical instruments, the focal length is determined by the required magnification of the instrument. In such a case, the resolution is basically determined by an aperture size of the lens. On the other hand, in optical microscopes, the aperture of objective lens is roughly equal to its focal length. In this case, the spatial resolution is solely determined by the wavelength of light.

4.3 Diffraction in a General Case: Kirchhoffs Diffraction Theory

The phenomenon of light diffraction was strictly treated by Kirchhoff in 1882–3. We will follow his theory using a geometry shown in Fig. 4.6. We consider a whole system is divided into three: a space containing a light source located at a point O, that containing a detector located at a point P, and the rest of the space, which is assumed to be enclosed by a surface C. It is also assumed that the first space is enclosed by a surface S, on which a slit or an opening is attached. The second space is enclosed by a spherical surface T

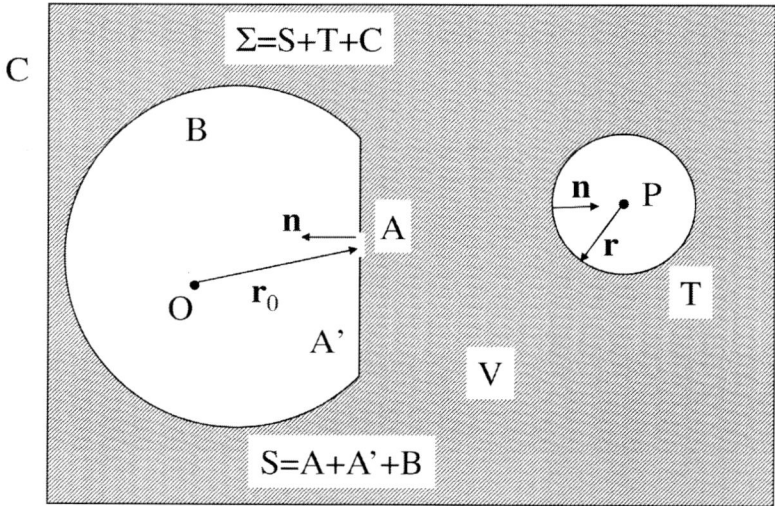

Figure 4.6 Geometry to explain Kirchhoff's diffraction theorem.

with the center placed at P. Let us denote a space without these two spaces by V, within which we consider a wave function associated with the diffraction of light as $\tilde{u}$. On the other hand, we define another wave function $\tilde{v} = (1/r_P)\exp[ikr_P]$ that is a function of r_P, a distance from a detecting point P, and is expressed in the form of outgoing spherical wave (see Sec. 6.3.1). Further, we define a unit vector **n** that is normal to the surfaces S and T, and is always directed outward when one sees from the space V.

Kirchhoff's diffraction theory leads to a general expression for the diffraction of light through the Green's theorem (see Appendix A.2), which states that two arbitrary scalar functions $\tilde{u}$ and $\tilde{v}$ defined within a volume V generally satisfy the following relation:

$$\int_V (\tilde{u}\nabla^2\tilde{v} - \tilde{v}\nabla^2\tilde{u})\mathrm{d}V = \int_\Sigma \left(\tilde{u}\frac{\partial\tilde{v}}{\partial\mathbf{n}} - \tilde{v}\frac{\partial\tilde{u}}{\partial\mathbf{n}} \right)\mathrm{d}\sigma, \qquad (4.11)$$

where the volume V is assumed to be enclosed by a surface Σ and a unit vector directed outward and normal to the surface is denoted as **n**. We put $\int_V$ and $\int_\Sigma$ as volume and surface integrals within a volume V and on a surface Σ, respectively. We further put

$$\frac{\partial}{\partial\mathbf{n}} = \mathbf{n} \cdot \nabla. \qquad (4.12)$$

Let us apply the Green's theorem to our case, where Σ will correspond to a sum of three surfaces expressed by $\Sigma = S + T + C$. We first transform the differential operator ∇ in the right-hand side of Eq. (4.12) into the polar coordinates as

$$\nabla = \mathbf{j}_r \frac{\partial}{\partial r} + \mathbf{j}_\theta \frac{1}{r} \frac{\partial}{\partial \theta} + \mathbf{j}_\phi \frac{1}{r \sin \theta} \frac{\partial}{\partial \phi}, \tag{4.13}$$

where $\mathbf{j}_r$, $\mathbf{j}_\theta$ and $\mathbf{j}_\phi$ are unit vectors directing along the r, θ and ϕ coordinates. Further, applying it to the surface of T and considering that $\mathbf{n}$ is directed to the center of the sphere, we obtain

$$\frac{\partial}{\partial \mathbf{n}} = \mathbf{n} \cdot \left(\mathbf{j}_{r_P} \frac{\partial}{\partial r_P} + \mathbf{j}_{\theta_P} \frac{1}{r_P} \frac{\partial}{\partial \theta_P} + \mathbf{j}_{\phi_P} \frac{1}{r_P \sin \theta_P} \frac{\partial}{\partial \phi_P} \right) = -\frac{\partial}{\partial r_P}, \tag{4.14}$$

where we consider the polar coordinates with their origins at P. For this purpose, we have added a suffix P to indicate the coordinates. Using this relation, the right-hand side of Eq. (4.11) becomes

$$\int_T \left(\tilde{v} \frac{\partial \tilde{u}}{\partial r_P} - \tilde{u} \frac{\partial \tilde{v}}{\partial r_P} \right) \mathrm{d}\sigma$$

$$= \int_T \left(\frac{1}{r_P} e^{ikr_P} \frac{\partial \tilde{u}}{\partial r_P} - \tilde{u} \left(-\frac{1}{r_P^2} + \frac{ik}{r_P} \right) e^{ikr_P} \right) r_P^2 \mathrm{d}\Omega_P$$

$$\rightarrow 4\pi \tilde{u}(P) \qquad (r_P \rightarrow 0), \tag{4.15}$$

for the surface of T, where we have replaced an infinitesimal areal element $\mathrm{d}\sigma$ by $r_P^2 \mathrm{d}\Omega_P$ with an infinitesimal solid angle $\mathrm{d}\Omega_P$. In addition, we have used the relations

$$\frac{\partial \tilde{v}}{\partial r_P} = \frac{\partial}{\partial r_P} \left(\frac{1}{r_P} e^{ikr_P} \right) = -\frac{1}{r_P^2} e^{ikr_P} + \frac{ik}{r_P} e^{ikr_P},$$

and $\int \mathrm{d}\Omega_P = 4\pi$. We have further assumed that $\tilde{u}$ is a smooth function of position around the point P and have taken the limit of $r_P \rightarrow 0$.

Next, we consider that $\tilde{u}$ and $\tilde{v}$ correspond to the spatial parts of wave functions within the volume V, which are expressed by $\tilde{u}_0(\mathbf{r}_P, t) = \tilde{u}(\mathbf{r}_P) \exp[-i\omega t]$ and $\tilde{v}_0(\mathbf{r}_P, t) = \tilde{v}(\mathbf{r}_P) \exp[-i\omega t]$, where $\mathbf{r}_P$ is a position vector with its origin at P. These two wave functions will satisfy the following wave equations:

$$\nabla^2 \tilde{u}_0(\mathbf{r}_P, t) = \frac{1}{c^2} \frac{\partial^2}{\partial t^2} \tilde{u}_0(\mathbf{r}_P, t),$$

$$\nabla^2 \tilde{v}_0(\mathbf{r}_P, t) = \frac{1}{c^2} \frac{\partial^2}{\partial t^2} \tilde{v}_0(\mathbf{r}_P, t).$$

Thus, from these relations, it is easy to derive the following expressions,

$$\nabla^2 \tilde{u} + k^2 \tilde{u} = 0, \tag{4.16}$$

$$\nabla^2 \tilde{v} + k^2 \tilde{v} = 0, \tag{4.17}$$

with $k^2 = \omega^2/c^2$.

Applying these relations to the left-hand side of Eq. (4.11), we obtain

$$\int_V (\tilde{u}\nabla^2\tilde{v} - \tilde{v}\nabla^2\tilde{u})\mathrm{d}V = \int_V -k^2(\tilde{u}\tilde{v} - \tilde{v}\tilde{u})\mathrm{d}V = 0, \tag{4.18}$$

and hence

$$\int_\Sigma \left(\tilde{u}\frac{\partial\tilde{v}}{\partial\mathbf{n}} - \tilde{v}\frac{\partial\tilde{u}}{\partial\mathbf{n}} \right) \mathrm{d}\sigma = 0. \tag{4.19}$$

If we consider[a] that the surface of the volume V consists of only S and T, that is, $\Sigma = S + T$, Eq. (4.19) is rewritten as $\int_T = -\int_S$ and hence

$$\tilde{u}(\mathrm{P}) = -\frac{1}{4\pi} \int_S \left(\tilde{u}\frac{\partial\tilde{v}}{\partial\mathbf{n}} - \tilde{v}\frac{\partial\tilde{u}}{\partial\mathbf{n}} \right) \mathrm{d}\sigma$$

$$= -\frac{1}{4\pi} \int_S \left[\tilde{u}\frac{\partial}{\partial\mathbf{n}} \left(\frac{1}{r_P}e^{ikr_P} \right) - \frac{1}{r_P}e^{ikr_P}\frac{\partial\tilde{u}}{\partial\mathbf{n}} \right] \mathrm{d}\sigma. \tag{4.20}$$

Since $\tilde{v}$ is a function of r_P, the following transformation holds,

$$\frac{\partial\tilde{v}}{\partial\mathbf{n}} = \mathbf{n}\cdot\nabla_P\tilde{v} = \frac{\partial\tilde{v}}{\partial r_P}\mathbf{n}\cdot\nabla_P r_P = \frac{\partial\tilde{v}}{\partial r_P}\mathbf{n}\cdot\frac{\mathbf{r}_P}{r_P} \equiv \frac{\partial\tilde{v}}{\partial r_P}\cos(\mathbf{n}, \mathbf{r}_P),$$

where $(\mathbf{n}, \mathbf{r}_P)$ indicates an angle made by two vectors of $\mathbf{n}$ and $\mathbf{r}_P$, and ∇_P is a differential operator with respect to $\mathbf{r}_P$.

---

[a]This assumption is not self-evident and theoretically the outer surface C can also contribute to the diffraction because if we integrate over the whole area of C, its contribution becomes finite. However, in Kirchhoff's diffraction theorem, C is not considered to contribute to the diffraction. This is based on the following qualitative inspection: If a light wave having a finite length is emitted from the light source and the distance from the light source to the surface C is sufficiently large, the secondary light wave generating on the outer surface will not interfere eventually with that generating on the inner surface S. In a similar manner, we will find a reason why only an opening on S contributes to the diffraction. The surface surrounding the light source is considered to consist of an opening A, its supporting substance A' and the rest of the surface B. However, as in a similar manner to the outer surface C, B is not considered to contribute to the diffraction, while light is assumed to be completely absorbed on the surface A'. Thus, only light diffracted at A and directly arriving at the detector contributes to the detection at P.

On the other hand, we assume that light is emitted from a point light source located at O and its spatial part of the wave function is expressed[a] as $\tilde{u} = (A/r_S)\exp[ikr_S]$ on the surface S, where r_S is a distance from O to a point on the surface S, while A, in general, varies only gently with the direction of emission and reflects the angular dependence of the light intensity. In a similar manner as in $\tilde{v}$, we obtain

$$\frac{\partial \tilde{u}}{\partial \mathbf{n}} = \frac{\partial \tilde{u}}{\partial r_S}\mathbf{n}\cdot\nabla_S r_S = \frac{\partial \tilde{u}}{\partial r_S}\cos(\mathbf{n},\mathbf{r}_S),$$

where $\mathbf{r}_S$ is a position vector with its origin O, and ∇_S is a differential operator with respect to $\mathbf{r}_S$. Inserting these two relations into Eq. (4.20), we can derive the following relation:

$$\tilde{u}(P) = -\frac{1}{4\pi}\int_S\left[\frac{A}{r_S}e^{ikr_S}\left(-\frac{1}{r_P^2}+\frac{ik}{r_P}\right)e^{ikr_P}\cos(\mathbf{n},\mathbf{r}_P)\right.$$
$$\left.-\frac{1}{r_P}e^{ikr_P}A\left(-\frac{1}{r_S^2}+\frac{ik}{r_S}\right)e^{ikr_S}\cos(\mathbf{n},\mathbf{r}_S)\right]d\sigma.$$

If we assume that r_S and r_P are sufficiently large as compared with the wavelength of light, the relations such that $1/r_j^2 \ll k/r_j = 2\pi/(\lambda r_j)$ hold with $j = S, P$. Thus, the terms having $1/r_j^2$ can be neglected and we obtain

$$\tilde{u}(P) = \frac{1}{2i\lambda}\int_S\frac{Ae^{ikr_S}}{r_S}\cdot\frac{e^{ikr_P}}{r_P}[\cos(\mathbf{n},\mathbf{r}_P)-\cos(\mathbf{n},\mathbf{r}_S)]\,d\sigma. \quad (4.21)$$

This relation is called *Kirchhoff diffraction theorem*. Further, if we define *inclination factor* as

$$B \equiv \frac{1}{2i\lambda}[\cos(\mathbf{n},\mathbf{r}_P)-\cos(\mathbf{n},\mathbf{r}_S)], \quad (4.22)$$

the above relation is expressed simply as

$$\tilde{u}(P) = \int_S\frac{Ae^{ikr_S}}{r_S}\cdot\frac{Be^{ikr_P}}{r_P}d\sigma, \quad (4.23)$$

where A expresses the intensity distribution of the light source, while B is regarded as the diffraction efficiency, both of which should be functions of position in a strict sense. Since B becomes maximum when both relations of $(\mathbf{n},\mathbf{r}_P) = 0$ and $(\mathbf{n},\mathbf{r}_S) = \pi$

---

[a]We can regard $\tilde{u}(\mathbf{r}_P)$ as a function of $\mathbf{r}_S$, since $\mathbf{r}_P$ is connected with $\mathbf{r}_S$ through a linear relation $\mathbf{r}_P = \mathbf{r}_S + \mathbf{r}_{PO}$, where $\mathbf{r}_{PO}$ is a vector directed to a point O from a point P.

are satisfied, the diffracted light intensity is maximized when light is incident on an opening normally to the surface and only the diffracted light propagating forward is detected. On the other hand, the diffracted light intensity becomes considerably weaker when the light is detected at right angles.

4.4 Fresnel and Fraunhofer Diffraction

When the positions of the light source and detector are far enough from the opening, r_S and r_P in the denominators do not depend largely on the selection of a point within the opening. Further, in this case, A and B are also considered to be independent of the selection of a point within the opening, then the following approximated expression will be obtained:

$$\tilde{u}(P) \approx \frac{AB}{R_S R_P} \int_S e^{ik(r_S + r_P)} d\sigma. \tag{4.24}$$

Further, as shown in Fig. 4.7, r_j $(j = S, P)$ can be expanded into

$$
\begin{aligned}
r_j &= |\mathbf{R}_j + \mathbf{r}| = R_j \left| \mathbf{e}_j + \mathbf{r}/R_j \right| \\
&= R_j \sqrt{1 + 2\mathbf{e}_j \cdot \mathbf{r}/R_j + r^2/R_j^2} \\
&= R_j \left(1 + \mathbf{e}_j \cdot (\mathbf{r}/R_j) + \frac{1}{2} \left(r^2 - (\mathbf{e}_j \cdot \mathbf{r})^2 \right) / R_j^2 + \cdots \right) \\
&= R_j \left(1 + \mathbf{e}_j \cdot (\mathbf{r}/R_j) + \frac{1}{2} \left(r^2/R_j^2 \right) \sin^2 \theta_j + \cdots \right),
\end{aligned}
$$

where $\mathbf{e}_j = \mathbf{R}_j/R_j$ and $\sin^2 \theta_j = 1 - (\mathbf{e}_j \cdot \mathbf{r}/r)^2$, where θ_j is shown in Fig. 4.7, and we have used an expansion $\sqrt{1+x} \approx 1 + x/2 - x^2/8 + \cdots$. Inserting this expression into Eq. (4.24) and considering the exponent up to the quadratic term of r, we obtain

$$\tilde{u}(P) \approx AB \frac{e^{ik(R_S + R_P)}}{R_S R_P} \int_S e^{i(\mathbf{k}_S - \mathbf{k}_P) \cdot \mathbf{r} + \frac{1}{2} ikr^2 (\sin^2 \theta_S/R_S + \sin^2 \theta_P/R_P)} d\sigma,$$

$$\tag{4.25}$$

where R_S and R_P are the distances shown in Fig. 4.7 and we put $\mathbf{k}_S = k\mathbf{R}_S/R_S$ and $\mathbf{k}_P = -k\mathbf{R}_P/R_P$. When an exponent in the above integrand is evaluated up to the quadratic term of r and the diffraction phenomenon expressed by this formula is called *Fresnel diffraction* (Exercise (1)).

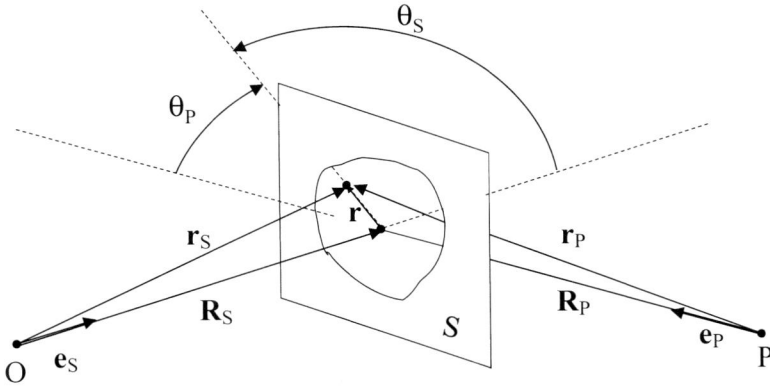

Figure 4.7   Geometry for the diffraction of light.

On the other hand, if the exponent is evaluated up to the linear term of r as in

$$\tilde{u}(P) \approx AB\frac{e^{ik(R_S+R_P)}}{R_S R_P} \int_S e^{i(\mathbf{k}_S-\mathbf{k}_P)\cdot\mathbf{r}}d\sigma = AB\frac{e^{ik(R_S+R_P)}}{R_S R_P} \int_S e^{-i\mathbf{K}\cdot\mathbf{r}}d\sigma,$$
(4.26)

it is called *Fraunhofer diffraction*, where $\mathbf{K} \equiv \mathbf{k}_P - \mathbf{k}_S$ is called *scattering vector*.

When we have to take into account the explicit shape of an opening on a surface S, it is convenient to introduce a function $\hat{A}(\mathbf{r})$ with $\hat{A}(\mathbf{r}) = 1$ inside the opening and $\hat{A}(\mathbf{r}) = 0$ otherwise, and to express Eq. (4.26) as

$$\tilde{u}(P) \approx AB\frac{e^{ik(R_S+R_P)}}{R_S R_P} \int_S \hat{A}(\mathbf{r})e^{-i\mathbf{K}\cdot\mathbf{r}}d\sigma.$$
(4.27)

which is essentially the same expression that we have derived in 1D or 2D diffraction case[a], although an additional terms are included in the present case. Thus, it gives a general expression for the light diffraction phenomena for arbitrary shapes of opening in 3D (three-dimensional) space.

---

[a]To accommodate the formula of the 2D diffraction to the present one, $\mathbf{k}$ in the last formula of Eq. (4.4) should be replaced by $\mathbf{k}_P$, which is further replaced by a scattering vector $\mathbf{K}$, since $\mathbf{K} \cdot \mathbf{r}' = \mathbf{k}_P \cdot \mathbf{r}'$ holds under normal incidence.

4.5 Diffraction Grating

4.5.1 *1D Diffraction Grating*

Double slit interference   Consider a case where two infinitely long slits separated by d are placed in parallel within a plane and a plane wave of light is incident normal to this plane. This configuration belongs to one of the typical cases of the well-known Young's interference experiment, in which diffraction and interference of light work together (see Sec. 3.1.2). The treatment of this type of diffraction/interference phenomenon is, however, completely the same as in case of 1D diffraction.

If the two slits have the same width of a, the amplitude of the diffracted light detected at a sufficiently distant position, $\mathbf{r}$, with the direction determined by an angle θ becomes a simple sum of the two contributions and is expressed as

$$
\begin{aligned}
\tilde{u}_\theta &\propto \left(\int_{-a/2}^{a/2} \mathrm{e}^{-\mathrm{i}k_x x}\,\mathrm{d}x + \int_{d-a/2}^{d+a/2} \mathrm{e}^{-\mathrm{i}k_x x}\,\mathrm{d}x \right) \cdot \mathrm{e}^{\mathrm{i}(\mathbf{k}\cdot\mathbf{r}-\omega t)} \\
&= \frac{\mathrm{e}^{\mathrm{i}k_x a/2} - \mathrm{e}^{-\mathrm{i}k_x a/2}}{\mathrm{i}k_x} \left(1 + \mathrm{e}^{-\mathrm{i}k_x d}\right) \mathrm{e}^{\mathrm{i}(\mathbf{k}\cdot\mathbf{r}-\omega t)} \\
&= \frac{a \sin(k_x a/2)}{k_x a/2} \left(1 + \mathrm{e}^{-\mathrm{i}k_x d}\right) \cdot \mathrm{e}^{\mathrm{i}(\mathbf{kr}-\omega t)},
\end{aligned}
$$

where $k_x = k \sin\theta$. The cycle-averaged intensity is then given as

$$
I = \frac{1}{2} |\tilde{u}_\theta|^2 = 2a^2 \left(\frac{\sin(k_x a/2)}{k_x a/2} \right)^2 \cos^2(k_x d/2), \qquad (4.28)
$$

which is expressed by the product of the contributions from a single-slit case and the $\cos^2$ term. Figures 4.8a and b show the calculated results for the diffraction due to a double slit, where the separation between the two slits is set at 10 times the slit width. It is clear that the diffracted light intensity for a single-slit case (shown as a thin line) is modulated by the $\cos^2$ term and takes the minimum values of zero at $k_x = (2m - 1)\pi/d$, which corresponds to a condition of $\cos^2(k_x d/2) = 0$.

Multi-slit interference   When N infinitely long slits are aligned equidistantly with the interval of d, the amplitude of the diffracted

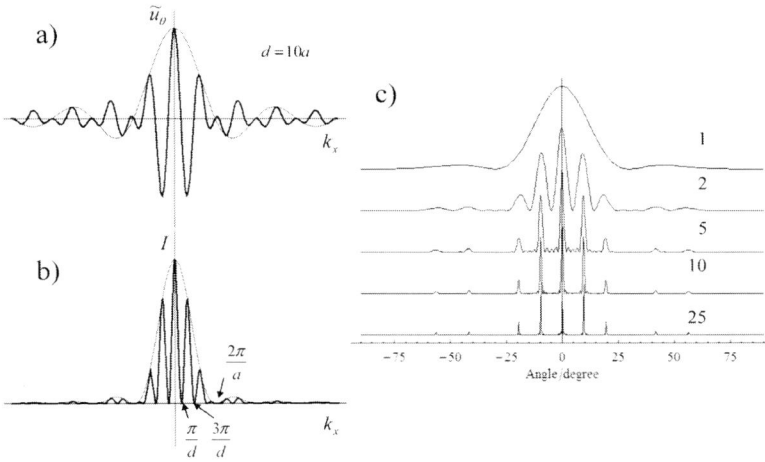

Figure 4.8 (a) Amplitude and (b) intensity of light diffracted by a double slit that is illuminated by a monochromatic plane wave under normal incidence. A thick line shows a case where the separation of the slits is set at 10 times the slit width, while a thin line shows a single slit case with the same slit width. (c) Angular dependence of diffracted light intensity for various numbers of slits, indicated in the figure. The widths of the slits are set at 1 μm, while their interval is set at 3 μm with the wavelength of 0.5 μm.

light will be given by analogy with double-slit interference as

$$\tilde{u}_\theta \propto \frac{a\,\sin(k_x a/2)}{k_x a/2}\left(1 + e^{-ik_x d} + \cdots + e^{-i(N-1)k_x d}\right)\cdot e^{i(\mathbf{k}\cdot\mathbf{r}-\omega t)}$$

$$= \frac{a\,\sin(k_x a/2)}{k_x a/2}\cdot\frac{1-e^{-iNk_x d}}{1-e^{-ik_x d}}\cdot e^{i(\mathbf{k}\cdot\mathbf{r}-\omega t)}.$$

Consequently, the cycle-averaged intensity is obtained as

$$I = \frac{a^2}{2}\left(\frac{\sin(k_x a/2)}{k_x a/2}\right)^2\cdot\frac{\sin^2(Nk_x d/2)}{\sin^2(k_x d/2)}. \tag{4.29}$$

It is clear that the result is also expressed by the product of single-slit diffraction with the width of a and the effect that the slits are aligned equidistantly with the interval of d.

The above argument is more generally described using a function $\hat{A}(\mathbf{r})$ that expresses the overall shape of aligned openings or structures that contribute to the diffraction of light. Consider a case where N openings or structures are periodically aligned in 1D.

The functional form of such a periodic structure can be generally expressed by

$$\hat{A}(\mathbf{r}) = \int d\mathbf{r}' \hat{a}(\mathbf{r} - \mathbf{r}') \sum_{j=0}^{N-1} \delta(\mathbf{r}' - jd\mathbf{v}), \qquad (4.30)$$

where $\hat{a}(\mathbf{r})$ is a function that depends on the shape of a single opening or structure, $\mathbf{v}$ a unit vector expressing the direction of the alignment, and d an interval. Notice $\mathbf{r}$ in this case indicates a position vector where openings or structures are placed. According to Eq. (4.27), the amplitude of the wave at a point P is generally expressed as

$$\tilde{u}(P) \propto \int d\mathbf{r} \, \hat{A}(\mathbf{r}) e^{-i\mathbf{K}\cdot\mathbf{r}},$$

which leads to

$$\begin{aligned}
\tilde{u}(P) &\propto \int d\mathbf{r} \, e^{-i\mathbf{K}\cdot\mathbf{r}} \int d\mathbf{r}' \, \hat{a}(\mathbf{r} - \mathbf{r}') \sum_{j=0}^{N-1} \delta(\mathbf{r}' - jd\mathbf{v}) \\
&= \int d\mathbf{r} \, \hat{a}(\mathbf{r}) e^{-i\mathbf{K}\cdot\mathbf{r}} \cdot \int d\mathbf{r}' \sum_{j=0}^{N-1} \delta(\mathbf{r}' - jd\mathbf{v}) e^{-i\mathbf{K}\cdot\mathbf{r}'} \\
&= \int d\mathbf{r} \, \hat{a}(\mathbf{r}) e^{-i\mathbf{K}\cdot\mathbf{r}} \cdot \sum_{j=0}^{N-1} e^{-i\mathbf{K}\cdot jd\mathbf{v}} \\
&= \int d\mathbf{r} \, \hat{a}(\mathbf{r}) e^{-i\mathbf{K}\cdot\mathbf{r}} \cdot \frac{1 - e^{-iK_v Nd}}{1 - e^{-iK_v d}}, \qquad (4.31)
\end{aligned}$$

where we put $K_v = \mathbf{K} \cdot \mathbf{v}$. Further, we have used the well-known relation of $\mathcal{F}(f * g) = \mathcal{F}f \cdot \mathcal{F}g$, where $\mathcal{F}$ indicates the Fourier transform operation with $f * g$ the convolution between the functions, f and g. Thus, the product rule described above is again confirmed.

For example, the diffraction due to an array of an infinitely long slit directed normal to the alignment is expressed as

$$\begin{aligned}
\tilde{u}(P) &\propto \int_{-\infty}^{\infty} dx \, \hat{A}(x) e^{-iK_v x} \\
&= \int_{-\infty}^{\infty} dx \, e^{-iK_v x} \int_{-\infty}^{\infty} dx' \, \hat{a}(x - x') \sum_{j=0}^{N-1} \delta(x' - jd) \\
&= \int_{-a/2}^{a/2} dx \, e^{-iK_v x} \cdot \frac{1 - e^{-iK_v Nd}}{1 - e^{-iK_v d}}, \qquad (4.32)
\end{aligned}$$

where an axis of x or x' is taken in parallel to the alignment and we have put $\hat{a}(x) = 1$ for $-a/2 < x < a/2$ and $\hat{a}(x) = 0$ otherwise. The intensity of diffracted light then becomes

$$I(P) = \frac{1}{2}|\tilde{u}(P)|^2 = \frac{a^2}{2} \left(\frac{\sin(K_v a/2)}{K_v a/2} \right)^2 \frac{\sin^2(K_v N d/2)}{\sin^2(K_v d/2)}. \qquad (4.33)$$

Thus, completely the same result as above is obtained except for a factor K_v, which eliminates the restriction with respect to the incident direction.

In Fig. 4.8c, we show typical examples of the diffraction from a single and multiple slits illuminated by a plane wave of light under normal direction. In a single slit, light is diffracted mostly toward a forward direction with small sidebands on both sides. Between main peak and sidebands, or between sidebands, null points are observed, the positions of which are determined by setting a numerator of Eq. (4.2) at zero and are obtained as $ka \sin \theta = 2m\pi$ with m an integer. With increasing number of slits, it is clear that another structure is superimposed on the diffraction pattern. The presence of these sharp lines is one of the characteristics of diffraction grating and is called *diffraction spots*, the positions of which are further determined by the relation of $kd \sin \theta = 2m'\pi$ with m' an integer.

Interference condition and zeroth-order grating   A general interference condition for diffraction grating is obtained from Eq. (4.29) or (4.33). Since the periodicity of structures only affects the intensity through the last term of this relation, the condition to set the denominator at zero will become that to show diffraction spots. Thus, the interference condition becomes $K_v d/2 = m\pi$, that is, $(\mathbf{k}_P - \mathbf{k}_S) \cdot \mathbf{v} d = 2m\pi$. By considering an experimental geometry shown in Fig. 4.9a, the interference condition is reduced to a simple form,

$$nd(\sin \theta + \sin \phi) = m\lambda, \qquad (4.34)$$

because $\mathbf{k}_S \cdot \mathbf{v} = -k \sin \theta$ and $\mathbf{k}_P \cdot \mathbf{v} = k \sin \phi$ with $k = 2n\pi/\lambda$, where n and λ are the refractive index of the medium and the wavelength of light, respectively. $m = 0, \pm 1, \pm 2, \cdots$ give the orders of diffraction spots. This relation is also derived from the following geometrical inspection. Namely, as shown in Fig. 4.10, the path length differences for incident and diffracted light between adjacent structural units

(a)   (b)

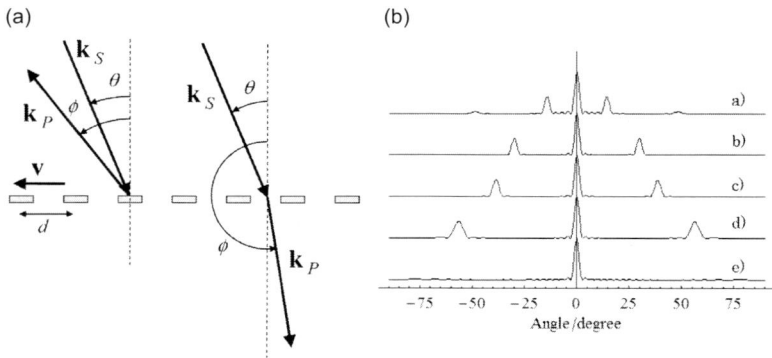

Figure 4.9 (a) Schematic diagram of (left) reflection- and (right) transmission-type diffraction grating. (b) Diffraction properties due to periodic multi-slits with reducing slit separation. A plane wave with the wavelength of 0.5 μm is assumed to illuminate the slits within a limited range of 10 μm in width. The ratio of the separation of the neighboring slits to the slit width is set at a constant value of 2. The separations of the neighboring slits are set at a) 2.0, b) 1.0, c) 0.8, d) 0.6 and e) 0.4 μm, respectively.

are easily obtained as $d \sin \theta$ and $d \sin \phi$, respectively, and the total optical path length difference thus becomes $nd(\sin \theta + \sin \phi)$. Since the interference condition is given in terms that the optical path length difference is equal to integer-multiplied wavelength, Eq. (4.34) is easily derived.

In Fig. 4.9b, we show the angular dependence of light diffraction from a diffraction grating with reducing interval of the periodic structures under normal incidence. Since the angles showing the diffraction spots are generally determined by the interference condition $nd(\sin \theta + \sin \phi) = m\lambda$, $\sin \phi$ is inversely proportional to the periodicity d under normal incidence of $\theta = 0$ and accordingly, angles of the diffraction spots increase with decreasing d. When d decreases much more and finally becomes less than the wavelength of light, diffraction spots other than zeroth-order disappear. Since diffracted light of the zeroth-order is nothing but specular reflection, the diffraction grating in this case behaves like a flat mirror, which is called *zeroth-order grating*. The typical example of the zeroth-order grating is seen in e) of Fig. 4.9b for $d = 0.4$ and $\lambda = 0.5$ μm.

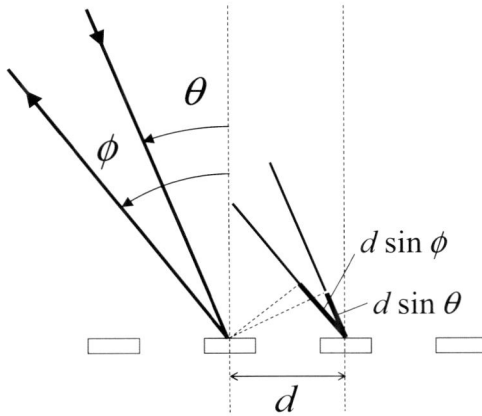

Figure 4.10 Geometrical inspection on the interference condition for diffraction grating.

The condition for the zeroth-order grating is more generally derived for an arbitrary angle of incidence by using the relation $\sin\theta + \sin\phi \leq 2$, where the equality holds for $\theta = \phi = \pi/2$. Thus, in order that diffraction spots appear, it is necessary that the relation $\sin\theta + \sin\phi = m\lambda/(nd) \leq 2$ anyway holds. This relation can be transformed into the condition for m as $m \leq 2nd/\lambda$, which states that for $2nd/\lambda < 1$, the inequality holds only for $m = 0$. Thus, this becomes a condition for the zeroth-order diffraction grating. In nature, the 2D zeroth-order diffraction grating appears as the moth-eye type anti-reflection structure, which does not produce specific reflection, but inhibits it, owing to the sophisticated surface structure.

Diffraction grating type spectrometer   Diffraction grating has been commonly employed as the most effective spectroscopic element in actual monochromators. This function is deeply related with a fact that the angle of higher-order diffraction spot changes with the wavelength of light.

Here, we will investigate the function of diffraction grating as a spectroscopic tool. Consider that a reflection-type diffraction grating is placed in a medium with the refractive index n, as shown in the left side of Fig. 4.9a. The optical-path-length difference for light diffracted at two adjacent structures becomes $nd(\sin\theta + \sin\phi)$, which is converted into time difference τ_g by dividing it by the light velocity of c as $\tau_g = nd(\sin\theta + \sin\phi)/c$.

The amplitude of light wave diffracted at N periodically aligned structures will be expressed as

$$\tilde{E}_g(t) = \tilde{E}(t) + \tilde{E}(t - \tau_g) + \tilde{E}(t - 2\tau_g) + \cdots + \tilde{E}(t - (N-1)\tau_g)$$
$$= \sum_{j=0}^{N-1} \tilde{E}(t - j\tau_g), \tag{4.35}$$

where $\tilde{E}(t)$ is the amplitude of diffracted light from a single structure placed at the left end. We consider a case where incident light is an ensemble of light with various angular frequencies and assume that its amplitude is expressed by the sum of components with the angular frequencies ω's as

$$\tilde{E}(t) = \int \tilde{E}_\omega e^{-i\omega t} d\omega. \tag{4.36}$$

Here, $\tilde{E}_\omega$ is a complex amplitude for a component with the angular frequency of ω. Then, the intensity of diffracted light from the diffraction grating is expressed as

$$I \propto \left| \sum_{j=0}^{N-1} \tilde{E}(t - j\tau_g) \right|^2 = \left| \sum_{j=0}^{N-1} \int d\omega \tilde{E}_\omega e^{-i\omega t + i\omega j\tau_g} \right|^2$$
$$= \left| \int d\omega \tilde{E}_\omega e^{-i\omega t} \left(1 + e^{i\omega\tau_g} + \cdots + e^{i(N-1)\omega\tau_g}\right) \right|^2$$
$$= \left| \int d\omega \tilde{E}_\omega e^{-i\omega t} \frac{1 - e^{i\omega N\tau_g}}{1 - e^{i\omega\tau_g}} \right|^2$$
$$= \left| \int d\omega \tilde{E}_\omega e^{-i\omega t + i(N-1)\omega\tau_g/2} \frac{\sin(N\omega\tau_g/2)}{\sin(\omega\tau_g/2)} \right|^2. \tag{4.37}$$

If the number of structures N is sufficiently large, the last term within $|\cdots|^2$ will give sharp peaks when ω satisfies the relation $\omega\tau_g/2 = m\pi$. Each peak is then approximated by a triangle with the height and base length of N and $4\pi/(N\tau_g)$, respectively. Since the area of this triangle becomes $2\pi/\tau_g$, this term can be replaced by a sum of delta functions such that $\sum_l (2\pi/\tau_g)a_l\delta(\omega - 2\pi l/\tau_g)$, where $a_l = 1$ for even l, while $a_l = -1$ for odd l.

Since we normally select a certain order, say l, in actual experiments, the intensity concerning the lth-order diffraction will become

$$
\begin{aligned}
I_l &\propto \left| \int d\omega \tilde{E}_\omega e^{-i\omega t + i(N-1)\omega\tau_g/2}\omega_1 a_l \delta(\omega - l\omega_1) \right|^2 \\
&= \left| \tilde{E}_{l\omega_1} e^{-il\omega_1 t + i(N-1)l\omega_1\tau_g/2}\omega_1 a_l \right|^2 \\
&= \omega_1^2 \left| \tilde{E}_{l\omega_1} \right|^2,
\end{aligned}
\tag{4.38}
$$

where we put $\omega_1 = 2\pi/\tau_g$. $\left| \tilde{E}_{l\omega_1} \right|^2$ is the intensity of the spectral component at $\omega = l\omega_1$. Since $\omega_1 = 2\pi/\tau_g = 2\pi c/\{nd(\sin\theta + \sin\phi)\}$, the change in ϕ causes that in ω_1 and hence the spectral distribution of the incident light wave is obtained.

The spectral resolution by using this type of grating is obtained from the last term in $|\cdots|$ of Eq. (4.37). Since this function shows sharp peaks at $\omega\tau_g/2 = m\pi$, their widths $\Delta\omega$ are evaluated as a full width at half maximum of a triangle-shaped peak and thus are determined as $\Delta\omega = 2\pi/(N\tau_g)$, independently of the diffraction order. Since $N\tau_g$ is the maximum optical-path-length difference of the grating, a large number and small pitch of grooves will give the high resolution. It is also noticed that since the diffraction spots appears at angles satisfying $nd(\sin\theta + \sin\phi) = m\lambda$ and hence $\sin\phi = m\lambda/(nd) - \sin\theta$, its derivative yields $\Delta\sin\phi = m/(nd)\Delta\lambda$ under the assumption of constant incident angle, which indicates that the angular resolution is related to the wavelength resolution via a sin function and is proportional to the order of grating.

In Fig. 4.11, we show a schematic diagram of a typical grating-type monochromator. In this monochromator, incident light is once focused on an entrance slit and then made parallel using a concave mirror to illuminate a flat diffraction grating. The diffracted light is focused again on an exit slit by another concave mirror. Since light having different angular frequency focuses on a different position around the exit slit, the monochromatization of light will be attained. The spectrum is normally obtained by rotating the grating so that θ and ϕ will simultaneously change in this case. If we put $\theta = \theta_0 + \alpha$ and $\phi = \phi_0 + \alpha$, the interference condition becomes $nd\{\sin(\theta_0 +$

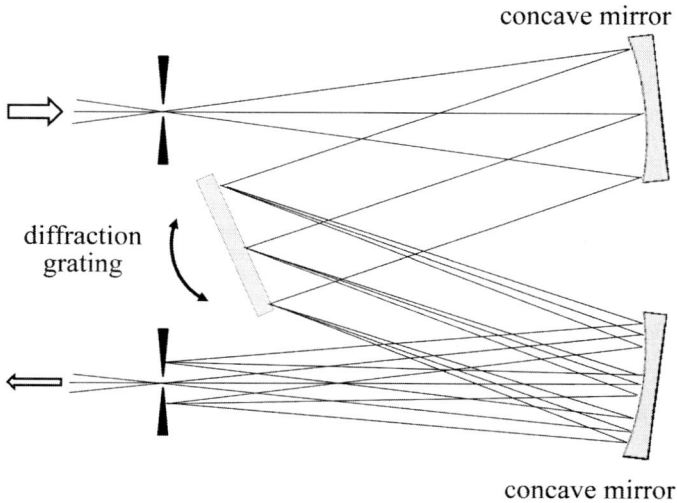

concave mirror

diffraction grating

concave mirror

Figure 4.11   Czerny-Turner type monochromator.

$\alpha) + \sin(\phi_0 + \alpha)\} = 2nd\,\sin\{(\theta_0 + \phi_0)/2 + \alpha\}\cos\{(\theta_0 - \phi_0)/2\} = m\lambda$, which states that if we change α so as to be linear to the wavelength, we can directly obtain a spectrum whose horizontal axis is linear to the wavelength (Exercise (2)). This type of monochromator is called *Czerny-Turner type* and has been most commonly employed in spectroscopy. Sometimes, two sets of this type of monochromator are used in tandem to increase the spectral resolution and to reduce the stray light, which is called *double monochromator*.

4.5.2 *2D Diffraction Grating*

If small particles are regularly distributed on a plane, a two-dimensional (2D) grating will be formed. The amplitude of the diffracted light in far field for $N \times N'$ 2D grating is expressed as

$$\tilde{u} \propto \sum_{m=0}^{N-1}\sum_{n=0}^{N'-1}\sum_{j} e^{-i\mathbf{K}\cdot(\mathbf{R}_{mn}+\mathbf{r}_j)} \qquad (4.39)$$

where $\mathbf{R}_{mn} = m\mathbf{a} + n\mathbf{b}$ with unit lattice vectors $\mathbf{a}$ and $\mathbf{b}$, and $\mathbf{r}_j$ is a vector expressing the position of a particle j in a unit cell. Here, for simplicity, we have omitted the term due to the diffraction from each

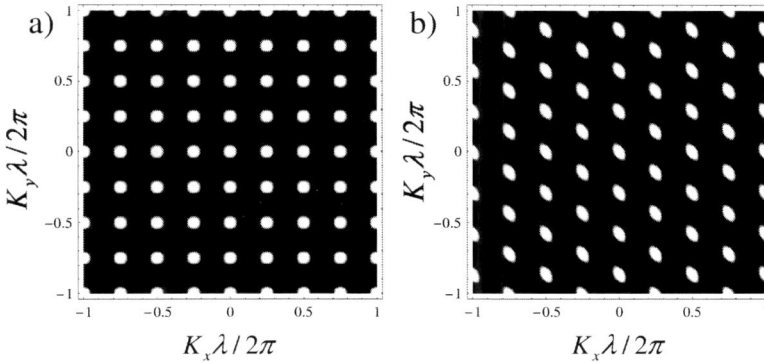

Figure 4.12 Diffraction patterns from 4×4 2D grating with (a) square and (b) triangle lattices. The nearest particle-particle distances are set at 2 μm in both cases and the wavelength of light at 0.5 μm.

particle. The intensity of the diffracted light is then proportional to

$$|\tilde{u}|^2 = \frac{\sin^2(N\mathbf{K} \cdot \mathbf{a}/2)}{\sin^2(\mathbf{K} \cdot \mathbf{a}/2)} \frac{\sin^2(N'\mathbf{K} \cdot \mathbf{b}/2)}{\sin^2(\mathbf{K} \cdot \mathbf{b}/2)} \left| \sum_j e^{-i\mathbf{K} \cdot \mathbf{r}_j} \right|^2. \tag{4.40}$$

A typical diffraction patterns obtained for a square and triangle lattice are shown in Fig. 4.12. The size of each diffraction spot is related with the total number of lattices and the distribution of spots directly depends on the lattice structure.

4.6 Colorations due to Diffraction Gratings

4.6.1 *Familiar Diffraction Gratings: CD and DVD*

Probably, the most familiar diffraction grating will be CD or DVD. In Fig. 4.13b, we show an example of diffraction spots obtained by illuminating a CD with a He-Ne laser (see Experiment I). We can clearly see the diffraction spots up to the second order in addition to the zeroth-order spot that corresponds to specular reflection.

We can easily understand why CD or DVD becomes a diffraction grating. Although several kinds of CD's are now available, here we will explain the structure of CD-ROM as an example, where digital data are written in during the fabrication process. CD-ROM

a)   b)

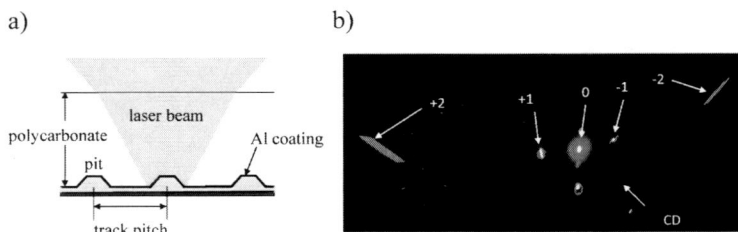

Figure 4.13 (a) Schematic illustration of the cross section of CD. (b) Diffraction pattern due to a CD illuminated by a He-Ne laser. The light beam is incident from a hole below the pattern and hits the CD at a point that is seen as a red spot. The diffraction spots with the orders from -2 to $+2$ are clearly seen.

is fabricated by reproducing a master disk by pressing a plastic material on it. The reproduced disk is then coated by aluminum reflector and covered by polycarbonate resin layer. Data are written on a spiral-shaped track in forms of *pit* and *land* (see Fig. 4.13a). The pit is a small projection that plays a role of light scatterer, while the land is a flat surface to induce specular reflection. When a focused light beam from a semiconductor laser is illuminated on a track, the intensity of reflected light is modulated by the presence of pit and land so that by rotating a disc and changing the focusing position along the movement of a spiral-shaped track, we can read out the data.

The pitch of the spiral is called *track pitch* and corresponds to a lattice constant in diffraction grating. The amount of data that can be written in is mainly determined by track pitch and minimum pit length. For example, they are 1.6 μm and 0.87 μm for CD (700 MB), while 0.74 μm and 0.4 μm for DVD (2.6 GB).

The positions of the diffraction spots observed in the above experiment directly reflect the track pitch of CD, which is easily measured by the experiment (see Experiment II). The interference condition for CD or DVD is generally obtained as

$$nd(\sin\theta + \sin\phi) = m\lambda,$$

where n is a refractive index of a resin layer, and θ and ϕ are angles of incidence and reflection within the resin layer with m an integer. Actually, we need not know the value of n, because refraction occurs at an air-resin interface. If we denote an angle of incidence from air to resin as θ' and a corresponding angle of reflection as ϕ', then the Snell's law gives the following relations:

$$\sin\theta' = n\sin\theta, \quad \sin\phi' = n\sin\phi.$$

Inserting these relations into the above interference condition, we obtain

$$d(\sin\theta' + \sin\phi') = m\lambda.$$

Thus, completely the same interference condition is obtained, in which we replace the angles within resin by those in air. From this relation, we can directly obtain the track pitch d. Newly developed blue-ray disc (BD; 27 GB) has a track pitch of 0.32 μm with the minimum pit length of 0.14 μm. Thus, in order to succeed an experiment on this disc with a laser pointer, the wavelength of the laser should be shorter than 0.64 μm and the incident beam is directed as obliquely as possible to a plane of the disc. Then, only the first-order diffraction spot will be observed around the direction of incidence. In near future, we will not be able to measure a track pitch by using such type of diffraction experiment.

4.6.2 *Structural Colors in Shells*

Diffraction grating can be one of the most familiar structural colors in nature. Surprisingly, in the animal kingdom, it is quite rare to find out the grating-type structural colors. One of the possible biological reasons may be that the diffraction grating does not show a specific color in itself, but gives only a mixture of various colors.

In this sense, pearl and mother-of-pearl will offer interesting topics on grating-type structural colors, though it is somewhat complicated. In the middle of the 19th century, Brewster first observed the optical phenomena concerning mother-of-pearl [Brewster (1845)]. Under the illumination of a candle, he observed that the candle images reflected from the surface of the shell showed several separate images on a screen, whose intensity distribution depended on whether the shell was polished or not. Further, if the surface of the shell was coated with bee's wax, the surface of the wax was also shining with prismatic colors after removing it from the shell. He

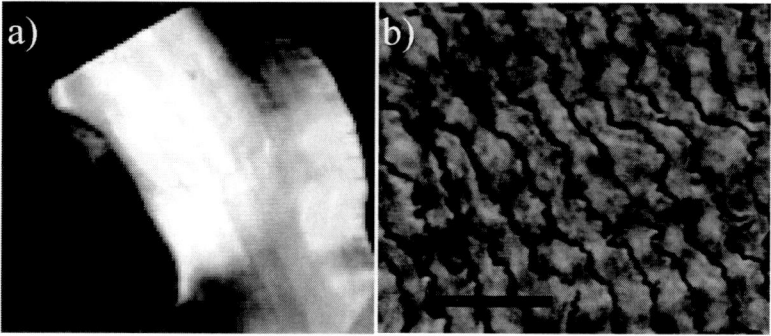

Figure 4.14 (a) Thin piece of polished turban shell (Courtesy of Saga Nomura, Kyoto, Japan) and (b) its surface observed under a microscope. Scale bar is 20 μm.

considered that the colors of mother-of-pearl was "communicated" to the wax surface and called "communicable" colors. He considered that the surface of the shell had a grooved structure of 3,000 lines per inch (118 lines/mm), which originated from the cross section of numerous strata within the shell.

Later, Lord Rayleigh (Robert John Strutt, 4th Baron Rayleigh) stressed the presence of a series of equally spaced parallel plates, which reflected distinguishable colors due to multilayer interference [Rayleigh (1923a)]. He measured a transmission spectrum through a thin plate of shell and found the presence of two dark bands located at 590 and 430 nm. He analyzed it in terms that the reflection from the stratified layer of the shell corresponded to the second- and third-order reflection bands, respectively.

In Fig. 4.14, we show a thin piece of a turban shell and its surface observed under a microscope. The surface of the shell is actually covered with the outcrops of stratified plates. These plates are known to consist of a single crystal of aragonite ($CaCO_3$) with a thickness of ~0.5 μm and a diameter of ~10 μm, which is cemented by organic materials called conchiolin (complex proteins) secreted from the epithelium with a thickness of ~25 nm. These layers are stratified one another to strengthen the body like a brick wall.

The origins of the structural colors in mother-of-pearl were, thus, conflicting among the theories based on diffraction grating,

(a)

(b)

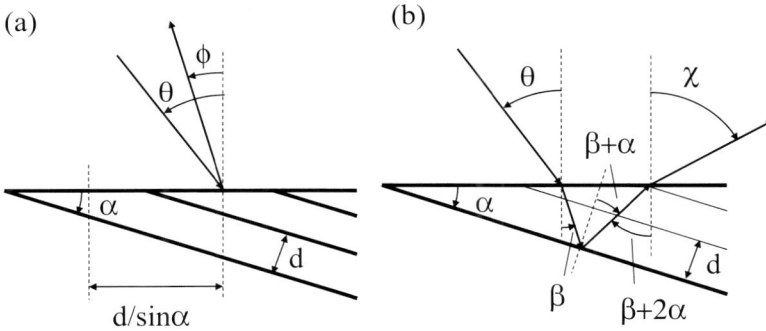

Figure 4.15 Geometries for (a) a diffraction grating appearing on the surface of a shell and (b) an oblique multilayer below the surface.

multilayer interference, and the combination of both. Raman played a decisive role on this problem by considering that the internal laminations within a shell were inclined to the external surface [Raman (1934b,c)]. He stressed the importance of a relationship between the diffraction grating on the surface and the multilayered structure below the surface, because the outcrops of the multilayered structure were just the origin of the diffraction grating.

From the inspection based on geometrical optics, he proved that the diffraction grating and the multilayer gave rise to the reflections into the same directions. Here, we will follow his proof, using geometries shown in Fig. 4.15. If a layered structure with the periodicity of d is inclined by an angle of α to the surface, a diffraction grating with the pitch of $d/\sin\alpha$ is naturally formed at the surface, as shown in Fig. 4.15a. When incident light is inclined by an angle of θ from the normal to the surface, the interference condition will be given as

$$d(\sin\theta + \sin\phi)/\sin\alpha = m\lambda, \qquad (4.41)$$

where ϕ is an angle of diffraction and m is an integer.

On the other hand, for the inclined multilayer below the surface, an incident angle to this oblique multilayer will be changed into $\alpha + \beta$ due to the refraction at the surface according to the Snell's law of $\sin\theta = n\sin\beta$ (see Fig. 4.15b). Thus, the multilayer interference occurs under a condition of

$$2nd\cos(\alpha + \beta) = m'\lambda, \qquad (4.42)$$

where n is the average refractive index for a sum of the $CaCO_3$ and cochiolin layers with m' a positive integer. Then, an angle of outgoing light at the surface should satisfy a relation $\sin \chi = n \sin(2\alpha + \beta)$.

By using the relations given above, the left-hand side of Eq. (4.42) can be transformed into

$$2nd \cos(\alpha + \beta) = nd \{\sin(2\alpha + \beta) - \sin \beta\} / \sin \alpha$$
$$= d(\sin \chi - \sin \theta)/ \sin \alpha = -d(\sin \theta - \sin \chi)/ \sin \alpha.$$

Under the fixed incident angle, χ should be constant, and then light is reflected only when the wavelength of light satisfies the condition $-d(\sin \theta - \sin \chi) = m'\lambda$ for each m' of a positive integer. On the other hand, in case of diffraction grating, ϕ is essentially variable so that diffraction spots are, in principle, observable for every wavelength, corresponding to the value of m, which takes a value of both positive and negative integers. Thus, if the conditions of $\chi = -\phi$ and $m' = -m$ are satisfied at a particular wavelength, the above relation of the multilayer reflection results in $d(\sin \theta + \sin \phi) = m\lambda$, which is completely the same as that of the diffraction grating.

In Fig. 4.16, we show the result of simulation for parameters of $n = 1.57$, $\alpha = 5.7°$ and $d = 0.5$ (μm) with $\theta = 30°$, which are typical values of a turbo. It is clear that above conditions are satisfied at the wavelengths of 0.715, 0.477 and 0.358 (μm), which correspond to $m' = -m = 2$, 3 and 4, respectively, with $\chi = 51.7°$. Thus, the diffraction at the surface and the multilayer reflection below the surface of the same order, but with opposite signs, are essentially indiscernible under monochromatic light illumination when the wavelength satisfies the above condition. On the contrary, under the illumination by a light source with a broad emission spectrum, the superposition of the two processes occurs at particular wavelengths, and thus within a visible region, a diffraction spot with a particular order (for example, $m = -3$ in Fig. 4.16) among various spots will be felt brighter.

It is also noticed that the form of the outcrop surface will strongly affect the intensity distribution of diffraction spots and also the incidence angle to the multilayer. Thus, the intensity profile of reflected light is strongly affected by a surface condition, whether it is as grown, polished or cracked. Further, relatively large lamellar thickness will give the higher-order interference, which will be able

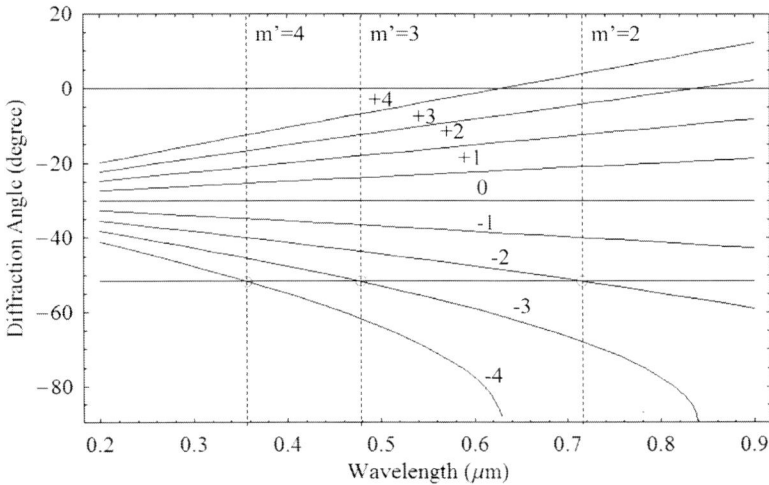

Figure 4.16 Diffraction angles ϕ calculated using Eq. (4.41) for various values of m, which are shown as numbers from -4 to $+4$ in this figure, with the parameters of $n = 1.57$, $d = 0.5$ (μm), $\alpha = 5.7°$ and $\theta = 30°$, which are typical values for turban shell. The conditions given by Eq. (4.42) are shown as dashed lines corresponding to each value of m'. The points of the intersection of these lines with the curves of diffraction spots offer the conditions of the superposition of these two optical processes. A horizontal line at $\phi = -51.7°$ corresponds to $\phi = -\chi$.

to give two reflection peaks in a visible region. In addition, thin layer of conchiolin causes a strongly non-ideal multilayer, which results in relatively sharp reflection peaks. As a result, a shell will display green/purple two-color iridescence more vividly than that was found in rock dove (see Sec. 3.6.5).

Raman and Krishnamurti investigated various shells and pearl using this analysis and found that in *Margaratifera*, the fourth-order diffraction spots superposed on the reflection from the layered structure, while in *Turbo* the third-order, and in *Nautilus* and *Haliotis*, the second-order [Raman and Krishnamurti (1954a,b)]. They also analyzed strong diffusion halo that was observed in the reflected light from a shell and in the transmitted light through a thin plate of shell. They assigned its origin partly as a diffraction effect due to a narrow width of a crystal plate and partly as due to the difference of the optical path lengths between a crystal and conchiolin when a light wave propagated in the oblique direction.

Recently, it was reported using a high-power electron microscope that the CaCO$_3$ crystal was not a single crystal but consisted of an ensemble of many nanocrystals [Takahashi *et al.* (2004)]. The nanocrystals had a diameter of 5–15 nm and were present within a network of polymers, whose orientations were equal within a block, but were different from those in the other block. If this is true, there is a possibility that the diffusion halo is deeply associated with nanoparticles that constitute the macroscopic structure, which reminds us of the case of opal as will be described in Sec. 5.3.1.2. Anyway, the diffusion halo in mother-of-pearl is essential for its pearly appearance and thus further research is strongly expected.

Exercises

(1) Equation (4.25) is called *Fresnel integral*. This integral can be solved geometrically using Cornu's spiral. Explain the method in a single-slit diffraction case.
(2) Explain a method how to obtain a spectrum that is linear to wavelength or energy in case of the Czerny-Turner type monochromator (see Fig. 4.11).

Experiments

I. Diffraction of Light by CD

Materials: CD or DVD, a laser pointer and a screen made of white paper with a hole at a center.

Procedure: Prepare a laser pointer and a screen made of white thick paper with a small hole at the center. Put a CD or DVD behind the screen and the laser pointer in front of it. Direct the laser beam to the hole of the screen so that a light beam can pass through the hole and then illuminate a plane of CD or DVD. Then, you will clearly see the diffraction spots on both sides of a spot due to specular

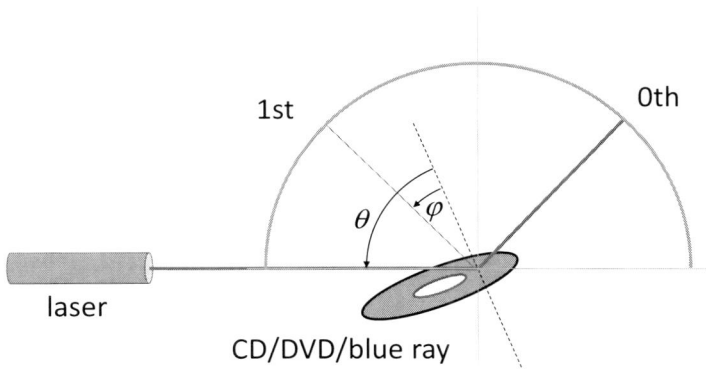

Figure 4.17 Experiment II. Experimental arrangement for the track pitch measurement.

reflection. The typical example is shown in Fig. 4.13c. These spots are called first- and second-order diffraction spots, while a spot due to specular reflection is sometimes called zeroth-order.

II. Measurement of Track Pitch

Materials: CD, DVD, MD, MO or BD, a laser pointer, a protractor or a scale to measure the angles and a screen.

Procedure: Place a laser pointer so as to be parallel to a desktop and put a CD or DVD vertical to it (see Fig. 4.17). Make sure that the height of the light beam coincides with just a center of the CD or DVD so that specular reflection and diffraction spots will form in line on a screen within a plane parallel to the horizontal plane. Then, measure the angles of incidence and diffraction accurately. It will be smart to use a semicylindrically curved screen and to put the markings of corresponding angles on it.

III. Making a Direct-vision Spectroscope

Materials: a sheet of black drawing paper of an octavo size, a sheet of transmission grating (500 lines/mm), a pair of scissors, and a glue or an adhesive tape.

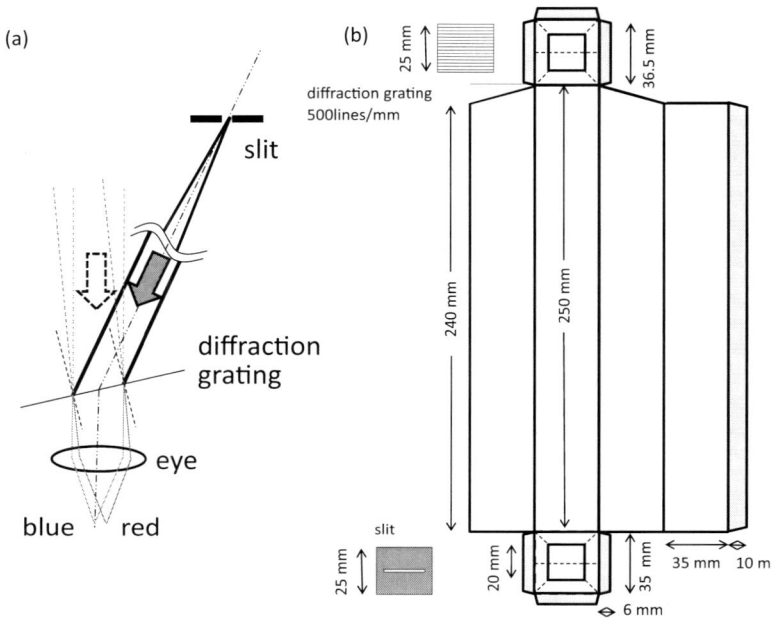

Figure 4.18 Experiment III. (a) Principle and (b) the development of a direct-vision spectroscope.

Procedure: Prepare a sheet of octavo-size black drawing paper and print a development of a spectroscope on it, which is shown in Fig. 4.18b. Prepare a sheet of transmission grating with 500 lines/mm. Cut the grating slightly larger than a square window shown in the upper end of the development, while a piece of black paper, within which a slender slit is notched, is also prepared and pasted on a square hole in the lower end. Construct a spectroscope of a quadrangular shape after pasting appropriate positions. Seeing a fluorescence lamp, LED, mercury lamp, sodium lamp, and so on through this spectroscope, you can find continuous and line spectra for these lamps. When you see the blue sky, you can find even Fraunhofer lines, which are due to light absorption within the earth's and sun's atmospheres.

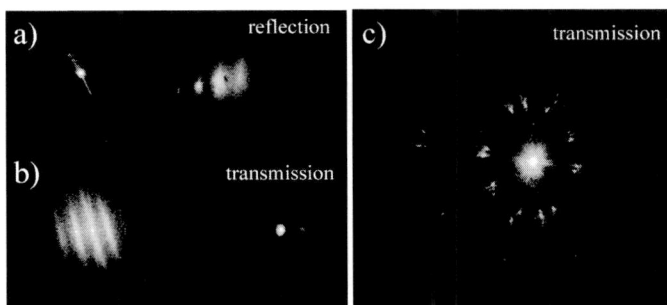

Figure 4.19 Experiment IV. The diffraction patterns due to a thin shell piece, observed on the (a) reflection and (b) transmission sides. c) The ring-shaped diffraction pattern due to cover scales of *Morpho didius*, which are scattered on a glass slide.

IV. Natural Diffraction Gratings

Materials: Butterfly scales, a thin polished shell piece with a thickness of ∼0.1 mm, a laser pointer, a lens (focal length of ∼10 cm), a screen and a glass slide.

Procedure: 1. Prepare butterfly scales and scatter them on a glass slide. Place a laser pointer in front of a screen made of a sheet of white paper. Then, illuminate the scales by a laser beam through a weakly focusing lens. You will find a ring pattern on the screen on the transmission side (Fig. 4.19c). The diameter of the ring is basically dependent on butterfly species and also the kinds of scales employed. Sometimes, you will find a double ring as shown in the figure.

A butterfly scale furnishes a lot of lines called *ridges* that run in parallel along a long axis of the scale, whose interval usually lies within a range of 0.7∼1.5 μm. Thus, each scale works as a transmission diffraction grating. Since the directions of scales are randomly distributed on a glass slide, diffraction spots will be generated in various directions, which causes a ring pattern. The clearest ring pattern corresponds to the first-order diffraction spot, from which we can calculate the interval of ridges.

2. Prepare a polished shell piece with a thickness of ~0.1 mm and conduct the same experiment as above. A polished shell piece gives the diffraction patterns both on the transmission and reflection sides (see Figs. 4.19a and b). The origin of the diffraction grating comes from multilayered crystalline layers within a shell piece, which consists of thin aragonite crystals. These thin crystals lie slightly obliquely to a surface. Thus, the edges of the crystalline layers appear like stairs with the interval of several μm (see Fig. 4.14b).

Chapter 5

Photonic Crystals

5.1 Fundamentals of Photonic Crystal

5.1.1 *What Is Photonic Crystal?*

If small and identical particles are regularly arranged like a crystal
as in Fig. 5.1c, whose lattice size lies on the order of the wavelength
of light, light scattered from each particle will interfere with each
other and will be reflected to a regular direction. Such a structure
is normally called *photonic crystal* and has attracted particular
attention in the fields of science and technology. Although this type
of structure seems to be a simple extension of 2D diffraction grating
to 3D system, the importance of photonic crystal lies in the fact that
light scattered at each particle interferes in a complicated way to
produce unique features in an optical region.

The typical examples of the photonic crystals are illustrated in
Fig. 5.1. The simplest case of the photonic crystals is a 1D regular
arrangement of layers as shown in Fig. 5.1a, which is nothing but
a multilayer described in the previous chapter. However, in the
previous case, it is implicitly assumed that the multilayer consists
of a finite number of layers, whereas in the present case, the number

Bionanophotonics: An Introductory Textbook
Shuichi Kinoshita
Copyright © 2013 Pan Stanford Publishing Pte. Ltd.
ISBN 978-981-4364-71-3 (Hardcover), 978-981-4364-72-0 (eBook)
www.panstanford.com

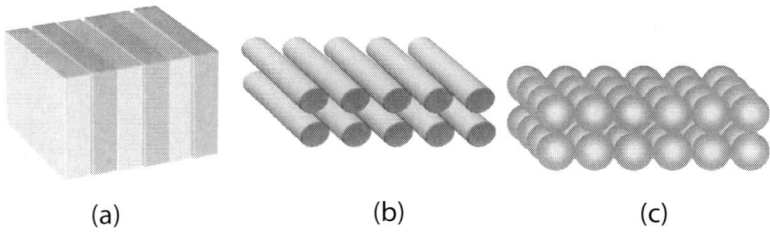

(a)   (b)   (c)

Figure 5.1 Examples of (a) one-, (b) two- and (c) three-dimensional photonic crystals.

of layers is, in principle, assumed to be infinite. Figure 5.1b shows an example of 2D photonic crystals, where cylinders with an infinite length arrange regularly to form a 2D lattice structure. Figure 5.1c is an example of 3D photonic crystals, where spherical particles are arranged like a crystal. In the natural world, photonic crystals of 1D, 2D and 3D types are all distributed widely in living and non-living systems, which have been offering us a lot of novel topics concerning their structures and functions. However, it is always necessary to remember that in the natural systems, photonic crystals are not complete and are finite in size, which differs considerably from ideal photonic crystals that will be treated here.

The theoretical treatment on the scattering of electromagnetic waves within such crystalline structures was first developed for atomic crystals in the X-ray diffraction method (see [James (1948)]). In this method, the theory considering only one-time scattering from each particle was called *kinematical theory*, whereas that considering multiple scattering was called *dynamical theory*. The latter was further classified into two according to the treatment of atomic dipoles. Ewald considered an ensemble of discrete dipoles and derived an approximate expression for the scattering using electromagnetic theory. On the other hand, Laue assumed a spatially continuous function for permittivity with a small periodic modulation due to the presence of lattice structure. Anyway, rigorous solution of the scattering problem was not generally obtainable and two-beam approximation, in which only incident and scattered waves were considered, was often employed.

Until recently, attention has not been paid to a system consisting of much larger particles, which will be soon triggered to open a new research field called *photonics*. One could say that the concept of photonics started from the papers published separately by Yablonovitch and John [Yablonovitch (1987); John (1987)]. In their works, they proposed a new idea to control light by regular structure on the order of the wavelength of light. Before that, light had been believed to be one of the most difficult object to be controlled. This idea resembles a case where electrical conduction due to electrons or holes are controlled within semiconductor materials that have the moderate width of energy band gap. By adding impurities and applying electric field, electrons are admirably controlled, which leads to new electronics predominating over the present world. Accordingly, the interaction of electromagnetic wave with periodic microstructure becomes an alternative that leads to new photonic technology, which will prevail over prosperous electronics based on semiconductor technology.

From a theoretical viewpoint, theoretical treatments to consider the interaction of microstructure with electromagnetic wave in photonic crystal are essentially different from those employed in the X-ray diffraction method applied to atomic crystal. This mainly originates from the difference of interaction strength between light and particle. In the former, the interaction is usually weak so that it is often convenient to express an atom as a point scatterer, whereas in photonic crystal, it is so strong that it is often necessary to solve the problem rigorously. This situation is equivalent to a case of scattering problem due to a small particle. That is, the Rayleigh scattering regarding a particle as a point scatterer is only applicable to an extremely small particle, while Mie scattering is applied to a much larger particle, whose size is comparable with the wavelength of light. Actually, the situation in photonic crystal is closed to that of electron in atomic crystal rather than X-ray diffraction, and hence, similar theoretical treatment has been often applied to photonic crystal.

In the present chapter, we will describe the fundamentals of photonic crystal and its interaction with light. Since the detailed descriptions of photonic crystal are easily available nowadays, we will confine ourselves only within simple cases of 1D and 2D systems

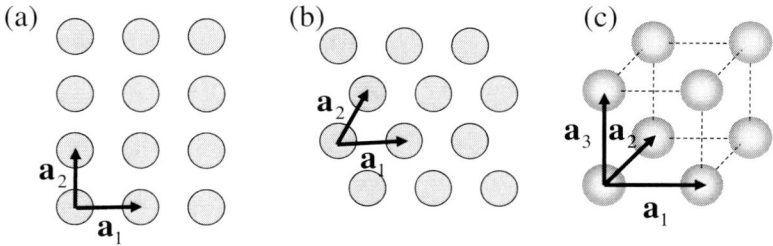

Figure 5.2 Primitive translation vectors for (a) 2D square lattice, (b) 2D hexagonal lattice and (c) 3D simple cubic lattice.

to aim at the intuitive understanding of the optical properties of photonic crystal.

5.1.2 *Crystal Structure and Reciprocal Vector*

Photonic crystal is similar to ordinary atomic crystal from a viewpoint of spatial periodicity, although their sizes differ by three orders of magnitude. A crystal structure is generally treated as an ensemble of a unit structure called *unit cell* that fills a whole 3D space by being regularly arranged. The unit cell has a form of a parallelepiped in 3D, a parallelogram in 2D and a line segment in 1D. In order to express the unit cell, *primitive translation vectors* are normally defined, whose number is naturally dependent on the dimension such as one in 1D, two in 2D space and three in 3D space. The typical examples of the primitive translation vectors in 2D and 3D are shown in Fig. 5.2. In this figure, (a) and (b) are examples of 2D lattice: (a) shows a square lattice, whose primitive translation vectors are expressed by two orthogonal vectors of $\mathbf{a}_1$ and $\mathbf{a}_2$, while (b) is a hexagonal lattice, whose vectors are expresses by two vectors that make an angle of 60°. On the other hand, (c) is a simple cubic lattice, whose primitive translation vectors, $\mathbf{a}_1$, $\mathbf{a}_2$ and $\mathbf{a}_3$, are orthogonal to each other. Since the periodicity is a basic feature of crystal, when a position vector corresponding to a certain point within a unit cell is denoted as $\mathbf{r}$, then an equivalent point in an arbitrary unit cell in crystal is generally expressed by $\mathbf{r} + \mathbf{T}$, where $\mathbf{T}$ is a translation vector defined as $\mathbf{T} = n_1\mathbf{a}_1 + n_2\mathbf{a}_2 + n_3\mathbf{a}_3$ with n_1, n_2 and n_3 integers (see Fig. 5.3).

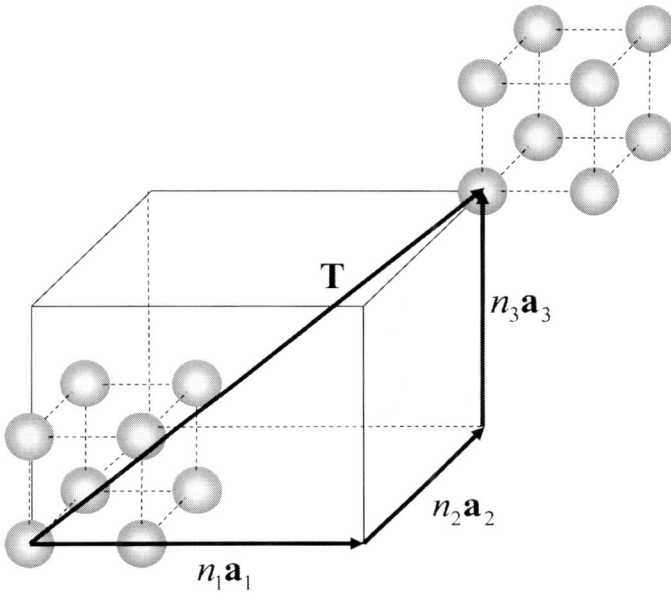

Figure 5.3 Translation vector, $\mathbf{T}$, expressed as $\mathbf{T} = n_1\mathbf{a}_1 + n_2\mathbf{a}_2 + n_3\mathbf{a}_3$.

Consider a case where light hits a crystal from one direction and is scattered to the other directions. For simplicity, we consider that light is scattered only once within a crystal as in a scattering problem of X-ray in atomic crystal. The scattered light waves will interfere with each other and will radiate emission to a specific direction according to the crystal structure. To formulate this, let us consider a point expressed by a position vector $\mathbf{r}$ and evaluate the phase difference of light wave scattered at this point as compared with that scattered at the origin. As shown in Fig. 5.4, the phase difference between these two is expressed by $\mathbf{k}_i \cdot \mathbf{r} - \mathbf{k}_s \cdot \mathbf{r}$ and then the amplitude of the scattered wave should have a factor of

$$e^{-i(\mathbf{k}_s - \mathbf{k}_i) \cdot \mathbf{r}}.$$

If many points are arranged regularly, the scattering field is expressed by the sum of the waves scattered at these points as in the following:

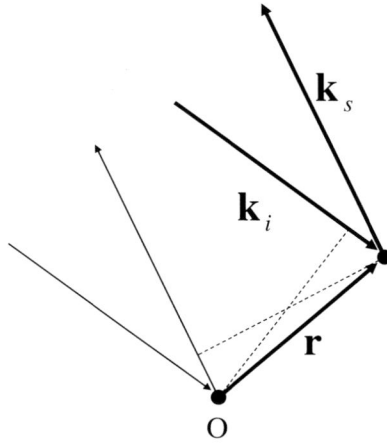

Figure 5.4 A method to calculate the phase difference in case where light with a wave vector of $\mathbf{k}_i$ is incident, which is scattered in the direction determined by a wave vector of $\mathbf{k}_s$.

$$\tilde{E} \propto \sum_{lmn} e^{-i(\mathbf{k}_s - \mathbf{k}_i) \cdot \mathbf{r}_{lmn}}$$

$$= \sum_l \sum_m \sum_n e^{-i\mathbf{K} \cdot (l\mathbf{a}_1 + m\mathbf{a}_2 + n\mathbf{a}_3)}$$

$$= \sum_l e^{-il\mathbf{K} \cdot \mathbf{a}_1} \sum_m e^{-im\mathbf{K} \cdot \mathbf{a}_2} \sum_n e^{-in\mathbf{K} \cdot \mathbf{a}_3},$$

where $\mathbf{K}$ is a *scattering vector* defined by $\mathbf{K} = \mathbf{k}_s - \mathbf{k}_i$ and $\mathbf{r}_{lmn}$ is a translation vector expressed as $\mathbf{r}_{lmn} = l\mathbf{a}_1 + m\mathbf{a}_2 + n\mathbf{a}_3$.

When M such points are arranged along each of three directions determined by the primitive translation vectors, the intensity of the scattered light will be expressed by

$$I \propto |\tilde{E}|^2$$

$$\propto \frac{\sin^2 M(\mathbf{K} \cdot \mathbf{a}_1/2)}{\sin^2(\mathbf{K} \cdot \mathbf{a}_1/2)} \cdot \frac{\sin^2 M(\mathbf{K} \cdot \mathbf{a}_2/2)}{\sin^2(\mathbf{K} \cdot \mathbf{a}_2/2)} \cdot \frac{\sin^2 M(\mathbf{K} \cdot \mathbf{a}_3/2)}{\sin^2(\mathbf{K} \cdot \mathbf{a}_3/2)}, \quad (5.1)$$

where we have used the relation $\sum_{l=0}^{M-1} \exp[-il\mathbf{K} \cdot \mathbf{a}_j] = (1 - \exp[-iM\mathbf{K} \cdot \mathbf{a}_j])/(1 - \exp[-i\mathbf{K} \cdot \mathbf{a}_j])$ with $j = 1, 2, 3$. Therefore, the interference conditions are obtained by putting each denominator to be zero. Thus, we obtain

$$\mathbf{K} \cdot \mathbf{a}_1 = 2\pi q, \quad \mathbf{K} \cdot \mathbf{a}_2 = 2\pi r, \quad \mathbf{K} \cdot \mathbf{a}_3 = 2\pi s, \quad (5.2)$$

where q, r and s are integers. If we can obtain **K** that simultaneously satisfies the above three conditions, we can solve a problem, to which direction light is scattered when a plane wave of light is incident on the crystal.

Reciprocal vector was introduced as the best method to solve this problem. If the primitive translational vectors, $\mathbf{a}_1$, $\mathbf{a}_2$ and $\mathbf{a}_3$, are orthogonal to each other, this problem is easily solved and **K** is obtained as follows:

$$\mathbf{K} = 2\pi \left(\frac{q}{a_1}\hat{\mathbf{a}}_1 + \frac{r}{a_2}\hat{\mathbf{a}}_2 + \frac{s}{a_3}\hat{\mathbf{a}}_3 \right), \tag{5.3}$$

where $\hat{\mathbf{a}}_1$, $\hat{\mathbf{a}}_2$ and $\hat{\mathbf{a}}_3$ are unit vectors directing to the primitive translational vectors, and $a_j = |\mathbf{a}_j|$ with $j = 1, 2, 3$. If we calculate the inner products with $\mathbf{a}_1, \mathbf{a}_2$ or $\mathbf{a}_3$, it is easy to show that **K** satisfies the above three conditions.

However, if $\mathbf{a}_1$, $\mathbf{a}_2$ and $\mathbf{a}_3$ are not orthogonal to each other, the problem will be somewhat complicated. In this case, it is convenient to consider that the scattering vector **K** is expressed by the sum of three vectors $\mathbf{G}_1$, $\mathbf{G}_2$ and $\mathbf{G}_3$ as

$$\mathbf{K} = q\mathbf{G}_1 + r\mathbf{G}_2 + s\mathbf{G}_3, \tag{5.4}$$

and to determine $\mathbf{G}_1$, $\mathbf{G}_2$ and $\mathbf{G}_3$ so as to satisfy the following conditions:

$$\mathbf{G}_1 \cdot \mathbf{a}_1 = 2\pi, \quad \mathbf{G}_1 \cdot \mathbf{a}_2 = 0, \quad \mathbf{G}_1 \cdot \mathbf{a}_3 = 0,$$
$$\mathbf{G}_2 \cdot \mathbf{a}_1 = 0, \quad \mathbf{G}_2 \cdot \mathbf{a}_2 = 2\pi, \quad \mathbf{G}_2 \cdot \mathbf{a}_3 = 0,$$
$$\mathbf{G}_3 \cdot \mathbf{a}_1 = 0, \quad \mathbf{G}_3 \cdot \mathbf{a}_2 = 0, \quad \mathbf{G}_3 \cdot \mathbf{a}_3 = 2\pi.$$

At first glance, $\mathbf{G}_1$ should be orthogonal to both $\mathbf{a}_2$ and $\mathbf{a}_3$, and in a similar manner, $\mathbf{G}_2$ is orthogonal to $\mathbf{a}_3$ and $\mathbf{a}_1$, and $\mathbf{G}_3$ is orthogonal to $\mathbf{a}_1$ and $\mathbf{a}_2$. Using a vector algebraic relation, $(\mathbf{a}_2 \times \mathbf{a}_3) \cdot \mathbf{a}_{2,3} = 0$, $\mathbf{G}_1$ is generally expressed in a form of

$$\mathbf{G}_1 = c_1 \mathbf{a}_2 \times \mathbf{a}_3,$$

where c_1 is a unknown constant. Using the relation,

$$\mathbf{G}_1 \cdot \mathbf{a}_1 = c_1 \mathbf{a}_1 \cdot \mathbf{a}_2 \times \mathbf{a}_3, \quad \mathbf{G}_1 \cdot \mathbf{a}_2 = \mathbf{G}_1 \cdot \mathbf{a}_3 = 0,$$

we can easily determine c_1 as

$$c_1 = 2\pi/(\mathbf{a}_1 \cdot \mathbf{a}_2 \times \mathbf{a}_3).$$

Since $\mathbf{a}_1 \cdot \mathbf{a}_2 \times \mathbf{a}_3$ expresses the volume of a unit cell, we put it as Ω and obtain

$$\mathbf{G}_1 = \frac{2\pi}{\Omega}\mathbf{a}_2 \times \mathbf{a}_3. \tag{5.5}$$

In a similar manner, we obtain[a]

$$\mathbf{G}_2 = \frac{2\pi}{\Omega}\mathbf{a}_3 \times \mathbf{a}_1, \tag{5.6}$$

$$\mathbf{G}_3 = \frac{2\pi}{\Omega}\mathbf{a}_1 \times \mathbf{a}_2. \tag{5.7}$$

Defining $\mathbf{G}_1$, $\mathbf{G}_2$ and $\mathbf{G}_3$ in this way, we can easily make the scattering vector $\mathbf{K}$ satisfy the interference condition as

$$\mathbf{K} \cdot \mathbf{a}_1 = q\mathbf{G}_1 \cdot \mathbf{a}_1 + r\mathbf{G}_2 \cdot \mathbf{a}_1 + s\mathbf{G}_3 \cdot \mathbf{a}_1 = 2\pi q,$$

$$\mathbf{K} \cdot \mathbf{a}_2 = q\mathbf{G}_1 \cdot \mathbf{a}_2 + r\mathbf{G}_2 \cdot \mathbf{a}_2 + s\mathbf{G}_3 \cdot \mathbf{a}_2 = 2\pi r,$$

$$\mathbf{K} \cdot \mathbf{a}_3 = q\mathbf{G}_1 \cdot \mathbf{a}_3 + r\mathbf{G}_2 \cdot \mathbf{a}_3 + s\mathbf{G}_3 \cdot \mathbf{a}_3 = 2\pi s.$$

Thus, if we define a reciprocal vector $\mathbf{G}$ as

$$\mathbf{G} = q\mathbf{G}_1 + r\mathbf{G}_2 + s\mathbf{G}_3, \tag{5.8}$$

then

$$\mathbf{K} = \mathbf{G}. \tag{5.9}$$

Namely, in order to obtain the relation between the directions of incident and scattered light, $\mathbf{G}_1$, $\mathbf{G}_2$ and $\mathbf{G}_3$ are first calculated for a given crystal structure using Eqs. (5.5)–(5.7) and then $\mathbf{K}$ is obtained through the relation (5.9), which contains arbitrary integers q, r and s. It is clear that $\mathbf{K}$ is related with the wave vectors of incident and scattered light through the relation, $\mathbf{K} = \mathbf{k}_s - \mathbf{k}_i$, and for elastic scattering, $|\mathbf{k}_i| = |\mathbf{k}_s|$ holds naturally. Thus, for a given reciprocal vector $\mathbf{G}$, the starting and end points of the wave vectors of incident and scattered light should lie on a bisector of the vector $\mathbf{G}$, as shown in Fig. 5.5a. In Fig. 5.5b, we show various reciprocal vectors and the corresponding bisectors, each of which guarantees the reflection of light due to the periodicity determined by the reciprocal vector.

---

[a]1D crystal is considered as a multilayer whose spatial range extends over a whole space. Thus, it has a periodic structure normal to the layers. We put the primitive translation vector as $\mathbf{a}_1$. On the contrary, for the other two directions, the primitive translation vectors are arbitrarily chosen within a plane perpendicular to a vector $\mathbf{a}_1$. Thus, according to Eq. (5.5), $\mathbf{G}_1$ is obtained as $\mathbf{G}_1 = 2\pi\mathbf{u}_2 \times \mathbf{u}_3/a_1 = 2\pi\mathbf{u}_1/a_1$, where we choose two unit vectors, $\mathbf{u}_2$ and $\mathbf{u}_3$, which are orthogonal to each other within a plane perpendicular to $\mathbf{a}_1$, while $\mathbf{u}_1$ is a unit vector parallel to $\mathbf{a}_1$. In a similar manner, for 2D crystal, $\mathbf{G}_1$ and $\mathbf{G}_2$ are obtained as $\mathbf{G}_1 = 2\pi\mathbf{a}_2 \times \mathbf{u}_3/(a_1 a_2)$ and $\mathbf{G}_2 = 2\pi\mathbf{u}_3 \times \mathbf{a}_1/(a_1 a_2)$ from Eqs. (5.5) and (5.6), where $\mathbf{u}_3$ is a unit vector defined as $\mathbf{u}_3 \equiv \mathbf{a}_1 \times \mathbf{a}_2/|\mathbf{a}_1 \times \mathbf{a}_2|$.

(a)

(b)

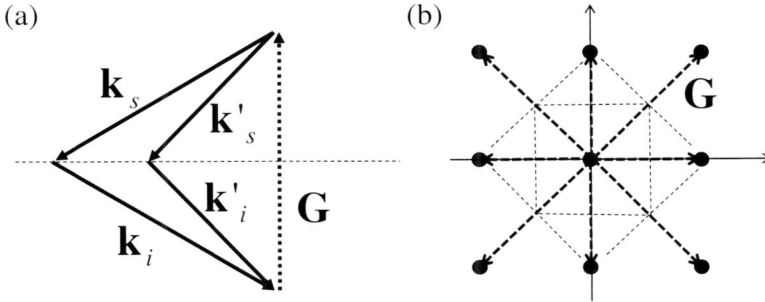

Figure 5.5 (a) Relation between wave vectors of incident and scattered light, which are associated with a given reciprocal vector of **G**. (b) Reciprocal vectors and their bisectors in case of a square lattice.

In a reciprocal space, we can view crystals from a different viewpoint from those in a real space. In Fig. 5.6a, we show examples of 2D lattice of square and hexagonal lattices, where we show the primitive translation vectors and the reciprocal vectors. In Fig. 5.6b, we show the corresponding structures viewed in reciprocal space. For a square lattice of a side a, the same square lattice is obtained for the reciprocal lattice with the side $2\pi/a$, while for a hexagonal lattice of a side a, 30° rotated hexagonal is obtained in the reciprocal space with the side $4\pi/(\sqrt{3}a)$.

As is described above, the light scattering process due to a periodic structure is associated with the bisectors of the reciprocal vectors. Since an infinite number of reciprocal points are distributed in the reciprocal space and hence numerous bisectors can be drawn in principle. Thus, there is a minimum area that is enclosed by the bisectors around the origin of the reciprocal space, which we call *first Brillouin zone*. In a similar manner, the second and third minimum spaces are called *second* and *third Brillouin zone*. On the other hand, the bisectors enclosing the Brillouin zone are called *zone boundary*. We show the first Brillouin zones for square and hexagonal lattices in Fig. 5.6b. For a square lattice of a side a, the first Brillouin zone is expressed as a square of a side $2\pi/a$, while for a hexagonal lattice of a side a, the first Brillouin zone is expressed as a hexagonal of a side $2\pi/(\sqrt{3}a)$.

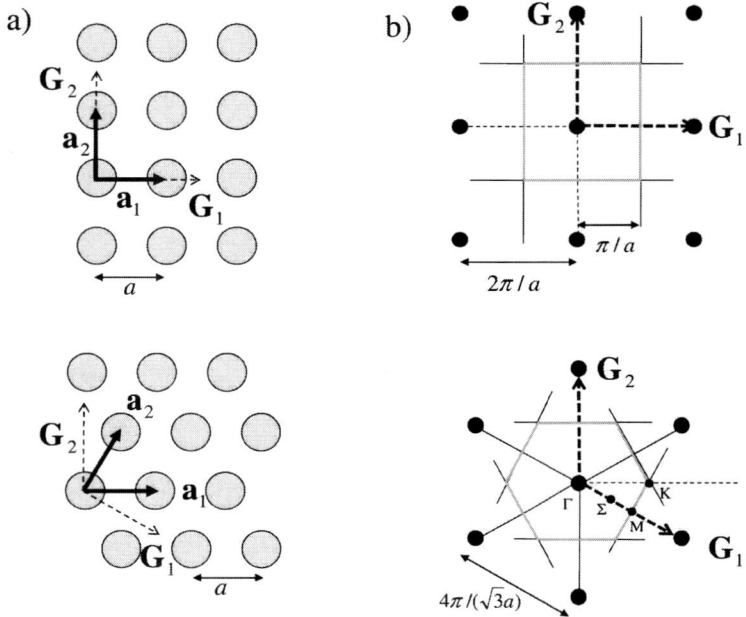

Figure 5.6 (a) Square (upper) and hexagonal (lower) lattices in real space with primitive translation vectors and the reciprocal vectors. (b) Square and hexagonal lattices in reciprocal space. The first Brillouin zone is expressed by a minimum area enclosed by several bisectors. In reciprocal space, it is a custom to name high-symmetry points using alphabetical or Greek letters. In general, alphabetical letters are used on a border of the first Brillouin zone, while Greek letters are used within the zone.

In a similar manner, we can write the first Brillouin zones for 3D crystals as shown in Fig. 5.7. In this figure, we illustrate the results for simple cubic (sc), face-centered cubic (fcc), body-centered cubic (bcc) and hexagonal close-packing (hcp) structures as examples. In order to explain a method to obtain reciprocal vectors in 3D crystal, we show bcc structure as an example (Exercise (1)). As shown in Fig. 5.7c, the primitive translation vectors in bcc structure are expressed as

$$\mathbf{a}_1 = (a/2)(\hat{\mathbf{x}} + \hat{\mathbf{y}} - \hat{\mathbf{z}}),$$
$$\mathbf{a}_2 = (a/2)(-\hat{\mathbf{x}} + \hat{\mathbf{y}} + \hat{\mathbf{z}}),$$
$$\mathbf{a}_3 = (a/2)(\hat{\mathbf{x}} - \hat{\mathbf{y}} + \hat{\mathbf{z}}),$$

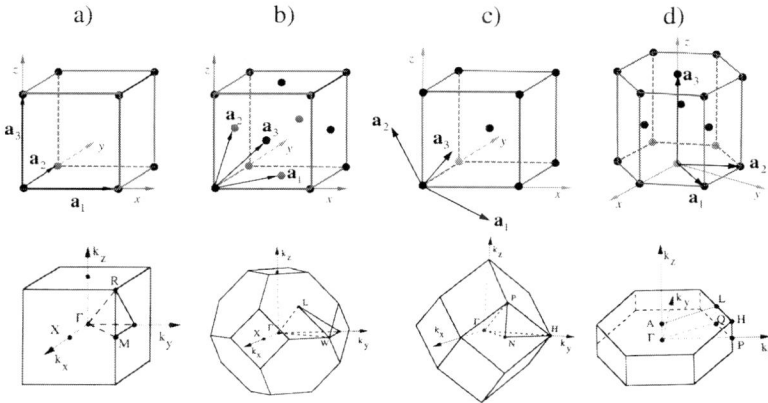

Figure 5.7 Crystal lattices expressed in real space and the corresponding first Brillouin zones. (a) Simple cubic (sc), (b) face-centered cubic (fcc), (c) body-centered cubic (bcc), and (d) hexagonal close-packed (hcp) structures.

where we put unit vectors along the x, y and z axes as $\hat{\mathbf{x}}$, $\hat{\mathbf{y}}$ and $\hat{\mathbf{z}}$, respectively. Inserting these relations into Eqs. (5.5), we obtain a reciprocal vector $\mathbf{G}_1$ as

$$
\begin{aligned}
\mathbf{G}_1 &= (2\pi/\Omega)(\mathbf{a}_2 \times \mathbf{a}_3) \\
&= \frac{2\pi}{\Omega} \left(\frac{a}{2}\right)^2 (-\hat{\mathbf{x}} + \hat{\mathbf{y}} + \hat{\mathbf{z}}) \times (\hat{\mathbf{x}} - \hat{\mathbf{y}} + \hat{\mathbf{z}}) \\
&= \frac{2\pi}{(a/2)(\hat{\mathbf{x}} + \hat{\mathbf{y}} - \hat{\mathbf{z}}) \cdot 2(a/2)^2(\hat{\mathbf{x}} + \hat{\mathbf{y}})} \cdot 2\left(\frac{a}{2}\right)^2 (\hat{\mathbf{x}} + \hat{\mathbf{y}}),
\end{aligned}
$$

which leads to

$$
\mathbf{G}_1 = (2\pi/a)(\hat{\mathbf{x}} + \hat{\mathbf{y}}).
$$

In a similar manner, using Eqs. (5.6) and (5.7), we obtain

$$
\mathbf{G}_2 = (2\pi/a)(\hat{\mathbf{y}} + \hat{\mathbf{z}}),
$$
$$
\mathbf{G}_3 = (2\pi/a)(\hat{\mathbf{z}} + \hat{\mathbf{x}}).
$$

These relations are nothing but primitive translation vectors for fcc structure (Fig. 5.7b). Thus, the reciprocal vectors for bcc are equivalent to primitive translation vectors for fcc. In fcc structure, the number of lattice points located at the distance of closest approach is totally 12 so that the first Brillouin zone is enclosed by 12 bisecting planes of these reciprocal vectors and becomes a regular dodecahedron as shown in Fig. 5.7c.

5.1.3 *Photonic Band Structure*

5.1.3.1 Basic equations

A starting point to analyze the features of photonic crystal is the Maxwell equations of Eqs. (2.21)–(2.24). In the previous treatment, we have put the permittivity ϵ as a constant and then have derived $\nabla \cdot \mathbf{E} = 0$ from the relation $\nabla \cdot \mathbf{D} = 0$ under the condition of $\rho = 0$. However, in photonic crystal, ϵ cannot be put as a constant and should be dependent on a position so that the above derivation is not applicable in principle. Hence, we put the permittivity as $\epsilon = \epsilon_0 \epsilon(\mathbf{r})$, where $\epsilon(\mathbf{r})$ is called *relative permittivity* and is generally expressed in terms of a 3×3 tensor. Thus, using Eqs. (2.21) and (2.22), and assuming $\mu = \mu_0$ and $\mathbf{j} = \mathbf{0}$, we obtain a general expression

$$\nabla \times (\nabla \times \mathbf{E}) = -\epsilon_0 \epsilon(\mathbf{r}) \mu_0 \frac{\partial^2 \mathbf{E}}{\partial t^2}, \qquad (5.10)$$

which becomes a basic equation for the electric field in photonic crystal. In a similar manner, we obtain an equation for magnetic flux density as

$$\nabla \times \left(\frac{1}{\mu_0 \epsilon_0 \epsilon(\mathbf{r})} \nabla \times \mathbf{B} \right) = -\frac{\partial^2 \mathbf{B}}{\partial t^2}. \qquad (5.11)$$

These equations become a basis to calculate the electromagnetic wave in photonic crystal and will essentially give the solutions in any case. However, in general, it is difficult to solve these equations analytically, because complicated boundary conditions should be applied at the surfaces of constituting particles. To obtain the general expressions concerning the photonic crystal, here we will focus on the periodicity of the structure, instead of imposing the complicated boundary conditions.

We first consider that the relative permittivity $\epsilon(\mathbf{r})$ is spatially periodic, according to the periodic nature of the crystal. Then, we can expand an inverse matrix of a tensor $\epsilon(\mathbf{r})$ using the reciprocal vectors as

$$\epsilon^{-1}(\mathbf{r}) = \sum_{\mathbf{G}} b_{\mathbf{G}} e^{i\mathbf{G} \cdot \mathbf{r}}, \qquad (5.12)$$

where $b_{\mathbf{G}}$ is an expansion coefficient of a 3×3 tensor. Assuming plane electromagnetic waves in crystal and using Bloch's theorem[a],

---

[a]See the proof in Appendix A.3.

we obtain

$$\mathbf{E}(\mathbf{r}, t) = \sum_G \mathbf{E_G} e^{i(\mathbf{k}+\mathbf{G})\cdot\mathbf{r}-i\omega t}. \tag{5.13}$$

Inserting these relations into Eq. (5.10), we obtain

$$\epsilon_0\mu_0\omega^2 \sum_G \mathbf{E_G} e^{i(\mathbf{k}+\mathbf{G})\cdot\mathbf{r}}$$

$$= \sum_{G'} b_{G'} e^{i\mathbf{G}'\cdot\mathbf{r}} \nabla \times \left(\nabla \times \sum_{G''} \mathbf{E_{G''}} e^{i(\mathbf{k}+\mathbf{G''})\cdot\mathbf{r}} \right)$$

$$= -\sum_{G'} b_{G'} e^{i\mathbf{G}'\cdot\mathbf{r}} \sum_{G''} (\mathbf{k}+\mathbf{G''}) \times \left\{ (\mathbf{k}+\mathbf{G''}) \times \mathbf{E_{G''}} e^{i(\mathbf{k}+\mathbf{G''})\cdot\mathbf{r}} \right\},$$

where we have used the relation $\nabla \times \mathbf{E_{G''}} \exp[i(\mathbf{k}+\mathbf{G''})\cdot\mathbf{r}] = i(\mathbf{k}+\mathbf{G''}) \times \mathbf{E_{G''}} \exp[i(\mathbf{k}+\mathbf{G''})\cdot\mathbf{r}]$. Now, let us compare the left- and right-hand sides of this equation. We notice $\mathbf{G}$ in the exponent of the left-hand side can be expressed by an arbitrary combination of $\mathbf{G}'$ and $\mathbf{G}''$ in the right-hand side, only if $\mathbf{G} = \mathbf{G}' + \mathbf{G}''$ is satisfied. Thus, putting $\mathbf{G}' = \mathbf{G} - \mathbf{G}''$ and comparing the coefficients of the exponential terms having the same exponents, we obtain

$$\frac{\omega^2}{c^2}\mathbf{E_G} = -\sum_{G''} b_{G-G''}(\mathbf{k}+\mathbf{G''}) \times \{(\mathbf{k}+\mathbf{G''}) \times \mathbf{E_{G''}}\}, \tag{5.14}$$

where c is the velocity of light in vacuum, which is expressed as $\epsilon_0\mu_0 = 1/c^2$.

In the same way, by putting

$$\mathbf{B}(\mathbf{r}, t) = \sum_G \mathbf{B_G} e^{i(\mathbf{k}+\mathbf{G})\cdot\mathbf{r}-i\omega t}, \tag{5.15}$$

an equation for magnetic flux density is obtained as

$$\epsilon_0\mu_0\omega^2 \sum_G \mathbf{B_G} e^{i(\mathbf{k}+\mathbf{G})\cdot\mathbf{r}} = \nabla \times \left(\sum_{G'} b_{G'} e^{i\mathbf{G}'\cdot\mathbf{r}} \nabla \times \sum_{G''} \mathbf{B_{G''}} e^{i(\mathbf{k}+\mathbf{G''})\cdot\mathbf{r}} \right).$$

Comparing the coefficients corresponding to the same exponents in both sides of the above relation, we obtain

$$\frac{\omega^2}{c^2}\mathbf{B_G} = -\sum_{G''} (\mathbf{k}+\mathbf{G}) \times \{b_{G-G''}(\mathbf{k}+\mathbf{G''}) \times \mathbf{B_{G''}}\}. \tag{5.16}$$

Since both of Eqs. (5.14) and (5.16) consist of only linear terms of $\mathbf{E_G}$ or $\mathbf{B_G}$, a condition for these equations to have non-zero solutions

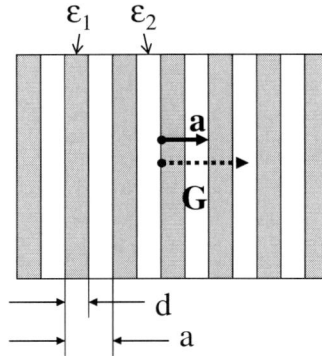

Figure 5.8 One-dimensional photonic crystal with a primitive translation vector **a** and a reciprocal vector **G**.

is to satisfy a condition that the determinants for their coefficient matrices should be zero, which are functions of ω and **k**. This type of equation is usually called *eigen equation*. Thus, by solving them, the relation between ω and **k** can be obtained. In the next section, we will follow this procedure for a simple 1D photonic crystal as an example.

5.1.4 *One-Dimensional Photonic Crystal*

5.1.4.1 Band structure

1D photonic crystal is considered as an extreme case of multilayer system, whose spatial range extends over a whole space. We consider a 1D photonic crystal shown in Fig. 5.8, where two layers having the widths of d and $a - d$ are alternately aligned. a thus denotes the interval of the repetitive layers. We set the permittivities of the layers at ϵ_1 and ϵ_2, respectively. Since the direction of the repetition of the layers is normal to the layers, the primitive translation vector **a** is expressed as a vector normal to the layers, whose magnitude is a, as shown in the figure. The corresponding reciprocal vector is parallel to **a** and is obtained from Eq. (5.5) as $\mathbf{G}_m = m\mathbf{G}_1 = 2m\pi\mathbf{u}_1/a$, where m is an arbitrary integer and $\mathbf{u}_1$ is a unit vector parallel to **a**.

In 1D photonic crystal shown in Fig. 5.8, Eqs. (5.12) is simply written in a form of

$$\epsilon^{-1}(x) = \sum_m b_m e^{iG_m x},$$

(5.17)

where we assume $\epsilon(x)$ and hence b_m's are scalar. On the other hand, from Eq. (5.15), we can put

$$B(x) = \sum_m B_m e^{i(k+G_m)x},$$

(5.18)

where $G_m = 2m\pi/a$. Thus, by using Eq. (5.16), the equation for the magnetic flux density becomes

$$\frac{\omega^2}{c^2} B_m = \sum_{m'} b_{m-m'}(k + G_m)(k + G_{m'}) B_{m'}.$$

(5.19)

Further, putting $b_{m-m'}(k + G_m)(k + G_{m'}) \equiv f_{mm'}$ and $(\cdots, B_{-1}, B_0, B_1, \cdots) = \mathbf{B}$, we rewrite the above equation in a matrix form:

$$\begin{pmatrix} \cdot & \cdot & \cdot & \cdot & \cdot \\ \cdot & f_{-1-1} & f_{-10} & f_{-11} & \cdot \\ \cdot & f_{0-1} & f_{00} & f_{01} & \cdot \\ \cdot & f_{1-1} & f_{10} & f_{11} & \cdot \\ \cdot & \cdot & \cdot & \cdot & \cdot \end{pmatrix} \begin{pmatrix} \cdot \\ B_{-1} \\ B_0 \\ B_{+1} \\ \cdot \end{pmatrix} = \frac{\omega^2}{c^2} \begin{pmatrix} \cdot \\ B_{-1} \\ B_0 \\ B_{+1} \\ \cdot \end{pmatrix},$$

or

$$F\mathbf{B} = \frac{\omega^2}{c^2} \mathbf{B},$$

(5.20)

where F is a matrix whose element is expressed by $\{f_{ij}\}$. This is a kind of eigen function so that by solving this, the dispersion relation between ω and k will be obtained. Although this equation, in principle, gives an infinite number of ω's for a given k, the equations are usually truncated to a finite number of equations.

We will show the actual process to obtain the explicit expression in case of Fig. 5.8 as an example. At first, the inverse of relative permittivity is expanded in a series of reciprocal vectors as

$$\epsilon^{-1}(x) = \sum_m b_m e^{iG_m x}.$$

The expansion coefficients are obtained by integrating over a range from 0 and a after multiplying the both sides by $\exp[-iG_{m'}x]$:

$$\int_0^a \frac{1}{\epsilon(x)} e^{-iG_{m'}x} dx = \sum_m b_m \int_0^a e^{iG_m x} e^{-iG_{m'}x} dx.$$

The calculation of the right-hand side becomes

$$\sum_m b_m \frac{e^{i(G_m - G_{m'})a} - 1}{i(G_m - G_{m'})} = 0,$$

for $m \neq m'$, where we have used the relation $G_m = 2m\pi/a$. On the other hand, it becomes

$$b_m \int_0^a dx = ab_m,$$

for $m = m'$. Thus, in case of $m \neq 0$, we obtain

$$
\begin{aligned}
b_m &= \frac{1}{a} \int_0^a \frac{1}{\epsilon(x)} e^{-iG_m x} dx \\
&= \frac{1}{a} \int_0^d \frac{1}{\epsilon_1} e^{-iG_m x} dx + \frac{1}{a} \int_d^a \frac{1}{\epsilon_2} e^{-iG_m x} dx \\
&= \frac{1}{a} \left\{ \frac{1}{\epsilon_1} \frac{e^{-iG_m d} - 1}{-iG_m} + \frac{1}{\epsilon_2} \frac{1 - e^{-iG_m d}}{-iG_m} \right\} \\
&= \frac{1}{a} \frac{1 - e^{-iG_m d}}{iG_m} \left(\frac{1}{\epsilon_1} - \frac{1}{\epsilon_2} \right) \\
&= \frac{1}{m\pi} e^{-i\pi md/a} \sin(\pi md/a) \left(\frac{1}{\epsilon_1} - \frac{1}{\epsilon_2} \right).
\end{aligned}
\tag{5.21}
$$

On the other hand, for $m = 0$, the coefficient becomes

$$b_0 = \frac{1}{a} \left(\frac{d}{\epsilon_1} + \frac{a-d}{\epsilon_2} \right). \tag{5.22}$$

In this way, all the expansion coefficients and hence all the elements of the eigen equation are determined. Then, the final equation is determined as

$$
\begin{vmatrix}
\cdot & \cdot & \cdot & \cdot & \cdot \\
\cdot & b_0(k + G_{-1})^2 - \omega^2/c^2 & b_{-1}(k + G_{-1})k & b_{-2}(k + G_{-1})(k + G_1) & \cdot \\
\cdot & b_1 k(k + G_{-1}) & b_0 k^2 - \omega^2/c^2 & b_{-1} k(k + G_1) & \cdot \\
\cdot & b_2(k + G_1)(k + G_{-1}) & b_1(k + G_1)k & b_0(k + G_1)^2 - \omega^2/c^2 & \cdot \\
\cdot & \cdot & \cdot & \cdot & \cdot
\end{vmatrix} = 0.
$$

$$(5.23)$$

In order to solve this equation, a numerical calculation is inevitable, although in a more simplified case, we will give the analytical solution, which will be described later (see Sec. 5.1.4.3).

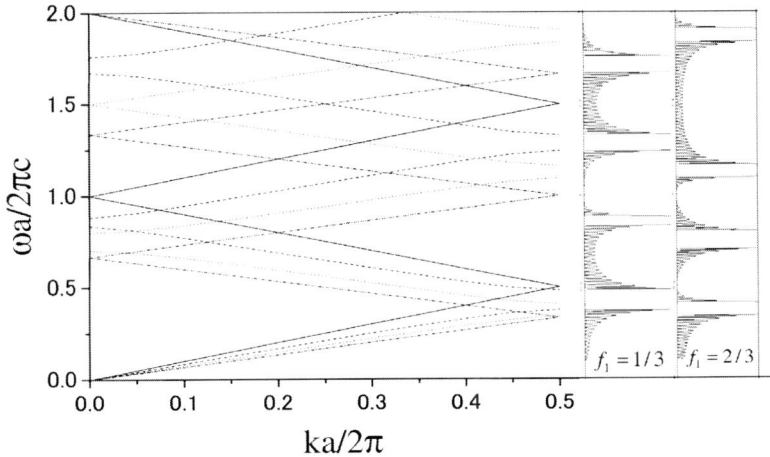

Figure 5.9 Band calculation result for 1D photonic crystal consisting of alternate layers with $\epsilon_1 = 2.25$ and the thickness of d, and $\epsilon_2 = 1.0$ and the thickness of $a - d$. We calculate the cases where the ratio of d and a, which is denoted as $d/a \equiv f_1$, takes values of 0, 1/3, 2/3 and 1, which appear as four curves from the top to bottom. Note that horizontal and vertical axes are multiplied by $a/(2\pi)$ and $a/(2\pi c)$ to make the both axes to be non-dimensional. (Courtesy of Dr. Y. Fujimura.) In the right side, the corresponding reflection spectra are calculated in cases of 21 layers of multilayer for $f_1 = 1/3$ and 2/3.

In Fig. 5.9, we show the results of the numerical calculations for $\epsilon_1 = 2.25$ and $\epsilon_2 = 1.0$. In this figure, we have defined $d/a \equiv f_1$, which corresponds to a filling factor in 2D or 3D photonic crystal, and have calculated in four cases, $f_1 = 0, 1/3, 2/3$ and 1, whose results appear as four curves from top to bottom, respectively. Note that horizontal and vertical axes are multiplied by $a/(2\pi)$ and $a/(2\pi c)$, respectively, to make both axes non-dimensional. If we look at the curves from the origin of this graph, they appear as almost linear lines with slightly different gradients. Then, they are folded back at $ka/(2\pi) = 0.5$, and are again so at $ka/(2\pi) = 0$ and so on. This representation is often called *reduced zone scheme*, where all the dispersion curves are expressed within the first Brillouin zone.

It is further noticed that except for $f_1 = 0$ and 1, there appear small gaps at $ka/(2\pi) = 0.5$. This type of gap is present at almost all the folded part of the curves for $f_1 = 1/3$ and $2/3$, while for $f_1 = 0$ and 1, the curves are connected continuously. In the right-hand side of this figure, we show the reflection spectra calculated for a multilayer consisting of totally 21 alternate layers with the refractive indices of $n_1 = 1.5$ and $n_2 = 1.0$ in cases of $f_1 = 1/3$ and $2/3$. It is easily noticed that the positions of the above gaps completely agree with the high-reflection bands in the reflection spectra. Such a gap appearing at $ka/(2\pi) = 0$ and 0.5 in reduced zone scheme is called *photonic band gap*. The photonic band gap appears only when the interaction of light and periodic structure is relatively large. On the other hand, in cased of $f_1 = 0$ and 1, the structure becomes homogeneous and no particular interaction will appear between light and the structure. It should be noted that oscillatory structure seen at both sides of reflection band in the reflection spectrum comes from the finiteness of the multilayer.

5.1.4.2 Origin of band gap

The origin of the band gap is qualitatively explained as follows: When a light wave propagates within an alternately layered structure with different permittivities, the light wave will partially reflect and partially transmit at the interface of these layers. The light waves thus generated will propagate to the right and left directions, and will interfere with each other to form a standing wave. At a zone boundary, the wavelength of light becomes twice the periodicity of the layers, because at a zone boundary of $k = \pi/a$, the wavelength and periodicity are connected with the relation, $\lambda/\langle n \rangle = 2\pi/k = 2a$ with $\langle n \rangle$ an average refractive index.

As shown in Fig. 5.10, if the loop of the standing wave thus created coincides with a high-permittivity part, the energy of electromagnetic wave will be higher, while in the opposite case, it will be lower. Thus, for a single wavelength, two values of energy will be realized, which is the origin of the band gap. When the width of one layer becomes narrower than the other, the overlap of the standing wave with the layered structure will be incomplete so that the energy separation will be smaller. Thus, the band gap width becomes also smaller. It is also expected that if the difference of the

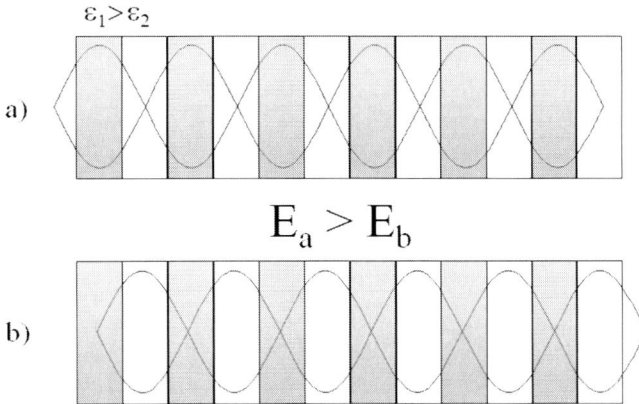

Figure 5.10 Schematic illustration to explain the origin of band gap at the zone boundary. The light waves propagating to the right and left interfere with each other to form a standing wave. At the zone boundary, the wavelength of the light wave is twice the periodicity of the structure. If the loop of the standing wave is coincident with the high-permittivity layer, the energy will be higher, while if it is coincident with the low-permittivity layer, the energy will be lower. Thus, the energy gap will be generated at the zone boundary.

permittivities becomes small, the band gap width will become small as well.

Careful readers will notice that the band gap widths shown in Fig. 5.9 are not constant, but are rather irregularly distributed, when one inspects the higher-order folded parts. For example, in case of $f_1 = 1/3$, the second-order band gap is somewhat narrower than the others, while in case of $f_1 = 2/3$, the fourth-order band gap is completely lost, and the third- and fifth-order gaps are slightly narrower. What is the origin of these variations?

Since 1D photonic crystal is essentially equivalent to a multilayer, it is convenient to consider multilayer interference instead of considering photonic band calculation [Fujimura (2009)]. As described in the previous section (Sec. 3.3), for multilayer interference, the interference conditions under normal incidence is simply expressed as

$$2(n_1 f_1 + n_2 f_2)a = m\lambda, \tag{5.24}$$

where we put $f_1 = d/a$ and $f_2 = 1 - f_1$ to accommodate the problem to the 1D photonic crystal. This condition comes from the consideration that the multilayer consists of two layers with high- and low-refractive indices of n_1 and n_2, and a thin-film interference condition of antireflection coating is applied to a sum of these two layers. On the other hand, if we consider only one layer, say 1 or 2, a soap-bubble condition should be applied. Thus, when the relation

$$2n_1 f_1 a = \left(m' - \frac{1}{2} \right) \lambda \qquad (5.25)$$

holds in addition to Eq. (5.24), the interference will be enhanced up to its maximum, while when the relation

$$2n_1 f_1 a = m' \lambda \qquad (5.26)$$

holds, they will be canceled.

The variation of photonic band gap widths is deeply connected with the above conditions of multilayer interference. In fact, Eq. (5.24) is easily transformed into

$$\frac{2n_1 f_1 a}{\lambda} = m - \frac{2n_2 f_2 a}{\lambda} = \frac{m\lambda - 2n_2 f_2 a}{\lambda}$$

$$= \frac{2n_1 f_1 a}{2(n_1 f_1 + n_2 f_2)a/m} = \frac{m}{1 + n_2 f_2/(n_1 f_1)}, \qquad (5.27)$$

which shows the term in the left-hand side is proportional to an integer m. In a similar way, Eqs. (5.25) and (5.26) are transformed into

$$\frac{2n_1 f_1 a}{\lambda} = m' - \frac{1}{2} \qquad (5.28)$$

$$\frac{2n_1 f_1 a}{\lambda} = m'. \qquad (5.29)$$

Comparing Eq. (5.27) with Eqs. (5.28) and (5.29), we notice that the terms in the left-hand sides of these relations are all equal to each other. Thus, if we plot this term against m or m', we can investigate the relation between the above two interference conditions.

In Fig. 5.11, we show a graph obtained in this way. In this figure, we plot $2n_1 f_1 a/\lambda$ in Eq. (5.27) against m for $f_1 = 1/3$ and $2/3$, which are shown as closed circles connected by two oblique lines. On the other hand, Eqs. (5.28) and (5.29) are given as horizontal solid and dashed lines, respectively. Thus, if the points

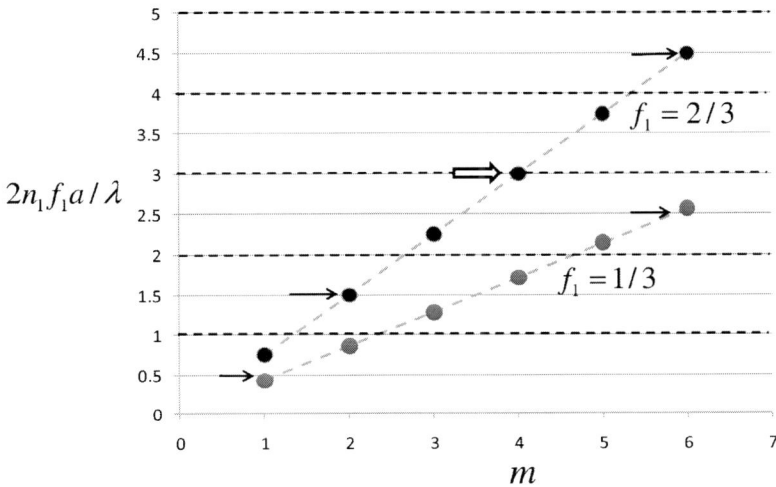

Figure 5.11 Condition to open a band gap in 1D photonic crystal. Closed circles indicate the values of $4n_1 f_1 a/\lambda$ calculated through Eq. (5.27) against m for $f_1 = 1/3$ and $2/3$, where we put $n_1 = 1.5$ and $n_2 = 1.0$ with $f_2 = 1 - f_1$ and $a/\lambda = 0.5$, which corresponds to the data in Fig. 5.9. Black arrow indicates a point where $4n_1 f_1 a/\lambda$ is almost coincident with a half-integer (Eq. (5.28)), while an arrow outline with a blank inside indicates a point where $4n_1 f_1 a/\lambda$ is almost coincident with an integer (Eq. (5.29)). The former corresponds to a condition for a widely opened band gap, while the latter is the condition for a band gap closed.

coincide with horizontal solid lines, the two reflection processes will constructively interfere and strong reflection will occur. On the contrary, if they coincide with horizontal dashed lines, the two processes will destructively interfere and the reflection will be strongly suppressed.

This speculation is well supported in the present case. Namely, for $f_1 = 1/3$, points corresponding to $m = 1$ and 6 almost completely overlap with the solid horizontal lines, while the other points are rather incomplete, which are in good agreement with the wider width of the first-order band observed in Fig. 5.10. In a similar manner, for $f_1 = 2/3$, the cases of $m = 2$ and 6 correspond to constructive interference, while the case of $m = 4$ to destructive interference, which explains the slightly wider width for the second-order band and also the disappearance of the fourth-

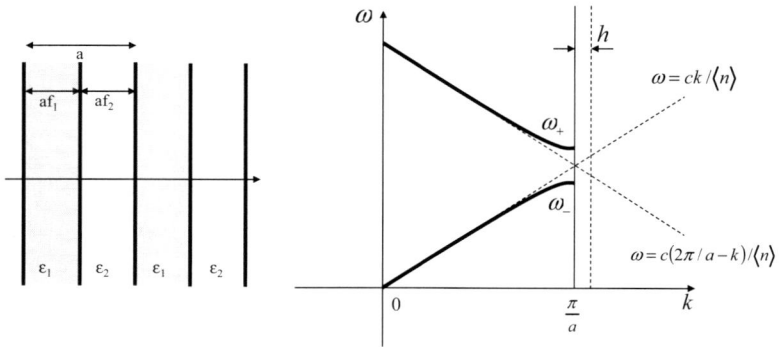

Figure 5.12 Illustration of a model to calculate the position and width of the first-order photonic band gap in 1D photonic crystal shown in the left. In this calculation, only the following two modes are considered, $\omega = ck/\langle n \rangle$ and $\omega = c(2\pi/a - k)/\langle n \rangle$ with $\langle n \rangle$ an average refractive index defined as $\langle n \rangle = 1/\sqrt{b_0}$.

order band. Thus, the magnitude of photonic band gap width is essentially determined by the degree of interference among the layers.

5.1.4.3 Method to determine the position and width of photonic band gap

When the difference of permittivities of the two layers is not so large in 1D photonic crystal, there is an easy way to estimate the position and width of the first-order band gap [Sakoda (2001)]. The first-order photonic band gap appears at a zone boundary of $k = \pi/a$, where only two modes are directly related to form a band gap, as shown in Fig. 5.12. These modes are associated with the reciprocal vectors of G_0 and G_{-1}, and are obtained by solving the diagonal terms of Eq. (5.23) after putting zero: $b_0(k + G_{-1})^2 - \omega^2/c^2 = 0$, which leads to $\omega = \pm c(k - 2\pi/a)/\langle n \rangle$, and $b_0 k^2 - \omega^2/c^2 = 0$, which leads to $\omega = \pm ck/\langle n \rangle$, where $\langle n \rangle = 1/\sqrt{b_0}$ is an average refractive index derived from Eq. (5.22). Thus, in order to treat the first-order band gap approximately, it is sufficient to consider only terms associated with G_0 and G_{-1}. This approximation greatly simplifies the determinant of Eq. (5.23) and leads to

$$\begin{vmatrix} b_0(k+G_{-1})^2 - \omega^2/c^2 & b_{-1}(k+G_{-1})k \\ b_1 k(k+G_{-1}) & b_0 k^2 - \omega^2/c^2 \end{vmatrix} = 0. \qquad (5.30)$$

The calculation of this determinant gives

$$\{b_0(k+G_{-1})^2 - \omega^2/c^2\}(b_0 k^2 - \omega^2/c^2) - b_1 b_{-1} k^2 (k+G_{-1})^2 = 0,$$

which leads to

$$\left(\frac{\omega^2}{c^2}\right)^2 - \left(\frac{\omega^2}{c^2}\right) b_0 \{k^2 + (k+G_{-1})^2\} + (b_0^2 - b_1 b_{-1})k^2 (k+G_{-1})^2 = 0.$$

This is a quadratic equation with respect to ω^2/c^2. Solving this, we obtain

$$\frac{\omega^2}{c^2} = \frac{1}{2} b_0 \left\{ k^2 + (k+G_{-1})^2 \right\}$$

$$\pm \frac{1}{2} \sqrt{ b_0^2 \left\{ k^2 - (k+G_{-1})^2 \right\}^2 + 4b_1 b_{-1} k^2 (k+G_{-1})^2}. \quad (5.31)$$

To investigate the behavior near a zone boundary ($k \sim \pi/a$), we put $k = \pi/a + h$ with $|h| \ll \pi/a$, insert it into the above solution and expand with respect to h as

$$\omega_\pm \approx \frac{\pi c}{a} \sqrt{b_0 \pm |b_1|} + \frac{ac}{2\pi \sqrt{b_0 \pm |b_1|}} \left(b_0 \pm \frac{2b_0^2 - |b_1|^2}{|b_1|} \right) h^2, \qquad (5.32)$$

where we show the result up to a quadratic term of h and use the relation $b_1 b_{-1} = |b_1|^2$ (Exercise (2)). The behavior around a zone boundary is illustrated in Fig. 5.12, where it is shown that both modes slightly bend near a zone boundary to form a band gap. The band center is easily obtained by putting $h = 0$ in the above solution and is given as

$$\omega_c = \pi \sqrt{b_0} c/a, \qquad (5.33)$$

while the band gap width is obtained by putting $h = 0$ and calculating the following difference of the angular frequencies as

$$\Delta \omega = \omega_+ - \omega_-. \qquad (5.34)$$

Since the expansion coefficient of the inverse permittivity is already given, putting $\epsilon_1 = n_1^2$ and $\epsilon_2 = n_2^2$ with $d/a = f_1$ and $(a-d)/a = f_2$, we obtain

$$b_0 = \frac{f_1}{n_1^2} + \frac{f_2}{n_2^2},$$

which results in

$$\omega_c = \frac{\pi c}{a} \sqrt{\frac{f_1}{n_1^2} + \frac{f_2}{n_2^2}}$$

$$= \frac{\pi c}{a} \sqrt{\frac{1}{n_2^2} \left\{ 1 + \frac{n_2^2 - n_1^2}{n_1^2} f_1 \right\}}. \qquad (5.35)$$

If we write the result in the form of wavelength, then it is given

$$\lambda_c \equiv \frac{2\pi c}{\omega_c}$$

$$= 2a \bigg/ \sqrt{\frac{1}{n_2^2} \left\{ 1 + \frac{n_2^2 - n_1^2}{n_1^2} f_1 \right\}}. \qquad (5.36)$$

Further, when the difference of the permittivities of the two layers is small enough ($n_1 \approx n_2$), the above expression is approximated as

$$1 \bigg/ \sqrt{\frac{1}{n_2^2} \left\{ 1 + \frac{n_2^2 - n_1^2}{n_1^2} f_1 \right\}} \approx n_2 \left(1 - \frac{1}{2} \frac{n_2^2 - n_1^2}{n_1^2} f_1 \right)$$

$$= n_2 \left(1 - \frac{1}{2} \frac{(n_2 + n_1)(n_2 - n_1)}{n_1^2} f_1 \right)$$

$$\approx n_2 \left(1 - \frac{n_2 - n_1}{n_1} f_1 \right)$$

$$\approx n_1 f_1 + n_2 f_2,$$

which leads to a well-known interference condition for multilayer of

$$\lambda_c = 2a(n_1 f_1 + n_2 f_2). \qquad (5.37)$$

On the other hand, the band gap width is derived as

$$\Delta\omega = \frac{\pi c}{a} \left(\sqrt{b_0 + |b_1|} - \sqrt{b_0 - |b_1|} \right)$$

$$= \frac{\pi c}{a} \frac{2|b_1|}{\sqrt{b_0 + |b_1|} + \sqrt{b_0 - |b_1|}}$$

$$\approx \frac{\pi c}{a} \frac{|b_1|}{\sqrt{b_0}}, \qquad (5.38)$$

where the last term is obtained by approximating under the assumption of $|b_1| \ll b_0$. When we write it in the form of wavelength, we obtain

$$\Delta\lambda \equiv \frac{2\pi c}{\omega_-} - \frac{2\pi c}{\omega_+} = \frac{2\pi c \Delta\omega}{\omega_+ \omega_-}$$

$$= \frac{2\pi c}{(\pi c/a)^2 \sqrt{b_0^2 - |b_1|^2}} \frac{(\pi c/a)|b_1|}{\sqrt{b_0}} \approx \frac{2a|b_1|}{(b_0)^{3/2}}.$$

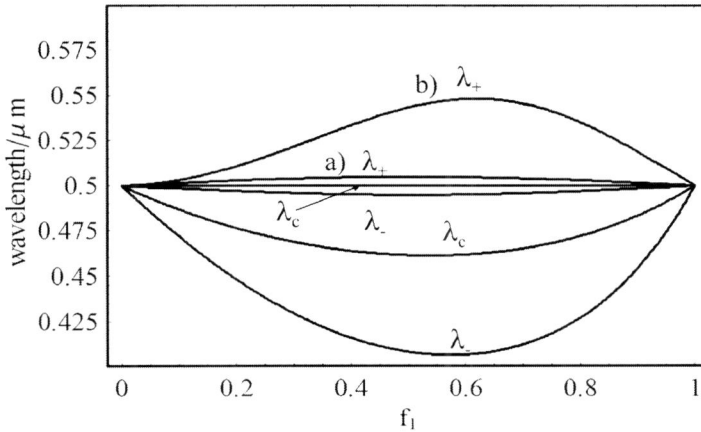

Figure 5.13 Filling factor, f_1, dependence of photonic band center and edges calculated on 1D photonic crystals in cases of (a) $n_1 = 1.60$ and $n_2 = 1.55$, and (b) $n_1 = 1.6$ and $n_2 = 1.0$. The calculation is performed using the relation Eq. (5.36), where a is varied so that the relation $a = \lambda/\{2(n_1 f_1 + n_2 f_2)\}$ will always hold for $\lambda = 500$ nm.

Further, under the condition of $f_1 \approx f_2$ and $n_1 \approx n_2$, this relation becomes

$$\Delta\lambda \approx 4a|n_2 - n_1|/\pi, \tag{5.39}$$

which means that when the differences of the refractive indices and widths between the two layers are small, the band gap width is proportional to the difference of the refractive indices (Exercise (3)).

In Fig. 5.13, we show the calculated results for λ_c, λ_- and λ_+ against a filling factor, f_1, for two combinations of the refractive indices. One is a case of $n_1 = 1.60$ and $n_2 = 1.55$, and the other is a case of $n_1 = 1.5$ and $n_2 = 1.0$. In this calculation, a is varied so as to always hold the relation $a = \lambda/\{2(n_1 f_1 + n_2 f_2)\}$ for $\lambda = 500$ (nm). It is found that when the difference of refractive indices is small, a narrow band gap opens without changing the band center. On the other hand, when the difference becomes large, a wide band gap opens with changing the position of the band center.

Figure 5.14 shows the comparison of the above approximate method with an exact band calculation. We compare the results in case of $n_1 = 1.5$ and $n_2 = 1.0$. Solid lines are obtained from

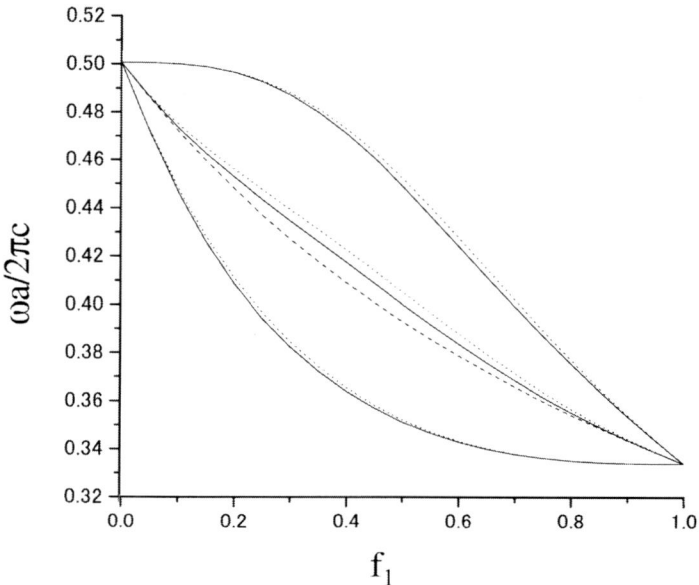

Figure 5.14 Filling factor, f_1, dependence of photonic band center and edges calculated on 1D photonic crystals in case of $n_1 = 1.5$ and $n_2 = 1.0$. Solid curves are obtained by a band calculation, while dotted lines are calculated using the relation Eq. (5.36) and a dashed line is calculated using the Maxwell-Garnett model. (Courtesy of Dr. Y. Fujimura.)

the band calculation, while dotted and dashed lines are obtained by an approximate method described above and the Maxwell-Garnett model described later (see Sec. 5.2.3). The agreement of the approximate method with the band calculation is fairly good even when the difference of refractive indices is considerably large, which confirms the validity of the approximation given above.

5.1.5 *2D Photonic Crystal*

In Fig. 5.15, we show an example of band calculation on 2D photonic crystal consisting of cylinders placed on square lattice points. Unlike the case of 1D photonic crystal, the band structure is largely dependent on the direction and polarization of incident light. In this figure, we show a case for *p*-polarization, that is, the polarization

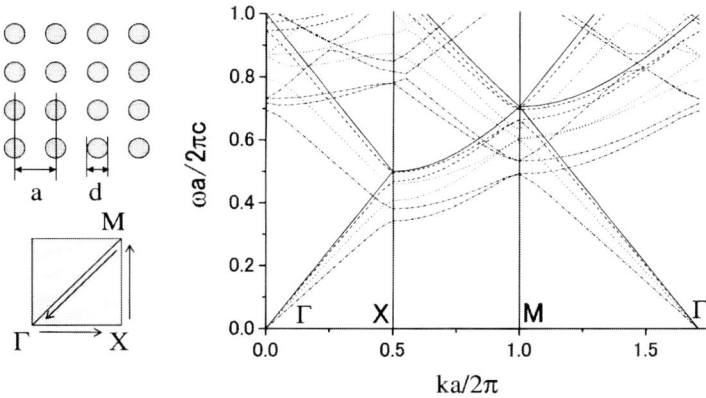

Figure 5.15 Band structure calculated for 2D photonic crystal consisting of cylinders arranged on square lattice points with $d/a = 0, 1/3, 2/3$ and 1 (from top to bottom). (Courtesy of Dr. Y. Fujimura.)

parallel to a plane of this page, and also show the variation of band structures for the magnitude and direction of **k**. Further, the band structures are calculated with changing ratios of the cylinder diameter d to the lattice constant a such as $d/a = 0, 1/3, 2/3$ and 1. At first, the magnitude of **k** is varied along a horizontal row of cylinders, which causes a translation from Γ $(k = 0)$ to X point $(k = \pi/a)$ in **k** space, then both the magnitude and direction of **k** are varied along a vertical zone boundary, causing a shift from X to M point, and then back to the Γ point along a diagonal line while only the magnitude of k is varied.

Solid line expresses a case of $d/a = 0$, which corresponds to a homogeneous medium with the refractive index of 1. Between Γ and X, a linear line that is folded at a zone boundary is observed as is usual, while between X and M, a curve that is convex downward is observed, which reflects that the magnitude of k varies according to the relation $k = \sqrt{(\pi/a)^2 + k_y^2}$, where we put y axis along a vertical direction. Finally, between M to Γ, k changes linearly from $\sqrt{2}(\pi/a)$ to 0. With increasing diameter, a gradient near the Γ point remarkably decreases and at a zone boundary, a curve show a clear photonic band gap. On the other hand, at the M point, a band gap does not open in every case. The band gap at X is open, even when the ratio takes a maximum value of $d/a = 1$. This is because for cylinders, they cannot fill a whole space and take a maximum filling factor of $\pi/4 \sim 0.785$ at $d/a = 1$.

Comparing the band structure with the structure in real space, we notice that X corresponds to the incident direction along a horizontal row of cylinders, where there is a clear spatial gap in the arrangement of vertical rows. Such a structure is expected to cause a large spatial modulation of refractive index. On the other hand, since the incident direction is along a diagonal line in case of M, $45°$-inclined rows of cylinders seem to be more densely packed, which will cause only a small spatial modulation of the refractive index. This intuitive speculation will partly explain why the band gap is not open at the M point. However, a simple analysis as has been done for the 1D crystal is generally difficult for the 2D crystal, because a lot of modes are more or less related with the zone boundary behavior [Fujimura (2009)].

5.2 Long-Wavelength Approximation and Average Refractive Index

5.2.1 *Behavior near a Γ Point*

In the preceding sections, we have shown typical band structures calculated for 1D and 2D photonic crystals. However, it seems rather difficult to know what information can be derived from these band structures and moreover to understand their physical meanings lying behind them. Of course, it is possible to obtain the information on where photonic band gaps appear and to what extent they are open. However, as described above, it is generally difficult to understand the origin of the band gap from the mode analysis as has been done in 1D photonic crystal, because in 2D and 3D crystals, so many modes are complicatedly interconnected to form a band gap.

In the present section, instead of analyzing the band gap, we will direct our attention to a behavior near the Γ point [Fujimura (2009)]. Near the Γ point, the characteristics of matter are essentially unrelated to the shape and configuration of composing microstructures, and only show a linear dispersion relation, whose gradient gives an average refractive index. This is natural because for a light wave having sufficiently long wavelength, it cannot discriminate the presence of microstructures and propagates as if the material consists of a uniform medium possessing the average material properties. Thus, under an approximation called *long-wavelength approximation*, the light wave can essentially probe the averaged properties of the material.

This concept of averaging is sometimes convenient to estimate the optical properties of the material that consists of random distribution of structure from a micrometer to nanometer size. For example, in liquid, the random distribution of liquid molecules appears as a uniform material for propagating light, whose property is essentially expressed by a single parameter of refractive index, although there is a small amount of scattering due to the random distribution of molecules. Here, we will consider the validity and range of the applicability of the averaging technique from a viewpoint of the behavior near the Γ point.

5.2.2 *Average Permittivity*

Obtaining the average permittivity of material, which is composed of several species of atoms, molecules or minute particles, has been one of the key problems for material sciences. In the present section, we will introduce several such theories and will discuss their physical meanings and the applicability to a system containing much larger particles.

At first, using Gauss's theorem (see App. A.1), let us derive the electric field induced by uniform charges on an extremely thin film extending over an infinite area. As shown in Fig. 5.16, we cut off the film by a cylinder perpendicular to the film with a cross section of A. If we consider the electric field on a sufficiently small area of the side face of the cylinder so that it can be regarded as a flat plane, the electric fields from the right and left sides of this plane will equally contribute to the electric field on this plane. Thus, the electric field normal to this plane will not be generated. On the other hand, the electric fields normal to the upper and lower faces of the cylinder will take the finite values.

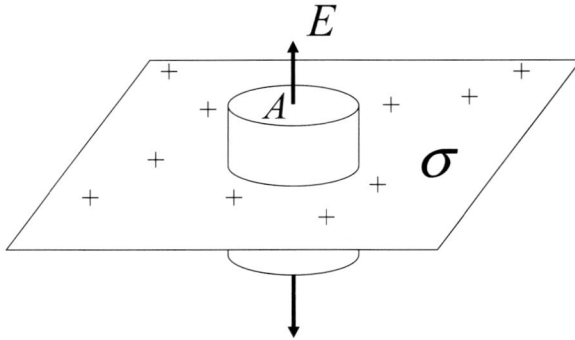

Figure 5.16 Method to calculate the electric field due to a plane carrying uniform charges.

Thus, by integrating the electric fields normal to the surface over all the surfaces of the cylinder, Gauss's theorem states $\int_S \mathbf{E} \cdot \mathbf{n}dS = \rho/\epsilon_0$, where $\mathbf{n}$ is a unit vector directing outside and normal to the surface, ρ a sum of charges within the cylinder, and ϵ_0 the permittivity of vacuum. dS denotes an infinitesimal area on the surfaces of the cylinder and the integration is performed over the whole surface S. Inserting $\mathbf{E} \cdot \mathbf{n} = E$ for the upper and lower surfaces, and $\rho = A\sigma$ into this formula, we obtain $2AE = A\sigma/\epsilon_0$, which leads to $E = \sigma/(2\epsilon_0)$. Therefore, as far as one considers a film with an infinite area, the electric field thus generated is normal to the plane of film and is independent of the distance from the plane.

Next, we consider a dielectric film of an infinite area with the thickness of d. It is assumed that a uniform electric field $\mathbf{E}_0$ is perpendicularly applied to this film. Since the film is dielectric, charges are induced on both surfaces of the film with opposite signs. We denote the density of the induced charge as σ. Using the above result, we can calculate the electric field within the film. Within the film, it is considered from a macroscopic viewpoint that the film is filled with continuous medium. However, from a microscopic viewpoint, the film is considered to consist of an ensemble of atoms or molecules with a vacuum filling the space other than the constituents. Thus, to eliminate the microscopic inhomogeneity, we hypothetically consider an extremely thin channel, within which it is assumed to be a vacuum as shown in Fig. 5.17a. The electric field

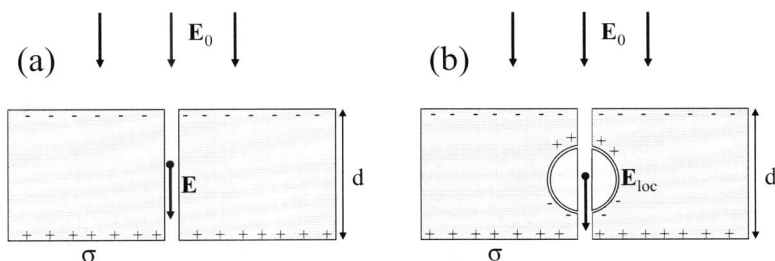

Figure 5.17 Method to calculate the electric field within a dielectric film (a) without and (b) with a hypothetical sphere.

inside the channel is obtained by the sum of the external field and the fields due to the charges induced on the surfaces. Although the latter is the sum of the contributions from the upper and lower surfaces, both give equal fields with the same signs, as has been derived above, so that we add them and obtain

$$E = E_0 - \frac{\sigma}{\epsilon_0}. \tag{5.40}$$

From the microscopic point of view, a lot of microscopic polarizations due to constituting atoms and molecules are to be distributed within the film. However, we consider that the effect of these microscopic polarizations will be canceled with each other and only that of the surface charges will remain. Thus, the macroscopic polarization P will be expressed as the product of the surface charge and the thickness of the film, and its direction is the same as that of the externally applied field. Therefore, if we consider a small volume determined by a surface area of A, the total amount of polarization within this area becomes $P\,Ad = \sigma\,Ad$, which leads to a simple result of $P = \sigma$. Inserting this relation into the above formula, we obtain

$$E = E_0 - \frac{P}{\epsilon_0},$$

which is transformed into

$$\epsilon_0 E_0 = \epsilon_0 E + P.$$

This is nothing but the continuity of electric flux density at the surface.

Next, we will take microscopic inhomogeneity into account. For this purpose, let us consider hypothetically a spherical hole insides the film, in addition to a hypothetical channel described above. The center of the hole is assumed to lie on a center line of the channel, as shown in Fig. 5.17b. Similar to the above case, charges will be induced on the surface of the spherical hole. Here, we will treat the material macroscopically outside the hole, while it is treated microscopically inside the hole. Thus, we have to take the contributions of the following four electric fields into account: (1) Externally applied field, (2) electric field due to the surface charges on the film, (3) that due to the surface charges on the spherical hole, and (4) that due to atoms or molecules within the spherical hole.

The first two are the same as those considered above. Since the surface charges on the spherical hole have an opposite sign as compared with those on the film, the induced electric field should have the same sign as the external field. The derivation of the magnitude of the electric field at the center of the hole is generally given in the textbook. Here, we will express it in terms of the ratio to the case of the thin film. The ratio L is usually called *depolarization coefficient*, which is generally dependent on the shape of hole and the direction of the electric field. We summarize the typical values of L in Table 5.1. In 3D system, a spherical hole is often employed to estimate the permittivity. In that case, the depolarization coefficient is put as $L = 1/3$. Hereafter, however, we will employ L as a parameter so that we can extend the discussion to more general cases having various shapes of holes.

On the other hand, the last term is concerned with the contribution from individual atoms (molecules) within the hole. Since the applied electric field will induce microscopic polarizations on the atoms (molecules), which are distributed homogeneously

Table 5.1 Depolarization coefficient L

Shape	Polarization[a]	L
sphere (3D)		1/3
cylinder (2D)	s	0
	p	1/2
film (1D)		1

[a]s and p indicate polarization directions of light, whose electric field vectors are parallel and perpendicular to the axis of cylinder.

within the hole, a net electric field evaluated at a center of the hole will then vanish on an average. Thus, the electric field at the center of the hole is expressed as

$$E_{loc} = E_0 - \frac{P}{\epsilon_0} + L\frac{P}{\epsilon_0}$$

$$= E + L\frac{P}{\epsilon_0}, \tag{5.41}$$

where we have used the relation $E = E_0 - P/\epsilon_0$.

The electric field, E_{loc}, called *local field*, is a microscopic electric field, to which every individual atom (molecule) within an interior of the film is susceptible. Thus, atoms (molecules) will be polarized under the influence of this local field. We denote the microscopic polarization thus generated as μ. If we define microscopic polarizability as α, then

$$\mu = \alpha E_{loc} = \alpha \left(E + L\frac{P}{\epsilon_0} \right)$$

holds naturally. Macroscopic polarization should be given by the sum of these microscopic polarizations. If we denote a number of atoms (molecules) within a unit volume as N, then the macroscopic polarization P is given as

$$P = N\alpha E_{loc} = N\alpha \left(E + L\frac{P}{\epsilon_0} \right).$$

Solving this with respect to P, we obtain

$$P = \frac{N\alpha E}{1 - L\frac{N\alpha}{\epsilon_0}}.$$

Further, defining the permittivity ϵ as $\epsilon_0 E + P \equiv \epsilon E$, the following relation is obtained:

$$\epsilon - \epsilon_0 = \frac{N\alpha}{1 - L\frac{N\alpha}{\epsilon_0}},$$

which leads to

$$(\epsilon - \epsilon_0) \left(1 - L\frac{N\alpha}{\epsilon_0} \right) = N\alpha,$$

and then

$$\epsilon - \epsilon_0 = \frac{N\alpha}{\epsilon_0} \{\epsilon_0 + (\epsilon - \epsilon_0)L\} = \frac{N\alpha}{\epsilon_0} \{\epsilon L + \epsilon_0(1 - L)\}.$$

From this relation, the following *Clausius-Mossotti relation* is derived:

$$\frac{\epsilon - \epsilon_0}{\epsilon + \frac{1-L}{L}\epsilon_0} = L\frac{N\alpha}{\epsilon_0}. \tag{5.42}$$

Consider a case that a material consists of a mixture of two kinds of atoms (molecules), a and b. If we denote their microscopic polarizations and polarizabilities as $\mu_{a,b}$ and $\alpha_{a,b}$, respectively, the macroscopic polarization is described as

$$\begin{aligned}P &= N_a\mu_a + N_b\mu_b\\ &= (N_a\alpha_a + N_b\alpha_b)E_{\text{loc}}\\ &= (N_a\alpha_a + N_b\alpha_b)\left(E + L\frac{P}{\epsilon_0}\right),\end{aligned}$$

where $N_{a,b}$ are numbers of a and b in a unit volume. Thus, we obtain

$$\frac{\epsilon - \epsilon_0}{\epsilon + \frac{1-L}{L}\epsilon_0} = L\frac{N_a\alpha_a + N_b\alpha_b}{\epsilon_0}. \tag{5.43}$$

If we consider that a material consists of either a or b and the permittivity in this case as $\epsilon_{a,b}$, the following relations will be obtained:

$$\frac{\epsilon_a - \epsilon_0}{\epsilon_a + \frac{1-L}{L}\epsilon_0} = L\frac{N\alpha_a}{\epsilon_0}, \qquad \frac{\epsilon_b - \epsilon_0}{\epsilon_b + \frac{1-L}{L}\epsilon_0} = L\frac{N\alpha_b}{\epsilon_0}.$$

Inserting these relations into Eq. (5.43), we obtain

$$\frac{\epsilon - \epsilon_0}{\epsilon + \frac{1-L}{L}\epsilon_0} = \frac{N_a}{N}\frac{\epsilon_a - \epsilon_0}{\epsilon_a + \frac{1-L}{L}\epsilon_0} + \frac{N_b}{N}\frac{\epsilon_b - \epsilon_0}{\epsilon_b + \frac{1-L}{L}\epsilon_0},$$

where we put $N \equiv N_a + N_b$. Further, by putting $f_{a,b} = N_{a,b}/N$, the following relation is obtained

$$\frac{\epsilon - \epsilon_0}{\epsilon + \frac{1-L}{L}\epsilon_0} = f_a\frac{\epsilon_a - \epsilon_0}{\epsilon_a + \frac{1-L}{L}\epsilon_0} + f_b\frac{\epsilon_b - \epsilon_0}{\epsilon_b + \frac{1-L}{L}\epsilon_0}. \tag{5.44}$$

Thus, if we consider a quantity on the left-hand side of the above relation, the additivity is proved to hold.

When a large amount of a is present, while an amount of b is small, it is sometimes convenient to consider a host medium is filled with a instead of considering vacuum as a host. Thus, replacing $\epsilon_0 \to \epsilon_a$ in Eq. (5.44), we obtain

$$\frac{\epsilon - \epsilon_a}{\epsilon + \frac{1-L}{L}\epsilon_a} = f_b\frac{\epsilon_b - \epsilon_a}{\epsilon_b + \frac{1-L}{L}\epsilon_a},$$

which is known as *Maxwell-Garnett model*. Further, by considering that a host medium is a mixture itself and replacing $\epsilon_0 \rightarrow \epsilon$ in Eq. (5.44), the *Bruggeman formula*

$$f_a \frac{\epsilon_a - \epsilon}{\epsilon_a + \frac{1-L}{L}\epsilon} + f_b \frac{\epsilon_b - \epsilon}{\epsilon_b + \frac{1-L}{L}\epsilon} = 0,$$

is derived.

These relations are anyway phenomenological so that their applicabilities should be always checked in actual systems before using them. Finally, we should emphasize that the above relations are only applicable that a material is actually regarded as completely homogeneous on a scale sufficiently smaller than the wavelength of light. Otherwise, the assumptions of the hypothetical sphere and additivity that we have made above cannot be justified.

5.2.3 Models for Average Permittivity and Long-Wavelength Approximation

Recently, the above models concerning the microscopic mixture of atoms (molecules) have been often extended to estimate the average permittivity or the average refractive index for a system containing much larger particles that are arranged regularly or irregularly, as shown in Fig. 5.18b. Although the above models are phenomenological in nature, the applicability of these models to this system is not simply justified as is easily seen from their derivations, because the above models originated essentially from the assumption of an optically homogeneous system (Fig. 5.18a). Thus, it is absolutely necessary to investigate the applicability and expandability of these phenomenological models before using them.

At first, let us qualitatively discuss the difference between a system consisting of microscopic mixture of atoms (molecules) and that containing much larger particles. It is clear that in the former, a spherical hole is anyway hypothetical and thus only a single hole is considered. Hence, it cannot be disturbed by any other hole around that hole. Furthermore, if we choose a certain point within the film, we can always consider a spherical hole hypothetically with that point at a center. Hence, the effect of induced microscopic polarizations created around that point will be always canceled owing to the symmetry property of the system.

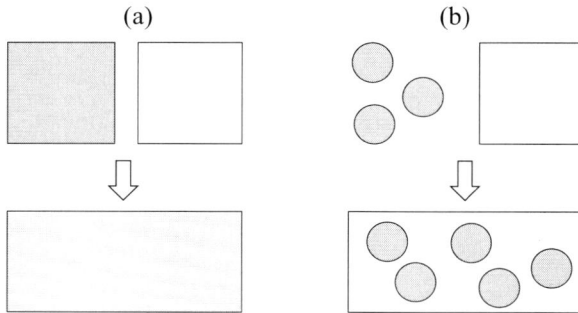

Figure 5.18 (a) Microscopic mixture of two kinds of atom (molecule) and (b) inclusion of much larger particles. The former is regarded as optically homogeneous, while the latter as optically inhomogeneous.

On the other hand, in the latter case, particles actually exist in the film. Then, larger the difference of the permittivity from that of the film becomes and/or more the number of particles increases, more remarkable the disturbance of the electric field around the particle becomes. If the density of particles increases, the distorted field around the particle will overlap with each other to make complicated field. Further, even if a spherical (or spheroidal) particle is considered so that the electric field within the particle becomes parallel to the external field, we should evaluate the electric field at each point within the particle, since the particle is not hypothetical but actually exists. Hence, the cancellation of the fields created by the other microscopic polarizations will not occur essentially without the center of the hole. The assumption made in a microscopic system is thus no longer applicable to this system and hence all the models considered above are, in principle, inapplicable. It is also expected that the above models will hold, in an approximate sense, only when spherical (or spheroidal) particles with small difference of permittivity from that of the medium are dispersed sparsely.

However, the above models have been often employed in actual systems containing larger particles or having complicated microstructures. Thus, we should have a correct understanding that these models are nothing but phenomenological ones so that their applicabilities should be carefully examined before using. Even then, it sometimes becomes a troubling problem which quantity is more appropriate, average permittivity or average refractive index. The long-wavelength approximation in the band calculation will give a hint to this problem.

Let us inspect the result of band calculation for a system containing large particles that are arranged regularly, and compare an average refractive index obtained under long-wavelength approximation with those obtained by phenomenological models. We will choose the Maxwell-Garnett and Bruggeman models in addition to the following expressions of average quantities:

$$n_{av}^{(1)} = \sum_j f_j n_j, \quad n_{av}^{(2)} = \frac{1}{\sum_j f_j (1/n_j)},$$

$$n_{av}^{(3)} = \sqrt{\sum_j f_j \epsilon_j}, \quad n_{av}^{(4)} = \frac{1}{\sqrt{\sum_j f_j (1/\epsilon_j)}}, \quad (5.45)$$

where f_j is a filling factor of the jth particle species, which corresponds to the volume fraction for 3D system and to the area fraction for 2D system. Thus, the upper two quantities are a refractive index and inverse refractive index averaged by a filling factor, while the lower two are a permittivity and inverse permittivity averaged by a filling factor.

Let us see a case of 2D photonic crystals, where cylinders or square prisms are arranged on square or hexagonal lattice points with the permittivity of 4, which are embedded in a medium with the permittivity of 1. The polarization of the incident light wave employed is *s* or *p*, which corresponds to an electric field parallel or perpendicular to the axis of cylinder or square prism. An example of the calculated results is shown in Fig. 5.15. It is soon understood that the dispersion curve is generally linear in the vicinity of the Γ point, which gradually bends to form a photonic band gap at a zone boundary of $k = \pi/a$. It is convenient to evaluate the gradient of the curve by a tangential line at the Γ point, which corresponds to the inverse of average refractive index through the relation $\{\omega a/(2\pi c)\}/\{ka/(2\pi)\} = 1/\langle n \rangle$.

The result calculated with changing filling factor is shown in Fig. 5.19, where the average permittivity calculated from $\langle n \rangle$ is plotted. In this figure, a) and c) correspond to the average permittivity for cylinders arranged in hexagonal and square lattices

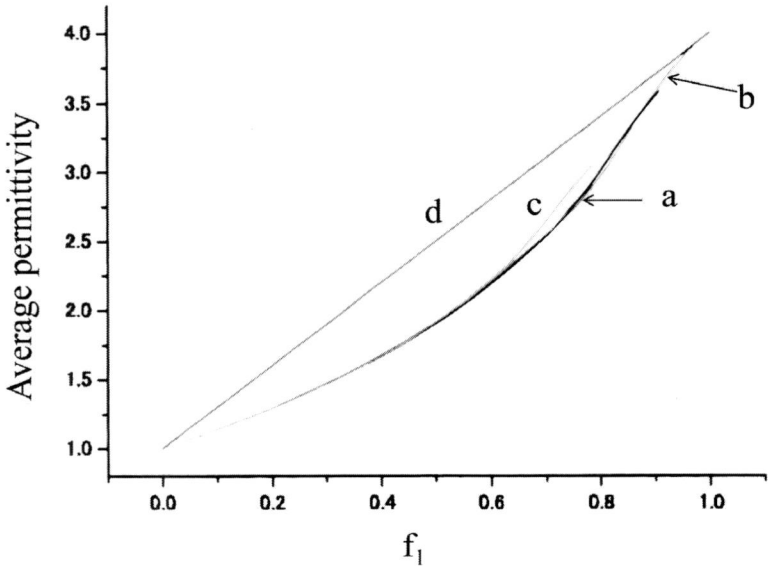

Figure 5.19 Average permittivities derived from long-wavelength approximation to band structures calculated for 2D photonic crystal. The average permittivity is evaluated from a gradient within a range of $0 < k < \pi/(500a)$, where k and a are the wave number of light and a lattice constant, respectively. The average permittivities thus calculated are plotted against a filling factor of cylinders or square prisms with the permittivity of 4, which are embedded in a medium with the permittivity of 1. a) Cylinders arranged in a hexagonal lattice (p-polarization), b) square prisms arranged in a square lattice (p-polarization), c) cylinders arranged in a square lattice (p-polarization), and d) cylinders or square prisms arranged in a square or hexagonal lattice under s-polarization. In a) and c), the maximum filling factors are determined by the relations of $(\pi a^2/8)/(\sqrt{3}a^2/4) \approx 0.9069$ and $\pi(a/2)^2/a^2 \approx 0.7853$, respectively. (Courtesy of Dr. Y. Fujimura.)

under p-polarization and b) that for square prisms arranged in a square lattice. d) is linear against the filling factor, which corresponds to cylinders or square prisms arranged in a square or hexagonal lattice under s-polarization. In c) of cylinders in a square lattice, it is found that the filling factor is limited within ~0.8, which is determined by the maximum areal fraction of $\pi(a/2)^2/a^2 = 0.7853$, where adjacent cylinders contact with each other.

First, we immediately notice that for *s*-polarization, the average permittivity is completely proportional to a filling factor. This is natural because if the electric field is directed parallel to the axis of cylinders or square prisms, the surface charge induced by the external field does not have, in principle, any effect on the electric field inside the cylinders or square prisms. Thus, the average permittivity is simply obtained by areal average of permittivity evaluated in the cross section. Second, for *p*-polarization, it is found that average permittivity is nonlinearly dependent on the areal fraction f_1, which is in contrast to a case of *s*-polarization. Further, it is also found that the dependence is rather independent of shape and lattice type particularly in a range of low areal fraction.

Next, let us compare the result with the phenomenological models of the Maxwell-Garnett and Bruggeman models in addition to various averaging procedures introduced in Eq. (5.45). In Fig. 5.20, we show the result for 2D photonic crystal, in which cylinders with permittivity of 2.25 (refractive index of 1.5) are embedded in a medium with the permittivity of 1.0 and are arranged in a square lattice. Compared with the average permittivity estimated from the long-wavelength approximation in the band calculation, simple mathematical averages of permittivity (a) and refractive index (b) clearly overestimate the averaging, while that derived from the mathematical averages of inverse permittivity (f) and refractive index (e) underestimate it. On the other hand, phenomenological models of the Maxwell-Garnett and Bruggeman give surprisingly good results. Therefore, although the agreements are not justified theoretically, these models actually become good expressions for the average permittivity even in a system containing large particles comparable to the wavelength of light.

It is generally believed that such averaging is valid only within a range where a particle size lies on an order of wavelength of light or less. We can investigate the effective range of averaging by inspecting the range of linearity in the dispersion curve. From the band calculation of photonic crystal, it is known that the dispersion curve shows a linear relation around the Γ point, which then bends near a zone boundary to form a photonic band gap. As far as the linearity holds, we should say that light regards a photonic crystal as a homogeneous medium having the average permittivity

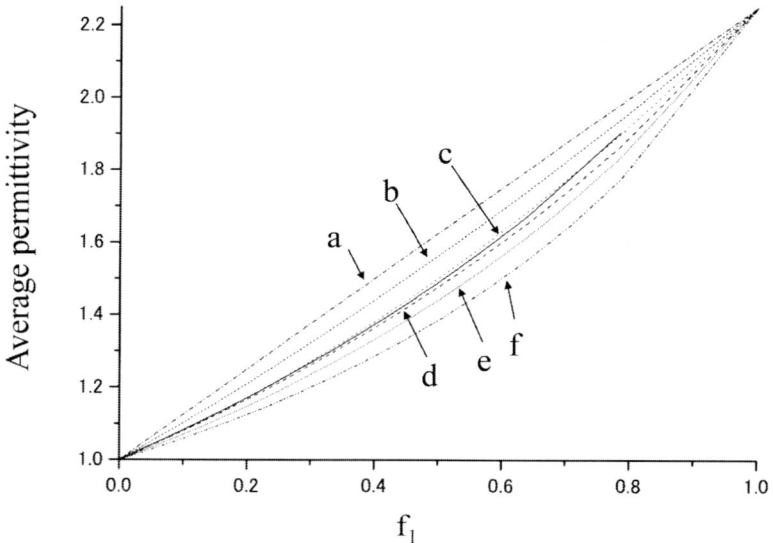

Figure 5.20 Comparison of average permittivities deduced from a band calculation under long-wavelength approximation and those obtained by phenomenological methods. Solid curve is obtained from a band calculation for 2D photonic crystal, where cylinders with the permittivity of 2.25 are arranged in a square lattice and are embedded in a medium with the permittivity of 1. The maximum filling factor in this case is determined by $\pi(a/2)^2/a^2 \approx 0.7853$, where a is a lattice constant. a)~f) are results of the following phenomenological models: a) average permittivity, b) average refractive index, c) Bruggeman model, d) Maxwell-Garnett model, e) average inverse refractive index, and f) average inverse permittivity. (Courtesy of Dr. Y. Fujimura.)

determined by its slope. In Fig. 5.21, we show the result of the above consideration for 2D crystal, in which cylinders with various permittivities are arranged in a square lattice. The vertical axis of this figure is a ratio of the lattice constant to the wavelength of light and several curves correspond to the difference of the permittivities. These curves show the upper limits, below which the averaging is effective.

For example, let us examine a curve numbered as 1. This curve corresponds to a system where cylinders with the permittivity of

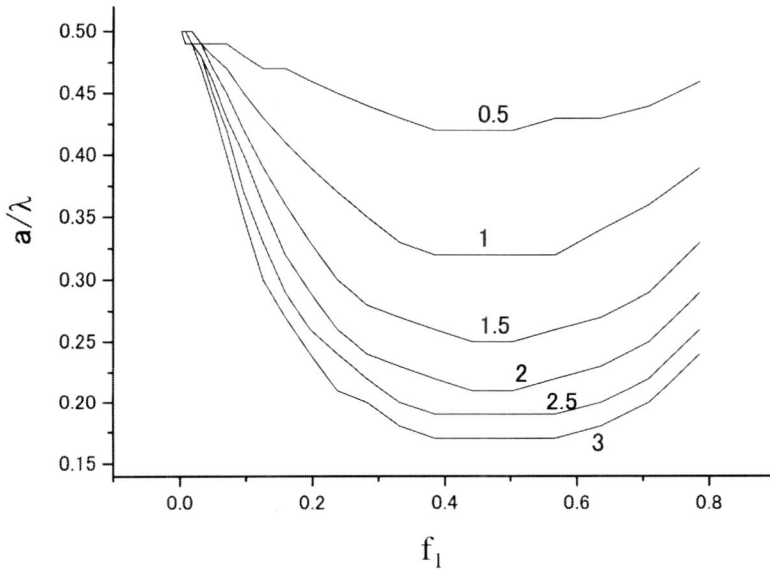

Figure 5.21 Effective range of average permittivity deduced from band calculations for 2D photonic crystals consisting of cylinders with various permittivities arranged in a square lattice. The average permittivity is effective below a solid line in each case. The number indicates the difference of permittivity of cylinder to that of an embedding medium. The effective range is estimated from a region of linear dispersion near a Γ point so as not to deviate from the linearity by 0.5%. (Courtesy of Dr. Y. Fujimura.)

2 are embedded in a medium with the permittivity of 1, and are arranged in a square lattice. The curve seems to be concave around a center and rises toward both ends. The minimum value lies at $a/\lambda \sim 0.32$ around $f_1 \sim 0.5$ with the lattice constant a. The upper limit of f_1 is again limited by the packing condition and takes a value of 0.7853. One may think that at $f_1 = 0$, the effective range should extend infinitely, because the system is completely homogeneous. In this figure, however, the range seems to be artificially restricted within a range of $a/\lambda = 0.5$. This is because we have evaluated the range of linearity within the first Brillouin zone of $0 < k < \pi/a$ and only a small amount of deviation from $f_1 = 0$ will cause a deviation from the linearity near a zone boundary. Thus, we can

actually evaluate the effective range only within a range of $0.7853 > f_1 > 0$ in this system.

From this figure, we can estimate a maximum diameter of cylinder, below which the concept of averaging is valid. By using the values of $a/\lambda = 0.32$ and $f_1 = 0.5$, the maximum diameter d_{max} is obtained as $d_{max} = 2a\sqrt{f_1/\pi} = 2 \cdot 0.32\lambda\sqrt{0.5/\pi} \sim 0.26\lambda$. This is the most strict condition for the permittivity difference of 1 and the restriction will be considerably relaxed as the filling factor deviates from 0.5.

Here, we will summarize the above results. The concept of the average permittivity or refractive index is very useful to analyze the optical properties of material containing particles of various shapes, whose sizes are comparable to or less than the wavelength of light. Though various phenomenological models have been employed so far, judging from the above evaluation using the long-wavelength approximation, the Maxwell-Garnett and Bruggeman models are found to give considerably good results, whereas the simple geometrical average of the refractive index or permittivity will overestimate the values. Deduced from the range of linearity in the dispersion curve, a limit for the applicability of this averaging procedure is also evaluated and is found to lie in a range considerably smaller than the wavelength of light, which is particularly severe when the filling factor lies in an intermediate range. The range of linearity is, however, largely affected by the formation of photonic band gap at a zone boundary. Thus, if a system does not have such a regular structure, the dispersion curve will be linear up to the zone boundary and the concept of averaging will be more extendable. The band calculations using a supercell method applied to randomly distributed cylinders clarified this expectation [Fujimura (2009)]. On the other hand, it was reported that the photonic band gap was also formed in amorphous photonic materials [Jin *et al.* (2001)], the extent of which was dependent on the range of short-range order [Yang *et al.* (2010)]. Thus, the relation between the expandability of the concept of averaging and the photonic band gap formation will become the next problem to be solved in future.

5.3 Photonic Crystals in the Natural World

5.3.1 *Colloidal Crystals and Opal*

5.3.1.1 Colloidal crystals

When particles of a size ranging from nm to µm are dispersed in fluid, they are called *colloid* or *colloidal particles*. Since an ensemble of colloidal particles takes various forms according to the interparticle interactions, they have attracted much attention for a long time and have been considered as one of the traditional research fields of chemistry called *colloid chemistry*. Colloidal particles of the same size are sometimes arranged simultaneously in solution to form a crystal in a macroscopic scale called *colloidal crystal*. In 1935, tobacco mosaic virus was found to form a crystalline shape in liquid, which was the first observation of colloidal crystal [Stanley (1935)]. In 1947, colloidal spheres of a uniform size were commercially available from Dow Chemical Co., which largely enhanced research on colloidal crystals.

Since a lattice constant of colloidal crystal is on the order of wavelength of light, light illumination on a colloidal crystal causes the Bragg reflection in a visible region, which shows iridescent color as shown in Fig. 5.22a. In general, when a colloidal particle is in solution, the functional groups existing on the surface of particle are dissociated to form surface charges. For example, in case of polystyrene sphere, sulfuric groups are present at the surface and will be almost completely dissociated into sulfuric ions having negative charges (see Fig. 5.23a). In case of silica sphere, silanol groups on the surface are partly dissociated to form weakly negative surface charges. Since even in solution that is sufficiently deionized, counter ions due to residual ions are inevitably present, some of these ions are firmly adsorbed on the surface of the sphere to form the *Stern layer*, while the other ions surround the sphere to screen the surface charges and form a *diffuse layer* (see Fig. 5.23b). The surface charges and the charges in these layers possess opposite signs with each other and constitute the layers called *electrical double layer*, while the screening due to the counter ions is called *Debye screening* and its effective distance is called *Debye's screening length*.

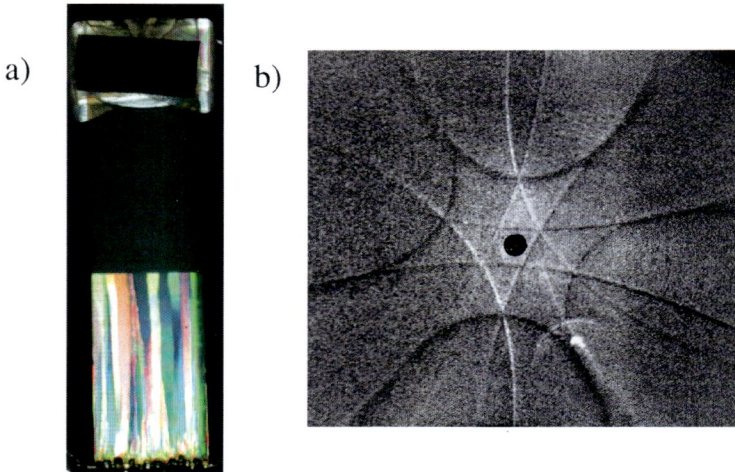

Figure 5.22 (a) Colloidal crystals grown in a columnar form and (b) Kossel lines. (Courtesy of Dr. H. Yamada.)

In general, when the concentration of residual ions becomes higher, the Debye's screening length becomes shorter because only a thin layer is sufficient to screen the surface charges. Thus, particles can come closer to each other so that they easily go into a working region of van der Waals' attractive force. Therefore, they tend to aggregate with each other. On the contrary, when the concentration becomes lower, the Debye's screening length becomes larger. Even when two such screened particles happen to come closer, a repulsive force will be naturally generated to exclude the overlap of their diffuse layers owing to the presence of osmotic pressure. In this sense, a colloidal particle is regarded as a particle wearing a diffuse layer of counter ions. Thus, the colloidal particles behave as if they can move independently, which should be called *liquid state*, as shown in Fig. 5.24.

When the concentration of residual ions becomes much lower, the apparent volume of a particle becomes extremely large and if it exceeds a certain limit, the particles will always touch with each other so that their free motions will be strongly restricted. Since the particles are originally affected by a random thermal force, under these conditions and under the presence of slight external force

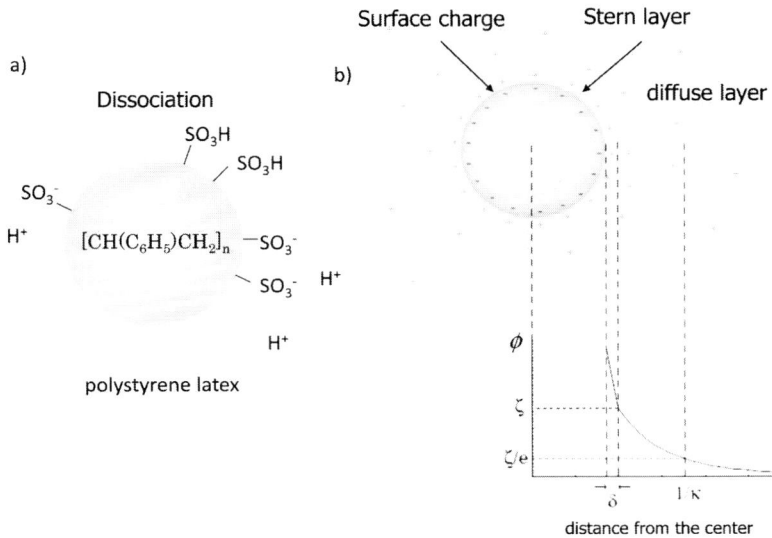

Figure 5.23 (a) Surface charges of a polystyrene latex, which are generated by dissociation of sulfuric groups. (b) Electrical double layer consisting of surface charges, a Stern layer, and a diffuse layer. Since counter ions involving in the latter two layers screen surface charges, an electric potential ϕ decreases with a distance from the center of particle. The electric potential measured at the surface of the Stern layer is called ζ-*potential*. The distance where the electric potential becomes $1/e$ of ζ-potential is called *Debye's screening length*, $1/\kappa$, which becomes a measure of the effective size of a particle. (Courtesy of Ms. R. Nakazawa.)

such as gravity and diffusion force due to concentration difference, they are arranged into a closest-packed structure of colloidal crystal. Thus, the physical quantities that determine the conditions of crystal formation will be the volume fraction of particles, the concentration of residual ions and the magnitude of surface charges, the latter two of which are related to the Debye's screening length.

Colloidal crystal is experimentally prepared as follows: (1) polystyrene or silica spheres of a uniform size are prepared and are dispersed in a solution. Then, this solution is dialyzed in ultrapure water to remove residual ions at least for several days. (2) The particles are then dispersed in ultrapure water after adding ion exchange resin and stored in a refrigerator. (3) Colloidal crystal is

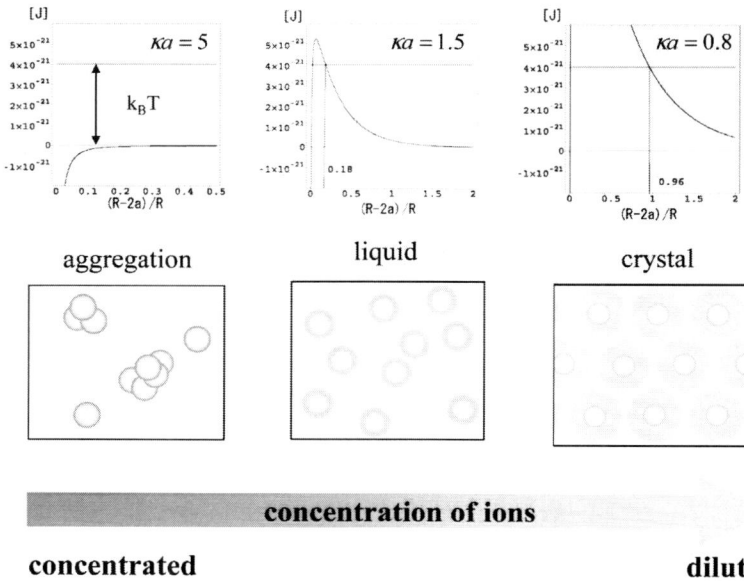

concentration of ions

concentrated   dilute

Figure 5.24 (Upper) Potential between two colloidal spherical particles in solution, and (lower) state of colloidal particles in solution. R is a distance between two colloidal spherical particles with a radius of a, and $1/\kappa$ is a Debye's screening length. Under high concentration of ions, colloidal particles tend to aggregate owing to van der Waals attraction, whereas under low concentration, they tend to form a colloidal crystal owing to increasing Debye's screening length. (Courtesy of Ms. R. Nakazawa.)

formed after pouring a stored sample into a solution containing a fixed concentration of residual ions. In order to grow large single crystals, ion exchange resins are placed at a bottom of a vessel to make a concentration gradient of residual ions.

The colloidal crystal thus grown usually takes a form of fcc structure and shows iridescent color under the illumination of light, as shown in Fig. 5.22a. In this figure, all the columnar crystals are single crystals grown from the bottom. Since their orientations are different with each other, they are observed as different colors. In ordinary atomic crystal, the orientation and structure of a crystal are investigated through the Bragg diffraction using X-ray as a light source. In colloidal crystal, we have to increase its scale by 10^3 times so that visible light is employed to investigate the orientation of the crystal.

The difference between atomic and colloidal crystals lies in a force to maintain their crystal structures. In atomic crystal, strong binding force works among neighboring atoms and maintains a crystal firmly, while in colloidal crystal, each particle floats in a solution and only weak electrostatic and chemical forces work between the particles. Thus, only a slight shake of the solution makes a crystal completely destroyed. On the other hand, within a colloidal crystal, the position of each particle fluctuates owing to Brownian motion so that under illumination of light, a clear halo is observed around a directional Bragg diffraction (see Fig. 6.14b). The halo directs in a wide angular range so that when its direction happens to coincide with the Bragg condition of the crystal, the Bragg diffraction again occurs, which deflects a part of the halo into a different direction. This phenomenon is easily observed if we project the reflected or transmitted light into a screen. As shown in Fig. 5.22b, the Bragg condition in this screen appears as several bright or dark lines, depending on the observation direction. These lines are called *Kossel lines*, and are used to analyze the crystal structure and orientation.

Although a lot of discussion have been made with regard to a fundamental question why a colloidal crystal is formed simultaneously, it is not easy to answer the question, because it is not yet completely known whether only a repulsive force will work between two like-charge particles in solution or not. The force operating between two like-charge particles was theoretically analyzed independently by Derjaguin and Landau [Derjaguin and Landau (1941)], and by Verwey and Overbeek [Verway and Overbeek (1948)]. These theories are now called *DLVO theory* and are used as a standard theory to analyze the interaction between colloidal particles. According to this theory, only a repulsive force should operate between two like-charge particles. However, under the presence of the other components, it is still not clear what happens actually. In fact, it was reported that an attractive force was actually present even between two like-charge particles when they were in the vicinity of a glass plate [Larsen and Grier (1997)].

Colloidal crystal is a system that is particularly attractive because the order formation process in a microscopic scale is directly observable under a microscope and can be quantitatively investigated in real time by various optical methods. In addition, it is easy to measure the fluctuation of each particle directly under a microscope, which promises the future development of this system to clarify a fundamental mechanism of self-organization process.

5.3.1.2 Opals

Natural opals A notable example of photonic crystals in the natural world is well-known opal. Opal has been prized as a jewel from ancient times and is now well known as a birthstone of October. Its variously colored appearance, which is peculiar to opal, is sometimes called *play of color*. Although many kinds of opal are known, those accompanying the play of color is especially called *precious opal* or *noble opal*, and are particularly prized. On the contrary, those not showing the play of color are called *common opal* or simply *potch*. Iridescent precious opal and common opal with beautiful ground color are frequently used as a jewel.

Opals are classified into several types according to their ground colors such as *white*, *black*, *water* and *fire opal*. In addition, they are distinguished according to thier formation processes: Those deposited in groundwater within sedimentary rocks are called *sandstone opal*, while those in hot water within volcanic rocks are called *mountain opal*. The main areas that produce opals are known as Australia and Mexico: The former mainly belongs to sandstone opal, while the latter belongs to mountain opal with red ground color.

Opal is a soft material composed of amorphous silica SiO_2 with high water content of 1–21%. The structure of opal was first discovered by Sanders *et al.* using an electron microscope [Jones *et al.* (1964); Sanders (1964)]. In their paper, two types of opal were investigated: one was a crystalline cristobalite-like type, which was a high-temperature polymorph of silica, and the other was an amorphous type. It was reported that the latter was composed of spheres of 150–350 nm diameter, which were packed in closest packing of fcc structure, while the former did not show any particular structure. They insisted that the unique color of opal was explained in term of 3D diffraction grating in addition to the refraction effect at the surface.

They estimated the maximum and minimum wavelengths, λ_{max} and λ_{min}, to cause the reflection of light by this microstructure: The maximum wavelength was derived by assuming that λ_{max} was obtained under normal incidence to a lattice spacing corresponding to {111} plane and was given as $\lambda_{max} = 2na_0/\sqrt{3}$, where a_0 and n were a lattice constant of fcc structure and an average refractive index, respectively.

On the other hand, the minimum wavelength was derived by considering that light was incident obliquely to the same plane and was refracted at a surface. The maximum angle of reflection, θ_{max}, measured inside the opal was obtained by the Snell's law of $n \sin \theta_{max} = 1$. The minimum wavelength was then derived as $\lambda_{min} = \lambda_{max}\sqrt{1 - 1/n^2}$, where the Bragg condition of $\lambda_{max} \cos\theta = \lambda$ was used after inserting $\theta = \theta_{max}$. They compared the estimated wavelengths with the observed ones and obtained good agreement.

Darragh *et al.* investigated the spheres more in detail using en electron microscope and found that each sphere had fine structure [Darragh *et al.* (1964)]. Within a sphere of 170–350 nm diameter, a central nucleus was present, which was surrounded by fine spheres of 30–40 nm diameter, arranging in a way of double or triple shell. The number of the shells depended on the color of opal: that of green to violet colors showed single or double shell, while that of red color did triple one. From these observations, they speculated that the production process of opal started with the aggregation of sol-like solution through the evaporation and condensation so that very fine spheres would be generated one after another. After these spheres grew up to 30–40 nm diameter, they aggregated again to form a larger particle. The particles thus produced gradually formed closest-packing structure during the slow sedimentation, while the fine spheres within a large particle gradually took a regular shell structure.

According to the above findings, it is considered that the play of color is nothing but interference effect due to regular arrangement of large spheres constituting an opal, which is also related to the irregular structure inherent to this type of photonic crystal. On the other hand, fine spheres involving in a large sphere will give so-called opalescence, which may be deeply connected with diffusion halo observed in shell and pearl (see Sec. 4.6.2).

Synthetic Opals   After the structure and coloration mechanism of natural opal were clarified through the 1960s into the 1970s, the motivations to fabricate synthetic opals had grown rapidly. In just the same year when Sanders published the first paper on the structure of opal, a group of CSIRO (The Commonwealth Scientific and Industrial Research Organisation, Australia) applied a patent on the fabrication of synthetic opal to Australia, USA and Great Britain, which were accepted within the years of 1968-1971.

From this patent, we can investigate the fabrication method of synthetic opal that they first proposed: They described that the fabrication process of synthetic opal was divided into three: (1) Preparing a suspension of spherical amorphous silica particles of a uniform size with the diameter of 150–450 μm, (2) packing the particles into an ordered close-packed array by sedimentation, and (3) stabilizing the spatial arrangement of the particles constituting the array.

The first process was carried out by deionizing a sodium silicate solution with ion-exchange resin for 30–300 hours at 100°C, which promoted the secondary colloidal particles by the aggregation of the primary ones. The second process was to select the particles of an appropriate size by drawing a part of sedimentation by means of pipetting or by repeating centrifuging treatment. These particles were then re-sedimented for another week or more, which resulted in a colored ordered array in a form of mechanically soft cake. The last process was to stabilize the array. This was performed by drying for weeks at 100°C to decrease the water content and to increase the extent of inter-particle welding, or by adding additional fresh silica sol before heating.

However, the earliest manufactured synthetic opal was white with almost no play-of-color and mechanically too soft and porous. Microscopically, the particles were not really spherical and not tightly packed. Thus, it was not actually merchandised. Stöber *et al.* reported a method to fabricate spherical silica particles of a uniform size under controlled growth [Stöber *et al.* (1968)]. They performed the hydrolysis of alkyl silicates to form the condensate of silicic acid in alcoholic solution under the presence of ammonia as a catalyst, which specifically promoted the formation of spherical particles. As a natural consequence, this finding greatly enhanced the commercialization of synthetic opal.

In gemology, man-made simulants of gemstones are classified into two: synthetic ones that are exact chemical equivalents of the natural material, and imitations that are simply look-alikes or chemically distinct from the natural material. The latter is further divided into two: those that have a very similar chemical composition and those that have an entirely different one. Since in case of synthetic opal, the following two modifications were anyway made during the manufacturing process, reducing or completely removing water and using stabilizing chemicals, it has been always a subject of fierce argument whether it is synthetic one or imitation. In general, man-made opals are usually classified into synthetic one when it is made mostly of silica, and imitation when it contains polymer impregnated into a silica array or is made entirely of polymer or glass.

In 1972, black and white Gilson synthetic opals were man-ufactured at the Gilson Laboratories and came into the market. This was further improved and various types of opal were actually merchandised. Although their detailed fabrication methods were unknown, this type of synthetic opal had a characteristic "lizard skin pattern" on the surface that were absent in the natural opal. In addition, the spheres did not show a trace of concentric growth and it revealed that Gilson synthetic black opal assumed the addition of small spheres of ZrO_2, the deliberate depletion of water and the presence of organic material.

In 1987, Kyocera applied a patent on synthetic opal, in which zirconium alkoxide $Zr(OR)_4$ with R alkyl group in solution was impregnated in the silica array so that it caused the formation of oxide or hydroxide with the amount of 0.005–8% when it came in contact with water in the pores. The resultant structure was then calcined at 1000–1300°C for 20–36 hours to obtain the final product. They described that the inclusion of ZrO_2 was important to impart the play of color equivalent to the natural opal and to impart the mechanical strength, weatherability, heat resistance, and chemical resistance.

Thus, synthetic opals do not possess exactly the same chemical compositions as the natural product. However, this is quite natural because the natural products are generally created under various chemical and physical operations of many year's duration, whereas synthetic ones are within several weeks or months. However, the long story of synthetic opal actually became beneficial to open up a new research field after the 1990s.

Opal as Photonic Material The era of photonics began with the papers of Yablonovitch [Yablonovitch (1987)] and John [John (1987)]. In these papers, it was predicted that within a material consisting of regular arrangement of dielectric microstructures, there are special energy ranges where light is not permitted to exist. This is just the same case as an electron in a semiconductor or insulator material, and hence such energy ranges for light are called *photonic band gap* in contrast to energy band gap for electron. In case of semiconductor that has an appropriate width of the energy band gap, man adds impurities to modify the band gap, which makes it possible to control electrons to propagate. In the similar manner, if the photonic band gap actually exists, it will be possible to control light.

In case a material shows a photonic band gap only when light is incident from particular directions, such a band is called *stop band*. For example, a multilayer shows a band gap only when the propagation of incident light has a component perpendicular to the layer, whereas it does not when light propagates along the layer. Therefore, in order to control light completely, it is necessary to search for materials that show a photonic band gap for any propagation direction. Thus, to fabricate a 3D photonic crystal is inevitable to attain complete photonic band gap. In this sense, opal is an ideal 3D material in an optical region.

At present, 3D or pseudo 3D photonic crystals are fabricated as in the following ways: One is a method of lithography, in which electron beam lithography or laser holography is often employed [Lin *et al.* (1998); Noda *et al.* (2000); Miklyaev *et al.* (2003)]. By using this method, various shapes of microstructure are fabricated quite accurately. However, this method is essentially based on 2D patterning so that the thickness of the structure is usually limited within several layers. It is also very time-consuming so that it is not convenient to fabricate a photonic crystal with a large area.

The other method is based on self-assembly of microstructures. Synthetic opal is involved in this category. According to the conventional method of fabricating synthetic opal, slow sedimentation and/or centrifugation of colloidal solution have been often employed as a typical fabrication method [Philipse (1989)]. Materials employed for this process are, however, limited in a case where uniform spherical particles are easily synthesized such as silica and polystyrene. After slow sedimentation, a crystalline structure thus sedimented is dried to form a photonic crystal. In case of silica, the crystal is further calcined to stabilize. Owing to this stabilization, the spheres are firmly fixed by forming a kind of bond between adjacent spheres. This method is generally easy to fabricate a 3D photonic crystal without using expensive instruments, whereas the control of crystal formation is usually difficult and the resulting crystal contains a considerable amount of disorders and domains.

Another one that is categorized into the self-assembly method is to fabricate an opaline film using a capillary force. This was first proposed by Denkov *et al.* to fabricate 2D crystallization [Denkov *et al.* (1993)]. When a small amount of colloidal solution are placed on a glass plate, particles involving in this solution will be condensed and form a crystalline arrangement during the solvent evaporation. Using this principle, various methods to fabricate an opaline film have been proposed [Jiang (1999); Meng (2000); Waterhouse and Waterland (2007)]. In this case, a substrate is usually placed vertically to the colloidal solution, which is sometimes called *vertical deposition*, and is left untouched or slowly drawn up. The slow solvent evaporation makes colloidal particles to be successively arranged on the substrate. The opaline film essentially consists of a multilayer of fcc form with the (111) plane faced on the substrate. The thickness of the multilayer depends on the concentration and diameter of colloidal particles, and several layers to several tens of layers are generally formed in a controlled way. It is possible to increase the thickness by repeating this process after drying the film. The opaline film thus fabricated is very uniform. It was also reported that an opaline film could be fabricated by applying shear flow to a colloidal solution sandwiched by two glass plates [Amos *et al.* (2000)].

The reflection and transmission spectral measurements were performed with these 3D crystals or opaline films as a sample, and the presence of a stop band was clarified, which gradually shifted with the change of the angle of incidence. However, since the colloidal particles of silica or polystyrene has an refractive index of ~1.5, the small refractive index contrast, defined as $n_{particle}/n_{void}$, makes it impossible to open a complete photonic band gap, where $n_{particle}$ and n_{void} are the refractive indices of particle and a medium filling voids among colloidal particles. In fact, the theoretical estimations showed that complete photonic band gap would be opened in case where the refractive index contrast was above 2.8 [Biswas *et al.* (1998); Busch and John (1998)].

On the other hand, opal can be regarded as porous structure, if one pays attention to voids created by arranged spherical particles. In this sense, opaline structure has attracted considerable attention as a support of catalyst and chromatographic separation. It is also considered to be useful for light-weight structural material with thermal, acoustic and electric insulator. As for the photonics, its porous structure has been attracting particular attention to modify the optical properties of opal [Yoshino *et al.* (1997)]. It was reported that infilling the voids by liquid crystal and applying the electric field to it, made it possible to change the refractive index of medium and consequently to tune the position of the stop band [Yoshino *et al.* (1997); Leonard *et al.* (2000); Kang *et al.* (2001)].

The new method to attain the high refractive index contrast was reported one after another in the late 1990s. In this method, voids are replaced by a hard material by making spherical particles as a template and then the templates are removed by chemical or thermal treatment [Velev (1997); Imhof and Pine (1997)]. The fabricated material consists of spherical air holes arranged in a crystalline form instead of spherical particles at lattice points, which is called *inverse opal*. So far various materials are reported to be introduced as a material of inverse opal through reactions in solution, or chemical vapor deposition, which include Si, SiO_2, TiO_2, CeO_2, ZrO_2, Al_2O_3, ZnS:Mn, CdSe, CdS, polymer, graphite and so on. Using this technique, various inverse opals with the refractive index contrast, which should be expressed as $n_{framework}/n_{void}$ in this case, larger than 2.8 was fabricated and the photonic band gap was investigated. Actually, Blanco *et al.* measured the reflection spectrum of Si inverse opal, which had a refractive index contrast of 3.5 [Blanco *et al.* (2000)].

Owing to the presence of photonic band gap, it is natural that light does not have any mode within this energy range so that fluorescent material cannot emit light within this energy range. This was first predicted by Yablonovitch, which was to cause a variation of spectral shape and to elongate the lifetime of the excited state [Yablonovitch (1987)]. The spectral variation was expressed by two processes, that is, the inhibition in the particular wavelength range and the redistribution to the other region, and the elongation of the lifetime, which appeared in the fluorescence decay measurement. These predictions were later confirmed in materials including fluorescent molecules in colloidal and opaline crystals, and semiconductor material in opaline crystals [Bogomolov *et al.* (1997); Romanov *et al.* (1997); Blanco *et al.* (1998); Yoshino *et al.* (1998)].

Further, various experiments have been performed concerning the photonic crystal using opaline crystal. For example, in the vicinity of photonic band gap, the dispersion curve will be largely changed, which will cause a large change in group velocity dispersion of light in the photonic crystal. This was confirmed by the experiment on light propagation experiment using a pulsed light, or by the measurement on the phase change using a stationary laser light [Vlasov *et al.* (1999); Imhof *et al.* (1999)]. On the other hand, under the X-ray illumination on atomic crystal, only the first-order Bragg reflection is observed. In contrast, owing the strong interaction between light and the crystal, spots due to the higher-order Bragg reflection are observed in photonic crystal [van Driel and Vos (2000)]. It is also predicted that superprism effect[a] can be observed in opaline crystal [Ochiai and Sánchez-Dehesa (2001)]. Thus, photonics based on opal is steadily progressing as a key material of leading-edge technology.

---

[a]An incident light beam into a photonic crystal shows an extraordinarily large angular dispersion

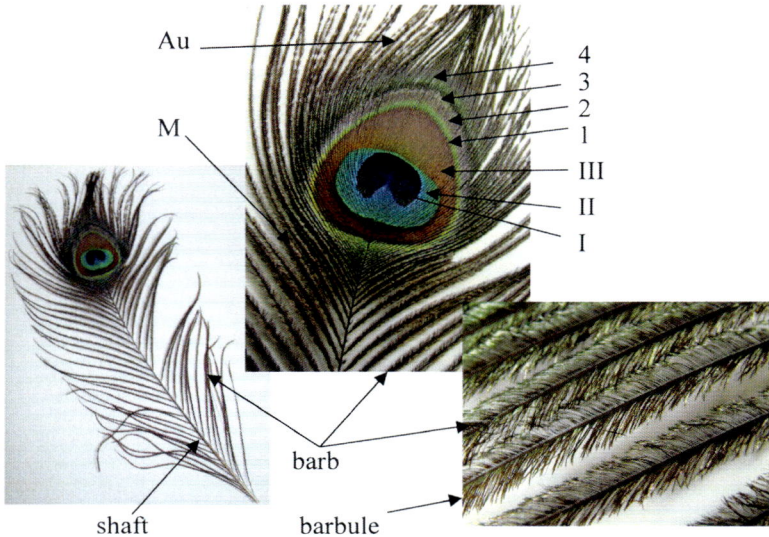

Figure 5.25 Feather of peacock.

5.3.2 *Photonic Crystals in Birds*

5.3.2.1 Structures of bird feather

Avian species are a vast treasure trove of photonic crystals that contribute to their structural colorations. In fact, various colorful birds are known to live everywhere, particularly in the tropics. As shown in Fig. 5.25, a feather of bird consists of many barbs sticking out from a main shaft and each barb has a lot of barbules. Most of the structural colors in birds originate from their barbs and barbules. Peacock, peasant, duck, hummingbird, trogon, paradise bird, and sunbird are the examples of the former, whereas the colors of kingfisher and parakeet have their origins in the barbs.

Structurally colored barbules in avian species are normally composed of well-ordered melanin granules. These granules originally add black, gray and brown colors to feathers, but play an important role to display structural colors by arranging themselves into regular structure and also to enhance the colors thus generated as a background absorber. On the other hand, structural colors in barbs are somewhat mysterious, because the barbs are filled with random network of sticks or air bubbles, which do not seem to contribute to light interference at first glance. In fact, this type of structure shows almost no or only weak dependence of viewing-angle-dependent color change and is called *noniridescence*. However, recent scientific approach has been gradually elucidating their mechanisms, which will be described in Sec. 6.5.2.

Durrer reported systematic investigations on iridescent barbules and classified the shapes and arrangements of melanin granules into several types [Durrer (1977, 1986)]. At first, he classified the shapes of melanin granules into 5 types, as shown in Fig. 5.26: (1) S-type (rod-shaped granule), (2) St-type (thin rod-shaped granule), (3) P-type (flattened stick), (4) R-type (hollow tube), and (5) K-type (platelet like a humming bird). Next, the arrangements of melanin granules were categorized into 5 types: (1) E-type (monolayer), (2) K-type (close packing), (3) S-type (multilayer), (4) G-type (lattice type) and (5) O-type (surface layer), as shown in Fig. 5.27. His elaborate work on totally 106 iridescent avian species revealed that the microstructures of these species were classified into 19 categories.

For example, peacock was categorized into StG-type (see Fig. 5.26b). Most of pheasants and ducks belonged to StK-type, while

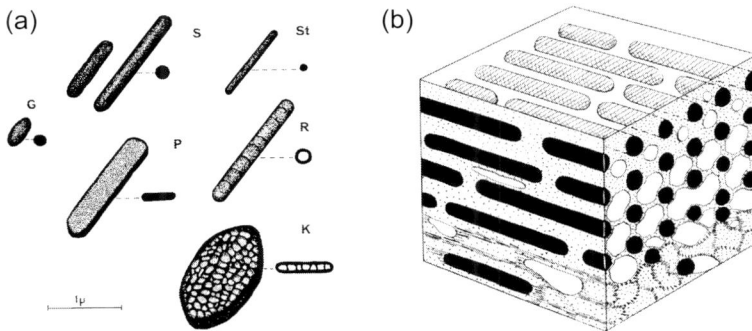

Figure 5.26 (a) Types of melanin granule used in the classification by Durrer. S: rod-shaped granule, St: thin rod-shaped granule, P: flattened stick, R: hollow tube, K: hollow platelet, G: basic form. (b) Schematic illustration of StG type. (Reproduced from [Durrer (1977)] with permission.)

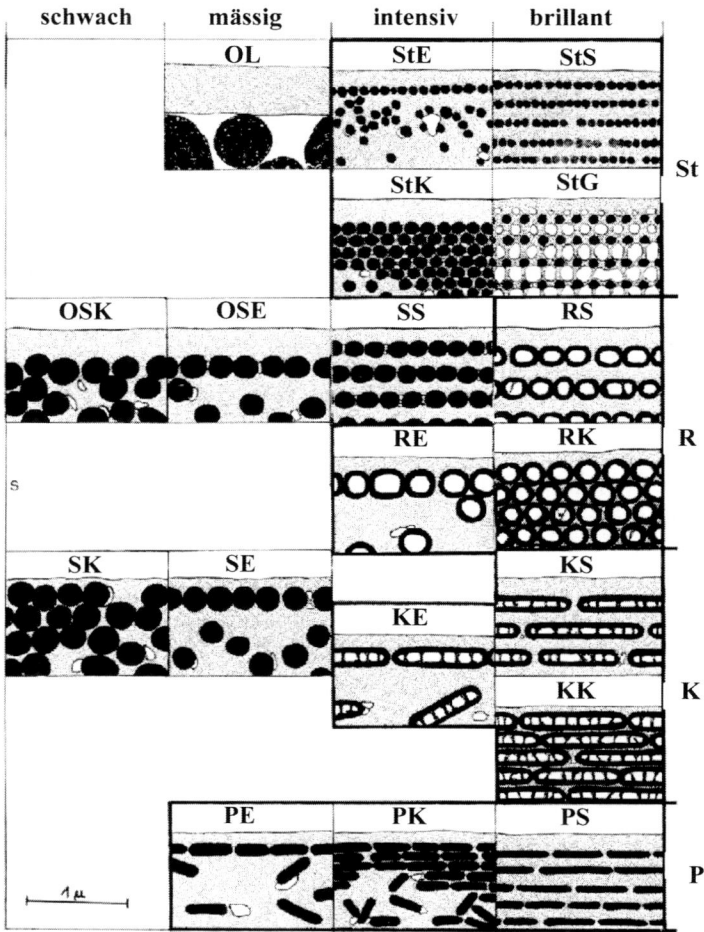

Figure 5.27 Type and arrangement of melanin granules in feather barbules. (Reproduced from [Durrer (1986)] with permission, where the names of the melanin types follow those employed in [Durrer (1977)].)

trogons to KS or RS-type, hummingbirds to KK-type, pigeons to StS-type, paradise birds to StS-type, and sunbirds to PS-type. Thus, at the present stage, we believe that the shapes and arrangements of melanin granules in barbules have been well systematized. However, their physical interpretations have not eventually proceeded since

Figure 5.28 (a,b) Feather barbule of peacock, and (c,d) its cross section and inside [Yoshioka and Kinoshita (2002, 2005)].

the 1960s, when the optical properties were mainly evaluated on the basis of a thin-layer interference theory.

5.3.2.2 Microstructures hidden in the feather of peacock

We will first show the barbules of peacock, one of the most famous avian species assuming structural colors. The color of peacock's feather has attracted a great deal of scientific attention for more than 300 years and is now known as one of typical examples of biological photonic crystals. Since the early observations by Hooke [Hooke (1665)] and Newton [Newton (1704)], it was in the 20th century that the detailed observation was first carried out by Mason [Mason (1923b)] and then by Durrer [Durrer (1962)].

As shown in Fig. 5.28a, a feather barbule of peacock is somewhat curved along its long axis and slightly twisted from the root. Each barbule has a shape of connected segments of a typical size of 20–30 μm, the surfaces of which are smoothly curved like a saddle. The sophisticated color-producing structure in the peacock's feather was first observed by Durrer using an electron microscope. He reported that below a cortex surface layer, totally 3–11 layers of melanin rods were arranged to form a 2D quasi-square lattice (Figs. 5.28c).

At a center of each square lattice, an air hole was present with increasing size toward the lower layers (see StG type in Fig. 5.26b). The medullary region of the barbule was filled with keratin filament with a small number of melanin rods and air holes. He found that the rod separation perpendicular to the surface correlated with the apparent color of the feather, while that parallel to the surface did not show such a correlation.

We have also observed a barbule using a SEM. The cross section of the barbule is crescent-shaped (Figs. 5.28b), and under high magnification, 8–12 layers consisting of periodically arrayed particles are observed beneath the surface layer (Fig. 5.28c), whose diameters range from 110 to 130 nm. The layer intervals are actually dependent on the color of the feather: for example, 140–150, 150, and 165–190 nm for blue, green, and yellow feathers, respectively. It is also noticed that the crescent-shaped barbule makes the lattice structure curved along the surface. In contrast to the transverse section, particles in the longitudinal direction have a long shape with the length of 0.7 µm and are rather randomly distributed (Fig. 5.28d).

Zi *et al.* reported the photonic band calculations on this complicated structure [Zi *et al.* (2003)]. They employed a barbule of green peafowl, *P. muticus*, and calculated a photonic band structure using the refractive indices of keratin and melanin of 1.54 and 2.0, respectively. They found a partial band gap along $\Gamma - X$ direction for two polarization directions. They also measured the reflection spectra for blue, green, yellow and brown barbules, and found sharp peaks located at 440 and 530 nm for the former two, respectively, while rather broad peaks with side peaks at a blue region for the latter two, indicating that the yellow and brown colors were non-spectral colors. They calculated the reflectance spectra for a finite number of lattice planes and found fairly good agreement with the experiment. Further, they paid special attention to a brown barbule and found even brown color was of a structural origin [Li *et al.* (2005)].

On the other hand, we measured the angular dependence of the reflection spectrum with changing size of the illuminated area [Yoshioka and Kinoshita (2002, 2005)]. When several barbs were illuminated, the reflection spectra were found to be quite smooth

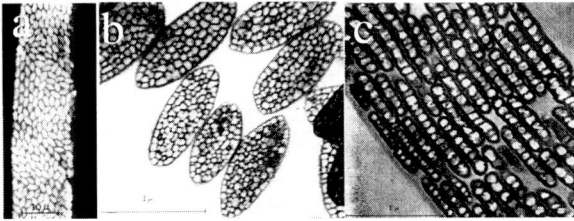

Figure 5.29 Platelets covering over a feather barbule of hummingbird, (a) *Heliangelus viola*, (b) the section containing a platelet plane, and (c) the cross section of a barbule of *Chrysolampis mosquitus*. (Reproduced from [Greenewalt *et al.* (1960a)] with permission.)

and tended to shift to blue when the angle of observation was changed, while the reflection spectra were rather irregular and the angle-dependent spectral shift was less conspicuous when a single barbule was illuminated. In order to investigate a relation between microscopic structure and actual appearance, we introduced the variation of lattice orientation by assuming a Gaussian distribution for the tilt angle of the lattice. The calculated results seemed to be in fairly good agreement with the experiment.

The tilt-angle distribution introduced here reflects partly a crescent-shaped cross section of a barbule, and partly a curved and twisted form of a barbule. Thus, macroscopic appearance of feather can be explained through statistical average of microscopic structures that are inevitably affected by some sort of irregularity. In general, it is absolutely important to consider both microscopic and macroscopic structures to truly reproduce the actual appearance.

5.3.2.3 Microstructures in feather of hummingbirds

Sophisticated structures were observed in a barbule of bird feather under an electron microscope by Greenewalt *et al.*, who found the presence of platelet mosaics on barbules (see Fig. 5.29a) in nearly 50 species of hummingbirds [Greenewalt *et al.* (1960a,b); Greenewalt (1960)]. They found that 7–15 layers of discrete platelets existed in a barbule in parallel with the surface. The platelet had an appearance of *an elliptical pancake made from dough to which baking soda had been liberally added* (see Fig. 5.29b) and the thickness ranged from

100 to 220 nm with the average of 150 nm. The inside of the platelet was not uniform and consisted of air voids sandwiched by melanin layers (Fig. 5.29c). The air voids were further partitioned into small chambers by structural reinforcements that were thick near the upper and lower ends, while they were thin in the middle. Viewed from the feather surface, air voids in a platelet formed irregular foam-like pattern (see Fig. 5.29a).

They speculated that a platelet constituted a half-wave interference plate by changing thickness of air voids to accommodate the specific color. In order to explain a characteristic narrow width observed in the reflection spectrum, they analyzed its structure as follows: a thin uniform keratin layer was located in the outer surface with a discontinuous rise of the refractive index at the interface between keratin and melanin layer, then the refractive index smoothly changed with decreasing toward the center of the platelet and then increasing again. Thus, periodic modulation in the refractive index occurred. The reason of this continuous change in the refractive index owed mostly to the shape of the melanin structural reinforcements described above. They solved the Riccati differential equation and attained good agreement with the experiment.

In spite of the first success in explaining the narrow reflection band using a model of continuously changing refractive index, however, no effort has been made to further explain the optical properties quantitatively. The angular dependence characterized by a relatively small color change with varying viewing angle, has not been clarified yet, although the shape of the platelet clearly contributes to the anisotropic diffraction effect and the melanin layer definitely plays a role of absorption contributing both to the spectral narrowing and background reduction. Thus, it is strongly expected to construct a new model to explain these properties in a unified manner.

5.3.3 *Photonic Crystals in Butterflies*

Natural photonic crystals are found in a more complete form in butterflies. First, we select a small lycaenid butterfly, *Callophrys rubi*, as an example. This butterfly is distributed widespread in Europe to

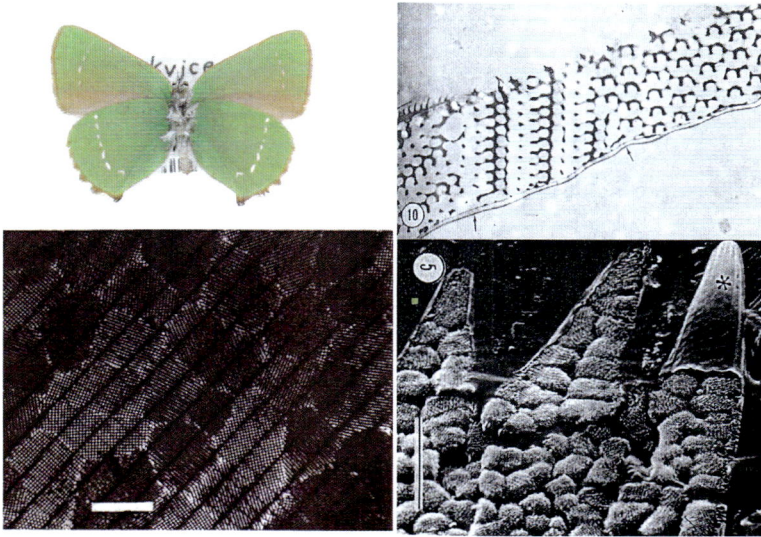

Figure 5.30 TEM (lower left) of green iridescent scale from *Callophrys rubi* showing many domains of regular structure (scale bar 5 μm), reproduced from [Morris (1975)] with permission. Longitudinal section of the iridescent cell (upper right) of *C. rubi* (upper right), showing photonic crystal with three domains, reproduced from [Ghiradella and Radican (1976)] with permission. SEM view (lower right) of the interior of a tip of cover scale of *Mitoura grynea*, showing many small domains of lattice structures (scale bar 10 μm), reproduced from [Ghiradella (1989)] with permission. (Specimen of the butterfly is in the possession of The Nature and Human Activities, Hyogo, Japan, and the photograph was taken by Dr. M. Kambe.)

Asia and North Africa, which was first received attention by Onslow [Onslow (1920, 1923)]. He noticed that a scale of this butterfly was covered with irregular polygonal dark patches when observed under transmitted illumination, which was later called *mosaic scale* by Schmidt [Schmidt (1943)]. Afterward, the microstructure of this mosaic scale was investigated through an electron microscope by Morris [Morris (1975)]. He observed that each polygonal grain, seen as a mosaic patch, was 5.4 μm in mean diameter and each consisted of a simple cubic network with hollow spheres at lattice points and the lattice constant of 0.257 μm.

Ghiradella and Radigan also made an electron microscopic observation of the scale and found hollow spheres were arranged in a closest-packed cubic structure (fcc) [Ghiradella and Radican (1976)]. Thus, the elaborated structure in the interior of a scale in *C. rubi* is now well known as a typical example of biological photonic crystal and mosaic patches observed about 100 years ago correspond to domain structure having different lattice orientations. The similar photonic structures have been reported in several other lycaenid butterflies.

Ghiradella also devoted to the developmental studies on a similar lycaenid butterfly, *Mitoura grynea*, and made electron microscopic observations during its pupal period [Ghiradella (1989, 1991)]. After the eighth day after pupation, the combination of membrane with cuticle formed a unit, which enclosed the network structure due to smooth endoplasmic reticulum as a nucleus, and formed quite regular arrangement of a closest-packed lattice structure. The enclosing membrane-cuticle units were connected with each other and also with the outer space. It was reported that the formation process of the lattices continued for 2 days and the structure became firmly fixed at that position as the cell died back and disappeared.

Similar photonic structure was also found in emerald-patched cattleheart, *Parides sesostris*, inhabiting in Central to South America, which showed a strong green patch in the fore wing. Ghiradella reported that the scale of this butterfly showed spectacularly regular lattice [Ghiradella (1984, 1991, 1994, 1999, 2005)]. The structure was composed of crystallites with a face-centered cubic lattice of granules (see Fig. 5.31), which would contribute to producing the green diffraction color. It was also noticed that very peculiar ridges were standing regularly on the lattice. Vukusic reported from the minute observation that the 3D lattices were divided into irregular but distinct domains of a few-micron size, which made it possible to offer a constant-color effect by spatial averaging [Vukusic and Sambles (2003)].

Recently, it has been reported that these photonic structures can be expressed by a gyroid structure, which is observable for block copolymer systems as a thermodynamically stable bicontinuous structure [Michielsen and Stavenga (2008); Saranathan *et al.*

Figure 5.31 SEM of a fractured green scale of *Parides sesostris* (left). Scale bar is 1 μm. (Reproduced from [Ghiradella (2005)] with permission. Specimen of the butterfly is in the possession of The Nature and Human Activities, Hyogo, Japan, and the photograph was taken by Dr. M. Kambe.)

(2010)]. A gyroid is an infinitely connected 3D curved surface expressed by a relation

$$g(x, y, z) = \sin x \cos y + \sin y \cos z + \cos x \sin z = t, \qquad (5.46)$$

which divides a 3D space into two according to a condition of $g(x, y, z) > t$ or $g(x, y, z) < t$ with t as a parameter. If a space defined by the latter condition corresponds to a cuticle network, while that by the former does to a remaining air space, the photonic crystal structures found in scales of various kinds of butterflies can be expressed fairly well by this formula. t in this case corresponds to a filling factor of cuticle network and it is reported that for *C. rubi*, t ranges for $-1.4 < t < -1.0$, corresponding to the filling factor of 0.16–0.26.

This proposal has been experimentally confirmed by the comparison of calculated cross sections with TEM or SEM images [Michielsen and Stavenga (2008)], by that of the reflection spectrum with the band calculation [Michielsen and Stavenga (2008); Poladian *et al.* (2009)] and the FDTD calculation [Michielsen *et al.* (2010)], and also by the diffraction patterns obtained in small-angle X-ray scattering studies [Saranathan *et al.* (2010)]. Although these comparisons are not complete at the present stage, it is expected to become a clue to solve a problem how these elaborate structures are formed during pupal period.

5.3.4 *Photonic Crystals in Beetles*

Among various scale-bearing beetles, weevils offer an interesting subject to the studies on structural colors. Michelson investigated a weevil called diamond beetle, *Estimus imperialis*, which possessed brilliant and exquisitely colored scales on the elytron [Michelson (1911)]. He considered that the color of this beetle was due to fine striations on the interior surface of the scale and had an unsymmetrical saw-tooth shape, which was intended to enhance a particular diffraction order. Mallock observed a polygonal pattern in the transmission view of a scale and investigated the effect of liquid immersion [Mallock (1911)]. On the other hand, Onslow found a well-defined crossed appearance like the stings of a tennis racquet when seen at the cross section of the scale, and opposed to Michelson's specially designed grating [Onslow (1920, 1923)]. He also described that a variety of colors among species might come from the irregularity of the periodicity.

Using an electron microscope, Ghiradella found very peculiar structures within a scale of local weevil, *Polydrusus sericeus* [Ghiradella (1984)]. The scale of this beetle possessed a lattice structure similar to that found in the scale of the butterfly, *Callophyrys rubi*. Parker *et al.* reported that a weevil, *Pachyrhynchus argus*, showed a metallic coloration visible from any direction owing to photonic crystal analogous to opal [Parker *et al.* (2003)]. They reported that within a weevil's scale, many transparent spheres of 250-nm diameter constituted a photonic crystal of hexagonal closest-packing and selectively reflected 530-nm light. Further, they considered that the domain structure found within a scale contributed to the omnidirectional color. Similar photonic crystals were also found in scales of long-horned beetles [Simonis and Vigneron (2011)].

5.3.5 *Miscellaneous*

One of the conspicuous findings in structural colors was also made in marine animals. Parker *et al.* reported that brilliantly arranged two-dimensional photonic crystal was found in a kind of sea mouse *Aphrodita* sp., whose lower part of the body was decorated by many iridescent hair and spines called *chaetae* [Parker (2001); McPhedran *et al.* (2001)]. Sea mice with 15–20 cm in length and 5 cm in width are known to live along continental shelves at depths of 1 m–2 km, whose natural history is hardly known. The first electron microscopic observation of the *Aphrodite aculeata* was performed by Lippert and Gentil, who observed numerous regularly arranged capillaries within a hair of this animal [Lippert and Gentil (1963)]. Parker *et al.* investigated thicker iridescent spines of a hollow cylinder shape and found that within a cylinder, a tremendous number of small hollow cylinders are hexagonally closest-packed like a photonic crystal fiber. They calculated the reflectivity by modeling this system and found a large reflection band around 650 nm. The photonic band calculation showed a partial photonic band gap of the first band appeared in an infrared region and the second band that were complicated by mixing with several modes, appeared in a visible region, which corresponded to the red reflection in the reflection spectrum.

Probably, marine animals will offer the most fascinating materials for the studies in the field of biophotonics, although only a few species have been investigated up to now. These may include jellyfishes, swimming crabs, squids, sea urchins, and marine planktons, which are now waiting for the future studies.

Exercises

(1) Calculate the reciprocal vectors for the fcc and hcp structures.
(2) Derive the approximate expression of Eq. (5.32) for the dispersion curves near the zone boundary of $k \sim \pi/a$ by putting $k = \pi/a + h$ with $|h| \ll \pi/a$.
(3) Derive Eq. (5.39) under the condition of $f_1 \approx f_2$ and $n_1 \approx n_2$.
(4) The photonic crystals are amply distributed in nature. Search for the examples and summarize their features.

Chapter 6

Light Scattering

6.1 Light Scattering

Light scattering is one of the simplest optical processes, which commonly appear in the fields of optics and spectroscopy. In general, it is classified into inelastic and elastic ones, according to whether the incident light loses/gains its energy or not. In the former, *Rayleigh*, *Brillouin*, and *Raman scatterings* are commonly known, whose differences are dependent on the amount of energy exchanged with the scatterers. In general, it is called Rayleigh scattering when light illuminating on particles floating in a fluid or a density and/or thermal fluctuation in fluid and solid causes inelastic scattering. Its frequency range is normally below MHz. To measure the spectrum due to Rayleigh scattering, photon correlation or optical beat spectroscopy is often employed. It is sometimes called dynamical light scattering to characterize the fluctuations of particles in liquid and those of a medium itself.

Brillouin scattering is that in an intermediate frequency range from MHz to THz, which results from inelastic scattering due to thermally excited sound waves and the fluctuations on a level of molecular ensemble [Brillouin (1922)]. In order to measure the spectrum in this frequency range, Fabry-Perot type interferometer

Bionanophotonics: An Introductory Textbook
Shuichi Kinoshita
Copyright © 2013 Pan Stanford Publishing Pte. Ltd.
ISBN 978-981-4364-71-3 (Hardcover), 978-981-4364-72-0 (eBook)
www.panstanford.com

is often employed. Since the sound waves or fluctuations are directly connected with the physical properties of matter such as elasticity, viscosity, and thermal and dielectric properties, it is often used as an excellent probe to characterize the polymer dynamics and phase transition phenomena.

Raman scattering is that in a high frequency range above THz, which is mainly caused by vibrations of atoms or chemical groups within a molecular framework or lattice vibrations in crystal [Raman (1928); Raman and Krishnan (1928)]. Raman scattering is normally measured using a grating-type monochromator and is popular to obtain the structural information on molecule and solid.

On the other hand, for elastic scattering, we can give *Rayleigh*[a] [Rayleigh (1871a,b)] and *Mie scatterings* [Mie (1908)] as examples. The difference between these scattering processes are only dependent on a size of scattering particles. The theoretical basis of Rayleigh scattering comes from light scattering due to a point scatterer, whereas that of Mie scattering is a strict mathematical solution based on the electromagnetic theory in the presence of a small particle. In this sense, Rayleigh scattering is one of the approximation of Mie scattering in the limit of small particles.

On the contrary, in the other limit of large particles, the usual geometrical optics such as reflection and refraction of light becomes dominant. Thus, Mie scattering is the intermediate case, where both wave nature of light and geometrical optics work together. Thus, it is completely dependent on the shape and distribution of the structure. On the other hand, mathematically soluble shapes are restricted within a single particle whose shape belongs to a cylinder and concentric cylinder in 2D and a sphere and concentric sphere in 3D. However, recently, owing to the development of personal computers, numerical methods such as FDTD (finite-difference time domain) [Yee (1966)] and RCWA (rigorous coupled wave analysis) [Moharam and Gaylord (1981)] are widespread and become one of the most common methodologies on a laboratory level, which have been breaking a wall of the above computational difficulties.

In the following, however, we will only concentrate on an analytical approach concerning elastic scattering, because most

[a]The name "Rayleigh scattering" is doubly used both in elastic and inelastic light scattering. Of course, they are due to different processes.

of the nanophotonic processes can be treated without inelastic processes. We will derive the expressions for the most fundamental scattering process of Rayleigh scattering due to a single point scatterer and Mie scattering due to a spherical particle. Since excellent books and papers are now available, readers who aim to study the details on the scattering problems should refer to these resources [van de Hulst (1957); Born and Wolf (1959); Bohren and Huffman (1983)].

6.2 Rayleigh Scattering

6.2.1 *Vector and Scalar Potentials*

From the Maxwell equation of $\nabla \cdot \mathbf{B} = 0$ and by applying a general vector algebraic relation $\nabla \cdot (\nabla \times \mathbf{A}) = 0$, it is easy to show that the relation

$$\mathbf{B} = \nabla \times \mathbf{A},$$

holds[a] for an arbitrary vector $\mathbf{A}$. Inserting this relation into the Maxwell equation of $\nabla \times \mathbf{E} + \partial \mathbf{B}/\partial t = 0$, we obtain

$$\nabla \times \left(\mathbf{E} + \frac{\partial \mathbf{A}}{\partial t} \right) = 0.$$

Further, we again apply another vector algebraic relation $\nabla \times \nabla \phi = 0$ for an arbitrary scalar function of ϕ and obtain

$$\mathbf{E} + \frac{\partial \mathbf{A}}{\partial t} = -\nabla \phi,$$

where we have added a minus sign to accommodate to a general definition of Coulomb potential. Thus, both of electric and magnetic fields are expressed by the derivatives of $\mathbf{A}$ and ϕ as follows:

$$\mathbf{E} = -\frac{\partial \mathbf{A}}{\partial t} - \nabla \phi, \tag{6.1}$$

$$\mathbf{B} = \nabla \times \mathbf{A}. \tag{6.2}$$

We call these arbitrary quantities $\mathbf{A}$ and ϕ *vector* and *scalar potentials*, respectively. It is sometimes convenient that instead of

---

[a] Actually, in the right-hand side of this relation, an arbitrary constant should be added. However, here we leave it untouched, because a potential always has such arbitrariness. Later, we will take account of this arbitrariness by considering a concept of gauge.

directly calculating the electric and magnetic fields, these potentials are first evaluated and then converted into the electric and magnetic fields by applying Eqs. (6.1) and (6.2).

Next, we will derive general relations for $\mathbf{A}$ and ϕ using the Maxwell equations. By using $\nabla \times \mathbf{H} = \mathbf{j} + \partial \mathbf{D}/\partial t$, it is easy to show

$$\nabla \times (\nabla \times \mathbf{A}) - \frac{1}{c^2} \frac{\partial}{\partial t} \left(-\frac{\partial \mathbf{A}}{\partial t} - \nabla\phi \right) = \mu_0 \mathbf{j},$$

where we have assumed that except for free charge and electric current, the remaining space is a vacuum so that the permittivity and permeability of the space are set at ϵ_0 and μ_0, respectively, with $c^2 = 1/(\epsilon_0\mu_0)$. Using a vector algebraic relation of $\nabla \times (\nabla \times \mathbf{A}) = \nabla(\nabla \cdot \mathbf{A}) - \nabla^2\mathbf{A}$, we obtain

$$\nabla(\nabla \cdot \mathbf{A}) - \nabla^2\mathbf{A} - \frac{1}{c^2} \left(-\frac{\partial^2 \mathbf{A}}{\partial t^2} - \nabla\frac{\partial\phi}{\partial t} \right) = \mu_0 \mathbf{j},$$

which leads to

$$\nabla \left(\nabla \cdot \mathbf{A} + \frac{1}{c^2} \frac{\partial\phi}{\partial t} \right) + \frac{1}{c^2} \frac{\partial^2 \mathbf{A}}{\partial t^2} - \nabla^2\mathbf{A} = \mu_0 \mathbf{j}. \tag{6.3}$$

On the other hand, the Maxwell equation of $\nabla \cdot \mathbf{D} = \rho$ can be expressed as $\nabla \cdot \mathbf{E} = \rho/\epsilon_0$ when charges are distributed in a vacuum. Thus, inserting Eq. (6.1) into this relation gives

$$-\nabla \cdot \left(\frac{\partial \mathbf{A}}{\partial t} + \nabla\phi \right) = \rho/\epsilon_0,$$

which leads to

$$-\nabla \cdot \left(\frac{\partial \mathbf{A}}{\partial t} \right) - \nabla^2\phi = \rho/\epsilon_0. \tag{6.4}$$

In these two relations, Eqs. (6.3) and (6.4), if the relation

$$\nabla \cdot \mathbf{A} + \frac{1}{c^2} \frac{\partial\phi}{\partial t} = 0 \tag{6.5}$$

holds, then Eq. (6.3) directly leads to the following equation

$$\frac{1}{c^2} \frac{\partial^2 \mathbf{A}}{\partial t^2} - \nabla^2\mathbf{A} = \mu_0 \mathbf{j}, \tag{6.6}$$

while Eq. (6.4) is transformed into

$$\frac{1}{c^2} \frac{\partial^2\phi}{\partial t^2} - \nabla^2\phi = \rho/\epsilon_0, \tag{6.7}$$

by inserting $\nabla \cdot \mathbf{A} = -(1/c^2)\partial\phi/\partial t$ into the first term of the left-hand side. These two equations correspond to wave equations with their source terms in the right-hand sides. The meaning of these equations is that when the electric current or charge change with time, its change will propagate as a wave of vector or scalar potential with the velocity of light.

Since both vector and scalar potentials are defined as their derivatives to express the electric and magnetic fields, they contain the arbitrariness within themselves. Is it possible to derive the above relation, Eq. (6.5), by using their arbitrariness? To investigate its possibility, let us go back to a starting point of their definitions. Since a vector relation $\nabla \cdot (\nabla \times \mathbf{A}) = 0$ holds for an arbitrary vector $\mathbf{A}$, it is possible to put $\mathbf{A}' = \mathbf{A} + \nabla u$ for an arbitrary scalar function u to satisfy the vector relation. In a similar manner, since $\nabla \times \nabla\phi = 0$ holds for an arbitrary function ϕ, it is possible to put $\phi' = \phi - \partial u/\partial t$ as a new scalar potential. Inserting these relations into Eqs. (6.1) and (6.2), we obtain

$$\mathbf{E} = -\frac{\partial \mathbf{A}'}{\partial t} - \nabla\phi' = -\frac{\partial \mathbf{A}}{\partial t} - \frac{\partial \nabla u}{\partial t} - \nabla\phi + \nabla\frac{\partial u}{\partial t} = -\frac{\partial \mathbf{A}}{\partial t} - \nabla\phi,$$

$$\mathbf{B} = \nabla \times \mathbf{A}' = \nabla \times \mathbf{A} + \nabla \times \nabla u = \nabla \times \mathbf{A},$$

which shows that adding the extra terms to $\mathbf{A}$ and ϕ does not change their functional forms.

Next, we have to investigate whether we can choose a function u so as to satisfy the relation $\nabla \cdot \mathbf{A}' + (1/c^2)\partial\phi'/\partial t = 0$. By using the newly defined potentials $\mathbf{A}'$ and ϕ', the left-hand side of Eq. (6.5) becomes

$$\nabla \cdot \mathbf{A}' + \frac{1}{c^2}\frac{\partial \phi'}{\partial t} = \nabla \cdot \mathbf{A} + \frac{1}{c^2}\frac{\partial \phi}{\partial t} + \nabla^2 u - \frac{1}{c^2}\frac{\partial^2 u}{\partial t^2}.$$

From this relation, if we can determine a function u for a given set of $\mathbf{A}$ and ϕ to satisfy

$$\nabla^2 u - \frac{1}{c^2}\frac{\partial u^2}{\partial t^2} = -\nabla \cdot \mathbf{A} - \frac{1}{c^2}\frac{\partial \phi}{\partial t},$$

by solving this equation, we can always attain a set of $\mathbf{A}'$ and ϕ', which satisfy the relation $\nabla \cdot \mathbf{A}' + (1/c^2)\partial\phi'/\partial t = 0$. Instead of performing this procedure each time, it is convenient to employ $\mathbf{A}'$ and ϕ' in advance with an additional condition of

$$\nabla \cdot \mathbf{A}' + \frac{1}{c^2}\frac{\partial \phi'}{\partial t} = 0. \tag{6.8}$$

This additional condition to combine $\mathbf{A}'$ and ϕ' is called *Lorentz condition*.

6.2.2 *Retarded Potential*

Rayleigh scattering is a kind of light scattering in which light is emitted by an oscillating dipole created within a small particle through light irradiation. To obtain a general formulation of the Rayleigh scattering, it is necessary to investigate vector and scalar potentials in far field when a charge or a current is oscillating at a point in vacuum. Thus, it is necessary to solve the wave functions with the source terms of Eqs. (6.6) and (6.7). For this purpose, we will utilize a fact that a function $f(\mathbf{r}, t)$ can be generally converted into $\hat{f}(\mathbf{r}, \omega)$ through the Fourier transform with respect to time as $f(\mathbf{r}, t) = \int_{-\infty}^{\infty} \hat{f}(\mathbf{r}, \omega)e^{-i\omega t}d\omega$. Applying this relation to $\mathbf{A}$, ϕ, ρ and $\mathbf{i}$, and inserting them into Eqs. (6.6) and (6.7), we obtain

$$\nabla^2 \hat{\mathbf{A}} + \frac{\omega^2}{c^2} \hat{\mathbf{A}} = -\mu_0 \hat{\mathbf{j}}, \tag{6.9}$$

$$\nabla^2 \hat{\phi} + \frac{\omega^2}{c^2} \hat{\phi} = -\hat{\rho}/\epsilon_0. \tag{6.10}$$

These equations can be solved by means of a Green function. First, we pay attention to the second equation. This equation can be regarded as a Helmholtz equation $(\nabla^2 + \omega^2/c^2)\phi(\mathbf{r}) = 0$ with respect to a function $\phi(\mathbf{r})$, to which a term $-\rho(\mathbf{r})/\epsilon_0$ is added to the right-hand side. Instead of solving Eq. (6.10) directly, let us consider the following equation,

$$\left(\nabla^2 + \frac{\omega^2}{c^2}\right) G(\mathbf{r}) = -\delta(\mathbf{r}), \tag{6.11}$$

which is obtained by replacing the term on the right-hand side of Eq. (6.10) with $-\delta(\mathbf{r})$, where a function $G(\mathbf{r})$ is called *Green function*.

In order to solve this equation, it is convenient to use the Fourier transform method by putting $G(\mathbf{r})$ and $\delta(\mathbf{r})$ in $\mathbf{k}$ space:

$$G(\mathbf{r}) = \int \hat{G}(\mathbf{k})e^{i\mathbf{k}\cdot\mathbf{r}}d\mathbf{k}, \tag{6.12}$$

$$\delta(\mathbf{r}) = \frac{1}{(2\pi)^3} \int e^{i\mathbf{k}\cdot\mathbf{r}}d\mathbf{k}. \tag{6.13}$$

Inserting these expressions into Eq. (6.11), we obtain

$$\left(-k^2 + \frac{\omega^2}{c^2}\right)\hat{G}\,(\mathbf{k}) = -\frac{1}{(2\pi)^3}. \tag{6.14}$$

The left-hand side of this equation becomes zero when $k^2 - \omega^2/c^2 = 0$, while the right-hand side is a nonzero constant. Thus, $\hat{G}\,(\mathbf{k})$ has singularities at the points satisfying $k^2 - \omega^2/c^2 = 0$. Putting $k_0 \equiv \omega/c$ and using a fact that a relation $(k^2 - k_0^2)\delta(k^2 - k_0^2) = 0$ always holds for δ-function, we can put

$$\left(k^2 - k_0^2\right)\hat{G}\,(\mathbf{k}) = \frac{1}{(2\pi)^3}\left\{1 + C\,(k^2 - k_0^2)\delta(k^2 - k_0^2)\right\},$$

where C is an arbitrary constant. Then, $\hat{G}\,(\mathbf{k})$ is obtained within a framework of hyperfunction[a] as

$$\hat{G}\,(\mathbf{k}) = \frac{1}{(2\pi)^3}\left\{\frac{\mathcal{P}}{k^2 - k_0^2} + C\,\delta(k^2 - k_0^2)\right\}, \tag{6.15}$$

where $\mathcal{P}$ is called *Cauchy's principal value*, which indicates the integration should be carried out without points where the denominator becomes zero.

By using the mathematical relation

$$\lim_{\epsilon \to 0}\frac{1}{x \pm i\epsilon} = \frac{\mathcal{P}}{x} \mp i\pi\,\delta(x), \tag{6.16}$$

and putting $C = \mp i\pi$, the above equation can be rewritten as

$$\hat{G}_{\pm}(\mathbf{k}) = \lim_{\epsilon \to 0}\frac{1}{(2\pi)^3}\frac{1}{k^2 - k_0^2 \pm i\epsilon}. \tag{6.17}$$

Although $\hat{G}\,(\mathbf{k})$ originally has singular points on the real axis, this formula indicates that $G\,(\mathbf{r})$ is obtainable by the Fourier transform of $\hat{G}\,(\mathbf{k})$ after shifting the singular point by $\mp i\epsilon$ and then making $\epsilon \to 0$. $\mp$ corresponds to two particular solutions of the Green function.

Inserting Eq. (6.17) into Eq. (6.12), we obtain

$$G_{\pm}(\mathbf{r}) = \lim_{\epsilon \to 0}\frac{1}{(2\pi)^3}\int \frac{1}{k^2 - k_0^2 \pm i\epsilon}e^{i\mathbf{k}\cdot\mathbf{r}}d\mathbf{k}$$

$$= \lim_{\epsilon \to 0}\frac{1}{(2\pi)^3}\int_0^\infty k^2 dk \int_0^\pi \sin\theta d\theta \int_0^{2\pi}d\phi\frac{1}{k^2 - k_0^2 \pm i\epsilon}e^{ikr\cos\theta},$$

[a]Hyperfunction is a function having the discontinuity in its nature such as δ-function or θ-function, and is used only as an integrand.

where we have employed a polar coordinate with a direction of $\mathbf{r}$ as z axis and have defined θ as an angle made by two vectors $\mathbf{r}$ and $\mathbf{k}$. Further, we change a variable by putting $\cos\theta = t$, which results in $-\sin\theta d\theta = dt$, and obtain

$$
\begin{aligned}
G_\pm(\mathbf{r}) &= \lim_{\epsilon\to 0} \frac{2\pi}{(2\pi)^3} \int_0^\infty k^2 dk \int_{-1}^{1} dt \frac{1}{k^2 - k_0^2 \pm i\epsilon} e^{ikrt} \\
&= \lim_{\epsilon\to 0} \frac{1}{(2\pi)^2} \int_0^\infty k^2 dk \frac{1}{k^2 - k_0^2 \pm i\epsilon} \frac{e^{ikr} - e^{-ikr}}{ikr} \\
&= \lim_{\epsilon\to 0} \frac{1}{(2\pi)^2 ir} \int_0^\infty \frac{k}{k^2 - k_0^2 \pm i\epsilon} \left(e^{ikr} - e^{-ikr}\right) dk \\
&= \lim_{\epsilon\to 0} \frac{1}{(2\pi)^2 ir} \int_{-\infty}^\infty \frac{k}{k^2 - k_0^2 \pm i\epsilon} e^{ikr} dk.
\end{aligned}
$$

To investigate the positions of singular points, we put $k^2 - k_0^2 \pm i\epsilon = 0$ and calculate k, which results in

$$
k = \pm\sqrt{k_0^2 \mp i\epsilon} \approx \pm k_0 \left(1 \mp \frac{i\epsilon}{2k_0^2}\right) = \pm(k_0 \mp i\mu),
$$

where we have assumed that ϵ is infinitely small and put $\mu \equiv \epsilon/(2k_0)$. Thus,

$$
\hat{G}_\pm(\mathbf{r}) = \lim_{\mu\to 0} \frac{1}{2(2\pi)^2 ir} \int_{-\infty}^\infty \left\{ \frac{1}{k - k_0 \pm i\mu} + \frac{1}{k + k_0 \mp i\mu} \right\} e^{ikr} dk.
$$

(6.18)

In this equation, all the combinations of double signs are possible.

To calculate the integration, we employ a complex integral on paths of integration shown in Fig. 6.1 and use the Cauchy's theorem. For example, consider the upper sign for each double sign. Then, the singular point for the first term in the integrand becomes $k = k_0 - i\mu$, while the second term does $k = -k_0 + i\mu$, which are shown as crosses in the figure. If we consider the integration only for the upper half of the space and take the paths of C_1, C_2 and C_3, then the Cauchy's theorem states $\int_{C_1} - \int_{C_2} + \int_{C_3} = 0$. The integration along C_3 becomes zero when the radius of the path becomes infinitely large. Therefore, a relation $\int_{C_1} = \int_{C_2}$ holds so that we obtain

$$
G_+(\mathbf{r}) = \frac{1}{2(2\pi)^2 ir} \cdot 2\pi i \cdot e^{-ik_0 r} = \frac{1}{4\pi} \frac{e^{-ik_0 r}}{r}.
$$

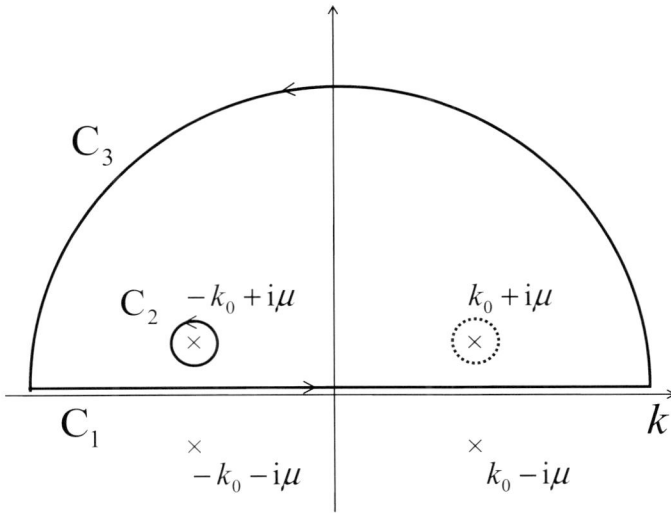

Figure 6.1 Paths of integration to calculate Eq. (6.18). Solid lines C_1, C_2 and C_3 are the integration paths to calculate the integral for the upper signs of double signs, while solid lines ($C1$ and $C3$) and a dashed line are those for the lower signs.

In a similar manner, if we take the lower sign for each double sign, then a dashed circle corresponding to the second term should be considered, which gives another solution,

$$G_-(\mathbf{r}) = \frac{1}{4\pi} \frac{e^{ik_0 r}}{r}.$$

Thus a general solution of the Green function is expressed by a sum of these particular solutions as

$$G(\mathbf{r}) = aG_+(\mathbf{r}) + bG_-(\mathbf{r}),$$

where a and b are arbitrary constants. Since the Fourier transform of both of $G_+(\mathbf{r})$ and $G_-(\mathbf{r})$, in addition to its sum $G(\mathbf{r})$, satisfy Eq. (6.14), a relation $a + b = 1$ holds naturally. Hence, a general solution for the Green function becomes

$$G(\mathbf{r}) = aG_+(\mathbf{r}) + (1-a)G_-(\mathbf{r})$$

$$= \frac{a}{4\pi} \frac{e^{-ik_0 r}}{r} + \frac{1-a}{4\pi} \frac{e^{ik_0 r}}{r}. \tag{6.19}$$

The solution of Eq. (6.10) is obtained as follows: In Eq. (6.10), we first change a variable $\mathbf{r} \to \mathbf{r} - \mathbf{r}'$ as

$$\left(\nabla^2 + \frac{\omega^2}{c^2} \right) G\left(\mathbf{r} - \mathbf{r}'\right) = -\delta(\mathbf{r} - \mathbf{r}').$$

Since the relation $\nabla_{\mathbf{r}-\mathbf{r}'} = \nabla_{\mathbf{r}}$ holds naturally, ∇ appearing on the left-hand side should be read as $\nabla_{\mathbf{r}}$. Next, we integrate the equation with respect to $\mathbf{r}'$ after multiplying $\hat{\rho}(\mathbf{r}', \omega)/\epsilon_0$ at both sides. Then, we obtain

$$\left(\nabla^2 + \frac{\omega^2}{c^2} \right) \int G\left(\mathbf{r} - \mathbf{r}'\right) \frac{\hat{\rho}(\mathbf{r}', \omega)}{\epsilon_0} d\mathbf{r}' = - \int \delta(\mathbf{r} - \mathbf{r}') \frac{\hat{\rho}(\mathbf{r}', \omega)}{\epsilon_0} d\mathbf{r}'$$

$$= - \frac{\hat{\rho}(\mathbf{r}, \omega)}{\epsilon_0}.$$

Thus, by comparing the above result with Eq. (6.10), the solution of Eq. (6.10) becomes

$$\hat{\phi}(\mathbf{r}, \omega) = \int G\left(\mathbf{r} - \mathbf{r}'\right) \frac{\hat{\rho}(\mathbf{r}', \omega)}{\epsilon_0} d\mathbf{r}'. \tag{6.20}$$

In order to confirm that Eq. (6.20) actually becomes a solution of Eq. (6.10), we will operate $(\nabla^2 + \omega^2/c^2)$ to the both sides of Eq. (6.20): The left-hand side becomes

$$\left(\nabla^2 + \frac{\omega^2}{c^2} \right) \hat{\phi}(\mathbf{r}, \omega),$$

while the right-hand side does

$$\int \left(\nabla^2 + \frac{\omega^2}{c^2} \right) G\left(\mathbf{r} - \mathbf{r}'\right) \frac{\hat{\rho}(\mathbf{r}', \omega)}{\epsilon_0} d\mathbf{r}'$$

$$= - \int \delta(\mathbf{r} - \mathbf{r}') \frac{\hat{\rho}(\mathbf{r}', \omega)}{\epsilon_0} d\mathbf{r}' = - \frac{\hat{\rho}(\mathbf{r}, \omega)}{\epsilon_0},$$

which confirms its justification. A constant a involved in $G(\mathbf{r})$ should be determined by the boundary condition. For example, for a scattering problem, a scattered wave will propagate as a spherical wave from the scatterer (see Sec. 6.3.1), thus the $\exp[ik_0 r]/r$ type expression is anyway necessary. Therefore, putting $a = 0$, we obtain

$$\hat{\phi}(\mathbf{r}, \omega) = \int \frac{1}{4\pi} \frac{e^{ik_0|\mathbf{r}-\mathbf{r}'|}}{|\mathbf{r} - \mathbf{r}'|} \frac{\hat{\rho}(\mathbf{r}', \omega)}{\epsilon_0} d\mathbf{r}'. \tag{6.21}$$

In a similar manner, we can calculate the expression for the vector potential. Thus, summarizing, we obtain

$$\hat{\mathbf{A}}(\mathbf{r}, \omega) = \frac{\mu_0}{4\pi} \int d\mathbf{r}' \frac{e^{\pm i(\omega/c)|\mathbf{r}-\mathbf{r}'|}}{|\mathbf{r} - \mathbf{r}'|} \hat{\mathbf{j}}(\mathbf{r}', \omega) \qquad (6.22)$$

$$\hat{\phi}(\mathbf{r}, \omega) = \frac{1}{4\pi\epsilon_0} \int d\mathbf{r}' \frac{e^{\pm i(\omega/c)|\mathbf{r}-\mathbf{r}'|}}{|\mathbf{r} - \mathbf{r}'|} \hat{\rho}(\mathbf{r}', \omega), \qquad (6.23)$$

where by custom, we have expressed the two particular solutions for $G(\mathbf{r})$ as a double sign. Further, we perform the Fourier transform of Eq. (6.22) and obtain

$$\mathbf{A}(\mathbf{r}, t) = \int_{-\infty}^{\infty} d\omega\, \hat{\mathbf{A}}(\mathbf{r}, \omega) e^{-i\omega t}$$

$$= \frac{\mu_0}{4\pi} \int_{-\infty}^{\infty} d\omega\, e^{-i\omega t} \int d\mathbf{r}' \frac{e^{\pm i(\omega/c)|\mathbf{r}-\mathbf{r}'|}}{|\mathbf{r} - \mathbf{r}'|} \cdot \frac{1}{2\pi} \int_{-\infty}^{\infty} dt'\mathbf{j}(\mathbf{r}', t')e^{i\omega t'}$$

$$= \frac{\mu_0}{4\pi} \int d\mathbf{r}' \int_{-\infty}^{\infty} dt' \frac{\mathbf{j}(\mathbf{r}', t')}{|\mathbf{r} - \mathbf{r}'|} \cdot \frac{1}{2\pi} \int_{-\infty}^{\infty} d\omega\, e^{-i\omega(t-t')\pm i(\omega/c)|\mathbf{r}-\mathbf{r}'|}$$

$$= \frac{\mu_0}{4\pi} \int d\mathbf{r}' \int_{-\infty}^{\infty} dt' \frac{\mathbf{j}(\mathbf{r}', t')}{|\mathbf{r} - \mathbf{r}'|} \delta\left(t - t' \mp |\mathbf{r} - \mathbf{r}'|/c\right),$$

where we have defined the inverse Fourier transform with respect to t as

$$\hat{f}(\omega) = \frac{1}{2\pi} \int f(t)e^{i\omega t} dt,$$

and have used

$$\delta(t) = \frac{1}{2\pi} \int_{-\infty}^{\infty} e^{-i\omega t} d\omega.$$

Thus, the final expression for a vector potential induced by a time variation of the electric current density becomes

$$\mathbf{A}(\mathbf{r}, t) = \frac{\mu_0}{4\pi} \int d\mathbf{r}' \frac{\mathbf{j}(\mathbf{r}', t')}{|\mathbf{r} - \mathbf{r}'|}. \qquad (6.24)$$

In a similar manner, a scalar potential induced by a time variation of the electric charge density becomes

$$\phi(\mathbf{r}, t) = \frac{1}{4\pi\epsilon_0} \int d\mathbf{r}' \frac{\rho(\mathbf{r}', t')}{|\mathbf{r} - \mathbf{r}'|}, \qquad (6.25)$$

where t' is expressed as $t' = t \mp |\mathbf{r} - \mathbf{r}'|/c$. A minus sign in t' means that the time variation of electric charge and current densities observed at $|\mathbf{r} - \mathbf{r}'|$ will be delayed by a amount of $|\mathbf{r} - \mathbf{r}'|/c$, which is equivalent to a time for light to propagate the same distance. These potentials are called *retarded potentials* and t' in this case is called *retarded time*, while for a plus sign, such potentials are called *advanced potentials*.

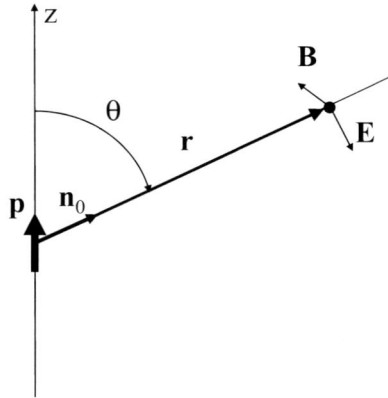

Figure 6.2 Geometry for dipole radiation.

6.2.3 *Dipole Radiation*

Consider a case where a point charge is oscillating at the origin of a coordinate shown in Fig. 6.2. Since the current density in this case is expressed by the motion of the point charge, Eq. (6.24) can be transformed into

$$\mathbf{A} = \frac{\mu_0}{4\pi} \frac{\dot{\mathbf{p}}(t - r/c)}{r},$$

where we have used the relation

$$\mathbf{j}(\mathbf{r}', t') = q\frac{d\mathbf{r}_0(t')}{dt'}\delta(\mathbf{r}' - \mathbf{r}_0) = \frac{d\mathbf{p}(t')}{dt'}\delta(\mathbf{r}' - \mathbf{r}_0) = \dot{\mathbf{p}}(t')\delta(\mathbf{r}' - \mathbf{r}_0).$$

Here, we denote an amount and position of charge as q and $\mathbf{r}_0$, and put $\mathbf{p}(t') = q\mathbf{r}_0(t')$. We further put $|\mathbf{r} - \mathbf{r}_0| \approx r$ and employ a minus sign for the retarded time. The magnetic flux density in this case becomes

$$\mathbf{B} = \nabla \times \mathbf{A}$$
$$= \frac{\mu_0}{4\pi} \nabla \times \frac{\dot{\mathbf{p}}(t - r/c)}{r}.$$

The actual calculation in the second expression is carried out as follows: First, we only consider the x component, which results in

$$\left(\nabla \times \frac{\dot{\mathbf{p}}(t - r/c)}{r}\right)_x = \frac{\partial}{\partial y}\left(\frac{\dot{p}_z(t - r/c)}{r}\right) - \frac{\partial}{\partial z}\left(\frac{\dot{p}_y(t - r/c)}{r}\right).$$

Further, we consider the first term on the right-hand side of the resulting relation, which is calculated as

$$\frac{\partial}{\partial y}\left(\frac{\dot{p}_z(t-r/c)}{r}\right) = \frac{\ddot{p}_z(t-r/c)(-1/c)(y/r)r - \dot{p}_z(t-r/c)(y/r)}{r^2}$$

$$= -\frac{(r/c)y\ddot{p}_z(t-r/c) + y\dot{p}_z(t-r/c)}{r^3},$$

where we have used $\partial r/\partial x_j = x_j/r$ with $x_j = x, y, z$. We notice that on the right-hand side of this result, $y\ddot{p}_z$ and $y\dot{p}_z$ are found to exist, which are considered as a component of vector products, $\mathbf{r} \times \ddot{\mathbf{p}}$ and $\mathbf{r} \times \dot{\mathbf{p}}$. Thus,

$$\nabla \times \frac{\dot{\mathbf{p}}(t-r/c)}{r} = -\frac{\mathbf{r} \times \{(r/c)\ddot{\mathbf{p}}(t-r/c) + \dot{\mathbf{p}}(t-r/c)\}}{r^3},$$

which leads to

$$\mathbf{B} = \frac{\mu_0}{4\pi}\frac{\dot{\mathbf{p}}(t-r/c) + (r/c)\ddot{\mathbf{p}}(t-r/c)}{r^3} \times \mathbf{r}. \qquad (6.26)$$

The scalar potential can be obtained using the Lorentz condition $\nabla \cdot \mathbf{A} + (1/c^2)\partial\phi/\partial t = 0$. First, we consider the first term of $\nabla \cdot \mathbf{A} = \partial A_x/\partial x + \partial A_y/\partial y + \partial A_z/\partial z$ and obtain

$$\frac{\mu_0}{4\pi}\frac{\partial}{\partial x}\left(\frac{\dot{p}_x(t-r/c)}{r}\right)$$

$$= \frac{\mu_0}{4\pi}\frac{\ddot{p}_x(t-r/c)(-1/c)(x/r)r - \dot{p}_x(t-r/c)(x/r)}{r^2}$$

$$= -\frac{\mu_0}{4\pi}\frac{(r/c)x\ddot{p}_x(t-r/c) + x\dot{p}_x(t-r/c)}{r^3}.$$

In this case, the right-hand side is expressed as the products, $x\ddot{p}_x$ and $x\dot{p}_x$, and is considered to be expressed by the inner products, $\mathbf{r} \cdot \ddot{\mathbf{p}}$ and $\mathbf{r} \cdot \dot{\mathbf{p}}$. Thus,

$$\nabla \cdot \mathbf{A} = -\frac{\mu_0}{4\pi}\frac{\dot{\mathbf{p}}(t-r/c) + (r/c)\ddot{\mathbf{p}}(t-r/c)}{r^3} \cdot \mathbf{r},$$

which results in

$$\phi = \frac{1}{4\pi\epsilon_0}\frac{\mathbf{p}(t-r/c) + (r/c)\dot{\mathbf{p}}(t-r/c)}{r^3} \cdot \mathbf{r}. \qquad (6.27)$$

The electric field induced by an oscillating point charge is derived by using a relation of $\mathbf{E} = -\nabla\phi - \partial\mathbf{A}/\partial t$. The similar calculation as above is applied to $\nabla\phi$ (Exercise (1)), and the following relations are obtained:

$$\nabla\left(\frac{\mathbf{p}(t-r/c) \cdot \mathbf{r}}{r^3}\right) = \frac{\mathbf{p}}{r^3} - \frac{\mathbf{r}(\dot{\mathbf{p}} \cdot \mathbf{r})}{cr^4} - \frac{3\mathbf{r}(\mathbf{p} \cdot \mathbf{r})}{r^5},$$

and

$$\nabla \left(\frac{\dot{\mathbf{p}}(t - r/c) \cdot \mathbf{r}}{r^2} \right) = \frac{\dot{\mathbf{p}}}{r^2} - \frac{\mathbf{r}(\ddot{\mathbf{p}} \cdot \mathbf{r})}{cr^3} - \frac{2\mathbf{r}(\dot{\mathbf{p}} \cdot \mathbf{r})}{r^4}.$$

Inserting these relations and adding the derivative of the vector potential, we obtain

$$\mathbf{E} = \frac{-1}{4\pi\epsilon_0 r^3} \left\{ \mathbf{p} - \frac{\mathbf{r}(\dot{\mathbf{p}} \cdot \mathbf{r})}{cr} - \frac{3\mathbf{r}(\mathbf{p} \cdot \mathbf{r})}{r^2} \right.$$
$$\left. + \frac{r}{c}\dot{\mathbf{p}} - \frac{1}{c^2}\mathbf{r}(\ddot{\mathbf{p}} \cdot \mathbf{r}) - \frac{2\mathbf{r}(\dot{\mathbf{p}} \cdot \mathbf{r})}{cr} \right\} - \frac{\mu_0}{4\pi}\frac{\ddot{\mathbf{p}}}{r}$$
$$= \frac{-1}{4\pi\epsilon_0 r^3} \left\{ \mathbf{p} + \frac{r}{c}\dot{\mathbf{p}} - \frac{3\mathbf{r}(\mathbf{p} \cdot \mathbf{r})}{r^2} \right.$$
$$\left. - \frac{r}{c}\frac{3\mathbf{r}(\dot{\mathbf{p}} \cdot \mathbf{r})}{r^2} + \frac{r^2\ddot{\mathbf{p}} - \mathbf{r}(\ddot{\mathbf{p}} \cdot \mathbf{r})}{c^2} \right\},$$

where we have used $\mu_0 = 1/(\epsilon_0 c^2)$. Further, putting $\mathbf{p}(t - r/c) + (r/c)\dot{\mathbf{p}}(t - r/c) \equiv \mathbf{p}_0(t - r/c)$, we have reached the final expression as

$$\mathbf{E}(\mathbf{r}, t) = \frac{-1}{4\pi\epsilon_0 r^3} \left[\mathbf{p}_0(t - r/c) - \frac{3\{\mathbf{p}_0(t - r/c) \cdot \mathbf{r}\}\mathbf{r}}{r^2} \right.$$
$$\left. - \frac{1}{c^2}\left\{ \ddot{\mathbf{p}}\left(t - \frac{r}{c}\right) \times \mathbf{r} \right\} \times \mathbf{r} \right], \tag{6.28}$$

where we have used a vector algebraic relation $\mathbf{A} \times (\mathbf{B} \times \mathbf{C}) = \mathbf{B}(\mathbf{A} \cdot \mathbf{C}) - \mathbf{C}(\mathbf{A} \cdot \mathbf{B})$.

In Eq. (6.28), the first two terms will decrease with a dependence of r^{-3} when a distance from the origin is increased, while the last term will decrease only with that of r^{-1}. Thus, at a sufficiently distant point, only the last term will remain as a major component. In a similar manner, in Eq. (6.26), the first term has r^{-2} dependence, while the last term has r^{-1} dependence. Thus, only the last term will remain as far field (Exercise (2)).

Both of these terms remaining even in far field have a form of the second derivative of $\mathbf{p}$ with respect to time and the radiation due to such terms is called *dipole radiation*. If we denote these terms as $\mathbf{E}^{(2)}$ and $\mathbf{B}^{(2)}$, then they are expressed as

$$\mathbf{E}^{(2)}(\mathbf{r}, t) = \frac{1}{4\pi\epsilon_0}\frac{(\ddot{\mathbf{p}}(t - r/c) \times \mathbf{r}) \times \mathbf{r}}{c^2 r^3}, \tag{6.29}$$

$$\mathbf{B}^{(2)}(\mathbf{r}, t) = \frac{\mu_0}{4\pi}\frac{\ddot{\mathbf{p}}(t - r/c) \times \mathbf{r}}{cr^2}. \tag{6.30}$$

Inserting the right-hand side of Eq. (6.30) into Eq. (6.29), we obtain

$$\mathbf{E}^{(2)}(\mathbf{r}, t) = c\mathbf{B}^{(2)}(\mathbf{r}, t) \times \mathbf{r}/r, \tag{6.31}$$

and also taking an inner product of Eq. (6.30) with $\mathbf{r}$, we obtain

$$\mathbf{B}^{(2)}(\mathbf{r}, t) \cdot \mathbf{r} = \frac{\mu_0}{4\pi} \frac{(\ddot{\mathbf{p}}(t - r/c) \times \mathbf{r}) \cdot \mathbf{r}}{cr^2} = 0, \tag{6.32}$$

where we have used a vector algebraic relation $\mathbf{r} \cdot (\mathbf{r} \times \mathbf{A}) = 0$. Further, calculating the vector product of Eq. (6.31) with $\mathbf{r}$, we obtain

$$
\begin{aligned}
\mathbf{E}^{(2)} \times \mathbf{r} &= \frac{c(\mathbf{B}^{(2)} \times \mathbf{r}) \times \mathbf{r}}{r} \\
&= \frac{c\{\mathbf{r}(\mathbf{r} \cdot \mathbf{B}^{(2)}) - r^2 \mathbf{B}^{(2)}\}}{r} \\
&= -cr\mathbf{B}^{(2)},
\end{aligned} \tag{6.33}
$$

where we have used Eq. (6.32). Thus, the relation

$$\mathbf{B}^{(2)} = -\frac{1}{cr}\mathbf{E}^{(2)} \times \mathbf{r}, \tag{6.34}$$

is also attained. From these relations, it is easily shown that $\mathbf{E}^{(2)}$, $\mathbf{B}^{(2)}$ and $\mathbf{r}$ are mutually orthogonal.

The intensity of electromagnetic wave radiated by the oscillating dipole can be obtained by calculating a cycle-averaged Poynting vector (Eq. (2.44)), which is expressed as

$$
\begin{aligned}
\bar{\mathbf{S}} &= \frac{1}{2\mu_0} \mathrm{Re} \left\{ \mathbf{E}^{(2)} \times \mathbf{B}^{(2)*} \right\} \\
&= -\frac{c}{2\mu_0 r} \mathrm{Re} \left\{ \mathbf{B}^{(2)*} \times \left(\mathbf{B}^{(2)} \times \mathbf{r} \right) \right\} \\
&= -\frac{c}{2\mu_0 r} \mathrm{Re} \left\{ \mathbf{B}^{(2)} \left(\mathbf{B}^{(2)*} \cdot \mathbf{r} \right) - \mathbf{r} \left(\mathbf{B}^{(2)*} \cdot \mathbf{B}^{(2)} \right) \right\},
\end{aligned}
$$

where we have used Eq. (6.33) and have again used a vector algebraic relation described above. Further, since the first term in the right-hand side vanishes owing to the orthogonality given by Eq. (6.32), the remaining term becomes

$$
\begin{aligned}
\bar{\mathbf{S}} &= \frac{c}{2\mu_0} \left(\frac{\mathbf{r}}{r} \right) (\mathbf{B}^{(2)*} \cdot \mathbf{B}^{(2)}) \\
&= \frac{c}{2\mu_0} \left(\frac{\mu_0}{4\pi} \right)^2 \frac{1}{(cr^2)^2} \left(\frac{\mathbf{r}}{r} \right) |\mathbf{r} \times \ddot{\mathbf{p}}(t - r/c)|^2,
\end{aligned}
$$

where we have used Eq. (6.30). Putting $\mathbf{r}/r \equiv \mathbf{n}_0$ (see Fig. 6.2), the Poynting vector is reduced to the following simple form:

$$\bar{\mathbf{S}} = \frac{1}{2(4\pi)^2\epsilon_0 c^3} \frac{|\mathbf{n}_0 \times \ddot{\mathbf{p}}(t-r/c)|^2}{r^2} \mathbf{n}_0. \qquad (6.35)$$

If we take the z-axis parallel to the direction of a vector $\ddot{\mathbf{p}}(t-r/c)$, it is proved through the relation $|\mathbf{n}_0 \times \ddot{\mathbf{p}}(t-r/c)|^2 = |\ddot{\mathbf{p}}(t-r/c)|^2 \sin^2\theta$ that the radiated intensity is proportional to $\sin^2\theta$, where θ is an angle between the directions of $\mathbf{r}$ and the z-axis, and is inversely proportional to a square of a distance from the oscillating dipole.

The total power radiated from the oscillating dipole is evaluated by integrating Eq. (6.35) over a whole surface of a sphere with a radius r and is given as

$$P = \iint \bar{\mathbf{S}} \cdot \mathbf{n}_0 r^2 \sin\theta d\theta d\phi$$

$$= \frac{1}{2(4\pi)^2\epsilon_0 c^3} \iint |\mathbf{n}_0 \times \ddot{\mathbf{p}}(t-r/c)|^2 \sin\theta d\theta d\phi$$

$$= \frac{1}{2(4\pi)^2\epsilon_0 c^3} \iint |\ddot{\mathbf{p}}(t-r/c)|^2 \sin^3\theta d\theta d\phi,$$

which is further calculated as

$$P = \frac{1}{2(4\pi)^2\epsilon_0 c^3} |\ddot{\mathbf{p}}(t-r/c)|^2 \int_0^\pi d\theta \int_0^{2\pi} d\phi \sin^3\theta$$

$$= \frac{1}{2(4\pi)^2\epsilon_0 c^3} |\ddot{\mathbf{p}}(t-r/c)|^2 2\pi \left[\frac{1}{12}\cos 3\theta - \frac{3}{4}\cos\theta \right]_0^\pi$$

$$= \frac{1}{12\pi\epsilon_0 c^3} |\ddot{\mathbf{p}}(t-r/c)|^2. \qquad (6.36)$$

Thus, the total intensity is simply expressed by a square of the second derivative of the oscillating dipole.

6.2.4 Rayleigh Scattering

Light scattering by a particle much smaller than the wavelength of light is called *Rayleigh scattering* and can be derived by using the electromagnetic theory of dipole radiation. Consider a plane wave of light with the wave vector and angular frequency of $\mathbf{k}_i$ and ω, respectively, is incident on a small particle located at a point of $\mathbf{r}_0$

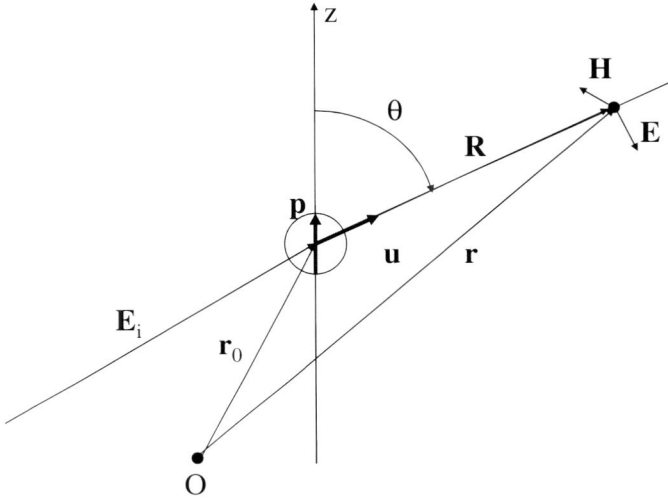

Figure 6.3 Geometry for the Rayleigh scattering.

(Fig. 6.3). Then, an oscillating dipole is induced on the particle, which is expressed as

$$\mathbf{p}(\mathbf{r}_0, t) = \alpha \mathbf{E}_0 e^{i(\mathbf{k}_i \cdot \mathbf{r}_0 - \omega t)} \equiv \mathbf{p}_0 e^{i(\mathbf{k}_i \cdot \mathbf{r}_0 - \omega t)}, \qquad (6.37)$$

where α is a polarizability tensor of the particle and $\mathbf{E}_0$ is the amplitude of the electric field of the incident light. The induced dipole is oscillating in synchronism with the incident field, and radiates the secondary emission of light.

According to Eq. (6.29), the electric field observed at a point $\mathbf{r}$ is described as

$$\mathbf{E}(\mathbf{r}, t) = \frac{\mathbf{u} \times (\mathbf{u} \times \ddot{\mathbf{p}}(\mathbf{r}_0, t'))}{4\pi \epsilon_m c_m^2 R}, \qquad (6.38)$$

where $\mathbf{u}$ and t' are a unit vector in the direction of $\mathbf{r} - \mathbf{r}_0$ and a retarded time expressed as $t' = t - R/c_m$ with the velocity of light in the medium c_m and $|\mathbf{r} - \mathbf{r}_0| \equiv R$. ϵ_m is the permittivity of the medium immersing the particle. Further, $\ddot{\mathbf{p}}(\mathbf{r}_0, t')$ is rewritten as

$$\begin{aligned}
\ddot{\mathbf{p}}(\mathbf{r}_0, t') &= -\omega^2 \mathbf{p}_0 \, e^{i\{\mathbf{k}_i \cdot \mathbf{r}_0 - \omega(t - R/c_m)\}} \\
&= -\omega^2 \mathbf{p}_0 \, e^{i(\mathbf{k}_i - \mathbf{k}_u) \cdot \mathbf{r}_0} \, e^{i(\mathbf{k}_u \cdot \mathbf{r} - \omega t)}, \qquad (6.39)
\end{aligned}$$

where $\mathbf{k}_u = (\omega/c_m)\mathbf{u}$ and we have used a relation $\mathbf{R} = \mathbf{r} - \mathbf{r}_0$, as shown in Fig. 6.3.

Next, we take the direction of the induced dipole as the z axis and employ a polar coordinate. Then, the components of the electric field become

$$E_\theta = -\frac{|\mathbf{p}_0|\omega^2}{4\pi\epsilon_m c_m^2 R}\sin\theta\ e^{i(\mathbf{k}_i - \mathbf{k}_u)\cdot\mathbf{r}_0}\ e^{i(\mathbf{k}_u\cdot\mathbf{r} - \omega t)},$$

$$E_\phi = 0, \tag{6.40}$$

where θ is a polar angle. The magnetic flux density is then obtained using Eq. (6.34) as

$$B_\theta = 0,$$

$$B_\phi = -\frac{|\mathbf{p}_0|\omega^2}{4\pi\epsilon_m c_m^3 R}\sin\theta\ e^{i(\mathbf{k}_i - \mathbf{k}_u)\cdot\mathbf{r}_0}\ e^{i(\mathbf{k}_u\cdot\mathbf{r} - \omega t)}. \tag{6.41}$$

Thus, both the electric and magnetic flux density vectors are orthogonal to each other and their directions guarantee the propagation of a transverse electromagnetic wave. The light intensity in the direction of $\mathbf{u}$ is directly calculated using these expressions. That is,

$$I \equiv \overline{|\mathbf{E}\times\mathbf{H}|} = \frac{1}{2\mu_0}\mathrm{Re}\left\{E_\theta B_\phi^*\right\} = \frac{\omega^4}{32\pi^2\epsilon_m c_m^3 R^2}|\mathbf{p}_0|^2\sin^2\theta, \tag{6.42}$$

where $\bar{A}$ means a cycle average of A.

The features of Rayleigh scattering are summarized as follows: (1) The scattering intensity is proportional to the fourth power of light frequency and then the fourth power of the reciprocal of wavelength. (2) The radiation is most intense in a plane perpendicular to the direction of an induced dipole and is dependent as $\sin^2\theta$ on the polar angle. (3) The scattering intensity is inversely proportional to the second power of a distance between an induced dipole and a point of observation. (4) The scattered light is completely polarized, if the polarizability is isotropic. As a result, the feature (1), i.e. blue light is more efficiently scattered than red light, holds generally, and thus the material containing many scatterers assumes bluish, even if it is illuminated with white light.

6.3 Mie Scattering

6.3.1 *Polar Coordinate and Spherical Wave*

When a particle size becomes comparable or larger than the wavelength of light, the above approximation of a point dipole no longer holds. This is because various optical phenomena will explicitly come out. For example, the diffraction of light due to the particle hinders the propagation of light wave, and the refraction and reflection of light take place at the surface. Thus, another and more strict theory is anyway necessary. The first attempt to solve this problem for a spherical particle was done by Mie, which appeared in his paper entitled "Contribution to the optics of turbid media, particularly of colloidal metal solutions" [Mie (1908)]. This theory has been further developed by many researchers so that various shapes of particles can be treated nowadays [van de Hulst (1957); Born and Wolf (1959); Kerker (1969); Bohren and Huffman (1983)]. Although the shape of a particle originally considered by Mie was spherical, light scattering due to a particle, whose size is comparable to the wavelength of light, is now generally called *Mie scattering* regardless of its shape.

The basis of Mie scattering theory is to solve a scattering problem in an exact way to consider the boundary condition at the surface of the particle and to express the solution using an expansion series of orthogonal functions. Although its formulation is quite complicated and yet its detailed derivations were described in various textbooks, here we will derive it again for the readers' convenience.

Before doing it, we will show various algebraic relations concerning a polar coordinate, since in Mie scattering theory, it is more convenient to use the polar coordinate instead of most commonly used Cartesian coordinate. At first, we will give algebraic rules to transform from the Cartesian to polar coordinate. The components of a vector **A** in the Cartesian coordinate are transformed into the polar coordinate as in the following way:

$$A_r = A_x \sin\theta \cos\phi + A_y \sin\theta \sin\phi + A_z \cos\theta,$$
$$A_\theta = A_x \cos\theta \cos\phi + A_y \cos\theta \sin\phi - A_z \sin\theta, \qquad (6.43)$$
$$A_\phi = -A_x \sin\phi + A_y \cos\phi,$$

and the inverse transformations are described as

$$A_x = A_r \sin\theta \cos\phi + A_\theta \cos\theta \cos\phi - A_\phi \sin\phi,$$
$$A_y = A_r \sin\theta \sin\phi + A_\theta \cos\theta \sin\phi + A_\phi \cos\theta, \quad (6.44)$$
$$A_z = A_r \cos\theta - A_\theta \sin\theta.$$

Further, various differential operators in the polar coordinate take the following forms:

$$\nabla \cdot \mathbf{A} = \frac{1}{r^2 \sin\theta} \left\{ \frac{\partial}{\partial r}(r^2 \sin\theta\, A_r) + \frac{\partial}{\partial\theta}(r \sin\theta\, A_\theta) + \frac{\partial}{\partial\phi}(r A_\phi) \right\},$$
$$(6.45)$$

$$\Delta V = \nabla^2 V = \frac{1}{r^2}\frac{\partial}{\partial r}\left(r^2 \frac{\partial V}{\partial r}\right) + \frac{1}{r^2 \sin\theta}\frac{\partial}{\partial\theta}\left(\sin\theta \frac{\partial V}{\partial\theta}\right)$$
$$+ \frac{1}{r^2 \sin^2\theta}\frac{\partial^2 V}{\partial\phi^2}, \quad (6.46)$$

$$(\nabla\times\mathbf{A})_r = \frac{1}{r \sin\theta}\left\{\frac{\partial}{\partial\theta}(\sin\theta\, A_\phi) - \frac{\partial A_\theta}{\partial\phi}\right\}, \quad (6.47)$$

$$(\nabla\times\mathbf{A})_\theta = \frac{1}{r}\left\{\frac{1}{\sin\theta}\frac{\partial A_r}{\partial\phi} - \frac{\partial}{\partial r}(r A_\phi)\right\}, \quad (6.48)$$

$$(\nabla\times\mathbf{A})_\phi = \frac{1}{r}\left\{\frac{\partial}{\partial r}(r A_\theta) - \frac{\partial A_r}{\partial\theta}\right\}. \quad (6.49)$$

Using these expressions, we will first transform the following wave equation described in the Cartesian coordinate into that in the polar coordinate:

$$\nabla^2\mathbf{A} - \frac{1}{c^2}\frac{\partial}{\partial t}\mathbf{A} = 0. \quad (6.50)$$

By using Eq. (6.46), the above equation is transformed into

$$\frac{1}{r^2}\frac{\partial}{\partial r}\left(r^2 \frac{\partial\mathbf{A}}{\partial r}\right) + \frac{1}{r^2 \sin\theta}\frac{\partial}{\partial\theta}\left(\sin\theta \frac{\partial\mathbf{A}}{\partial\theta}\right) + \frac{1}{r^2 \sin^2\theta}\frac{\partial^2\mathbf{A}}{\partial\phi^2} - \frac{1}{c^2}\frac{\partial^2}{\partial t^2}\mathbf{A} = 0.$$
$$(6.51)$$

In order to intuitively interpret this expression, it is convenient to consider a simple case where $\mathbf{A}$ is expressed only by a distance r from the origin of the coordinate. In this case, the above wave function is simplified considerably into the following form:

$$\frac{1}{r^2}\frac{\partial}{\partial r}\left(r^2 \frac{\partial\mathbf{A}}{\partial r}\right) - \frac{1}{c^2}\frac{\partial^2}{\partial t^2}\mathbf{A} = 0.$$

Further utilizing the relation,

$$\frac{1}{r^2}\frac{\partial}{\partial r}\left(r^2\frac{\partial \mathbf{A}}{\partial r}\right) = \frac{2}{r}\frac{\partial \mathbf{A}}{\partial r} + \frac{\partial^2 \mathbf{A}}{\partial r^2} = \frac{1}{r}\frac{\partial^2}{\partial r^2}(r\mathbf{A}),$$

we obtain

$$\frac{\partial^2}{\partial r^2}(r\mathbf{A}) - \frac{1}{c^2}\frac{\partial^2}{\partial t^2}(r\mathbf{A}) = 0, \qquad (6.52)$$

which is nothing but a wave equation whose amplitude is expressed by $r\mathbf{A}$. Hence, its solution should include a scalar function expressed by $f(r - ct) + g(r + ct)$ and thus $\mathbf{A}$ should include a scalar function, $\{f(r - ct) + g(r + ct)\}/r$. $f(r - ct)/r$ expresses a wave that is generated at the origin and propagates spherically with decreasing amplitude by a factor of $1/r$, while $g(r + ct)/r$ is a wave that spherically converges to the origin, both of which are called *spherical wave*.

6.3.2 *Derivation of General Expression for Mie Scattering*

6.3.2.1 Polar coordinate and boundary conditions

In the following, we will consider the light scattering due to a dielectric sphere that is illuminated by a plane wave of light, and will derive a general expression for this scattering problem. Basically, we will follow the derivation given in the famous book *Principles of Optics*, written by Born and Wolf [Born and Wolf (1959)]. However, in order to simplify it and to avoid misunderstandings, a few modifications will be made, particularly to the notations.

Consider a case where a linearly polarized, a plane wave of light propagates along the z axis and illuminates a dielectric sphere with a radius of a fixed within a uniform medium, as is shown schematically in Fig. 6.4. We take the x axis to agree with the polarization direction so that the direction of the magnetic field becomes parallel to the y axis. We denote the spaces outside and inside the sphere as I and II, where we consider the materials constituting these spaces have different permittivities and permeabilities.

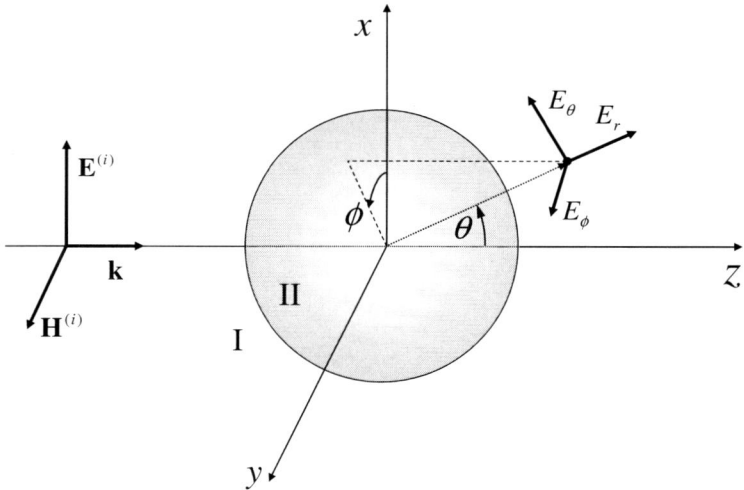

Figure 6.4 Geometry for Mie scattering.

We begin with the following Maxwell equations concerning electric and magnetic fields:

$$\nabla \times \mathbf{H} = \mathbf{j} + \frac{\partial \mathbf{D}}{\partial t}, \tag{6.53}$$

$$\nabla \times \mathbf{E} = -\frac{\partial \mathbf{B}}{\partial t}. \tag{6.54}$$

We consider a x-polarized monochromatic wave propagating to the $+z$ direction with an angular frequency ω as an incident light wave, which is generally expressed in a form of

$$E_x^{(i)} = E_{x0}^{(i)} e^{i(kz-\omega t)}.$$

Then, we incorporate the current density $\mathbf{j}$ into a complex permittivity as has been done in Sec. 2.2.2. Since the time dependence of all the quantities concerning the scattering of this wave is anyway expressed in a form of $\exp[-i\omega t]$, we will omit it and use the same symbols such as $\mathbf{H}$ and $\mathbf{E}$. Then, the above Maxwell equations become

$$\nabla \times \mathbf{H} = -i\omega\epsilon\mathbf{E}, \tag{6.55}$$

$$\nabla \times \mathbf{E} = i\omega\mu\mathbf{H}, \tag{6.56}$$

where ϵ in this case is obtained by replacing $(\epsilon + i\sigma/\omega) \to \epsilon$ with σ conductivity. The electric and magnetic fields of the x-polarized incident wave of light are then expressed as

$$E_x^{(i)} = e^{ikz},$$ (6.57)

$$H_y^{(i)} = \frac{k}{\omega\mu} e^{ikz},$$ (6.58)

where we put the amplitude of the incident electric field to be unity for simplicity.

When a spherical particle is fixed in a medium and a plane wave of light is incident on it, the scattering wave will be generated in the outer space, while the complex electromagnetic field will be generated inside the particle. Therefore, the field outside the particle is expressed by a sum of the incident (i) and scattered (s) waves, which are different from the field (w) inside the particle. Thus the electric fields in this system will be expressed as

$$\mathbf{E} = \begin{cases} \mathbf{E}^{(i)} + \mathbf{E}^{(s)} & \cdots\cdots r > a, \\ \mathbf{E}^{(w)} & \cdots\cdots r < a. \end{cases}$$ (6.59)

Then, we rewrite the components of Eqs. (6.55) and (6.56) into a polar coordinate

$$-i\omega\epsilon E_r = \frac{1}{r^2 \sin\theta} \left\{ \frac{\partial(r H_\phi \sin\theta)}{\partial\theta} - \frac{\partial(r H_\theta)}{\partial\phi} \right\},$$ (6.60)

$$-i\omega\epsilon E_\theta = \frac{1}{r \sin\theta} \left\{ \frac{\partial H_r}{\partial\phi} - \frac{\partial(r H_\phi \sin\theta)}{\partial r} \right\},$$ (6.61)

$$-i\omega\epsilon E_\phi = \frac{1}{r} \left\{ \frac{\partial(r H_\theta)}{\partial r} - \frac{\partial H_r}{\partial\theta} \right\},$$ (6.62)

$$i\omega\mu H_r = \frac{1}{r^2 \sin\theta} \left\{ \frac{\partial(r E_\phi \sin\theta)}{\partial\theta} - \frac{\partial(r E_\theta)}{\partial\phi} \right\},$$ (6.63)

$$i\omega\mu H_\theta = \frac{1}{r \sin\theta} \left\{ \frac{\partial E_r}{\partial\phi} - \frac{\partial(r E_\phi \sin\theta)}{\partial r} \right\},$$ (6.64)

$$i\omega\mu H_\phi = \frac{1}{r} \left\{ \frac{\partial(r E_\theta)}{\partial r} - \frac{\partial E_r}{\partial\theta} \right\},$$ (6.65)

where we have used Eqs. (6.47)–(6.49).

In order to solve the equation, it is better to find two independent particular solutions and to find a general solution by taking a sum of these solutions. It is a custom to find the solutions for the following two special cases: One is a case where a longitudinal component of the magnetic field is absent, which is called *TM (transverse magnetic) mode*. The other is a case where a longitudinal component of the electric field is absent, which is called *TE (transverse electric) mode*. According to Born and Wolf, we put TM and TE modes as symbols of *e* and *m*, respectively. Thus, from the definitions, the following relations are assumed to be satisfied:

$$^eE_r = E_r, \quad ^eH_r = 0, \quad ^mE_r = 0, \quad ^mH_r = H_r,$$

We first exemplify a case of the TM mode. By putting $H_r = 0$ in Eqs. (6.61) and (6.62), we obtain

$$i\omega\epsilon\,^eE_\theta = \frac{1}{r\sin\theta}\frac{\partial(r\,^eH_\phi\sin\theta)}{\partial r},$$

$$-i\omega\epsilon\,^eE_\phi = \frac{1}{r}\frac{\partial(r\,^eH_\theta)}{\partial r},$$

which lead to

$$i\omega\mu\,^eH_\theta = \frac{1}{r\sin\theta}\left\{\frac{\partial\,^eE_r}{\partial\phi} - \frac{\partial}{\partial r}\left(-\frac{1}{i\omega\epsilon}\frac{\partial(r\,^eH_\theta)}{\partial r}\sin\theta\right)\right\},$$

by inserting the latter relation into Eq. (6.64). Transforming further, we will reach a simple form:

$$\left(\frac{\partial^2}{\partial r^2} + k^2\right)(r\,^eH_\theta) = -\frac{i\omega\epsilon}{\sin\theta}\frac{\partial\,^eE_r}{\partial\phi}, \tag{6.66}$$

where we put $k^2 = \omega^2\epsilon\mu$. On the other hand, if we insert the former relation into Eq. (6.65), we obtain

$$i\omega\mu\,^eH_\phi = \frac{1}{r}\left\{\frac{\partial}{\partial r}\left(\frac{1}{i\omega\epsilon}\frac{\partial(r\,^eH_\phi)}{\partial r}\right) - \frac{\partial\,^eE_r}{\partial\theta}\right\},$$

which leads to

$$\left(\frac{\partial^2}{\partial r^2} + k^2\right)(r\,^eH_\phi) = i\omega\epsilon\frac{\partial\,^eE_r}{\partial\theta}. \tag{6.67}$$

In order to solve these equations, it is necessary to obtain all the components of the electric and magnetic fields except for one component that has been put to be zero. To simplify this procedure, it is convenient to find out scalar functions which finally lead to the electric and magnetic field vectors. For this purpose, we will direct our attention to Eq. (6.63), the left-hand side of which becomes zero in TM mode. Thus, a relation

$$\frac{\partial(r\,^eE_\phi\sin\theta)}{\partial\theta} = \frac{\partial(r\,^eE_\theta)}{\partial\phi},$$

is always satisfied in TM mode. This relation can be expressed by using a scalar function U as

$$^eE_\phi = \frac{1}{r\sin\theta}\frac{\partial U}{\partial\phi}, \qquad ^eE_\theta = \frac{1}{r}\frac{\partial U}{\partial\theta}. \tag{6.68}$$

Further, we will express U as a product of another scalar function u and r, such that

$$U = \frac{\partial(ru)}{\partial r}, \tag{6.69}$$

which is then inserted into Eq. (6.68). Thus, we obtain

$$^eE_\theta = \frac{1}{r}\frac{\partial^2(ru)}{\partial r\partial\theta}, \tag{6.70}$$

$$^eE_\phi = \frac{1}{r\sin\theta}\frac{\partial^2(ru)}{\partial r\partial\phi}. \tag{6.71}$$

On the other hand, if we put $^eH_r = 0$ in Eq. (6.61) and insert Eq. (6.70), we obtain

$$-i\omega\epsilon\frac{1}{r}\frac{\partial^2(ru)}{\partial r\partial\theta} = -\frac{1}{r\sin\theta}\frac{\partial(r\,^eH_\phi\sin\theta)}{\partial r},$$

which leads to[a]

$$^eH_\phi = \frac{i\omega\epsilon}{r}\frac{\partial(ru)}{\partial\theta}. \tag{6.72}$$

In a similar manner, using Eq. (6.62), we obtain

$$^eH_\theta = -\frac{i\omega\epsilon}{r\sin\theta}\frac{\partial(ru)}{\partial\phi}. \tag{6.73}$$

Further, inserting Eqs. (6.72) and (6.73) into Eq. (6.60), we obtain

$$^eE_r = -\frac{1}{r\sin\theta}\left\{\frac{\partial}{\partial\theta}\left(\sin\theta\frac{\partial u}{\partial\theta}\right) + \frac{1}{\sin\theta}\frac{\partial^2 u}{\partial\phi^2}\right\}. \tag{6.74}$$

If we insert Eqs. (6.72), (6.73) and (6.74) into Eqs. (6.66) and (6.67), the following relations are obtained:

$$\left(\frac{\partial^2}{\partial r^2}+k^2\right)\frac{\partial(ru)}{\partial\phi} = \frac{\partial}{\partial\phi}\left\{-\frac{1}{r\sin\theta}\left[\frac{\partial}{\partial\theta}\left(\sin\theta\frac{\partial u}{\partial\theta}\right)+\frac{1}{\sin\theta}\frac{\partial^2 u}{\partial\phi^2}\right]\right\},$$

$$\left(\frac{\partial^2}{\partial r^2}+k^2\right)\frac{\partial(ru)}{\partial\theta} = \frac{\partial}{\partial\theta}\left\{-\frac{1}{r\sin\theta}\left[\frac{\partial}{\partial\theta}\left(\sin\theta\frac{\partial u}{\partial\theta}\right)+\frac{1}{\sin\theta}\frac{\partial^2 u}{\partial\phi^2}\right]\right\}.$$

[a]Here, it is natural to consider that a constant term $C(\theta,\phi)$ with respect to r should be added to this relation. Even if such a term is present, we can newly define a function ru' so as to satisfy the relation $(i\omega\epsilon/r)\partial(ru)/\partial\theta - C(\theta,\phi)/(r\sin\theta) = (i\omega\epsilon/r)\partial(ru')/\partial\theta$.

We notice that these two relations are obtainable by differentiating the following function with respect to θ or ϕ, and then setting the result to be zero:

$$\left(\frac{\partial^2}{\partial r^2} + k^2\right)(ru) + \frac{1}{r\sin\theta}\left[\frac{\partial}{\partial\theta}\left(\sin\theta\frac{\partial u}{\partial\theta}\right) + \frac{1}{\sin\theta}\frac{\partial^2 u}{\partial\phi^2}\right]. \quad (6.75)$$

This fact leads to the following simple relation after some considerations:[a]

$$\frac{1}{r}\frac{\partial^2(ru)}{\partial r^2} + \frac{1}{r^2\sin\theta}\frac{\partial}{\partial\theta}\left(\sin\theta\frac{\partial u}{\partial\theta}\right) + \frac{1}{r^2\sin^2\theta}\frac{\partial^2 u}{\partial\phi^2} + k^2 u = 0, \quad (6.76)$$

which indicates that a function u satisfies a wave function expressed in a polar coordinate (compare with Eq. (6.51)). Further, using this, Eq. (6.74) reduces to

$$^eE_r = \frac{\partial^2(ru)}{\partial r^2} + k^2 ru. \quad (6.77)$$

As for the TE mode, the following replacements are applicable:

$$e \longleftrightarrow m,$$
$$-i\omega\epsilon \longleftrightarrow i\omega\mu,$$
$$E \longleftrightarrow H,$$
$$u \longleftrightarrow v,$$

where we consider a scalar function v instead of u in the TM mode. The general solutions are obtained by adding the solutions for the

---

[a]The fact that a function becomes zero after differentiating with respect to two independent variables of θ and ϕ means that the function is expressed by a function of r. Thus, we put $(\partial^2/\partial r^2 + k^2)(ru) + F(r; u) = C(r)$, where $C(r)$ is an arbitrary function of r and $F(r; u)$ is the remaining part of Eq. (6.75). To incorporate $C(r)$ into u, we define a new function u' so as to satisfy a relation $(\partial^2/\partial r^2 + k^2)(ru) - C(r) = (\partial^2/\partial r^2 + k^2)(ru')$. Further, we assume that we can determine a function $D(r)$ to satisfy $C(r) = -(\partial^2/\partial r^2 + k^2)D(r)$. Thus, a relation $(\partial^2/\partial r^2 + k^2)(ru + D(r)) = (\partial^2/\partial r^2 + k^2)(ru')$ holds. Hence, the new function ru' can be expressed as $ru' = ru + D(r)$, and using ru', we can put $(\partial^2/\partial r^2 + k^2)(ru') + F(r; u') = 0$, since $F(r; u') = F(r; u)$.

TM and TE modes, and are expressed as

$$E_r = \frac{\partial^2(ru)}{\partial r^2} + k^2 ru, \tag{6.78}$$

$$E_\theta = \frac{1}{r}\frac{\partial^2(ru)}{\partial r\partial\theta} + i\omega\mu\frac{1}{r\sin\theta}\frac{\partial(rv)}{\partial\phi}, \tag{6.79}$$

$$E_\phi = \frac{1}{r\sin\theta}\frac{\partial^2(ru)}{\partial r\partial\phi} - i\omega\mu\frac{1}{r}\frac{\partial(rv)}{\partial\theta}, \tag{6.80}$$

$$H_r = k^2 rv + \frac{\partial^2(rv)}{\partial r^2}, \tag{6.81}$$

$$H_\theta = -i\omega\epsilon\frac{1}{r\sin\theta}\frac{\partial(ru)}{\partial\phi} + \frac{1}{r}\frac{\partial^2(rv)}{\partial r\partial\theta}, \tag{6.82}$$

$$H_\phi = i\omega\epsilon\frac{1}{r}\frac{\partial(ru)}{\partial\theta} + \frac{1}{r\sin\theta}\frac{\partial^2(rv)}{\partial r\partial\phi}. \tag{6.83}$$

The boundary conditions at the surface of the sphere are expressed in terms that the tangential components of the electric and magnetic fields, that is, E_θ, E_ϕ, H_θ and H_ϕ, inside and outside the sphere agree with each other. To this end, the following four functions should connect continuously at the surface.

$$\epsilon ru, \quad \mu rv, \quad \partial(ru)/\partial r, \quad \partial(rv)/\partial r. \tag{6.84}$$

6.3.2.2 Separation of variables and series expansion

Both the functions u and v satisfy a wave equation of a form of Eq. (6.76). It will be possible to solve this equation by using a method of the separation of variables. For this purpose, we put

$$u = R(r)\Theta(\theta)\Phi(\phi), \tag{6.85}$$

and insert it into Eq. (6.76), which results in

$$\frac{1}{r}\frac{\partial^2(rR)}{\partial r^2}\Theta\Phi + \frac{1}{r^2\sin\theta}\frac{\partial}{\partial\theta}\left(\sin\theta\frac{\partial\Theta}{\partial\theta}\right)R\Phi$$
$$+ \frac{1}{r^2\sin^2\theta}\frac{\partial^2\Phi}{\partial\phi^2}R\Theta + k^2 R\Theta\Phi = 0.$$

Dividing both sides of this equation by $R\Theta\Phi/r^2$ and arranging the terms containing r to the left-hand side and those containing θ and ϕ to the right-hand side, we obtain

$$\frac{r}{R}\frac{\partial(rR)}{\partial r^2} + k^2 r^2 = -\frac{1}{\Theta\sin\theta}\frac{\partial}{\partial\theta}\left(\sin\theta\frac{\partial\Theta}{\partial\theta}\right) - \frac{1}{\Phi\sin^2\theta}\frac{\partial^2\Phi}{\partial\phi^2} \equiv \alpha.$$

Since this equation is expressed in terms that the left-hand side is only a function of r, while its right-hand side is that of θ and ϕ, it holds only when both the left- and right-hand sides are expressed by a common constant, which we have set at α in the above equation.

From this procedure, we obtain an equation with respect to r as

$$\frac{d^2(rR)}{dr^2} + \left(k^2 - \frac{\alpha}{r^2}\right)rR = 0, \tag{6.86}$$

and that with respect to θ and ϕ as

$$-\frac{\sin\theta}{\Theta}\frac{\partial}{\partial\theta}\left(\sin\theta\frac{\partial\Theta}{\partial\theta}\right) - \alpha\sin^2\theta = \frac{1}{\Phi}\frac{\partial^2\Phi}{\partial\phi^2} \equiv -\beta,$$

where we have again arranged the terms so that the left-hand side is only a function of θ, while the right-hand side is that of ϕ. In a similar way as above, this equation holds only when the functions in the left- and right-hand sides become a constant, which we have set at $-\beta$. This procedure further divides the equation into two as shown below:

$$\frac{1}{\sin\theta}\frac{d}{d\theta}\left(\sin\theta\frac{d\Theta}{d\theta}\right) + \left(\alpha - \frac{\beta}{\sin^2\theta}\right)\Theta = 0, \tag{6.87}$$

$$\frac{d^2\Phi}{d\phi^2} + \beta\Phi = 0. \tag{6.88}$$

The last equation, Eq. (6.88), is easily solved and its general solution is expressed as

$$\Phi = a\cos(\sqrt{\beta}\phi) + b\sin(\sqrt{\beta}\phi),$$

where a and b are arbitrary constants, which should be determined by the boundary conditions. Since the function $\Phi(\phi)$ should be a periodic function of ϕ with a period of 2π, β should take a value of $\beta = m^2$, where m is an integer. Thus, we obtain

$$\Phi = a_m\cos(m\phi) + b_m\sin(m\phi), \tag{6.89}$$

where a_m and b_m are constants corresponding to m.

Equation (6.87) can be transformed into

$$\frac{d}{d\xi}\left\{(1-\xi^2)\frac{d\Theta}{d\xi}\right\} + \left(\alpha - \frac{m^2}{1-\xi^2}\right)\Theta = 0, \tag{6.90}$$

by changing a variable as $\xi = \cos\theta$ and using $d\xi = -\sin\theta\, d\theta$. This equation corresponds to an associated Legendre function by putting $\alpha = l(l+1)$. Thus,

$$\Theta = P_l^m(\xi) = P_l^m(\cos\theta), \tag{6.91}$$

where l is an integer satisfying $l > |m|$.

Finally, Eq. (6.86) is transformed into

$$\frac{1}{k^2}\left(2\frac{dR}{dr} + r\frac{d^2R}{dr^2}\right) + \left\{1 - \frac{l(l+1)}{k^2r^2}\right\}rR = 0,$$

after dividing the both sides by k^2. Further, changing a variable by putting $kr = \rho$, we obtain

$$\frac{d^2R}{d\rho^2} + \frac{2}{\rho}\frac{dR}{d\rho} + \left\{1 - \frac{l(l+1)}{\rho^2}\right\}R = 0. \tag{6.92}$$

This equation is known as a kind of Bessel differential equations, whose solution is called spherical Bessel function. The general solution to this equation is expressed by a sum of arbitrary two spherical Bessel functions among the following four:

$$j_n(\rho) = \sqrt{\frac{\pi}{2\rho}}J_{n+\frac{1}{2}}(\rho), \tag{6.93}$$

$$n_n(\rho) = \sqrt{\frac{\pi}{2\rho}}N_{n+\frac{1}{2}}(\rho), \tag{6.94}$$

$$h_n^{(1)}(\rho) = \sqrt{\frac{\pi}{2\rho}}H_{n+\frac{1}{2}}^{(1)}(\rho) = j_n(\rho) + in_n(\rho), \tag{6.95}$$

$$h_n^{(2)}(\rho) = \sqrt{\frac{\pi}{2\rho}}H_{n+\frac{1}{2}}^{(2)}(\rho) = j_n(\rho) - in_n(\rho). \tag{6.96}$$

Here, $J_{n+\frac{1}{2}}$ and $N_{n+\frac{1}{2}}$ denote Bessel and Neumann functions, and $H_{n+\frac{1}{2}}^{(1)}$ and $H_{n+\frac{1}{2}}^{(2)}$ are Hankel functions of the first and second kind, respectively.

In our case, since the solution is anyway in a form of spherical wave, rR, it is convenient to define functions as

$$\psi_l(\rho) = \rho j_l(\rho) = \sqrt{\frac{\pi\rho}{2}}J_{l+\frac{1}{2}}(\rho),$$

$$\chi_l(\rho) = -\rho n_l(\rho) = -\sqrt{\frac{\pi\rho}{2}}N_{l+\frac{1}{2}}(\rho). \tag{6.97}$$

Thus, the general solution to Eq. (6.86) is expressed as

$$r R = c_l \psi_l(kr) + d_l \chi_l(kr). \qquad (6.98)$$

Combining these solutions, we finally obtain a general solution to the wave function, Eq. (6.76), with respect to ru, which is expr essed as follows:

$$ru = \sum_{l=0}^{\infty} \sum_{m=-l}^{l} \{c_l \psi_l(kr) + d_l \chi_l(kr)\} P_l^m(\cos\theta)\{a_m \cos(m\phi) + b_m \sin(m\phi)\}.$$

$$(6.99)$$

Next, we will express the incident wave in terms of similar series expansions as above. For this purpose, we express the components of the electric and magnetic fields of the incident wave in a polar coordinate. For example, by using Eq. (6.43), the radial part of the electric field (Eq. (6.57)) is expressed as

$$E_r^{(i)} = e^{ik^{(l)}r\cos\theta} \sin\theta \cos\phi,$$

where $k^{(l)}$ is a wave vector of the incident light in the medium l. In order to obtain a scalar function $u^{(i)}$ for this component, we use Eq. (6.78) and obtain

$$e^{ik^{(l)}r\cos\theta} \sin\theta \cos\phi = \frac{\partial^2(ru^{(i)})}{\partial r^2} + k^{(l)2}ru^{(i)}, \qquad (6.100)$$

which should be solved to obtain $u^{(i)}$. For this purpose, we use the following mathematical relation,

$$e^{ik^{(l)}r\cos\theta} = \sum_{l=0}^{\infty} i^l(2l+1)j_l(k^{(l)}r)P_l(\cos\theta) \qquad (6.101)$$

$$= \sum_{l=0}^{\infty} i^l(2l+1)\frac{\psi_l(k^{(l)}r)}{k^{(l)}r}P_l(\cos\theta). \qquad (6.102)$$

Further, using the relation,

$$e^{ik^{(l)}r\cos\theta}\sin\theta = -\frac{1}{ik^{(l)}r}\frac{\partial}{\partial\theta}e^{ik^{(l)}r\cos\theta}, \qquad (6.103)$$

we obtain

$$e^{ik^{(l)}r\cos\theta}\sin\theta\cos\phi = -\frac{1}{ik^{(l)}r}\sum_{l=0}^{\infty} i^l(2l+1)\frac{\psi_l(k^{(l)}r)}{k^{(l)}r}\frac{\partial}{\partial\theta}P_l(\cos\theta)\cos\phi$$

$$= \frac{1}{(k^{(l)}r)^2}\sum_{l=1}^{\infty} i^{l-1}(2l+1)\psi_l(k^{(l)}r)P_l^1(\cos\theta)\cos\phi,$$

$$(6.104)$$

where we have used the relation $(\partial/\partial\theta)P_l(\cos\theta) = -P_l^1(\cos\theta)$ with an associated Legendre's function $P_l^m(\cos\theta)$ of degree l and order m, and have further removed a term for $l = 0$ using a relation of $P_0^1(\cos\theta) = 0$.

To solve the differential equation, Eq. (6.100), we employ an expansion series as a trial function, which has a similar form as Eq. (6.104):

$$ru^{(i)} = \frac{1}{k^{(I)2}} \sum_{l=1}^{\infty} \alpha_l \psi_l(k^{(I)}r) P_l^1(\cos\theta)\cos\phi,$$

where α_l is a constant to be determined. Inserting this function into the right-hand side of Eq. (6.100) and comparing the terms in both sides, we obtain

$$\alpha_l \left\{ k^{(I)2}\psi_l(k^{(I)}r) + \frac{d^2\psi_l(k^{(I)}r)}{dr^2} \right\} = i^{l-1}(2l+1)\frac{\psi_l(k^{(I)}r)}{r^2}, \quad (6.105)$$

where we have to determine a constant α_l for any value of r, which formally has a form of differential equation with respect to $\psi_l(k^{(I)}r)$. As shown in Eq. (6.98), a general solution to rR satisfying a wave function in medium I should be expressed by $rR^{(I)} = c_l\psi_l(k^{(I)}r) + d_l\chi_l(k^{(I)}r)$. Since the incident wave also satisfies the wave function and is expressed only by $\psi_l(k^{(I)}r)$, it is natural to take a combination of $c_l = 1$ and $d_l = 0$, and hence $rR^{(I)} = \psi_l(k^{(I)}r)$. On the other hand, from Eq. (6.86), $rR^{(I)}$ should satisfy

$$\frac{d^2(rR^{(I)})}{dr^2} + \left(k^{(I)2} - \frac{l(l+1)}{r^2}\right)rR^{(I)} = 0. \quad (6.106)$$

Transforming Eq. (6.105) into the following form,

$$\frac{d^2}{dr^2}\psi_l(k^{(I)}r) + \left(k^{(I)2} - \frac{i^{l-1}(2l+1)}{\alpha_l r^2}\right)\psi_l(k^{(I)}r) = 0, \quad (6.107)$$

and comparing it with the above formula, we immediately find a relation

$$\alpha_l = \frac{i^{l-1}(2l+1)}{l(l+1)}. \quad (6.108)$$

Thus, we obtain the expression for $ru^{(i)}$ as follows:

$$ru^{(i)} = \frac{1}{k^{(I)2}} \sum_{l=1}^{\infty} i^{l-1}\frac{(2l+1)}{l(l+1)}\psi_l(k^{(I)}r)P_l^1(\cos\theta)\cos\phi. \quad (6.109)$$

Next, we will use Eq. (6.81) to obtain $rv^{(i)}$. For this purpose, we transform Eq. (6.58) into a polar coordinate as

$$H_r^{(i)} = \frac{k^{(I)}}{\omega\mu^{(I)}} e^{ik^{(I)}r\cos\theta} \sin\theta \sin\phi.$$

In a similar manner as in the electric field, we obtain

$$rv^{(i)} = \frac{1}{k^{(I)2}} \sum_{l=1}^{\infty} i^{l-1} \frac{k^{(I)}}{\omega\mu^{(I)}} \frac{(2l+1)}{l(l+1)} \psi_l(k^{(I)}r) P_l^1(\cos\theta) \sin\phi$$

$$= \frac{i}{k^{(I)2}i\omega\mu^{(I)}} \sum_{l=1}^{\infty} i^{l-1} k^{(I)} \frac{(2l+1)}{l(l+1)} \psi_l(k^{(I)}r) P_l^1(\cos\theta) \sin\phi.$$

(6.110)

Thus, both $ru^{(i)}$ and $rv^{(i)}$ are obtained for the incident wave.

From the boundary conditions, it is expected that ϕ dependence for $ru^{(s,w)}$ and $rv^{(s,w)}$ are the same as that for $ru^{(i)}$ and $rv^{(i)}$. Further, we have to choose the spherical Bessel functions for an inner wave so as not to diverge at the center of the sphere and for a scattering wave to become a spherical wave in its asymptotic form. Thus, for the former, ψ_l is considered to be appropriate, while for the latter, $h_l^{(1)}$ seems to be appropriate, because its asymptotic form is expressed as $\exp[ikr]/r$ in a far field. Thus, the scalar functions corresponding to the inner and scattering fields are expressed as

$$ru^{(w)} = \frac{1}{k^{(II)2}} \sum_{l=1}^{\infty} {}^eA_l \psi_l(k^{(II)}r) P_l^1(\cos\theta) \cos\phi, \qquad (6.111)$$

$$rv^{(w)} = \frac{i}{k^{(II)}i\omega\mu^{(II)}} \sum_{l=1}^{\infty} {}^mA_l \psi_l(k^{(II)}r) P_l^1(\cos\theta) \sin\phi, \quad (6.112)$$

$$ru^{(s)} = \frac{1}{k^{(I)2}} \sum_{l=1}^{\infty} {}^eB_l \zeta_l^{(1)}(k^{(I)}r) P_l^1(\cos\theta) \cos\phi, \qquad (6.113)$$

$$rv^{(s)} = \frac{i}{k^{(I)}i\omega\mu^{(I)}} \sum_{l=1}^{\infty} {}^mB_l \zeta_l^{(1)}(k^{(I)}r) P_l^1(\cos\theta) \sin\phi, \quad (6.114)$$

where we put $\zeta_l^{(1)}(\rho) = \psi_l(\rho) - i\chi_l(\rho) = \rho h_l^{(1)}(\rho)$, and ${}^eA_l$, ${}^mA_l$, ${}^eB_l$ and ${}^mB_l$ are unknown constants that should be determined from now.

Using Eqs. (6.109)–(6.110) and Eqs. (6.111)–(6.114), we have to impose the condition that the functions indicated in Eq. (6.84) are continuous at the surface of the sphere. Under this condition, we can determine the unknown constants, which will give the general expressions for the inner and scattering fields in the presence of a sphere. We first consider a function $\epsilon r u$ and compare the lth terms for the incident and scattered fields with that for the inner field at the surface of the sphere $r = a$ as[a]

$$\frac{\epsilon^{(I)}}{k^{(I)2}}\, {}^{e}B_{l}\zeta_{l}(q) + \frac{\epsilon^{(I)}}{k^{(I)2}}i^{l-1}\frac{2l+1}{l(l+1)}\psi_{l}(q) = \frac{\epsilon^{(II)}}{k^{(II)2}}\, {}^{e}A_{l}\psi_{l}(\hat{n}q),$$

where we put $q \equiv k^{(I)}a$. Further, we define the ratio of refractive index for a medium II to I as $\hat{n} = k^{(II)}/k^{(I)} = n^{(II)}/n^{(I)}$, and put $k^{(II)}a = \hat{n}q$. In a similar manner, we obtain the condition for $\partial(ru)/\partial r$ as

$$\frac{1}{k^{(I)}}\, {}^{e}B_{l}\zeta_{l}'(q) + \frac{1}{k^{(I)}}i^{l-1}\frac{2l+1}{l(l+1)}\psi_{l}'(q) = \frac{1}{k^{(II)}}\, {}^{e}A_{l}\psi_{l}'(\hat{n}q),$$

where $\zeta_{l}'(q)$ and $\psi_{l}'(q)$ are the derivative of the functions with respect to q.

Further, using the relation $k^2 = \omega^2 \epsilon \mu$ and performing the similar calculations, we can obtain the following four relations

$$\begin{aligned}
{}^{e}B_{l}\frac{1}{k^{(I)}}\zeta_{l}^{(1)'}(q) + \frac{1}{k^{(I)}}i^{l-1}\frac{2l+1}{l(l+1)}\psi_{l}'(q) &= {}^{e}A_{l}\frac{1}{k^{(II)}}\psi_{l}'(\hat{n}q),\\
{}^{m}B_{l}\frac{1}{\mu^{(I)}}\zeta_{l}^{(1)'}(q) + \frac{1}{\mu^{(I)}}i^{l-1}\frac{2l+1}{l(l+1)}\psi_{l}'(q) &= {}^{m}A_{l}\frac{1}{\mu^{(II)}}\psi_{l}'(\hat{n}q),\\
{}^{e}B_{l}\frac{1}{\mu^{(I)}}\zeta_{l}^{(1)}(q) + \frac{1}{\mu^{(I)}}i^{l-1}\frac{2l+1}{l(l+1)}\psi_{l}(q) &= {}^{e}A_{l}\frac{1}{\mu^{(II)}}\psi_{l}(\hat{n}q),\\
{}^{m}B_{l}\frac{1}{k^{(I)}}\zeta_{l}^{(1)}(q) + \frac{1}{k^{(I)}}i^{l-1}\frac{2l+1}{l(l+1)}\psi_{l}(q) &= {}^{m}A_{l}\frac{1}{k^{(II)}}\psi_{l}(\hat{n}q).
\end{aligned}$$

These equations are nothing but two sets of linear equations with respect to ${}^{e}A_l$ and ${}^{e}B_l$, and ${}^{m}A_l$ and ${}^{m}B_l$, which will be solved using the Cramer's rule. The solutions to ${}^{e}B_l$ and ${}^{m}B_l$ corresponding to scattering field are expressed as

$$
{}^{e}B_{l} = i^{l+1}\frac{2l+1}{l(l+1)}\frac{\begin{vmatrix} \frac{1}{k^{(II)}}\psi_{l}'(\hat{n}q) & \frac{1}{k^{(I)}}\psi_{l}'(q) \\ \frac{1}{\mu^{(II)}}\psi_{l}(\hat{n}q) & \frac{1}{\mu^{(I)}}\psi_{l}(q) \end{vmatrix}}{\begin{vmatrix} \frac{1}{k^{(II)}}\psi_{l}'(\hat{n}q) & \frac{1}{k^{(I)}}\zeta_{l}^{(1)'}(q) \\ \frac{1}{\mu^{(II)}}\psi_{l}(\hat{n}q) & \frac{1}{\mu^{(I)}}\zeta_{l}^{(1)}(q) \end{vmatrix}},
\tag{6.115}
$$

---

[a] To accommodate the notations to those given in "Principle of Optics", replace ϵ and μ appearing in the following expressions by k_1 and k_2 through the relations $k_1 \equiv i\omega\epsilon$ and $k_2 \equiv i\omega\mu$.

$$
{}^{m}B_{l} = i^{l+1} \frac{2l+1}{l(l+1)} \frac{\begin{vmatrix} \frac{1}{\mu^{(II)}}\psi_{l}'(\hat{n}q) & \frac{1}{\mu^{(I)}}\psi_{l}'(q) \\ \frac{1}{k^{(II)}}\psi_{l}(\hat{n}q) & \frac{1}{k^{(I)}}\psi_{l}(q) \end{vmatrix}}{\begin{vmatrix} \frac{1}{\mu^{(II)}}\psi_{l}'(\hat{n}q) & \frac{1}{\mu^{(I)}}\zeta_{l}^{(1)'}(q) \\ \frac{1}{k^{(II)}}\psi_{l}(\hat{n}q) & \frac{1}{k^{(I)}}\zeta_{l}^{(1)}(q) \end{vmatrix}}, \tag{6.116}
$$

which are explicitly expressed as

$$
\begin{aligned}
{}^{e}B_{l} &= i^{l+1} \frac{2l+1}{l(l+1)} \frac{\mu^{(I)}k^{(II)}\psi_{l}'(q)\psi_{l}(\hat{n}q) - \mu^{(II)}k^{(I)}\psi_{l}'(\hat{n}q)\psi_{l}(q)}{\mu^{(I)}k^{(II)}\zeta_{l}^{(1)'}(q)\psi_{l}(\hat{n}q) - \mu^{(II)}k^{(I)}\psi_{l}'(\hat{n}q)\zeta_{l}^{(1)}(q)} \\
&= i^{l+1} \frac{2l+1}{l(l+1)} \frac{(\mu^{(I)}/\mu^{(II)})\hat{n}\psi_{l}'(q)\psi_{l}(\hat{n}q) - \psi_{l}'(\hat{n}q)\psi_{l}(q)}{(\mu^{(I)}/\mu^{(II)})\hat{n}\zeta_{l}^{(1)'}(q)\psi_{l}(\hat{n}q) - \psi_{l}'(\hat{n}q)\zeta_{l}^{(1)}(q)},
\end{aligned} \tag{6.117}
$$

$$
\begin{aligned}
{}^{m}B_{l} &= i^{l+1} \frac{2l+1}{l(l+1)} \frac{\mu^{(I)}k^{(II)}\psi_{l}(q)\psi_{l}'(\hat{n}q) - \mu^{(II)}k^{(I)}\psi_{l}'(q)\psi_{l}(\hat{n}q)}{\mu^{(I)}k^{(II)}\zeta_{l}^{(1)}(q)\psi_{l}'(\hat{n}q) - \mu^{(II)}k^{(I)}\psi_{l}(\hat{n}q)\zeta_{l}^{(1)'}(q)} \\
&= i^{l+1} \frac{2l+1}{l(l+1)} \frac{(\mu^{(I)}/\mu^{(II)})\hat{n}\psi_{l}(q)\psi_{l}'(\hat{n}q) - \psi_{l}'(q)\psi_{l}(\hat{n}q)}{(\mu^{(I)}/\mu^{(II)})\hat{n}\zeta_{l}^{(1)}(q)\psi_{l}'(\hat{n}q) - \psi_{l}(\hat{n}q)\zeta_{l}^{(1)'}(q)}.
\end{aligned} \tag{6.118}
$$

Further, inserting Eqs. (6.113) and (6.114) into Eqs. (6.78)–(6.83), we can obtain the expressions for the scattering fields. We first consider Eq. (6.78). By inserting Eq. (6.113) into Eq. (6.78), we obtain

$$
\begin{aligned}
E_{r}^{(s)} &= \frac{\partial^{2}(ru^{(s)})}{\partial r^{2}} + k^{(I)2}ru^{(s)} \\
&= \frac{1}{k^{(I)2}} \sum_{l=1}^{\infty} {}^{e}B_{l} \left\{ \frac{\mathrm{d}^{2}}{\mathrm{d}r^{2}}\zeta_{l}^{(1)}(k^{(I)}r) + k^{(I)2}\zeta_{l}^{(1)}(k^{(I)}r) \right\} P_{l}^{1}(\cos\theta)\cos\phi.
\end{aligned}
$$

The formula in the parenthesis is further transformed into

$$
\begin{aligned}
\{\ \} &= k^{(I)2}\frac{\mathrm{d}^{2}}{\mathrm{d}\rho^{2}}\rho h_{l}^{(1)}(\rho) + k^{(I)2}\rho h_{l}^{(1)}(\rho) \\
&= k^{(I)2}\rho \left\{ \frac{2}{\rho}\frac{\mathrm{d}h_{l}^{(1)}}{\mathrm{d}\rho} + \frac{\mathrm{d}^{2}h_{l}^{(1)}}{\mathrm{d}\rho^{2}} + h_{l}^{(1)} \right\} = k^{(I)2}\frac{l(l+1)}{\rho^{2}}\rho h_{l}^{(1)},
\end{aligned}
$$

where we have changed the variable as $k^{(I)}r \to \rho$, and have used the relations $\zeta_{l}^{(1)}(\rho) = \rho h_{l}^{(1)}(\rho)$ and

$$
\frac{\mathrm{d}^{2}h_{l}^{(1)}}{\mathrm{d}\rho^{2}} + \frac{2}{\rho}\frac{\mathrm{d}h_{l}^{(1)}}{\mathrm{d}\rho} + \left\{ 1 - \frac{l(l+1)}{\rho^{2}} \right\}h_{l}^{(1)} = 0.
$$

Thus, we obtain

$$E_r^{(s)} = \frac{1}{k^{(I)2}} \frac{\cos\phi}{r^2} \sum_{l=1}^{\infty} l(l+1) \, {}^eB_l \zeta_l^{(1)}(k^{(I)}r) P_l^1(\cos\theta). \quad (6.119)$$

Next, we consider Eq. (6.79). Inserting Eqs. (6.113) and (6.114) into Eq. (6.79), we obtain

$$
\begin{aligned}
E_\theta^{(s)} &= \frac{1}{r} \frac{\partial^2(ru^{(s)})}{\partial r \partial\theta} + i\omega\mu^{(I)} \frac{1}{r\sin\theta} \frac{\partial(rv^{(s)})}{\partial\phi} \\
&= \frac{1}{k^{(I)2}r} \sum_{l=1}^{\infty} {}^eB_l \frac{\partial}{\partial r} \zeta_l^{(1)}(k^{(I)}r) \frac{\partial}{\partial\theta} P_l^1(\cos\theta)\cos\phi \\
&\quad + i\omega\mu^{(I)} \frac{1}{r\sin\theta} \frac{i}{k^{(I)}i\omega\mu^{(I)}} \sum_{l=1}^{\infty} {}^mB_l \zeta_l^{(1)}(k^{(I)}r) P_l^1(\cos\theta)\cos\phi,
\end{aligned}
$$

which gives

$$
\begin{aligned}
E_\theta^{(s)} = -\frac{1}{k^{(I)}} \frac{\cos\phi}{r} \sum_{l=1}^{\infty} \Big\{ &{}^eB_l \zeta_l^{(1)'}(k^{(I)}r) P_l^{1'}(\cos\theta)\sin\theta \\
&- i \, {}^mB_l \zeta_l^{(1)}(k^{(I)}r) P_l^1(\cos\theta)\frac{1}{\sin\theta} \Big\}. \quad (6.120)
\end{aligned}
$$

In a similar manner, we obtain (Exercise (3))

$$
\begin{aligned}
E_\phi^{(s)} = -\frac{1}{k^{(I)}} \frac{\sin\phi}{r} \sum_{l=1}^{\infty} \Big\{ &{}^eB_l \zeta_l^{(1)'}(k^{(I)}r) P_l^1(\cos\theta)\frac{1}{\sin\theta} \\
&- i \, {}^mB_l \zeta_l^{(1)}(k^{(I)}r) P_l^{1'}(\cos\theta)\sin\theta \Big\}, \quad (6.121)
\end{aligned}
$$

$$H_r^{(s)} = \frac{1}{k^{(I)}\omega\mu^{(I)}} \frac{\sin\phi}{r^2} \sum_{l=1}^{\infty} l(l+1) \, {}^mB_l \zeta_l^{(1)}(k^{(I)}r) P_l^1(\cos\theta), (6.122)$$

$$
\begin{aligned}
H_\theta^{(s)} = \frac{i}{\omega\mu^{(I)}} \frac{\sin\phi}{r} \sum_{l=1}^{\infty} \Big\{ &{}^eB_l \zeta_l^{(1)}(k^{(I)}r) P_l^1(\cos\theta)\frac{1}{\sin\theta} \\
&+ i \, {}^mB_l \zeta_l^{(1)'}(k^{(I)}r) P_l^{1'}(\cos\theta)\sin\theta \Big\}, \quad (6.123)
\end{aligned}
$$

$$
\begin{aligned}
H_\phi^{(s)} = -\frac{i}{\omega\mu^{(I)}} \frac{\cos\phi}{r} \sum_{l=1}^{\infty} \Big\{ &{}^eB_l \zeta_l^{(1)}(k^{(I)}r) P_l^{1'}(\cos\theta)\sin\theta \\
&+ i \, {}^mB_l \zeta_l^{(1)'}(k^{(I)}r) P_l^1(\cos\theta)\frac{1}{\sin\theta} \Big\}. \quad (6.124)
\end{aligned}
$$

6.3.2.3 Simple expression for scattered wave in far field

The expression for the scattered wave obtained in the preceding section can be simplified when we consider only the far field. At first, looking at the expressions for $E_r^{(s)}$ and $H_r^{(s)}$ in Eqs. (6.119) and (6.122), we find that both components express a longitudinal wave and have the dependence of r^{-2}, while the other ones express a transverse wave and have the dependence of r^{-1}. Thus, if we consider a wave sufficiently far from the sphere, the longitudinal wave will vanish and only the transverse wave will remain. Hence, we will consider only the transverse wave, which are expressed by the following four components, $E_\theta^{(s)}$, $E_\phi^{(s)}$, $H_\theta^{(s)}$ and $H_\phi^{(s)}$.

Since a function $\zeta_l^{(1)}(\rho)$ is defined as

$$\zeta_l^{(1)}(\rho) = \sqrt{\frac{\pi\rho}{2}} H_{l+\frac{1}{2}}^{(1)}(\rho) = \rho h_l^{(1)}(\rho),$$

$\zeta_l^{(1)}(\rho)$ and its derivative are approximately expressed as

$$\zeta_l^{(1)}(\rho) \sim (-i)^{l+1} e^{i\rho}, \tag{6.125}$$

$$\zeta_l^{(1)'}(\rho) \sim (-i)^l e^{i\rho}, \tag{6.126}$$

in far field, because it is known that a spherical Bessel function $h_l^{(1)}(\rho)$ takes an asymptotic form of

$$h_l^{(1)}(\rho) \sim (-i)^{l+1} e^{i\rho}/\rho,$$

in the limit of $\rho \to \infty$. Further, following van der Hulst [van de Hulst (1957)], we replace the functions and constants as follows:

$$\pi_l(\cos\theta) = P_l^1(\cos\theta)/\sin\theta, \tag{6.127}$$

$$\tau_l(\cos\theta) = \frac{d}{d\theta} P_l^1(\cos\theta) = -\sin\theta\, P_l^{1'}(\cos\theta), \tag{6.128}$$

and

$$^eB_l = i^{l+1} \frac{2l+1}{l(l+1)} a_l, \tag{6.129}$$

$$^mB_l = i^{l+1} \frac{2l+1}{l(l+1)} b_l. \tag{6.130}$$

Under these replacements, $E_\theta^{(s)}$ is expressed as

$$E_\theta^{(s)} = \frac{\cos\phi}{k^{(l)}r} e^{ik^{(l)}r} \sum_{l=1}^{\infty} (-i)^l \left\{ {}^eB_l \tau_l(\cos\theta) + {}^mB_l \pi_l(\cos\theta) \right\}$$

$$= \frac{i\cos\phi}{k^{(l)}r} e^{ik^{(l)}r} \sum_{l=1}^{\infty} \frac{2l+1}{l(l+1)} \left\{ a_l \tau_l(\cos\theta) + b_l \pi_l(\cos\theta) \right\}.$$

Therefore, putting

$$S_2(\theta) \equiv \sum_{l=1}^{\infty} \frac{2l+1}{l(l+1)} \{a_l \tau_l(\cos\theta) + b_l \pi_l(\cos\theta)\}, \qquad (6.131)$$

we obtain

$$E_\theta^{(s)} = \frac{\mathrm{i}\cos\phi}{k^{(l)}r} e^{\mathrm{i}k^{(l)}r} S_2(\theta). \qquad (6.132)$$

In a similar manner, we also obtain the expression for $H_\phi^{(s)}$ as

$$H_\phi^{(s)} = \frac{\mathrm{i}\cos\phi}{\omega\mu^{(l)}r} e^{\mathrm{i}k^{(l)}r} S_2(\theta). \qquad (6.133)$$

Using the above functions and constants, we can express $E_\phi^{(s)}$ as

$$E_\phi^{(s)} = -\frac{\sin\phi}{k^{(l)}r} e^{\mathrm{i}k^{(l)}r} \sum_{l=1}^{\infty} (-\mathrm{i})^l \{{}^e B_l \pi_l(\cos\theta) + {}^m B_l \tau_l(\cos\theta)\}$$

$$= -\frac{\mathrm{i}\sin\phi}{k^{(l)}r} e^{\mathrm{i}k^{(l)}r} \sum_{l=1}^{\infty} \frac{2l+1}{l(l+1)} \{a_l \pi_l(\cos\theta) + b_l \tau_l(\cos\theta)\}.$$

Thus, if we put

$$S_1(\theta) \equiv \sum_{l=1}^{\infty} \frac{2l+1}{l(l+1)} \{a_l \pi_l(\cos\theta) + b_l \tau_l(\cos\theta)\}, \qquad (6.134)$$

we can simply express both $E_\phi^{(s)}$ and $H_\theta^{(s)}$ as

$$E_\phi^{(s)} = -\frac{\mathrm{i}\sin\phi}{k^{(l)}r} e^{\mathrm{i}k^{(l)}r} S_1(\theta), \qquad (6.135)$$

$$H_\theta^{(s)} = \frac{\mathrm{i}\sin\phi}{\omega\mu^{(l)}r} e^{\mathrm{i}k^{(l)}r} S_1(\theta). \qquad (6.136)$$

Further, these relations lead to the following simple relations between the electric and magnetic fields

$$H_\phi^{(s)} = \frac{k^{(l)}}{\omega\mu^{(l)}} E_\theta^{(s)} = \sqrt{\frac{\epsilon^{(l)}}{\mu^{(l)}}} E_\theta^{(s)}, \qquad (6.137)$$

$$H_\theta^{(s)} = -\frac{k^{(l)}}{\omega\mu^{(l)}} E_\phi^{(s)} = -\sqrt{\frac{\epsilon^{(l)}}{\mu^{(l)}}} E_\phi^{(s)}, \qquad (6.138)$$

where a minus sign in Eq. (6.138) comes from a fact that we have employed the right-hand system for the coordinate (see Fig. 6.5). We have also used a relation of $k^{(l)} = \omega\sqrt{\epsilon^{(l)}\mu^{(l)}}$. Thus, the electric and magnetic fields of scattered wave in far field are expressed by Eqs. (6.132), (6.133), (6.135) and (6.136), which are eventually expressed by two newly defined functions of $S_1(\theta)$ and $S_2(\theta)$.

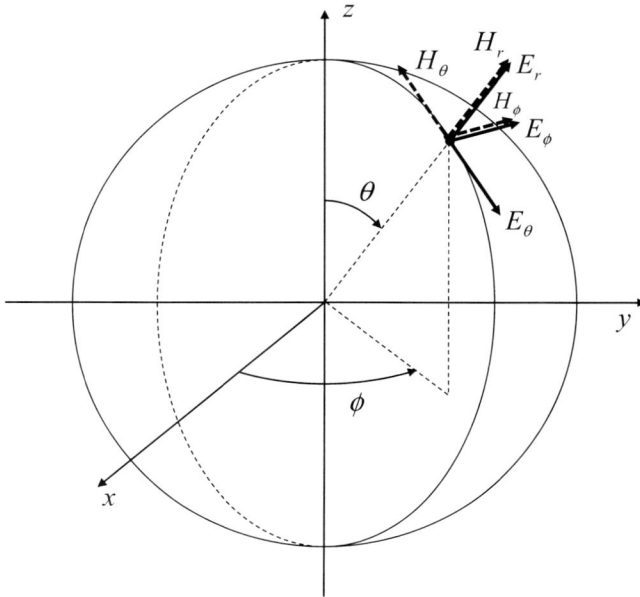

Figure 6.5 Geometry for a polar coordinate.

6.3.3 *Extinction and Scattering Cross Sections*

6.3.3.1 Extinction cross section

Extinction cross section expresses a quantity how the intensity of incident light is reduced owing to the scattering and absorption. The cross section expresses such loss of intensity in terms of an effective area that a plane wave of light illuminates. Thus, to obtain the cross section, it is evident that the decrease of the intensity of light that propagates along the incident direction should be calculated [van de Hulst (1957); Born and Wolf (1959); Bohren and Huffman (1983)].

Since in the medium *I*, only incident and scattered waves exist, and the magnetic field is connected with the electric field through the relations Eqs. (6.137) and (6.138), the intensity of light in far field is generally evaluated by

$$(\mathbf{E}^{(i)} + \mathbf{E}^{(s)}) \cdot (\mathbf{E}^{(i)*} + \mathbf{E}^{(s)*}).$$

In addition, we have assumed that the incident wave is polarized in the direction parallel to the *x* axis. The *x* components of $\mathbf{E}^{(i)}$ and $\mathbf{E}^{(s)}$

are expressed as

$$E_x^{(i)} = e^{ik^{(l)}z},$$

$$E_x^{(s)} = \frac{i}{k^{(l)}r}e^{ik^{(l)}r}\{S_2(\theta)\cos\theta\cos^2\phi + S_1(\theta)\sin^2\phi\}, \quad (6.139)$$

where we have used the inverse transformation from the polar to Cartesian coordinate according to Eq. (6.44) such that

$$E_x^{(s)} = E_\theta\cos\theta\cos\phi - E_\phi\sin\phi.$$

Then, the decrease in intensity of the incident wave can be evaluated only for the polarization directed to the x axis:

$$(\mathbf{E}^{(i)} + \mathbf{E}^{(s)})\cdot(\mathbf{E}^{(i)*} + \mathbf{E}^{(s)*})$$
$$= |E_x^{(i)}|^2 + 2\mathrm{Re}\{E_x^{(i)}E_x^{(s)*}\} + \mathbf{E}^{(s)}\cdot\mathbf{E}^{(s)*}.$$

Further, we assume that a position to observe the scattered light lies along the propagation direction of the incident wave and is sufficiently distant from the sphere. This assumption leads to the following approximate expression for r under the condition of $z \gg x, y$: Namely, we replace r according to the approximation $r = \sqrt{x^2 + y^2 + z^2} \approx z\{1 + (x^2 + y^2)/(2z^2)\} = z + (x^2 + y^2)/(2z)$ for the exponent in Eq. (6.139), while $r \approx z$ for the denominator. Thus, we obtain

$$(\mathbf{E}^{(i)} + \mathbf{E}^{(s)})\cdot(\mathbf{E}^{(i)*} + \mathbf{E}^{(s)*})$$
$$\approx 1 + 2\mathrm{Re}\left[\frac{-i}{k^{(l)}z}e^{-ik^{(l)}(x^2+y^2)/(2z)}\left\{S_2(\theta)\cos\theta\cos^2\phi + S_1(\theta)\sin^2\phi\right\}\right],$$

where we have neglected the term $|\mathbf{E}^{(s)}|^2$, because it includes a factor of z^{-2}. By putting $\theta \to 0$, the above expression reduces to

$$\to 1 + \frac{2}{k^{(l)}z}\mathrm{Re}\left[\frac{1}{i}e^{-ik^{(l)}(x^2+y^2)/(2z)}\left\{S_2(0)\cos^2\phi + S_1(0)\sin^2\phi\right\}\right].$$
$$(6.140)$$

Thus, we have to evaluate the values of $S_1(0)$ and $S_2(0)$. In these functions, the factors concerning θ are only included in the forms of $P_l^1(\cos\theta)/\sin\theta$ and $(d/d\theta)P_l^1(\cos\theta)$. It is known that an associated Legendre function $P_l^m(\cos\theta)$ is expressed in a power series as

$$P_l^m(\cos\theta) = \frac{\sin^m\theta}{2^m}\sum_{r=0}^{l-m}(-1)^r\frac{(l+m+r)!}{r!(m+r)!(l-m-r)!}\sin^{2r}\frac{\theta}{2}.$$

Therefore, putting $m = 1$ and taking a limit of $\theta \to 0$, we obtain

$$\frac{P_l^1(\cos\theta)}{\sin\theta} \to \frac{1}{2}\frac{(l+1)!}{0!1!(l-1)!} = \frac{1}{2}l(l+1).$$

On the other hand, $(d/d\theta)P_l^1(\cos\theta)$ is transformed as follows:

$$\frac{d}{d\theta}P_l^1(\cos\theta)$$

$$= \frac{\cos\theta}{2}\sum_{r=0}^{l-1}(-1)^r\frac{(l+1+r)!}{r!(1+r)!(l-1-r)!}\sin^{2r}\frac{\theta}{2}$$

$$+ \frac{\sin\theta}{2}\sum_{r=1}^{l-1}(-1)^r\frac{(l+1+r)!}{r!(1+r)!(l-1-r)!}\cdot 2r\sin^{2r-1}\frac{\theta}{2}\cdot\cos\frac{\theta}{2}\cdot\frac{1}{2}$$

$$\to \frac{1}{2}\frac{(l+1)!}{0!1!(l-1)!} = \frac{1}{2}l(l+1).$$

Using these results, we obtain

$$\pi_l(\cos 0) = \tau_l(\cos 0) = \frac{1}{2}l(l+1),$$

which leads to

$$S_1(0) = S_2(0) = \frac{1}{2}\sum_{l=1}^{\infty}(2l+1)(a_l+b_l). \tag{6.141}$$

Thus, we obtain

$$(\mathbf{E}^{(i)}+\mathbf{E}^{(s)})\cdot(\mathbf{E}^{(i)*}+\mathbf{E}^{(s)*}) \to 1 + \frac{2}{k^{(l)}z}\mathrm{Re}\left\{\frac{1}{i}e^{-ik^{(l)}(x^2+y^2)/(2z)}S_2(0)\right\}. \tag{6.142}$$

Assuming that the condition $z \gg x, y$ holds, let us integrate the light intensity within an area A including z axis on a plane of $z = z$. Integrating both sides of Eq. (6.142) within A, we obtain

$$\iint_A dx dy(\mathbf{E}^{(i)}+\mathbf{E}^{(s)})\cdot(\mathbf{E}^{(i)*}+\mathbf{E}^{(s)*})$$

$$= \iint_A dx dy\left[1 + \frac{2}{k^{(l)}z}\mathrm{Re}\left\{\frac{1}{i}e^{-ik^{(l)}(x^2+y^2)/(2z)}S_2(0)\right\}\right]$$

$$= A - \frac{4\pi}{k^{(l)2}}\mathrm{Re}\{S_2(0)\}, \tag{6.143}$$

where we have used the relation,

$$\iint_A e^{-ik^{(l)}(x^2+y^2)/(2z)}dx dy \approx \left(\int_A e^{-ik^{(l)}x^2/2z}dx\right)^2$$

$$\approx \left(\sqrt{\frac{2\pi z}{ik^{(l)}}}\right)^2 = \frac{2\pi z}{ik^{(l)}}.$$

The last relation is calculated under the condition that the area A is rectangular and the range of the integration is extended to infinity. Further, we have put $\iint_A 1 \cdot \mathrm{d}x\mathrm{d}y = A$. Equation (6.143) means that when a plane wave of light illuminates a sphere, the decrease of intensity will be observed, which appears as the second term on the right-hand side of this relation.

The decrease of the light intensity corresponds to the scattering and absorption of light by illuminating the sphere, and is called *extinction of light*. The situation is very likely to a case where a small particle hits a disturbing particle, which is usually expressed in terms of collision cross section. In the present case, such a cross section is called *extinction cross section*. The extinction cross section C_{ext} can be then expressed as

$$C_{\text{ext}} = \frac{4\pi}{k^{(I)2}}\operatorname{Re}\{S_2(0)\} = \frac{2\pi}{k^{(I)2}}\sum_{l=1}^{\infty}(2l+1)\operatorname{Re}\{a_l + b_l\}, \qquad (6.144)$$

which will be transformed into a dimensionless quantity after dividing it by a geometrical cross section of πa^2 as

$$Q_{\text{ext}} = C_{\text{ext}}/(\pi a^2) = \frac{2}{(k^{(I)}a)^2}\sum_{l=1}^{\infty}(2l+1)\operatorname{Re}\{a_l + b_l\}. \qquad (6.145)$$

6.3.3.2 Scattering cross section

The scattering cross section is obtained by integrating the intensity of scattering wave, which propagates as transverse wave in far field, within all the solid angles. By using Eqs. (6.132), (6.133), (6.135) and (6.136), the total scattering intensity is obtained from the Poynting vector of the scattering wave and is expressed as

$$
\begin{aligned}
I^{(s)} &= \frac{1}{2}\int_0^\pi r^2 \sin\theta\,\mathrm{d}\theta\int_0^{2\pi}\mathrm{d}\phi\left(E_\theta^{(s)}H_\phi^{(s)*} - E_\phi^{(s)}H_\theta^{(s)*}\right)\\
&= \frac{1}{2}\int_0^\pi \sin\theta\,\mathrm{d}\theta\int_0^{2\pi}\mathrm{d}\phi\\
&\quad \times\sqrt{\frac{\epsilon^{(I)}}{\mu^{(I)}}}\frac{1}{k^{(I)2}}\left\{|S_2(\theta)|^2\cos^2\phi + |S_1(\theta)|^2\sin^2\phi\right\}\\
&= \frac{1}{2}\sqrt{\frac{\epsilon^{(I)}}{\mu^{(I)}}}\frac{\pi}{k^{(I)2}}\int_0^\pi\left\{|S_2(\theta)|^2 + |S_1(\theta)|^2\right\}\sin\theta\,\mathrm{d}\theta, \quad (6.146)
\end{aligned}
$$

where the relation

$$\int_0^{2\pi} \cos^2 \phi d\phi = \int_0^{2\pi} \sin^2 \phi d\phi = \pi,$$

and Eqs. (6.137) and (6.138) are used, and a factor of 1/2 comes from the cycle-averaging.

Before calculating Eq. (6.146), we will derive the relations concerning the functions of $\pi_l(\cos\theta)$ and $\tau_l(\cos\theta)$, which are included in the functions of $S_1(\theta)$ and $S_2(\theta)$. From the definitions, Eqs. (6.127) and (6.128), the following relations are derived:

$$\int_0^\pi \pi_l(\cos\theta)\tau_{l'}(\cos\theta)\sin\theta d\theta$$

$$= \int_0^\pi P_l^1(\cos\theta)\frac{d}{d\theta}P_{l'}^1(\cos\theta)d\theta$$

$$= \left[P_l^1(\cos\theta)P_{l'}^1(\cos\theta)\right]_0^\pi - \int_0^\pi \frac{d}{d\theta}P_l^1(\cos\theta)\cdot P_{l'}^1(\cos\theta)d\theta.$$

Further, it is easy to show that $P_{l,l'}^1(\cos\theta) = 0$ for $\theta = 0, \pi$ from the expression for an associated Legendre function expanded in a power series. Therefore, we obtain

$$\int_0^\pi \{\pi_l\tau_{l'} + \pi_{l'}\tau_l\}\sin\theta d\theta = 0 \qquad (6.147)$$

where, for simplicity, we have abbreviated the expressions such that $\pi_l(\cos\theta) = \pi_l$.

Using a well-known relation of

$$\int_0^\pi \left\{\frac{m^2}{\sin\theta}P_n^m(\cos\theta)P_l^m(\cos\theta) + \sin\theta\frac{d}{d\theta}P_n^m(\cos\theta)\frac{d}{d\theta}P_l^m(\cos\theta)\right\}d\theta$$

$$= \begin{cases} 0 & n \neq l, \\ \dfrac{2(n+m)!n(n+1)}{(n-m)!(2n+1)} & n = l, \end{cases}$$

we obtain

$$\int_0^\pi (\pi_l\pi_{l'} + \tau_l\tau_{l'})\sin\theta d\theta = \begin{cases} 0 & l \neq l', \\ \dfrac{2l^2(l+1)^2}{2l+1} & l = l'. \end{cases} \qquad (6.148)$$

Using these relations, let us advance the calculation of Eq. (6.146). Since it is immediately proved that the following relation holds,

$$\int_0^\pi \left\{ |S_1(\theta)|^2 + |S_2(\theta)|^2 \right\} \sin\theta \, d\theta$$

$$= \sum_l \sum_{l'} \frac{2l+1}{l(l+1)} \frac{2l'+1}{l'(l'+1)} \int_0^\pi \left\{ (a_l \pi_l + b_l \tau_l)(a_{l'}^* \pi_{l'} + b_{l'}^* \tau_{l'}) \right.$$

$$\left. + (a_l \tau_l + b_l \pi_l)(a_{l'}^* \tau_{l'} + b_{l'}^* \pi_{l'}) \right\} \sin\theta \, d\theta,$$

the calculation of the integration gives

$$\int_0^\pi \left\{ (a_l a_{l'}^* + b_l b_{l'}^*)(\pi_l \pi_{l'} + \tau_l \tau_{l'}) + (a_l b_{l'}^* + a_{l'}^* b_l)(\pi_l \tau_{l'} + \pi_{l'} \tau_l) \right\} \sin\theta \, d\theta$$

$$= \frac{2l^2(l+1)^2}{2l+1}(|a_l|^2 + |b_l|^2)\delta_{ll'}.$$

Inserting this into the above relation, we obtain a simple result of

$$\int_0^\pi \left\{ |S_1(\theta)|^2 + |S_2(\theta)|^2 \right\} \sin\theta \, d\theta = 2 \sum_l (2l+1)(|a_l|^2 + |b_l|^2).$$

$$(6.149)$$

Thus, the scattering cross section is expressed as

$$C_{sca} \equiv I^{(s)}/I^{(i)} = \frac{2\pi}{k^{(l)2}} \sum_l (2l+1)(|a_l|^2 + |b_l|^2), \qquad (6.150)$$

where $I^{(i)}$ is the intensity of the incident light. Since we put $|E_x^{(i)}| = 1$, it is given as $I^{(i)} = (1/2)E_x^{(i)} H_y^{(i)*} = (1/2)\sqrt{\epsilon^{(l)}/\mu^{(l)}}$. Further, we define the dimensionless quantity Q_{sca}, which is obtained by dividing the scattering cross section by the geometrical cross section πa^2, as

$$Q_{sca} = \frac{C_{sca}}{\pi a^2} = \frac{2}{(k^{(l)}a)^2} \sum_{l=1}^\infty (2l+1)(|a_l|^2 + |b_l|^2). \qquad (6.151)$$

The absorption cross section is obtained by subtracting the scattering cross section from the extinction cross section as

$$Q_{abs} = Q_{ext} - Q_{sca}. \qquad (6.152)$$

6.3.4 Characteristics of Mie Scattering

By using the above expressions, various characteristics of Mie scattering can be derived. In Fig. 6.6, we show the angular dependence of the Mie scattering due to dielectric spheres with

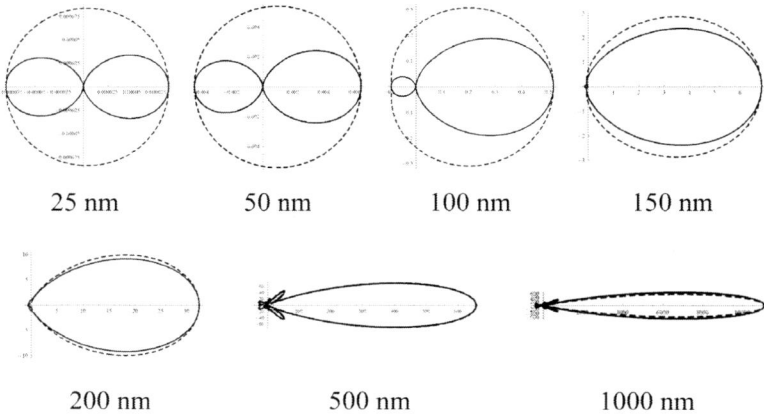

25 nm   50 nm   100 nm   150 nm

200 nm   500 nm   1000 nm

Figure 6.6 Angular dependence of Mie scattering intensities calculated for various radii of dielectric spheres. The scattering intensity is expressed by a distance from the origin of the coordinate axes for each scattering angle. The ratio of the refractive indices of the sphere to the surrounding medium is set at 1.5, with the wavelength of light 0.5 µm. Dashed and solid lines indicate the scattering intensities for the incident polarization perpendicular and parallel to a plane of this page. The plane wave of light is incident from the left of each figure.

various radii when they are illuminated by a linearly polarized plane wave of light with the wavelength of 500 nm. In case of small radii, it is found that the scattering is almost completely isotropic when the polarization direction of incident light is perpendicular to a plane of this page, whereas it shows a strongly constricted part along the direction perpendicular to the incident direction, when the polarization direction is parallel to the plane. This phenomenon is well described as Rayleigh scattering and is explained in terms that an oscillating dipole does not emit light toward the direction parallel to that of the dipole. In fact, as shown in Fig. 6.7, the angular dependence for both polarization directions almost completely reproduces that of Rayleigh scattering when the radius is 10 nm, while a slight enhancement in the forward direction is observed when the radius becomes 25 nm.

With increasing radius, the scattering tends to be considerably biased to the forward direction as shown in the figure (radius of 100 nm), which implies that a model based on the dipole radiation

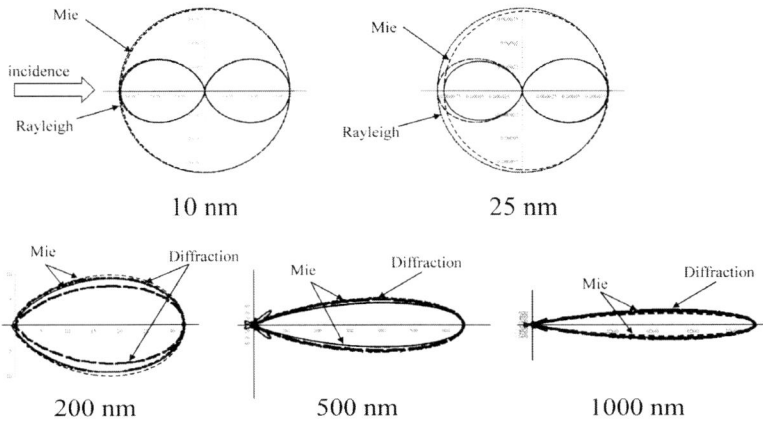

Figure 6.7 Angular dependence of the Mie scattering intensity for dielectric spheres with various radii, which are compared with simple models of Rayleigh scattering (upper) and 2D diffraction due to a circular opening (lower). The ratio of the refractive indices of the sphere to the surrounding medium is set at 1.5, with the wavelength of light 0.5 μm. Dashed and solid lines indicate the scattering intensities for the incident polarization perpendicular and parallel to a plane of this page in case of the Mie scattering, while long and short dashed lines are those for the models. For Rayleigh scattering, the angular dependence is calculated only by putting $I_{S,P} \propto A_{S,P}^2$, where I_S and I_P are the scattering intensities for the polarization of incident light, perpendicular and parallel to a plane of this page, while $A_{S,P}$ is the angular factor for Rayleigh scattering defined by $A_S = 1$ and $A_P = \cos\theta$, respectively. θ is a polar angle measured from the propagation direction of the incident wave. For the diffraction due to a circular opening, we employ a formula, $I_{S,P} \propto \{J_1(ka\sin\theta)/(ka\sin\theta)\}^2 B^2 A_{S,P}^2$, where B is an inclination factor expressed as $B = 1/(2i\lambda) \cdot (\cos\theta + 1)$ in this case.

is no longer applicable. In case of the radius of 150–200 nm, the backscattering component almost completely vanishes and further its polarization difference becomes less conspicuous, while it still possesses a wide angular range in the forward direction. At the radii larger than 500 nm, Mie scattering is characterized by strongly forward and pointed scattering, and its tendency is accentuated with increasing radius.

One may think that such a behavior is similar to that of diffraction phenomenon. To investigate its possibility, we calculate the angular dependence of 2D diffraction due to a circular opening with a radius a using the relation of $I_{S,P} \propto (J_1(ka\sin\theta)/ka\sin\theta)^2 \, B^2 A_{S,P}^2$, where $A_S = 1$ and $A_P = \cos\theta$ for the polarizations perpendicular and parallel to a plane of the page, respectively, and B is an inclination factor calculated as $B = 1/(2i\lambda) \cdot (\cos\theta + 1)$ in this case (see Eq. (4.22)).

As shown in Fig. 6.7, the angular dependence of Mie scattering is comparable to that of the diffraction when the radius of the sphere is 200 nm, while it is almost completely reproducible above 1000 nm. This means that the strong forward scattering appearing in Mie scattering is mostly interpretable as the diffraction effect due to a circular opening whose radius is the same as that of a sphere. Thus, the peculiarity found in the angular dependence of Mie scattering is explained in terms of Rayleigh scattering when the radius is much smaller than the wavelength of light, while it is explained in terms of 2D diffraction phenomenon due to a circular opening when the radius is comparable or larger than the wavelength.

In Fig. 6.8, we show the angular dependence of Mie scattering on changing refractive index of the sphere and also changing wavelength of illuminating light. It is found that the angular dependence does not change much when the refractive index is varied, while it changes more remarkably when the wavelength is varied. In the latter, the variation observed here is quite similar to that observed when the radius of the sphere is varied. This is natural because the scattering property is characterized by two functions of $S_1(\theta)$ and $S_2(\theta)$, which are then characterized by a factor of $q = k^{(I)}a = 2\pi n^{(I)}a/\lambda$. Thus, the variables of a and λ are not independent and they generally cause the same change in the scattering properties, but in the opposite directions.

In Fig. 6.9, we show the effect of the imaginary part on the angular dependence of Mie scattering for the radius of a sphere of 100 nm under the illumination of light with the wavelength of 500 nm. It is interesting that with increasing imaginary part, the angular dependence is considerably affected for the parallel polarization, which appears such that the constricted part becomes less pronounced. On the other hand, the perpendicular side is hardly affected and is almost isotropic.

a=100 nm, λ=500 nm

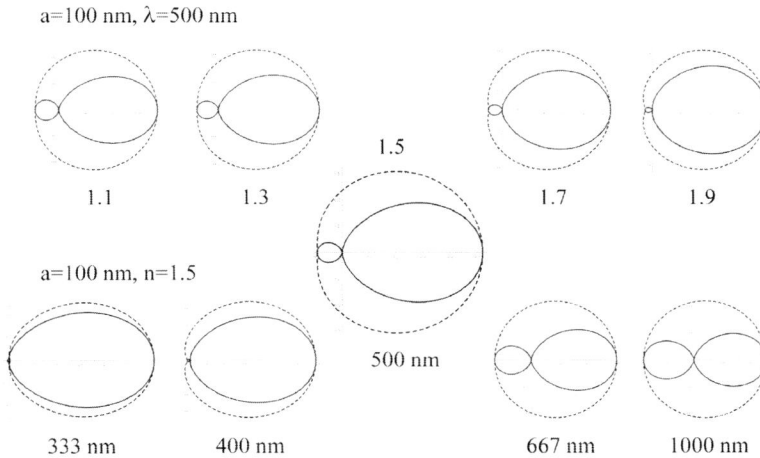

1.5

1.1   1.3   1.7   1.9

a=100 nm, n=1.5

500 nm

333 nm   400 nm   667 nm   1000 nm

Figure 6.8 Angular dependence of Mie scattering on changing refractive index (upper) and wavelength of light (lower). The upper part is calculated for the radius and wavelength of 100 nm and 500 nm, repsectively, while changing the refractive index from 1.1 to 1.9. The lower part is calculated for the radius and refractive index of 100 nm and 1.5, while changing the wavelength of light from 333 to 1000 nm. The wavelengths are chosen so as to make the interval of the photon energies to be constant.

In Fig. 6.10, we will show the results for various materials including metals, semiconductor and insulators. Some of them are characterized by their large imaginary parts and some are by their large real parts. If we compare these data with the data of BK7 as a standard, the results for metals are similar to those obtained

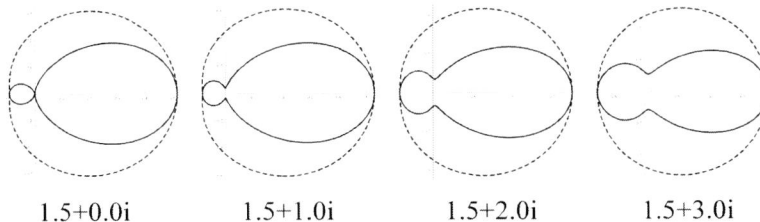

1.5+0.0i   1.5+1.0i   1.5+2.0i   1.5+3.0i

Figure 6.9 Angular dependence of Mie scattering calculated with increasing imaginary part of the refractive index. The radius and the wavelength of light are 100 nm and 500 nm, respectively.

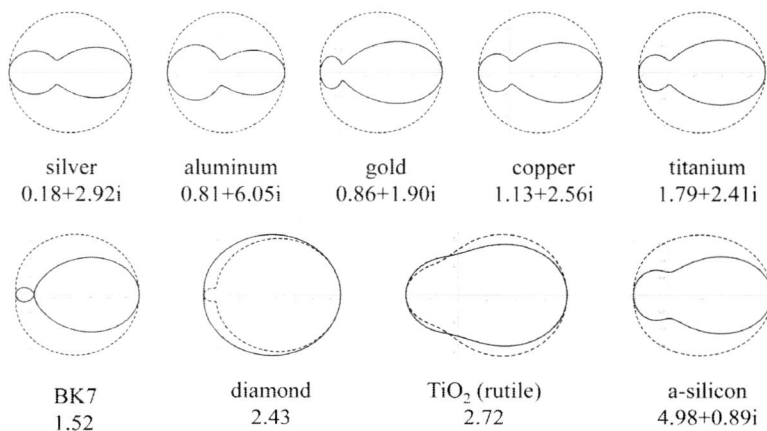

silver	aluminum	gold	copper	titanium
0.18+2.92i	0.81+6.05i	0.86+1.90i	1.13+2.56i	1.79+2.41i

BK7	diamond	TiO_2 (rutile)	a-silicon
1.52	2.43	2.72	4.98+0.89i

Figure 6.10 Angular dependence of Mie scattering for various spherical particles with the radius of 100 nm and the wavelength of light of 500 nm. The refractive indices employed are taken from those in Refractive Index Database (http://refractiveindex.info/).

in Fig. 6.9, which are essentially characterized by less pronounced constrictions. On the other hand, with increasing real part of the refractive index, large differences are observed as shown in the data of diamond and rutile. However, when we see the result for amorphous silicon that shows a large real part of the refractive index with a small amount of imaginary part, the result is rather similar to those obtained in metals. Thus, the presence of the imaginary part will play a role of suppressing the variations due to the presence of large real part.

In Fig. 6.11, the extinction cross sections normalized by geometrical ones, Eq. (6.145), are calculated for spheres with various radii. In general, the extinction cross sections tend to increase with decreasing wavelength. As shown in the right-hand side of the figure, when the radius is sufficiently small (see the result of $a = 10$ nm, for example), the extinction cross section is quite well expressed by a function of $1/\lambda^4$, where λ is the wavelength of light, indicating that its property is well described by Rayleigh scattering. It is also noticed that with increasing radius, the deviation from the Rayleigh scattering is gradually remarkable, particularly at shorter wavelengths. At the radius comparable or larger than

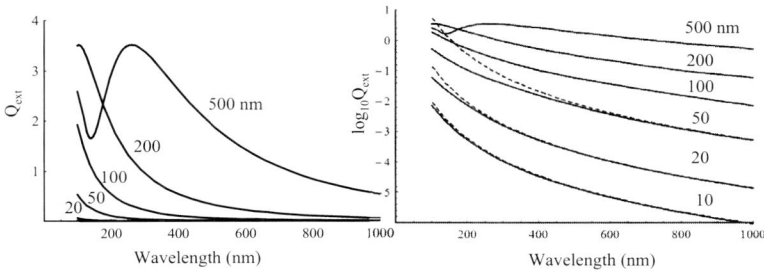

Figure 6.11 Left: Incidence-wavelength dependence of the extinction cross sections calculated for spheres with various radii, having the refractive index of 1.5, which are embedded in a medium with the refractive index of 1.33. Right: Logarithmic plot of the left figure. Dashed lines express curves proportional to $1/\lambda^4$, indicating that the scattering due to a sphere with the radius of 10 nm is almost completely expressed by the Rayleigh scattering, while the deviation from it becomes gradually larger when the radius is increased.

the wavelength of light, a peak appears in a visible region, which shifts to the red when the radius is increased. Thus, Mie scattering actually contributes to coloration, which is dependent on the size of sphere.

Next, we show a calculated result when the refractive index of a medium is varied, while the radius of a sphere remains constant. The result is shown in Fig. 6.12. With decreasing refractive index of the medium, it is natural that the relative index, $\hat{n} \equiv n_2/n_1$, is increased, which causes the shift of the peak to a longer wavelength. In addition, the extinction cross section more or less assumes complicated modulations, particularly at short wavelengths. The modulation consists of slowly changing part ranging to longer wavelengths and of rapidly changing part at shorter wavelengths. Such slow and rapid modulations are called *interference* and *ripple structures*, whose origins have been discussed in the literatures [Bohren and Huffman (1983)] in terms of the features of modified spherical Bessel functions appearing in the formulas of a_l and b_l, Eqs. (6.129) and (6.130).

As described above, it has been shown that extinction cross section Q_{ext} is a function of only three parameters: the wavelength of light λ, the radius of sphere a, and the relative refractive index $\hat{n}$.

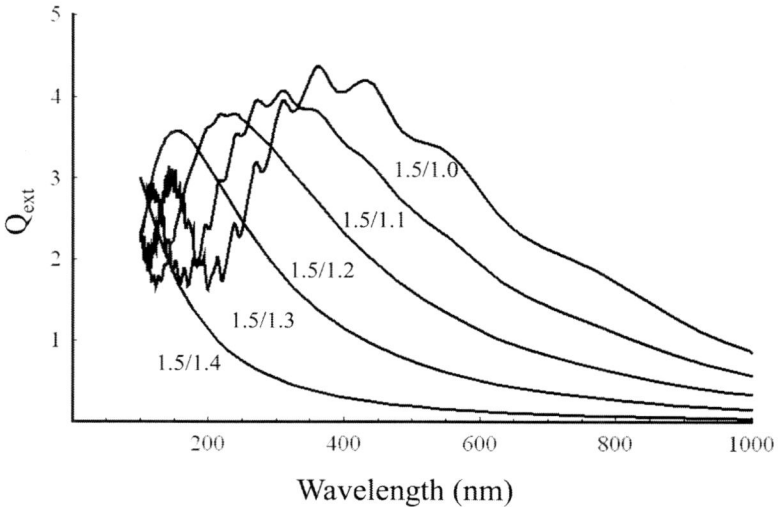

Figure 6.12 plot: y-axis Q_{ext} from 1 to 5, x-axis Wavelength (nm) from 200 to 1000. Curves labeled 1.5/1.0, 1.5/1.1, 1.5/1.2, 1.5/1.3, 1.5/1.4.

Wavelength (nm)

Figure 6.12 Incidence-wavelength dependence of the extinction cross sections for various ratios of the refractive index of the sphere to that of the medium. The radius of the sphere is set at 250 nm.

van de Hulst proposed an approximate expression for the extinction cross section using a comprehensive parameter of $\rho \equiv 2q|\hat{n} - 1|$, which includes the above three parameters in the form of $q = k^{(l)}a = 2\pi a/\lambda$ and $|\hat{n} - 1|$ [van de Hulst (1957)]. The expression is very simple as shown in the following:

$$Q_{\text{ext}} \approx 2 - \frac{4}{\rho}\sin\rho + \frac{4}{\rho^2}(1 - \cos\rho) \equiv Q_{\text{vdH}}, \qquad (6.153)$$

which is known as one of the most useful formulas among complicated formulations appearing in the Mie theory. The usefulness and a range of application of this approximate expression can be investigated easily. We have calculated Q_{ext} for various values of $\hat{n}$, and have compared them with Q_{vdH}. The result is shown in Fig. 6.13. Although slight deviations are seen, the extinction cross sections are fairly well expressed by Q_{vdH}, which shows that a scaling of Q_{ext} with respect to a parameter ρ is applicable in an approximate sense.

The treatment of Mie scattering has been extensively developed since the first paper presented by Mie and the scattering due to a particle with a variety of shapes can be calculated in the same

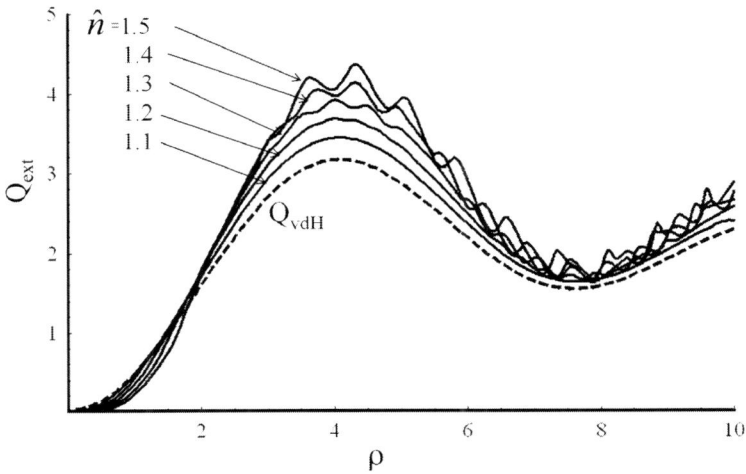

Figure 6.13 Comparison of the extinction cross sections (solid lines) calculated for various values of $\hat{n}$ with that of an approximate expression (dashed line) given by van de Hulst, Eq. (6.153).

theoretical framework. This includes a coated sphere or hollow sphere, a multi-coated sphere [Kerker (1969)], a cylinder with infinite length [van de Hulst (1957); Bohren and Huffman (1983)], a coated cylinder or hollow cylinder, a multi-coated cylinder [Kerker and Matijević (1961); Kerker (1969)] and so on. Since the number of pages in the present book is limited, readers should refer to each paper or book for their further studies.

In view of the coloration mechanisms, the scattering of light is completely different from other mechanisms such as thin-layer interference and photonic crystal, because it does not require a regular arrangement of microstructures. The coloration due to light scattering is also known as non-iridescence in contrast to iridescence due to the others.

The light scattering plays an important role in the natural colorations. For example, the origin of blue sky was first explained by Lord Rayleigh as due to light scattering by small particles in atmosphere [Rayleigh (1871a,b)]. The color of suspension involving small colloidal particles is called *Tyndall blue*. The blue colors in these cases are essentially based on a fact that the scattering cross sections are dependent on the fourth power of the frequency of light. With increasing size of particle, Mie scattering occurs, whose wavelength dependence differs considerably from Rayleigh scattering. Thus, the different coloration should appear.

It has been long believed that light scattering also plays a major role in non-iridescent blue colorings in animal world such as in birds and dragonflies. However, recent studies cast doubt on this thought, because most of them originate from a dense distribution of particles and structural elements so that multiple scattering will considerably affect the coloration, and also because a clear reflection peak usually appears in spite of scatterers of a small size, which is definitely different from that expected from Rayleigh and Mie scatterings. In Sec. 6.5.2, we will overview these phenomena that are widely distributed in the natural world.

6.4 Methods to Analyze Disordered Structures

6.4.1 *Deviation from an Average Structure*

A method to analytically treat the optical properties of disordered structures has not yet been established so that numerical methods have been often employed to elucidate the coloration mechanisms. However, to truly understand the essence of this phenomenon, it is inevitable to construct simple models, while remaining their essential characteristics, and to consider it in an analytical way. In the present section, we will show a few such models to permit analytical approach. The examples that will be shown from now contain a structure having regularity on an average with a small amount of irregularity posed on the structure, that possessing a short-range order but not possessing long-range one, and that consisting of completely irregular structure.

We first consider a case where a structure is characterized by regularity on an average, whereas it is subject to static or dynamical fluctuation, which causes deviations from the average. This is a case of colloidal crystal in liquid, where silica or polystyrene spheres are arranged in a crystalline form owing to electrostatic interaction between the spheres. When light hits on this crystal, the Bragg

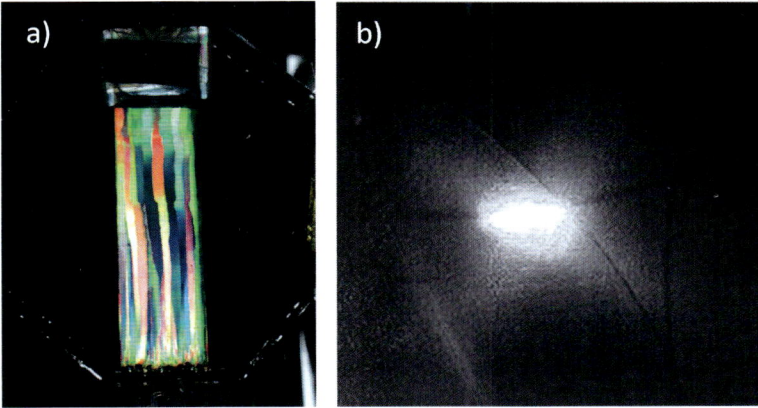

Figure 6.14 (a) Directionally growing colloidal crystal consisting of silica spheres. Each stripe corresponds to a single crystal. (b) Bragg reflection from a colloidal crystal, which accompanies halo around a strong light spot. Black curves in the background are known as Kossel lines. (Courtesy of Dr. H. Yamada.)

reflection occurs when the Bragg condition, $2nd\cos\theta = m\lambda$, is satisfied with n average refractive index, d distance between the arranged layers, θ angle of incidence, and λ wavelength of light. In Fig. 6.14b, we show an example of the Bragg reflection observed from a single colloidal crystal under illumination of laser light. A strong Bragg reflection spot is clearly seen, which is slightly elongated along one of the Kossel lines due to a focusing effect of the incident laser. However, on closer inspection, one notices the presence of halo around the Bragg-deflected light spot. The first step of our approach is to investigate the origin of the halo and to bring out physical information from this experiment.

We consider the simplest case where particles are regularly arranged in a line with a particle-particle distance of a and a plane wave of light illuminates on them, as shown in Fig. 6.15. The wave vector of incident light is denoted as $\mathbf{k}_i$, while that of scattered light due to particles is denoted as $\mathbf{k}_s$. The electric field observed sufficiently far from the jth scatter is calculated as

$$E_s(\mathbf{r}) = e^{i\mathbf{k}_i \cdot \mathbf{r}_j} e^{i\mathbf{k}_s \cdot (\mathbf{r} - \mathbf{r}_j)} = e^{i\mathbf{k}_s \cdot \mathbf{r}} e^{i(\mathbf{k}_i - \mathbf{k}_s) \cdot \mathbf{r}_j} = e^{i\mathbf{k}_s \cdot \mathbf{r}} e^{-i\mathbf{K} \cdot \mathbf{r}_j}, \quad (6.154)$$

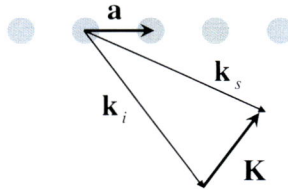

Figure 6.15 Light scattering due to particles regularly arranged in a line. **a** is a vector directing to a nearest neighboring particle. $\mathbf{k}_i$, $\mathbf{k}_s$ and $\mathbf{K}$ are wave vectors of incident and scattered light, and a scattering vector, respectively.

where we only consider a phase factor of the electric field. $\mathbf{r}_j$ and $\mathbf{r}$ are position vectors of the jth particle and an observing point that is assumed to be sufficiently far from the particle. Further, we put a scattering vector as $\mathbf{K} = \mathbf{k}_s - \mathbf{k}_i$.

Since the particles are arranged in 1D, we can immediately calculate the electric field of scattered light due to all the particles as

$$E_s(\mathbf{r}) = e^{i\mathbf{k}_s \cdot \mathbf{r}} \sum_{j=0}^{N-1} e^{-i\mathbf{K} \cdot \mathbf{r}_j} = e^{i\mathbf{k}_s \cdot \mathbf{r}} e^{-i\mathbf{K} \cdot \mathbf{r}_0} \sum_{j=0}^{N-1} e^{-ij\mathbf{K} \cdot \mathbf{a}}$$

$$= e^{i\mathbf{k}_s \cdot \mathbf{r}} e^{-i\mathbf{K} \cdot \mathbf{r}_0} \frac{1 - e^{-iN\mathbf{K} \cdot \mathbf{a}}}{1 - e^{-i\mathbf{K} \cdot \mathbf{a}}}, \tag{6.155}$$

where N is the number of particles. Further, we put a position vector of a particle located at one end as $\mathbf{r}_0$ and also define a lattice vector according to $\mathbf{r}_{j+1} - \mathbf{r}_j = \mathbf{a}$. Thus, the position vector of the jth particle becomes $\mathbf{r}_j = \mathbf{r}_0 + j\mathbf{a}$. Then, the scattering intensity is calculated as

$$|E_s(\mathbf{r})|^2 = \frac{\sin^2(N\mathbf{K} \cdot \mathbf{a}/2)}{\sin^2(\mathbf{K} \cdot \mathbf{a}/2)}, \tag{6.156}$$

which agrees completely with that for multi-slit interference and corresponds to a curve of $\alpha = 1$ in Fig. 6.16. Sharp peaks are found to appear under a condition of $\mathbf{K} \cdot \mathbf{a}/2 = m\pi$ with m an integer and are well known as diffraction spots in 1D diffraction grating.

Next, we assume that these particles slightly deviate from their regular positions. We denote the deviation for the jth particle as $\delta \mathbf{r}_j$. Then, the scattering amplitude will be expressed as

$$E_s(\mathbf{r}) = e^{i\mathbf{k}_s \cdot \mathbf{r}} \sum_{j=0}^{N-1} e^{-i\mathbf{K} \cdot (\mathbf{r}_j + \delta \mathbf{r}_j)}. \tag{6.157}$$

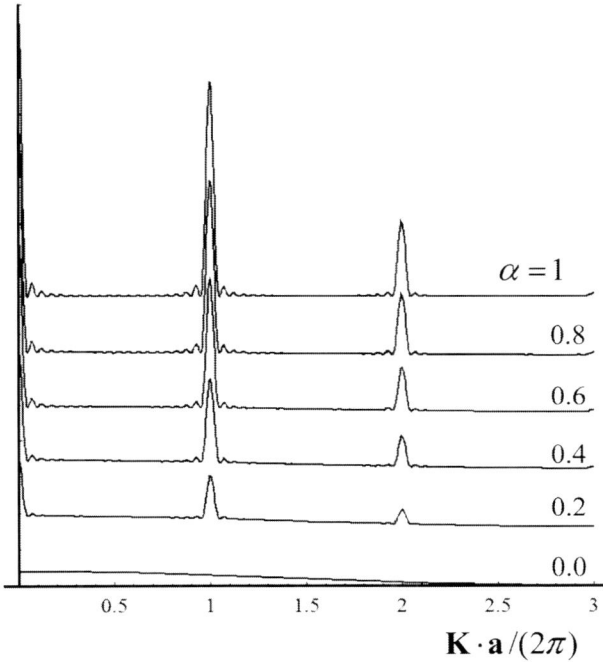

Figure 6.16 Light intensity diffracted by a diffraction grating consisting of particles arranged in 1D, which are subject to slight deviations from their original positions. The degree of deviations is expressed by a parameter α (see the text). $\alpha = 1$ means a completely regular grating, while $\alpha = 0$ is completely irregular grating. It is assumed that totally 20 particles are arranged with an interval of 1 μm. The diffraction due to each particle is assumed to be the same as that for 1D slit with the width of 0.3 μm.

If we treat such deviations in a statistical way, then we take a statistical average of the scattering intensity and express it as

$$\langle |E_s|^2 \rangle = \left\langle \left| \sum_{j=1}^{N-1} e^{-i\mathbf{K}\cdot(\mathbf{r}_j + \delta \mathbf{r}_j)} \right|^2 \right\rangle$$

$$= \sum_{j,l} e^{-i(j-l)\mathbf{K}\cdot\mathbf{a}} \langle e^{-i\mathbf{K}\cdot(\delta \mathbf{r}_j - \delta \mathbf{r}_l)} \rangle, \qquad (6.158)$$

where $\langle \cdots \rangle$ means the statistical average, which can be calculated essentially in the following two limiting cases: One is a case where the number of particles is sufficiently large and the other is a case where their positions fluctuate dynamically and the average is taken over for a sufficiently long time.

Moreover, if the deviations of the jth and lth particles have no correlation, the above relation becomes separable and leads to

$$\langle |E_s|^2 \rangle = \sum_{j,l(j \neq l)} e^{-i(j-l)\mathbf{K}\cdot\mathbf{a}} \langle e^{-i\mathbf{K}\cdot\delta\mathbf{r}_j} \rangle \langle e^{i\mathbf{K}\cdot\delta\mathbf{r}_l} \rangle + \sum_j 1$$

$$= \sum_{j,l(j \neq l)} e^{-i(j-l)\mathbf{K}\cdot\mathbf{a}} \left| \langle e^{-i\mathbf{K}\cdot\delta\mathbf{r}_j} \rangle \right|^2 + N,$$

where we have calculated the cases of $j \neq l$ and $j = l$ separately. We also assume the deviation or fluctuation does not depend on particle and hence replace the term $|\langle \exp[-i\mathbf{K}\cdot\delta\mathbf{r}_j] \rangle|^2$ by a constant parameter α as

$$|\langle e^{-i\mathbf{K}\cdot\delta\mathbf{r}_j} \rangle|^2 \equiv \alpha. \tag{6.159}$$

Then, the scattering intensity is derived as

$$\langle |E_s|^2 \rangle = \left(\left| \sum_j e^{-ij\mathbf{K}\cdot\mathbf{a}} \right|^2 - N \right)\alpha + N$$

$$= \alpha \left| \sum_j e^{-ij\mathbf{K}\cdot\mathbf{a}} \right|^2 + N(1-\alpha)$$

$$= \alpha \frac{\sin^2(N\mathbf{K}\cdot\mathbf{a}/2)}{\sin^2(\mathbf{K}\cdot\mathbf{a}/2)} + N(1-\alpha), \tag{6.160}$$

where $\alpha = 1$ means a completely regular case, while $\alpha = 0$ does completely irregular one. The first term in the right-hand side indicates the presence of diffraction spots, while the second term does the sum of diffracted light waves due to individual particles.

We plot the angular dependence of scattering intensity in Fig. 6.16 with changing α. In the above treatment, we only consider a phase factor so that no angular dependence appears due to the diffraction of an individual particle. For readers' convenience, we take account of angular dependence due to single-slit diffraction. It is soon understood that with decreasing α, the intensities of sharp diffraction spots decrease quickly and a broad background increases gradually. In the limit of $\alpha = 0$, the sharp spots completely disappear and only the broad background due to diffraction of individual particles remains.

Thus, this type of irregularity causes the coexistence of regular and irregular parts, whose fraction varies with the degree of irregularity. It is soon noticed that α contains only the inner product of $\mathbf{K}$ and $\delta\mathbf{r}_j$ so that only the fluctuation or deviation along the direction of $\mathbf{K}$ effectively affects the scattering intensity.

If the fluctuation obeys the Gaussian random process, it is known that the following relation holds:

$$\langle e^{ix} \rangle = e^{i\langle x \rangle - \frac{1}{2}\langle x^2 \rangle}, \qquad (6.161)$$

where x is a random variable. Thus, α will be simplified further into

$$\alpha = \left| \left\langle e^{-i\mathbf{K}\cdot\delta\mathbf{r}} \right\rangle \right|^2 = e^{-\langle (\mathbf{K}\cdot\delta\mathbf{r})^2 \rangle}. \qquad (6.162)$$

This relation indicates that α decreases with increasing amplitude of the fluctuation along a Gaussian curve. The decreasing rate of α is roughly expressed by a characteristic amplitude of the fluctuation $|\delta\mathbf{r}| \approx |\mathbf{K}|^{-1}$. Thus, the fluctuation whose magnitude is comparable to the wavelength of light becomes a key parameter for this phenomenon. It is also noticed that α remarkably decreases when the fluctuation has a component along the direction of $\mathbf{K}$, while it does not change when $\mathbf{K} \cdot \delta\mathbf{r} = 0$. Hence, the angle-dependent scattering experiment will offer information on anisotropic fluctuation.

Although we consider only a 1D case, we can speculate that a halo observed in colloidal crystal is a manifestation of the irregularity caused by the positional fluctuation of particles around the regular average positions. Thus, the intensity ratio of a Bragg diffraction spot to a background halo will give a degree of fluctuation of particles that consist of colloidal crystal in liquid. Furthermore, in colloidal crystal, the positions of particles are expected to fluctuate dynamically so that the time-course experiment on the background halo surely offers the information on the temporal fluctuation of particles, whereas the Bragg spots are due to an average structure that does not change essentially with time.

6.4.2 *Method to Utilize Spatial Correlation Function*

Next, we consider a case where a system does not have a long-range order, but has only a short-range order. We start with a system consisting of a lot of particles. As shown in the preceding section, the sum of the electric fields scattered by these particles is expressed as

$$E_s(\mathbf{r}) = e^{i\mathbf{k}_s \cdot \mathbf{r}} \sum_j e^{-i\mathbf{K} \cdot \mathbf{r}_j},$$

where we neglect multiple scattering among particles. We rewrite the above expression in an integral form as

$$E_s(\mathbf{r}) = e^{i\mathbf{k}_s \cdot \mathbf{r}} \int \sum_j f_j \delta(\mathbf{r} - \mathbf{r}_j) e^{-i\mathbf{K} \cdot \mathbf{r}} d\mathbf{r}. \tag{6.163}$$

Here, f_j is newly introduced to express the scattering efficiency including angular dependence of the jth particle. If we assume all the particles have the same scattering properties, we can put $f_j \equiv f$. Thus, putting a number density as $\rho(\mathbf{r}) \equiv \sum_j \delta(\mathbf{r} - \mathbf{r}_j)$, we obtain the scattering intensity as

$$|E_s(\mathbf{r})|^2 = |f|^2 \left| \int d\mathbf{r}\, \rho(\mathbf{r}) e^{-i\mathbf{K} \cdot \mathbf{r}} \right|^2$$

$$= |f|^2 \int d\mathbf{r} \int d\mathbf{r}'\, \rho(\mathbf{r})\rho(\mathbf{r}') e^{-i\mathbf{K} \cdot (\mathbf{r} - \mathbf{r}')}.$$

We change the valuable as $\mathbf{r} - \mathbf{r}' \equiv \mathbf{r}''$ and obtain

$$|E_s(\mathbf{r})|^2 = |f|^2 \int d\mathbf{r}'' \int d\mathbf{r}'\, \rho(\mathbf{r}')\rho(\mathbf{r}' + \mathbf{r}'') e^{-i\mathbf{K} \cdot \mathbf{r}''}$$

$$= |f|^2 \int d\mathbf{r}''\, G(\mathbf{r}'') e^{-i\mathbf{K} \cdot \mathbf{r}''} \cdot \int d\mathbf{r}' \rho^2(\mathbf{r}'),$$

where we have defined a normalized spatial correlation function

$$G(\mathbf{r}'') \equiv \frac{\int d\mathbf{r}' \rho(\mathbf{r}')\rho(\mathbf{r}' + \mathbf{r}'')}{\int d\mathbf{r}' \rho^2(\mathbf{r}')}. \tag{6.164}$$

Putting the total number of particles as N and since $\int d\mathbf{r}' \rho^2(\mathbf{r}') = N$, we have arrived at the final expression:

$$|E_s(\mathbf{r})|^2 = N \cdot |f|^2 \cdot \int d\mathbf{r}''\, G(\mathbf{r}'') e^{-i\mathbf{K} \cdot \mathbf{r}''}. \tag{6.165}$$

The right-hand side of this expression consists of three terms: The first term is a number of particles, while the second and third terms express the scattering properties for a single particle and the Fourier transform of the spatial correlation function. Thus, the scattering intensity is directly connected with the spatial correlation function of the particle arrangement: When the particles are only

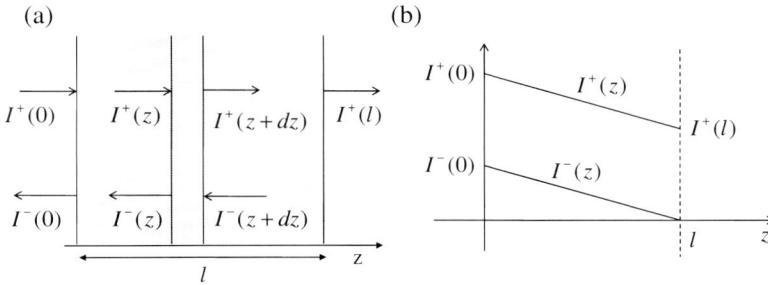

Figure 6.17 (a) Geometry of light propagation in a random medium. Light is assumed to completely lose the propagation direction within a medium and to propagate along the z axis on an average. (b) Intensity distribution within a random medium for the right- and left-propagating light.

locally arranged, i.e. a short-range order, the scattering vector cannot be settled, whereas it will be settled when the particles are arranged in a considerably regular way. Therefore, if the arrangement of the particle is random as a whole, but has a short-range order, it is possible that light scattered by such systems show the interference effect that reflects the spatial correlation function. The above treatment implicitly assumes a single scattering process, and hence neglects the effect of multiple scattering. However, it is expected that the multiple scattering will cause the localization of light [John (1987)] and will become an origin of the photonic band-like structure.

6.4.3 Reflection and Transmission of Light due to a Completely Random Medium

Medium without light absorption Consider an extreme case where light is multiply scattered by passing through a completely random medium and thoroughly loses its memory of propagation direction. We assume a random medium of the length l, as shown in Fig. 6.17a, to which diffuse light is incident from the left and propagates to the positive z direction on an average. We denote the intensity of light propagating to the positive direction along the z axis on an average as I^+, while that to the negative direction is denoted as I^-. We consider a region with an infinitesimal length along the z direction, which is denoted as dz, and assume that the forward-propagating light is subject to back scattering within this region, whose fraction is denoted as αdz.

The intensities of light transmitting through this region from the left to right and from the right to left are expressed, respectively, as

$$I^+(z+dz) - I^+(z) = -\alpha I^+(z)dz + \alpha I^-(z+dz)dz,$$
$$I^-(z) - I^-(z+dz) = -\alpha I^-(z+dz)dz + \alpha I^+(z)dz.$$

These relations are easily rewrite in the form of differential equations as

$$\frac{dI^+}{dz} = -\alpha(I^+ - I^-), \tag{6.166}$$

$$\frac{dI^-}{dz} = -\alpha(I^+ - I^-). \tag{6.167}$$

The subtraction of the second equation from the first one results in $d(I^+(z) - I^-(z))/dz = 0$, which leads to

$$I^+(z) - I^-(z) = C. \tag{6.168}$$

C is a constant that will be determined from the boundary condition such that

$$I^+(0) - I^-(0) = I^+(l) = C,$$

where we have assumed that the total intensity of light propagating in this medium is invariant.

Using this relation, we can solve the differential equation Eq. (6.166) by putting

$$\frac{dI^+}{dz} = -\alpha(I^+ - I^-) = -\alpha I^+(l),$$

which leads to

$$I^+(z) = I^+(0) - \alpha I^+(l)z. \tag{6.169}$$

In a similar way,

$$\frac{dI^-}{dz} = -\alpha(I^+ - I^-) = -\alpha I^+(l),$$

gives the solution as

$$I^-(z) = \alpha I^+(l)(l - z), \tag{6.170}$$

where we assume $I^-(l) = 0$, because light is only incident from the left.

The result is schematically shown in Fig. 6.17b. It is found that the intensity of light propagating within a random medium will vary linearly with the propagation length. Using the above results and the relations of $I^+(l) = I^+(0) - \alpha I^+(l)l$ and $I^-(0) = \alpha I^+(l)l$, we can define the reflectivity R and transmittance T as

$$I^+(l) = \frac{1}{1+\alpha l} I^+(0) \Rightarrow T \equiv \frac{I^+(l)}{I^+(0)} = \frac{1}{1+\alpha l}, \quad (6.171)$$

$$I^-(0) = \frac{\alpha l}{1+\alpha l} I^+(0) \Rightarrow R \equiv \frac{I^-(0)}{I^+(0)} = \frac{\alpha l}{1+\alpha l}, \quad (6.172)$$

which shows that both transmittance and reflectivity are determined by a product of the scattering efficiency α and the propagation length l. The wavelength dependence should be included in α. Hence, when the length is sufficiently large as compared with α^{-1}, then $\alpha l \gg 1$, which means T comes close to null, while R approaches unity. Thus, the wavelength dependence of the reflectivity will completely vanish so that light reflected by such a random medium becomes white. On the other hand, when $\alpha l \ll 1$, T will approach unity, while R is approximated as $R \approx \alpha l$, which means R is proportional both to α and l. Thus, although the reflectivity itself will decrease, instead the wavelength dependence will appear remarkably. Therefore, the coloration due to the scattering phenomenon needs the appropriate value of αl.

In Fig. 6.18, we show a result of simulation in case where light from black body radiation at 6000 K is incident on a random medium, which scatters light according to the fourth power of the frequency of light. For a small value of αl, the spectrum leans to shorter wavelengths with a scattering peak located around 280 nm, and then gradually decreases toward the longer wavelengths. With increasing αl, the scattering peak gradually shifts to the longer wavelengths and finally comes close to that of the incident light, which is shown as a dashed line. Thus, when one observes the random medium, it faintly assumes blue to violet at first, then the color shifts to blue, greenish and whitish-yellow with decreasing saturation.

In the blue sky, α will be overwhelmingly small, thus essentially very large l is necessary to increase the reflectivity. This may be one of the reasons why the blue sky cannot be reproduced in a

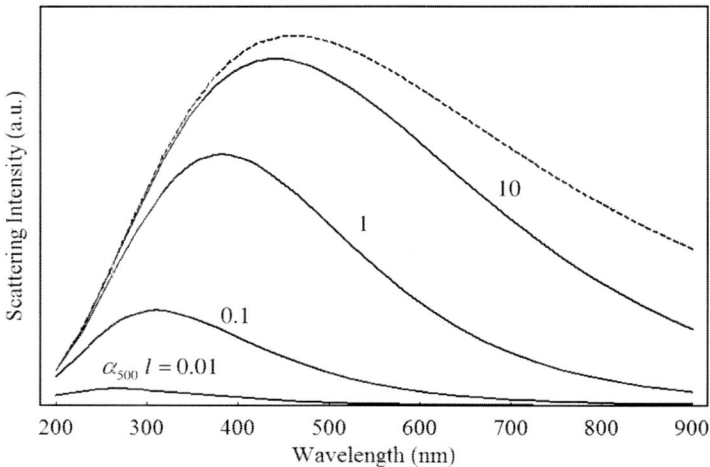

Figure 6.18 Scattering spectra for various values of αl when light from black body radiation at 6000 K is incident on a random medium, which is assumed to scatter the light according to the fourth-power of the light frequency. α_{500} means α at 500 nm, which is used as a normalization parameter. Dashed line is a spectrum of incident light. The calculation is performed according to a relation $S(\lambda) = (8\pi v^2/c^3) \cdot hv/(\exp[hv/(k_B T)] - 1) \cdot (l\alpha)/(1+l\alpha) \cdot (c/\lambda^2)$, where $\alpha = (v/v_{500})^4$ with v_{500} the frequency of light at 500 nm. T, k_B, c, and h are the absolute temperature, Boltzmann constant, light velocity in vacuum, and Planck constant, respectively.

laboratory. It is generally mentioned that to make colors effectively in a small-scale experiment, the mechanism of Rayleigh scattering will be insufficient and the interference effect to produce specific colorations is anyway necessary. The random structures possessing a short-range order in blue feather of a common kingfisher and blue body of dragonfly will be one of such devices that nature has been producing during a long evolutionary process.

Absorptive medium   Even when a medium is absorptive, the similar treatment is possible. Consider a case where light with the intensity of I^+ is incident on a thin plate with a thickness of dz and a fraction of the light is reflected, which is expressed as RI^+, while the other is transmitted, which is expressed as TI^+, as shown in Fig. 6.19. Then, without absorption, an energy conservation rule guarantees a relation $R + T = 1$. On the other hand, if the plate is absorptive, the

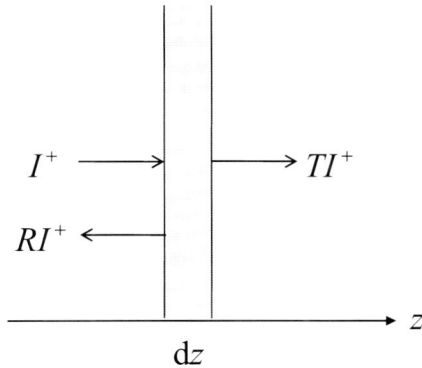

Figure 6.19 Light incident on a thin plate is partly transmitted and partly reflected with the fractions expressed by T and R.

fraction of transmission should be reduced according to a relation $T = 1 - R - A$, where A is a fraction of absorption. Under an assumption of an extremely thin plate, both R and A should be proportional to a thickness of the plate so that we can put $R = \alpha_1 dz$ and $A = \alpha_2 dz$.

Thus, we obtain

$$I^+(z + dz) - I^+(z) = -(\alpha_1 + \alpha_2)I^+(z)dz + \alpha_1 I^-(z + dz)dz,$$
$$(6.173)$$

$$I^-(z) - I^-(z + dz) = -(\alpha_1 + \alpha_2)I^-(z + dz)dz + \alpha_1 I^+(z)dz.$$
$$(6.174)$$

Then, the differential equations become

$$\frac{dI^+}{dz} = -(\alpha_1 + \alpha_2)I^+ + \alpha_1 I^-, \qquad (6.175)$$

$$\frac{dI^-}{dz} = (\alpha_1 + \alpha_2)I^- - \alpha_1 I^+. \qquad (6.176)$$

These equations are easily solved by putting $I^+ = X(z) \exp[-(\alpha_1 + \alpha_2)z]$ and $I^- = Y(z) \exp[(\alpha_1 + \alpha_2)z]$. Inserting these formulas into Eqs. (6.175) and (6.176), we obtain

$$X' = \alpha_1 Y e^{2(\alpha_1 + \alpha_2)z},$$
$$Y' = -\alpha_1 X e^{-2(\alpha_1 + \alpha_2)z}.$$

From the first relation, we obtain $Y = (1/\alpha_1)X' \exp[-2(\alpha_1 + \alpha_2)z]$, and insert it into the second relation, which leads to

$$(1/\alpha_1)\left\{ X''e^{-2(\alpha_1+\alpha_2)z} - 2(\alpha_1 + \alpha_2)X'e^{-2(\alpha_1+\alpha_2)z} \right\} = -\alpha_1 X e^{-2(\alpha_1+\alpha_2)z}.$$

Thus we obtain

$$X'' - 2(\alpha_1 + \alpha_2)X' + \alpha_1^2 X = 0, \qquad (6.177)$$

which is a homogeneous equation with respect to X and can be solved by a method of a characteristic equation.

Putting $X \sim \exp[\lambda z]$ and inserting it into the above equation, we obtain a quadratic equation with respect to λ such that

$$\lambda^2 - 2(\alpha_1 + \alpha_2)\lambda + \alpha_1^2 = 0,$$

whose solutions are given as

$$\lambda = \alpha_1 + \alpha_2 \pm \sqrt{\alpha_2^2 + 2\alpha_1\alpha_2}.$$

In case of $\alpha_2 \neq 0$, two independent solutions are obtained such that

$$\lambda_1 = \alpha_1 + \alpha_2 + \sqrt{\alpha_2^2 + 2\alpha_1\alpha_2},$$
$$\lambda_2 = \alpha_1 + \alpha_2 - \sqrt{\alpha_2^2 + 2\alpha_1\alpha_2},$$

with $\lambda_1 > \lambda_2$.

Hence, the general solutions to X and Y are expressed, using these solutions, as

$$X = Ae^{\lambda_1 z} + Be^{\lambda_2 z},$$
$$Y = \frac{1}{\alpha_1}\left(A\lambda_1 e^{-\lambda_2 z} + B\lambda_2 e^{-\lambda_1 z} \right),$$

where A and B are arbitrary constants that should be determined from the boundary conditions, and we have used the relations of $Y = (1/\alpha_1)X' \exp[-2(\alpha_1 + \alpha_2)z]$, $\lambda_1 - 2(\alpha_1 + \alpha_2) = -\lambda_2$ and $\lambda_2 - 2(\alpha_1 + \alpha_2) = -\lambda_1$. Inserting these results into I^+ and I^-, we obtain

$$I^+ = \left(Ae^{\lambda_1 z} + Be^{\lambda_2 z} \right)e^{-(\alpha_1+\alpha_2)z} = Ae^{\lambda z} + Be^{-\lambda z}, \quad (6.178)$$
$$I^- = (1/\alpha_1)\left(A\lambda_1 e^{-\lambda_2 z} + B\lambda_2 e^{-\lambda_1 z} \right)e^{(\alpha_1+\alpha_2)z}$$
$$= (1/\alpha_1)\left(A\lambda_1 e^{\lambda z} + B\lambda_2 e^{-\lambda z} \right), \qquad (6.179)$$

where we put $\lambda \equiv \sqrt{\alpha_2^2 + 2\alpha_1\alpha_2}$.

From the boundary conditions of $I^+(0) = I^+(0)$ and $I^-(l) = 0$, the following relations are obtained:

$$A + B = I^+(0),$$
$$A\lambda_1 e^{\lambda l} + B\lambda_2 e^{-\lambda l} = 0.$$

Solving these equations with respect to A and B, we can determine the constants as

$$A = -\frac{\lambda_2 e^{-\lambda l}}{\lambda_1 e^{\lambda l} - \lambda_2 e^{-\lambda l}} I^+(0),$$

$$B = \frac{\lambda_1 e^{\lambda l}}{\lambda_1 e^{\lambda l} - \lambda_2 e^{-\lambda l}} I^+(0).$$

Using these expressions, we obtain the transmittance and reflectance as follows:

$$T = \frac{I^+(l)}{I^+(0)} = \frac{2\lambda}{\lambda_1 e^{\lambda l} - \lambda_2 e^{-\lambda l}}, \tag{6.180}$$

$$R = \frac{I^-(0)}{I^+(0)} = \frac{\alpha_1 \left(e^{\lambda l} - e^{-\lambda l}\right)}{\lambda_1 e^{\lambda l} - \lambda_2 e^{-\lambda l}}, \tag{6.181}$$

where we have used a relation $\lambda_1 \lambda_2 = \alpha_1^2$.

These expressions are approximated at the following two extremes: One is a case where α_2 is sufficiently small, that is, a case where the effect of absorption is negligibly small. In this limit, the denominators of the both formulas are approximated as

$$\lambda_1 e^{\lambda l} - \lambda_2 e^{-\lambda l} \approx \lambda_1(1 + \lambda l) - \lambda_2(1 - \lambda l)$$
$$= (\lambda_1 - \lambda_2) + \lambda l(\lambda_1 + \lambda_2) = 2\lambda + 2\lambda l(\alpha_1 + \alpha_2)$$
$$\approx 2\lambda(1 + \alpha_1 l),$$

and thus $T = 1/(1 + \alpha_1 l)$, which completely agrees with the expression obtained without absorption, Eq. (6.171). In this derivation, we have used the relations, $\lambda_1 - \lambda_2 = 2\lambda$, $\lambda_1 + \lambda_2 = 2(\alpha_1 + \alpha_2)$ and $\lambda = \sqrt{(2\alpha_1 + \alpha_2)\alpha_2} \approx \sqrt{2\alpha_1\alpha_2} \ll 1/l$. On the other hand, using a relation, $\exp[\lambda l] - \exp[-\lambda l] \approx 2\lambda l$, we obtain an approximate expression for the reflection as $R = \alpha_1 l/(1 + \alpha_1 l)$, which again confirms the expression derived before, Eq. (6.172).

The other extreme is a case where l is sufficiently large. In this case, we can approximate the exponential terms appearing in a numerator and denominator of Eqs. (6.180) and (6.181) as

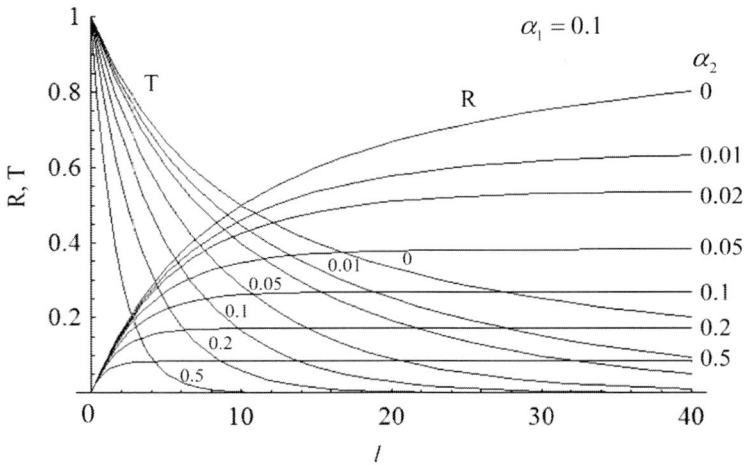

Figure 6.20 Transmittance and reflectance of light incident on an absorptive random media with the thickness of l for various values of the absorption efficiency α_2, while the backscattering efficiency α_1 is kept constant at $\alpha_1 = 0.1$.

$\exp[-\lambda l] \approx 0$, and hence we obtain $T \approx (2\lambda/\lambda_1)\exp[-\lambda l]$ and $R \approx \lambda_2/\alpha_1$. Thus, the transmittance through an absorptive random medium will vanish when the thickness is sufficiently large, while the reflection will remain constant, which is essentially determined by the ratio of scattering and absorption efficiencies. We show the simulated results in Fig. 6.20, where we can see that the transmission steeply decreases with increasing α_2, while the reflectivity tends to stay asymptotically at a constant value that decreases with increasing α_2. These behaviors completely agree with those we have estimated above.

The meaning of the constant value found in the reflectivity is roughly interpreted as follows. The value of λ_2/α_1 is approximated as

$$R \approx \frac{\lambda_2}{\alpha_1} = \frac{\alpha_1 + \alpha_2 - \sqrt{\alpha_2^2 + 2\alpha_1\alpha_2}}{\alpha_1} = \frac{\alpha_1}{\alpha_1 + \alpha_2 + \sqrt{\alpha_2^2 + 2\alpha_1\alpha_2}}$$

$$= \frac{\alpha_1/\alpha_2}{1 + \alpha_1/\alpha_2 + \sqrt{1 + 2\alpha_1/\alpha_2}} \approx \frac{\alpha_1/\alpha_2}{2(1 + \alpha_1/\alpha_2)}, \quad (6.182)$$

where the final relation is derived by assuming $\alpha_1/\alpha_2 \ll 1$. If we compare the result with a case without absorption of $R = \alpha_1 l/(1 +$

$\alpha_1 l$), we can conclude that apart from a numerical factor, the two expressions completely agree with each other when we put $1/\alpha_2 = l$. Thus, the effect of absorption on the reflectivity is to reduce the effective thickness of a medium, which is eventually determined by the inverse of the absorption efficiency.

6.5 Colorations Due to Light Scattering

6.5.1 *Origin of the Blue Sky*

Lord Rayleigh proposed a hypothesis that the blue sky originates from light scattering due to microparticles in the atmosphere, whose sizes were much smaller than the wavelength of light [Rayleigh (1871a,b)]. Before that, it was considered as due to the reflection at surfaces of water foams existing in the atmosphere. Using a dimensional analysis, he proved that the efficiency of light scattering was inversely proportional to the fourth power of wavelength.

When light is reflected by a thin film, dimensional quantities that determine the efficiency of reflection, η, will be the thickness of a film, d, and the wavelength of light, λ. Since both quantities have the dimension of length, while η is a non-dimensional quantity, η should be expressed by a ratio of these quantities. Further, the amplitude of light reflected by a thin film will be proportional to the thickness of the film in case that it is sufficiently thin. Thus, the efficiency should be proportional to d^2 which leads to a relation $\eta \propto d^2/\lambda^2$.

On the other hand, when light is scattered by a microparticle, the scattering amplitude will be proportional to the volume of the particle[a], V. In addition, it will be inversely proportional to a distance from the particle, r. Thus, the non-dimensional scattering efficiency, η, should be expressed by a relation, $\eta \propto V^2/(r^2\lambda^4)$, since V is expressed by a cube of length.

This consideration was generally accepted among scientists so that light scattering due to microparticles is now called *Rayleigh scattering*. His result shows that the light scattering due to

---

[a]This relation is derived under a consideration that a microparticle consists of N scattering points. Thus, the scattering amplitude should be simply expressed by that proportional to N and hence to a volume.

microparticles such as molecules and suspending particles in the atmosphere favors light with shorter wavelengths, while a radiation spectrum of the sun has a peak ranging from a blue to near infrared region. The product of these spectra gives a peak around blue-green region of 500 nm. In human eye, however, the enhancement of blue region as compared with white light due to the direct sunlight makes us to perceive blue as an impressive color.

Lord Rayleigh also noticed that the scattered light observed at right angles was largely polarized. This is easily confirmed when one observes the blue sky under a linear polarization analyzer. This phenomenon can be explained in term of the radiation due to an oscillating dipole induced in the particle: If we observe an oscillating dipole at right angles to the incident direction, it is expected that the radiation will be completely linearly polarized irrespective of the polarization direction of the incident light.

To tell the truth, this is not actually true. I performed the experiment once with my student on the top of a mountain. At that time, the degree of polarization was found to amount to at most 70%, even when we measured at right angles to the incident direction of the sun light, though the clear angular dependence was observed as was expected. The reason for this reduction of polarization is not clear up to now. First, we have to take account of the effect of anisotropy in the shape of scatterer. The light scattering in the atmosphere is considered to be mostly due to nitrogen and oxygen molecules. Since these molecules are anisotropic in shape, the degree of polarization will be deteriorated even when we observe at right angles. However, it will be soon found that their anisotropies do not give a large contribution to the polarization reduction.

Next, we consider the effect of multiple scattering. Lord Rayleigh also mentioned in his paper that the multiple scattering would affect the polarization. However, when multiple scattering becomes significant to affect the polarization of light, considerable deviation from the fourth power law will also appear. This is because even when white light illuminates scatterers, once it is scattered, the scattered light is no more white but becomes bluish. Thus, the multiple scattering will enhance the blue component even more. Actually, the truly blue sky is beautifully expressed by the fourth power law. Thus, it seems to need more quantitative study on the optical properties of the blue sky including the spectrum, polarization and their angular dependence.

On the other hand, the red evening sun is explained in terms that the sunlight passes through the atmosphere that is close to the surface of the earth. The thick atmospheric layer scatters off the blue component of the white sunlight and hence the remaining color will be red. This effect is easily experienced when one observes a turbid solution that is illuminated by white light. If the solution is observed from the side, we will feel somewhat bluish, while it becomes reddish when the light source is directly observed through the solution. Although the origin of this coloration is, of course, due to Rayleigh scattering, the scattered light from a solution is often called *Tyndall scattering* or *Tyndall blue*. It is soon noticed experimentally that it is rather difficult to reproduce the blue sky by this type of experiment, because if one wants to strengthen the blue by adding more amount of scatterers, it will be whitish rather than blue. In the case of the blue sky, microparticles are not so concentrated in the atmosphere to cause multiple scattering, but yet the number of scatterers is sufficiently large to cause the blue. Further, the darkness of space will play a role of background absorber, which will enhance the blue by completely removing the unwanted background colors.

6.5.2 *Non-iridescent Structural Colors*

The scattering of light can become one of the most important sources of structural colors, whose mechanism is completely different from that of thin-layer or photonic-crystal type, because it does not need regular structures in principle. Furthermore, an induced oscillating dipole uniformly emits scattered light so that it is basically non-iridescent, i.e. the color does not change with viewing angles. Thus, the light scattering has been thought as a cause of bluish colors with the lack of directionality, which are generally found in a wide variety of random media. As described above, the origin of the blue sky is explained as due to the light scattering by atmospheric microparticles. The pale bluish color of a suspension involving small colloidal particles is known as Tyndall blue. The blue or bluish color in these cases is anyway based on a fact that the scattering cross section is dependent on the fourth power of light frequency.

In the past, the Tyndall blue frequently appeared in the literatures as a typical example of non-iridescent structural colors. Mason reported that the colors of some birds and insects were due to the Tyndall blue and exemplified the feather of blue jay and the body/wing of dragonfly [Mason (1926)]. He considered six necessary conditions for a medium to show the Tyndall blue: (1) inhomogeneities in the refractive index, (2) dimensions comparable to the wavelength of light, (3) blue scattering and red transmission, (4) size-dependent depth of blue, (5) polarized scattering, and (6) inversely proportional to the fourth power of wavelength.

Frank and Ruska applied the first commercially available electron microscope to investigating the feathers of ivory-breasted pitta (*Pitta maxima*) and turquoise tanager (*Tangara mexicana*) that were known to display the Tyndall blue [Frank and Ruska (1939)]. Within a barb of the blue feather, they found a spongy structure consisting of keratin and air at the inner wall of the medullary cell with a characteristic size of 0.1–0.25 μm. Later, more detailed observation on the spongy structure was made by Schmidt and Ruska on Eurasian jay (*Garrulus glandarius*), purple-breasted cotinga (*Cotinga cotinga*) and blue-and-yellow macaw (*Ara ararauna*) [Schmidt and Ruska (1962)]. They reported that the spongy structure sometimes showed a structure consisting of numerous spherical or oval vacuoles, which is now called *spherical type*, while they sometimes showed a network structure consisting of hard sticks and air holes, which is called *channel type*.

The spongy structure are now known to be widely distributed in avian species (see Fig. 6.21b). Kingfisher, parakeet, cotinga, jay, manakin and bluebird are the well-known examples of this type. Their structural colors are apparently characterized by (1) weak or no change in color with varying view angles, and (2) rather low reflectivity with the lack of directionality. Most of the spongy structures are found in the medulla of feather barb, which consists of several medullary cells running along a barb, and is covered with a cortex. They lie in the inner part of medullary cells, but sometimes occupies the whole barb. Near the center of the barb, some vacuoles

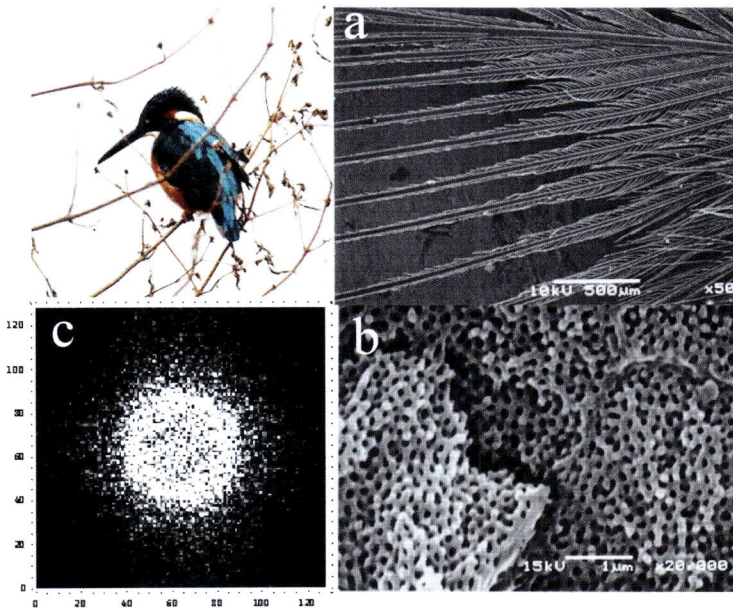

Figure 6.21 (a) Barbs and (b) the cross section of a barb of a common kingfisher, *Alcedo atthis*, and (c) 2D Fourier image of the TEM image of its spongy structure [Kinoshita and Yoshioka (2005b)].

are distributed, which correspond to the nuclei of medullary cells, around which small melanin granules are distributed.

Owing to the above detailed observations using electron microscopy, it was generally believed that the Tyndall blue was an only cause for the blue colorations in these spongy structures. Yet, there had been no report to measure accurately the wavelength and angular dependence of the scattering cross section, and its polarization characteristics, which were the necessary conditions for the Tyndall blue as were proposed by Mason. Thus, a problem whether pure Rayleigh/Tyndall scattering determined the colors of animals or not was actually unknown at that time.

However, it was gradually recognized that light scattering was somehow modified and contributed effectively to animal colors. The first implication on the presence of interference phenomenon on this problem was made by Raman, who threw doubt on the Tyndall scattering in the feather of Indian roller (*Coracias indica*) [Raman (1934a)]. He found that the barb involved a series of cells and the color varied individually from cell to cell. Further, he noticed that when the barb was fully damp, the dark blue color changed into bright green, while light blue into light red. He compared these results with light scattering phenomena known for particles of various sizes and concluded the inapplicability of light scattering theory. He suggested, without a final decision, the possibility of diffraction or interference for this phenomenon.

Later, Dyck performed detailed anatomical and optical investigations on the feathers of rosy-faced love bird (*Agapornis roseicollis*) and plum-throated cotinga (*Cotinga maynana*) [Dyck (1971)]. He found the inner part of a barb was completely filled up with a spongy structure, which consisted of 3D network of connecting keratin rods of 0.1 μm wide that supported empty spaces of 0.1 μm in diameter. He investigated the spongy structures in various avian species and found that the widths of the rods and air spaces were strikingly uniform in size, while their orientations were definitely random.

He measured the reflection spectra of violet, blue and green barbs of structurally colored three avian species and found a clear peak from ultraviolet to visible regions, which were in good accordance with the apparent colors of the feathers. Further, he found that the sizes of keratin rods and air spaces varied strongly in correlation with the reflection color. Thus, it was clear that a simple model based on Tyndall scattering was no longer applicable. He considered that the small structural units such as rods and spaces contributed to the interference of light and proposed a "hollow cylinder" model. Namely, the spongy structure was assumed to consist of randomly oriented hollow cylinders with the diameters of 200–400 nm. He compared a wavelength giving the maximum reflectivity with that deduced from the model, and found the relatively good correlation between them.

Along this stream, Prum *et al.* conducted a decisive analysis on this problem. They employed a spatial Fourier transformation of TEM image of the medullary spongy structure in blue feather barbs of plum-throated cotinga (*Cotinga maynana*), and found a clear ring structure around the origin in wave vector space [Prum *et al.* (1998)] (see Fig. 6.21c for example). If the spongy structures with a wide variety of sizes were distributed, a Gaussian-like distribution would be obtained around the origin. The presence of a ring structure clearly indicated that the uniform characteristic size of the matrix actually existed, which presumably corresponded to the reflection maximum.

They simulated the reflection spectrum using the 2D Fourier power spectrum by averaging it radially and converting into a real space using the average refractive index. The result was in fairly good agreement with the experiment. They applied this method to the feathers of various avian species and obtained the similar results [Prum *et al.* (1999b); Prum (2006)]. They also conducted the analysis on avian skins, which appeared as quasi-ordered arrays of parallel collagen fibers, and found the similar spatial order [Prum *et al.* (1999a); Prum and Torres (2003)]. Thus, it is now clear that the spongy structures generally possess a characteristic length that actually contributes to the generation of well-defined reflection peaks through constructive interference.

Prum *et al.* also performed an experiment on a familiar bluet damselfly (*Enallagma civile*) and a common green darner (*Anax junius*) [Prum *et al.* (2004)]. They measured the reflection spectra of the bodies of these two species and found that the reflection spectra showed clear peaks. Since the Tyndall blue should follow the fourth power law, the presence of the peak clearly contradicted to this mechanism.

Within a pigmentary cell located below chitinous cuticle, nanospheres with 200–300 nm in diameter and ommochrome spheres with a much larger size were known to be distributed (see Fig. 6.22) [Veron *et al.* (1974)]. They found that the nanospheres were packed densely with the average distance of 322 nm, which were truly spherical and highly consistent in diameter. Further, the inside of a sphere was heterogeneous with a central spot stained darkly and a clear peripheral boundary. They analyzed using the 2D Fourier transform of a transmission electron microgram and found the similar ring structure implying the presence of interference effect. They considered the origin of the blue reflection in terms of the quasi-ordered nanospheres or the heterogeneity within the sphere. However, both models failed to explain quantitatively the peak wavelength of the reflection spectrum.

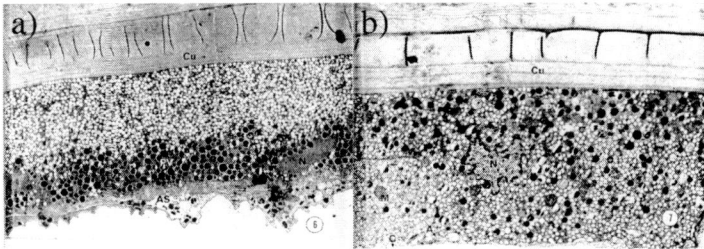

Figure 6.22 TEM image of the dorsal abdominal integument of Australian damselfly, (*Austrolestes annulosus*, blue ringtail), in (a) blue and (b) dark color phase. Cu: cuticle, Gr: refractive granules, PV: pigment vesicles, N: nucleus, AS: air sac, M: mitochondria, A: nerve axons, and C: collagen. (Reproduced from [Veron *et al.* (1974)] with permission.)

It is now evident that a simple model based on Rayleigh/Tyndall scattering is no longer applicable to explaining non-iridescent blue or bluish color, since a definite peak is usually observed in the reflection spectrum and also a clear spatial correlation exists in seemingly random structures. It also becomes clear that these structures are widely distributed in the animal kingdom, particularly in birds and insects. Their structural units appear sometimes in a form of a network and sometimes of randomly distributed spheres. These findings are quite important because the structure actually possesses a characteristic length and hence shows the short-range order, which induces the interference of light with enforcing reflection in a particular wavelength range.

Recently, Cao's group has extensively studied on this problem. Dufresne *et al.* employed small-angle X-ray scattering (SAXS) to analyze the spongy structures in various feather barbs and found similar ring structures both for channel and spherical types, as shown in Fig. 6.23 [Dufresne *et al.* (2009)]. The ring structure generally consisted of a definite ring, which was followed by a tail or a weak second ring outside the main ring. Since the scattering cross section of X-ray is very small, the structural information is directly obtainable without being disturbed by multiple scattering, which will be thus comparable with the Fourier analysis performed by Prum *et al.*

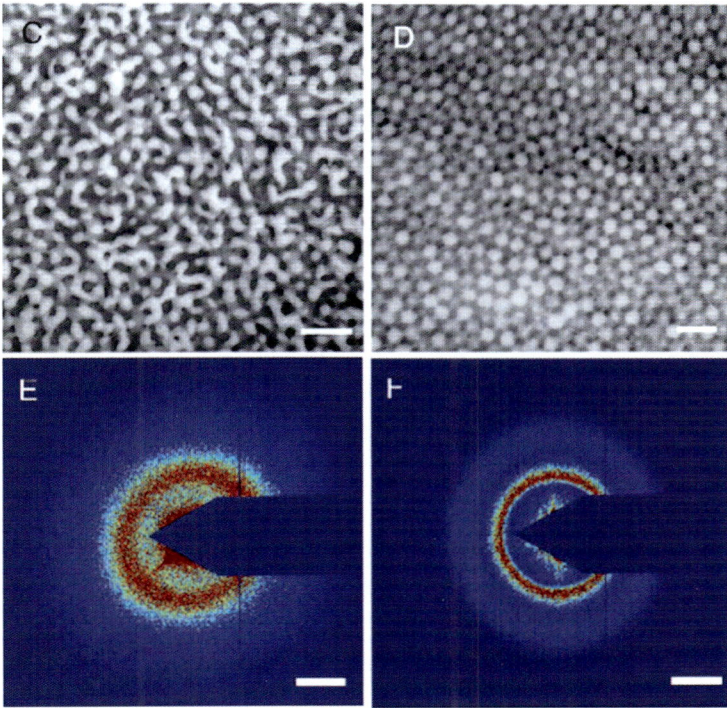

Figure 6.23 Electron micrograms of feather barbs of C) channel-type male easter bluebird (*Sialia sialis*) and D) sphere-type male plum-throated cotinga (*Cotinga maynana*), and the corresponding small-angle X-ray scattering patterns, E) and F). Scale bars are 500 nm for C) and D), and 0.025 cm^{-1} for E) and F). ([Dufresne *et al.* (2009)] - Reproduced by permission of The Royal Society of Chemistry.)

Further, Noh *et al.* measured the reflection spectra under various optical configurations and found that under directional lighting, the feather barbs were surprisingly iridescent, while under omni-directional illumination, they were actually non-iridescent [Noh *et al.* (2010a)]. They found that the reflection spectrum was eventually independent of an angle of incidence, if the refraction at the surface was properly corrected, but was only dependent on an angle between the directions of incidence and reflection. Further, they attempted the reproduction of the reflection spectrum from the

SAXS pattern, and obtained good agreement with respect to the main ring when the average refractive index of 1.25 was employed.

On the other hand, they noticed that the feather of plum-throated cotinga (*Cotinga mayanana*) showed a double peak with a secondary peak at 365 nm in addition to a primary peak at 534 nm. They noticed in the θ-2θ experiment[a] that the secondary peak was considerably depolarized and with increasing incident angle, θ, its peak position shifted oppositely as compared with the primary peak that shifted toward shorter wavelengths. They considered double-scattering mechanism to explain the origin of the secondary peak and obtained quantitative agreement [Noh *et al.* (2010b)]. Cao's group further performed an experiment using biomimetic nanostructures to confirm the above speculation and also discussed the origin of spongy structure in terms of phase separation [Liew *et al.* (2011)].

As was described in the papers, their results are easily understandable if we assume that the ring structure is expressed simply by a delta-function-like ring described as $f(\mathbf{q}) \equiv \delta(|\mathbf{q}| - q_0)$, where $\mathbf{q}$ is a scattering vector with a radius of the ring, q_0. At first, we will explain the iridescent and non-iridescent nature under directional and omni-directional illuminations, respectively. Figure 6.24a shows a scattering process under directional illumination, where $\mathbf{k}_i$ and $\mathbf{k}_s$ are the wave vectors of incident and scattered waves. Since the magnitude of $\mathbf{q}$ is fixed, the relation between the directions of incidence and reflection is easily obtained under an assumption of elastic scattering with $|\mathbf{k}_i| = |\mathbf{k}_s| \equiv k$ and a wave-vector conservation rule, $\mathbf{k}_s = \mathbf{k}_i + \mathbf{q}$. Thus, the change of $\mathbf{k}_i$, while its direction is fixed, inevitably causes the change of the angle of reflection. Hence, the iridescence is perceived to the eye under directional illumination by white light when a viewing angle is changed.

On the other hand, under omni-directional illumination, the shortest wave vector of incident light that satisfies the conservation

---

[a]θ-2θ method is often employed for the structural analysis of crystal in X-ray region, in which both the sample and detector are rotated with respect to the incident direction with the amount of $\Delta\theta$ for the sample and $2\Delta\theta$ for the detector. This measurement is used to probe only a specular component among various reflection components.

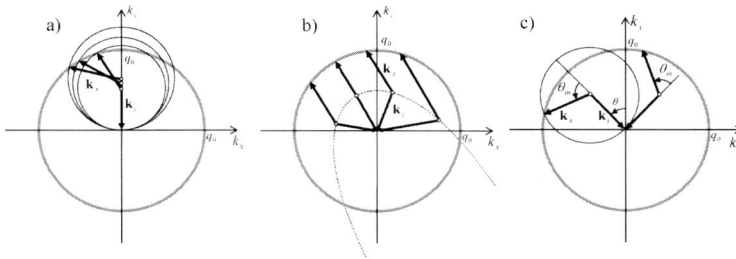

Figure 6.24 Scattering of light expressed in wave vector space when the ring structure is expressed as $f(\mathbf{q}) = \delta(|\mathbf{q}| - q_0)$ with $\mathbf{q}$ a scattering vector. (a) is a scattering process when a viewing angle is changed under directional illumination (from normal direction) and (b) that under omni-directional illumination. (c) is a scattering process with a fixed angle between incidence and scattering, θ_m, while an incident angle, θ, is changed. Thin solid circles in (a) and (c) indicate those with a radius of a wave number of incident light, k, while a dashed curve in (b) is a function of $k_y = (q_0^2 - 2k_x^2)/\left(2\sqrt{q_0^2 - k_x^2}\right)$, which is slightly inclined to accommodate to the scattering direction.

rule is always determined by a relation of $k = q_0/2$ irrespectively of a viewing angle. Thus, under white light illumination, the reflection spectrum is almost invariable with the longest wavelength of $4\pi/q_0$. This explains the non-iridescent nature under omni-directional illumination.

Next, we will explain that the reflection properties are only determined by an angle between incident and reflected light, and not by an angle of incidence. The schematic explanation is shown in Fig. 6.24c. Since the centrosymmetric nature of $f(\mathbf{q})$, the wave-vector conservation rule guarantees the presence of a fixed angle between incidence and reflection, which leads to a conclusion that the optical properties are essentially independent of the incidence angle, if the refraction at a sample surface is properly corrected.

Finally, we will explain their result of θ-2θ experiment on the feather of plum-throated cotinga. In this experiment, they considered double-scattering mechanism to explain a secondary peak appearing in a shorter wavelength region, which showed an opposite behavior as compared with a primary peak when the incidence angle is changed. The single and double scatterings under a delta-function-like ring structure are restricted within processes

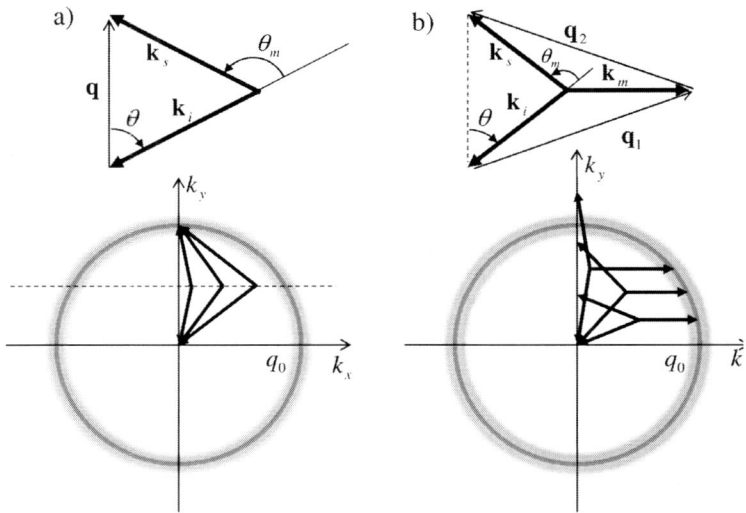

Figure 6.25 Scattering of light expressed in momentum space when the scattering vectors are expressed as $\delta(|\mathbf{q}| - q_0)$. (a) and (b) are the schematic illustration of single and double scattering processes in the θ-2θ experiment for three incident angles. θ and θ_m are an angle of incidence, and that between incident and scattered light wave, respectively.

illustrated in Fig. 6.25, where the incident and scattered wave vectors are accommodated to θ-2θ experiment, in which the angles of incidence and scattering are taken to be symmetric with respect to the normal to the surface.

As shown in Fig. 6.25a, the single scattering processes with varying k are simply expressed by bent double-pointed arrows with changing bent angles. On the other hand, the double scattering process includes a scattering characterized by a scattering vector of $\mathbf{q}_1$, which is further scattered through $\mathbf{q}_2$ to accommodate finally to the θ-2θ experiment (Fig. 6.25b). As is clear from this figure, with increasing θ, the magnitude of k decreases, which is just opposite to the case of single scattering process. Further, under normal incidence, the magnitude of wave vector in double scattering is generally larger than that in single scattering process, which suggests the presence of a secondary peak at a shorter wavelength.

As described above, most of the optical properties are now explained by this simplified model, which essentially originates from a ring-like order in the wave-vector space. Actually, the presence of short-range order without long-range one implies a considerable deviation from a delta-function-like ring, which will cause the broadening of the reflection spectrum and will fade out their optical processes considerably. However, one could say that the elucidation of the mechanism of non-iridescent color has brilliantly progressed owing to their extensive studies. We expect that such a methodology will play a key role to solve physical and biological meanings of the cooperation of regularity and irregularity, which is believed to be caused as the natural consequence.

Exercises

(1) Calculate $\nabla\phi$ using the expression described in Eq. (6.27).
(2) Confirm that Eqs. (6.28) and (6.26) actually give static electric and magnetic fields when $c \rightarrow \infty$, which proves that the electromagnetic wave cannot be generated from the static electric and magnetic fields.
(3) Derive Eqs. (6.121)–(6.124).

Experiments

I. Reproduction of the blue sky and the red evening sun

Materials: a transparent water tank, water, light scatterers such as a drop of milk and polystyrene or silica spheres with a diameter below 100 nm, a sheet of linear polarizer and an electric bulb of a high color temperature

Procedure: Pore water into a water tank with an appropriate amount of light scatterers in it. Put an electric bulb on one end of the tank and see it from a side and at the other end, then you can see slightly bluish color at the side of the tank and the reddish bulb when you see through the tank. Confirm that the scattered light viewed from a side of the tank is strongly polarized (see Fig. 6.26). In this

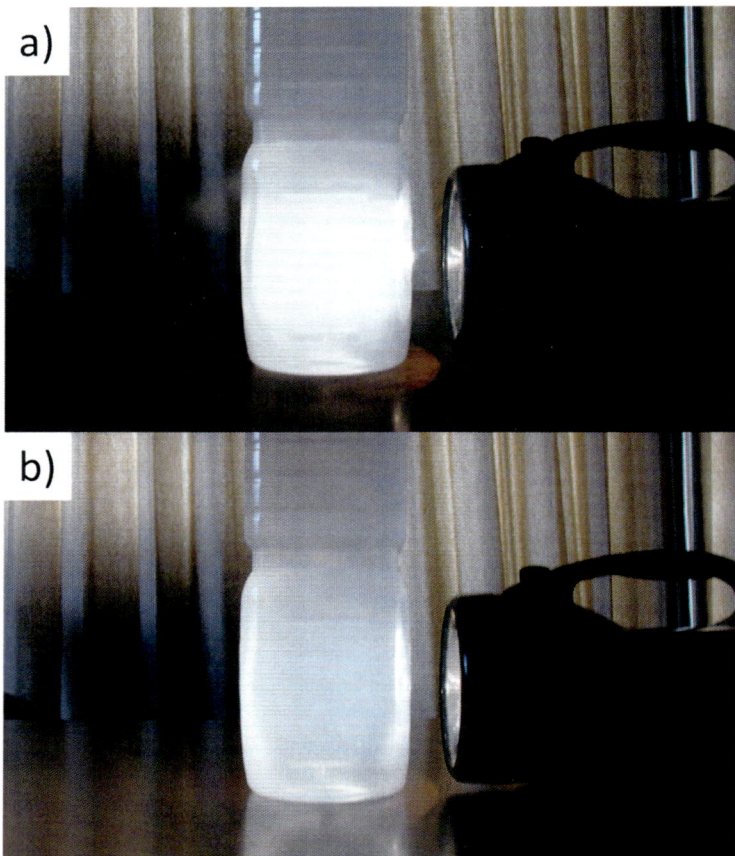

Figure 6.26 Experiment I. Polarization dependence of the scattered light by water including a small amount of bath oil, which is illuminated by a flashlight from the side. The polarization directions are (a) perpendicular and (b) parallel to the desktop.

experiment, the concentration of the scatters is important. If it is too low, you cannot see any color, while it is too high, you can perceive only white color in both cases. Anyway, you will find it quite difficult to reproduce the blue sky in the laboratory.

Chapter 7

FDTD Method and Its Applications

7.1 FDTD Method

7.1.1 *Method of Finite Differences*

Finite-difference time-domain (*FDTD*) method is known as a popular way to perform computational electrodynamics. This method was originally devised to analyze electromagnetic field distribution around an antenna. Recently, its outstanding usefulness has manifested itself with the development of nanotechnology, because near and far electromagnetic fields resulting from the interaction of light with complicated microstructure are easily simulated, which are often difficult to obtain in actual measurements. Although a lot of excellent papers and books describing its principle, algorithm and applications have been published so far, here we will briefly describe its principle of calculation for the reader's convenience. For this purpose, we will follow an excellent book written by Uno, where the comprehensive description of the FDTD method is given [Uno (1998)].

FDTD is basically a method to solve the Maxwell equation in the present of matter by means of finite differences. Thus, a starting

Bionanophotonics: An Introductory Textbook
Shuichi Kinoshita
Copyright © 2013 Pan Stanford Publishing Pte. Ltd.
ISBN 978-981-4364-71-3 (Hardcover), 978-981-4364-72-0 (eBook)
www.panstanford.com

point of this method is the general Maxwell equations given below:

$$\nabla \times \mathbf{E}(\mathbf{r}, t) = -\frac{\partial \mathbf{B}(\mathbf{r}, t)}{\partial t}, \tag{7.1}$$

$$\nabla \times \mathbf{H}(\mathbf{r}, t) = \mathbf{j}(\mathbf{r}, t) + \frac{\partial \mathbf{D}(\mathbf{r}, t)}{\partial t}, \tag{7.2}$$

$$\nabla \cdot \mathbf{D}(\mathbf{r}, t) = \rho(\mathbf{r}, t), \tag{7.3}$$

$$\nabla \cdot \mathbf{B}(\mathbf{r}, t) = 0, \tag{7.4}$$

where all the variables appearing in these formulas are assumed to be functions of position $\mathbf{r}$ and time t. Further, we assume the general relationships between these variables:

$$\mathbf{B}(\mathbf{r}, t) = \mu(\mathbf{r}, t)\mathbf{H}(\mathbf{r}, t), \quad \mathbf{D}(\mathbf{r}, t) = \epsilon(\mathbf{r}, t)\mathbf{E}(\mathbf{r}, t), \quad \text{and}$$

$$\mathbf{j}(\mathbf{r}, t) = \sigma(\mathbf{r}, t)\mathbf{E}(\mathbf{r}, t),$$

where $\mu(\mathbf{r}, t)$, $\epsilon(\mathbf{r}, t)$ and $\sigma(\mathbf{r}, t)$ are material parameters corresponding to permeability, permittivity and electric conductivity, respectively. All these parameters are also assumed to be functions of position and time.

In the FDTD method, only the upper two equations among the four are usually employed for the calculation. This is because the lower two equations can be included into the upper two under an appropriate assumption. Operating a differential operator ∇ on Eqs. (7.1) and (7.2) from the left of the both sides, we obtain

$$\nabla \cdot \{\nabla \times \mathbf{E}(\mathbf{r}, t)\} = -\frac{\partial}{\partial t} \nabla \cdot \mathbf{B}(\mathbf{r}, t),$$

$$\nabla \cdot \{\nabla \times \mathbf{H}(\mathbf{r}, t)\} = \nabla \cdot \mathbf{j}(\mathbf{r}, t) + \frac{\partial}{\partial t} \nabla \cdot \mathbf{D}(\mathbf{r}, t).$$

Using a vector algebraic relation $\nabla \cdot (\nabla \times \mathbf{A}) = 0$ that holds for an arbitrary vector $\mathbf{A}$, we notice that the left-hand sides of the both equations should vanish and also the conservation of charge, which is expressed by $\nabla \cdot \mathbf{j}(\mathbf{r}, t) + \partial \rho(\mathbf{r}, t)/\partial t = 0$, gives a set of the following equations:

$$\frac{\partial}{\partial t} \nabla \cdot \mathbf{B}(\mathbf{r}, t) = 0,$$

$$\frac{\partial}{\partial t} \{\nabla \cdot \mathbf{D}(\mathbf{r}, t) - \rho(\mathbf{r}, t)\} = 0.$$

Usually, the FDTD method assumes the presence of a source for electromagnetic wave generation such as incident electromagnetic wave, which will then interact with structures. Thus, all the electric and magnetic quantities are assumed to obey the four Maxwell equations at least at an initial time t_0, i.e. $\nabla \cdot \mathbf{B}(\mathbf{r}, t)|_{t=t_0} = 0$ and $\nabla \cdot \mathbf{D}(\mathbf{r}, t)|_{t=t_0} = \rho(\mathbf{r}, t_0)$. This condition and the above two relations guarantee that the relations $\nabla \cdot \mathbf{B}(\mathbf{r}, t) = 0$ and $\nabla \cdot \mathbf{D}(\mathbf{r}, t) = \rho(\mathbf{r}, t)$ hold even at a later time of $t > t_0$. Thus, as long as Eqs. (7.1) and (7.2) are assumed to hold, the two other equations hold automatically.

Equations (7.1) and (7.2) are anyway differential equations of the first order with respect to position and time, which will be approximated by applying the following *central difference* such that

$$\frac{\partial F(x, y, z)}{\partial x} \approx \frac{F(x + \Delta x/2, y, z) - F(x - \Delta x/2, y, z)}{\Delta x}. \quad (7.5)$$

Hence, we have to divide a space into small cells whose sizes are determined by the side lengths of Δx, Δy and Δz. Further, a time is also digitized into a step of Δt. Thus, an arbitrary spatiotemporal lattice point is generally expressed by $(x, y, z, t) = (i_1 \Delta x, i_2 \Delta y, i_3 \Delta z, n \Delta t)$, where i_1, i_2, i_3 and n are integers[a]. Using this relation, we will simplify the notation as

$$F(x, y, z, t) \equiv F^n(i_1, i_2, i_3),$$

for the later convenience. By employing this notation, the central differences around a spatiotemporal lattice point (i_1, i_2, i_3, n) with respect to x and t become

$$\frac{\partial F}{\partial x} \approx \frac{F^n(i_1 + \frac{1}{2}, i_2, i_3) - F^n(i_1 - \frac{1}{2}, i_2, i_3)}{\Delta x},$$

$$\frac{\partial F}{\partial t} \approx \frac{F^{n+\frac{1}{2}}(i_1, i_2, i_3) - F^{n-\frac{1}{2}}(i_1, i_2, i_3)}{\Delta t}.$$

Thus, it is natural that we should provide lattice points indexed by half integers in addition to those indexed by integers.

In computational electrodynamics, electric and magnetic fields should be both evaluated using difference equations derived from the original Maxwell equations of Eqs. (7.1) and (7.2). For this purpose, it becomes a problem how to deliver the lattice points to the two fields. This problem was first solved by Yee [Yee (1966)], and the programming based on this principle is now called *Yee algorithm*.

---

[a]By convention, we will use n to make a mark on a temporal lattice point. Do not confuse it with a refractive index that has been used in the previous chapters.

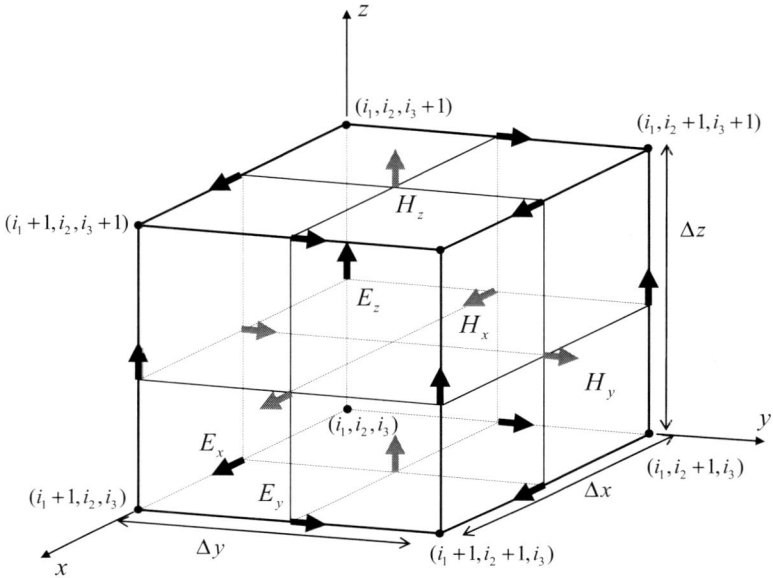

Figure 7.1 Yee cell model.

In Fig. 7.1, we show a typical example of the *Yee cell model*, where we see the electric and magnetic fields are delivered to various lattice points indicated by integer and half-integer indices.

Difference with respect to time Let us make a set of difference equations corresponding to Eqs. (7.1) and (7.2) according to this algorithm. We first rewrite the equations using material parameters given above as

$$\frac{\partial \mathbf{E}}{\partial t} = -\frac{\sigma}{\epsilon}\mathbf{E} + \frac{1}{\epsilon}\nabla \times \mathbf{H}, \tag{7.6}$$

$$\frac{\partial \mathbf{H}}{\partial t} = -\frac{1}{\mu}\nabla \times \mathbf{E}. \tag{7.7}$$

In order to derive the difference equations with respect to time, we deliver the temporal lattice points with integer indices for the electric field, while those of half-integer ones for the magnetic field. Thus, without considering the spatial coordinates, an electric field is generally written as $\mathbf{E}^n$, while a magnetic field as $\mathbf{H}^{n\pm\frac{1}{2}}$. Therefore, using the central differences, Eq. (7.7) is approximated with respect to time around an integer lattice point n as

$$\frac{\mathbf{H}^{n+\frac{1}{2}} - \mathbf{H}^{n-\frac{1}{2}}}{\Delta t} = -\frac{1}{\mu}\nabla \times \mathbf{E}^n.$$

In a similar manner, Eq. (7.6) is approximated around a half integer lattice point $n - \frac{1}{2}$ as

$$\frac{\mathbf{E}^n - \mathbf{E}^{n-1}}{\Delta t} = -\frac{\sigma}{\epsilon}\mathbf{E}^{n-\frac{1}{2}} + \frac{1}{\epsilon}\nabla \times \mathbf{H}^{n-\frac{1}{2}}.$$

However, since the electric field is only delivered on integer lattice points, $\mathbf{E}^{n-\frac{1}{2}}$ of the first term in the right-hand side should be anyway expressed by the fields on integer lattice points. Usually, this is attained by employing an approximation given as

$$\mathbf{E}^{n-\frac{1}{2}} \approx \frac{\mathbf{E}^n + \mathbf{E}^{n-1}}{2}.$$

Thus, we obtain

$$\frac{\mathbf{E}^n - \mathbf{E}^{n-1}}{\Delta t} = -\frac{\sigma}{\epsilon}\frac{\mathbf{E}^n + \mathbf{E}^{n-1}}{2} + \frac{1}{\epsilon}\nabla \times \mathbf{H}^{n-\frac{1}{2}},$$

which leads to

$$\mathbf{E}^n\left(\frac{1}{\Delta t} + \frac{\sigma}{2\epsilon}\right) = \mathbf{E}^{n-1}\left(\frac{1}{\Delta t} - \frac{\sigma}{2\epsilon}\right) + \frac{1}{\epsilon}\nabla \times \mathbf{H}^{n-\frac{1}{2}}.$$

Rewriting this expression, we finally obtain a difference equation for an electric field as

$$\mathbf{E}^n = \frac{1 - \sigma\,\Delta t/(2\epsilon)}{1 + \sigma\,\Delta t/(2\epsilon)}\mathbf{E}^{n-1} + \frac{\Delta t/\epsilon}{1 + \sigma\,\Delta t/(2\epsilon)}\nabla \times \mathbf{H}^{n-\frac{1}{2}}. \qquad (7.8)$$

On the other hand, a difference equation for a magnetic field is simply given as

$$\mathbf{H}^{n+\frac{1}{2}} = \mathbf{H}^{n-\frac{1}{2}} - \frac{\Delta t}{\mu}\nabla \times \mathbf{E}^n. \qquad (7.9)$$

Difference with respect to space   The differences with respect to the spatial coordinates are attained according to the Yee cell model, illustrated in Fig. 7.1. For simplicity, let us consider a 2D case, where we only consider the following two special cases: One is a case where the electric field is perpendicular to the 2D plane, and the other is a case where the magnetic field is perpendicular to it. The former and

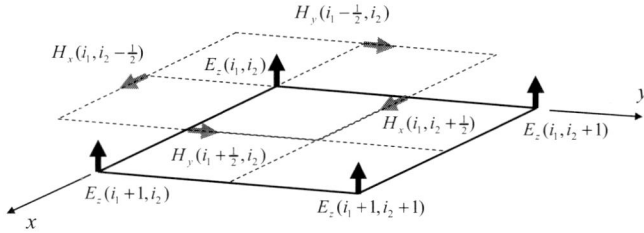

Figure 7.2 Calculation of difference with respect to 2D space coordinates in TM mode.

latter are usually called *TM* and *TE modes*, respectively[a]. An arbitrary electromagnetic field can be expressed by a linear combination of these two special cases.

We first consider a TM mode, which corresponds to a plane of $z = (i_3 + \frac{1}{2})\Delta z$ in Fig. 7.1. We have selected the lattice points necessary to calculate the differences in Fig. 7.2. We start from the time difference equation derived in Eq. (7.8) and rewrite it as follows:

$$\mathbf{E}^n = c_1 \mathbf{E}^{n-1} + c_2 \nabla \times \mathbf{H}^{n-\frac{1}{2}}, \qquad (7.10)$$

where $c_1 = \{1 - \sigma \Delta t/(2\epsilon)\} / \{1 + \sigma \Delta t/(2\epsilon)\}$ and $c_2 = (\Delta t/\epsilon)/ \{1 + \sigma \Delta t/(2\epsilon)\}$. Since the electric field directs to a positive direction of the z axis, we only consider the z component of the above equation and obtain

$$E_z^n = c_1 E_z^{n-1} + c_2 \left(\frac{\partial H_y^{n-\frac{1}{2}}}{\partial x} - \frac{\partial H_x^{n-\frac{1}{2}}}{\partial y} \right). \qquad (7.11)$$

Figure 7.2 shows that the electric fields are delivered to the lattice points with integer indices, while the magnetic fields to those with half-integer indices. According to this rule, we calculate differences with respect to x and y for the magnetic field, which results in

$$\frac{\partial H_y^{n-\frac{1}{2}}}{\partial x} \approx \frac{H_y^{n-\frac{1}{2}}(i_1 + \frac{1}{2}, i_2) - H_y^{n-\frac{1}{2}}(i_1 - \frac{1}{2}, i_2)}{\Delta x},$$

$$\frac{\partial H_x^{n-\frac{1}{2}}}{\partial y} \approx \frac{H_x^{n-\frac{1}{2}}(i_1, i_2 + \frac{1}{2}) - H_x^{n-\frac{1}{2}}(i_1, i_2 - \frac{1}{2})}{\Delta y}.$$

[a]TM and TE modes denote *transverse magnetic* and *transverse electric modes*, respectively.

Inserting these relations into Eq. (7.11), we obtain

$$
\begin{aligned}
E_z^n(i_1, i_2) = & \; c_1(i_1, i_2) E_z^{n-1}(i_1, i_2) \\
& + \frac{c_2(i_1, i_2)}{\Delta x} \left\{ H_y^{n-\frac{1}{2}}(i_1 + \tfrac{1}{2}, i_2) - H_y^{n-\frac{1}{2}}(i_1 - \tfrac{1}{2}, i_2) \right\} \\
& - \frac{c_2(i_1, i_2)}{\Delta y} \left\{ H_x^{n-\frac{1}{2}}(i_1, i_2 + \tfrac{1}{2}) - H_x^{n-\frac{1}{2}}(i_1, i_2 - \tfrac{1}{2}) \right\}.
\end{aligned}
$$

$$(7.12)$$

In a similar manner, time difference equations for the magnetic field are obtained from Eq. (7.9), whose x and y components are expressed as

$$
H_x^{n+\frac{1}{2}} = H_x^{n-\frac{1}{2}} - c_3 \frac{\partial E_z^n}{\partial y},
\tag{7.13}
$$

$$
H_y^{n+\frac{1}{2}} = H_y^{n-\frac{1}{2}} + c_3 \frac{\partial E_z^n}{\partial x},
\tag{7.14}
$$

with $c_3 = \Delta t / \mu$. For the former, the difference is calculated for a space point $(i_1, i_2 + \frac{1}{2})$ and is obtained as

$$
\begin{aligned}
H_x^{n+\frac{1}{2}}(i_1, i_2 + \tfrac{1}{2}) = & \; H_x^{n-\frac{1}{2}}(i_1, i_2 + \tfrac{1}{2}) \\
& - \frac{c_3(i_1, i_2 + \tfrac{1}{2})}{\Delta y} \left\{ E_z^n(i_1, i_2 + 1) - E_z^n(i_1, i_2) \right\}.
\end{aligned}
$$

$$(7.15)$$

On the other hand, for the latter, the difference is calculated for a space point $(i_1 + \frac{1}{2}, i_2)$ and is given as

$$
\begin{aligned}
H_y^{n+\frac{1}{2}}(i_1 + \tfrac{1}{2}, i_2) = & \; H_y^{n-\frac{1}{2}}(i_1 + \tfrac{1}{2}, i_2) \\
& + \frac{c_3(i_1 + \tfrac{1}{2}, i_2)}{\Delta x} \left\{ E_z^n(i_1 + 1, i_2) - E_z^n(i_1, i_2) \right\}.
\end{aligned}
$$

$$(7.16)$$

Thus, Eqs. (7.12), (7.15) and (7.16) are the final results for the FDTD expressions for the TM mode in 2D space.

Next, we consider a TE mode, where the direction of the magnetic field is perpendicular to a 2D plane. The lattice points concerned in this case is shown in Fig. 7.3, which correspond to those in a plane of $z = i_3 \Delta z$ in Fig. 7.1. In this case, the magnetic field has only the z

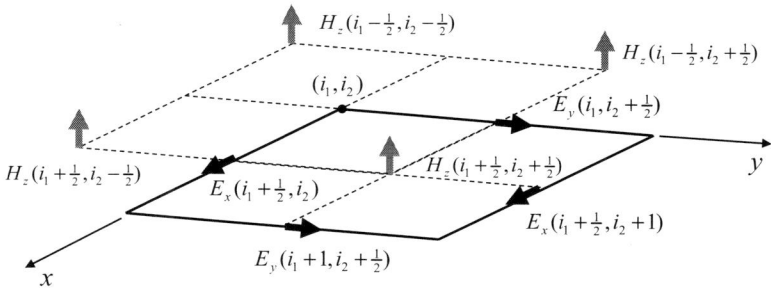

Figure 7.3 Calculation of difference with respect to 2D space coordinates in TE mode.

component and the same time difference equation of Eq. (7.10) for the electric field is used, whose x and y components are given as

$$E_x^n = c_1 E_x^{n-1} + c_2 \frac{\partial H_z^{n-\frac{1}{2}}}{\partial y},$$

$$E_y^n = c_1 E_y^{n-1} - c_2 \frac{\partial H_z^{n-\frac{1}{2}}}{\partial x}.$$

Since the magnetic field is delivered to half-integer lattice points, the x and y components of the corresponding electric field should be also delivered to space points of $(i_1 + \frac{1}{2}, i_2)$ and $(i_1, i_2 + \frac{1}{2})$, respectively. Inserting the expressions

$$\frac{\partial H_z^{n-\frac{1}{2}}}{\partial y} \approx \frac{H_z^{n-\frac{1}{2}}\left(i_1 + \frac{1}{2}, i_2 + \frac{1}{2}\right) - H_z^{n-\frac{1}{2}}\left(i_1 + \frac{1}{2}, i_2 - \frac{1}{2}\right)}{\Delta y},$$

$$\frac{\partial H_z^{n-\frac{1}{2}}}{\partial x} \approx \frac{H_z^{n-\frac{1}{2}}\left(i_1 + \frac{1}{2}, i_2 + \frac{1}{2}\right) - H_z^{n-\frac{1}{2}}\left(i_1 - \frac{1}{2}, i_2 + \frac{1}{2}\right)}{\Delta x},$$

into the above two time difference equations, we obtain

$$E_x^n\left(i_1 + \tfrac{1}{2}, i_2\right) = c_1\left(i_1 + \tfrac{1}{2}, i_2\right) E_x^{n-1}\left(i_1 + \tfrac{1}{2}, i_2\right)$$
$$+ \frac{c_2\left(i_1 + \tfrac{1}{2}, i_2\right)}{\Delta y}\left\{H_z^{n-\frac{1}{2}}\left(i_1 + \tfrac{1}{2}, i_2 + \tfrac{1}{2}\right) - H_z^{n-\frac{1}{2}}\left(i_1 + \tfrac{1}{2}, i_2 - \tfrac{1}{2}\right)\right\},$$

$$(7.17)$$

$$E_y^n\left(i_1, i_2 + \tfrac{1}{2}\right) = c_1\left(i_1, i_2 + \tfrac{1}{2}\right) E_x^{n-1}\left(i_1, i_2 + \tfrac{1}{2}\right)$$
$$- \frac{c_2\left(i_1, i_2 + \tfrac{1}{2}\right)}{\Delta x}\left\{H_z^{n-\frac{1}{2}}\left(i_1 + \tfrac{1}{2}, i_2 + \tfrac{1}{2}\right) - H_z^{n-\frac{1}{2}}\left(i_1 - \tfrac{1}{2}, i_2 + \tfrac{1}{2}\right)\right\}.$$

$$(7.18)$$

In the same way, the time difference equations for the magnetic field is obtained from Eq. (7.9) as

$$H_z^{n+\frac{1}{2}} = H_z^{n-\frac{1}{2}} - c_3 \left(\frac{\partial E_y^n}{\partial x} - \frac{\partial E_x^n}{\partial y} \right). \qquad (7.19)$$

Since the magnetic field is delivered to half-integer lattice points, putting it as $(i_1 + \frac{1}{2}, i_2 + \frac{1}{2})$, we obtain the differences for the partial differences of the electric field as

$$\frac{\partial E_y^n}{\partial x} \approx \frac{E_y^n \left(i_1 + 1, i_2 + \frac{1}{2} \right) - E_y^n \left(i_1, i_2 + \frac{1}{2} \right)}{\Delta x},$$

$$\frac{\partial E_x^n}{\partial y} \approx \frac{E_x^n \left(i_1 + \frac{1}{2}, i_2 + 1 \right) - E_x^n \left(i_1 + \frac{1}{2}, i_2 \right)}{\Delta y},$$

which results in

$$H_z^{n+\frac{1}{2}} \left(i_1 + \tfrac{1}{2}, i_2 + \tfrac{1}{2} \right) = H_z^{n-\frac{1}{2}} \left(i_1 + \tfrac{1}{2}, i_2 + \tfrac{1}{2} \right)$$

$$- \frac{c_3 \left(i_1 + \frac{1}{2}, i_2 + \frac{1}{2} \right)}{\Delta x} \left\{ E_y^n \left(i_1 + 1, i_2 + \tfrac{1}{2} \right) - E_y^n \left(i_1, i_2 + \tfrac{1}{2} \right) \right\}$$

$$+ \frac{c_3 \left(i_1 + \frac{1}{2}, i_2 + \frac{1}{2} \right)}{\Delta y} \left\{ E_x^n \left(i_1 + \tfrac{1}{2}, i_2 + 1 \right) - E_x^n \left(i_1 + \tfrac{1}{2}, i_2 \right) \right\}.$$

$$(7.20)$$

Thus, these difference equations, Eqs. (7.17), (7.18) and (7.20), are the final results for the TE mode.

In 3D space, the similar calculations should be performed as above. For example, if one would obtain a difference equation for E_z, one should focus on a plane of $z = (i_3 + \frac{1}{2})\Delta z$ and follow the same procedure as that in the TM mode in 2D space. Thus, completely the same expressions are obtained if one replace indices in 2D space by those in 3D one, e.g. $(i_1, i_2) \rightarrow (i_1, i_2, i_3 + \frac{1}{2})$. We can obtain all the components of the electric and magnetic fields in a similar manner.

Material parameters The material information is introduced through material parameters ϵ, μ and σ, which are involved in the coefficients of c_1, c_2 and c_3. It is generally attained by simply putting $\epsilon(i_1, i_2, i_3)$, $\mu(i_1, i_2, i_3)$ and $\sigma(i_1, i_2, i_3)$ for each integer and half-integer lattice point. However, near the material boundary, as shown in Fig. 7.4, some averaging procedure is usually necessary: For example, in case of 2D space shown in the figure, a material parameter $\alpha(i_1, i_2)$ is obtained by areal averaging indicated as $\alpha(i_1, i_2) = (A_1\alpha_1 + A_2\alpha_2)/(A_1 + A_2)$, where $\alpha = \epsilon$, μ and σ, and A_1 or A_2 is an area in each region.

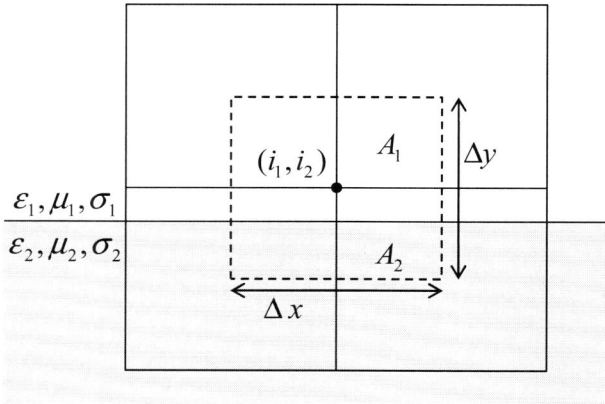

Figure 7.4 Determination of material parameters near the border.

7.1.2 *Boundary Condition*

The selection of a boundary condition at a surface surrounding an analyzing space is very important. The boundary is mainly provided from a computational requirement, and thus is anyway virtual. Moreover, the unsuitable selection will be fatal because an unwanted artificial effect will affect the calculation owing to the reflection at the boundary. In the FDTD method, this boundary condition is generally called *absorbing boundary condition* and is classified into two according to whether or not a virtual material is placed at the boundary. *Differential-based absorbing boundary condition* is a method to give the condition derived from a differential equation, while *material-based absorbing boundary condition* is that to place a material hypothetically to attenuate a propagating wave rapidly at the boundary. *Mur, Higdon* and *Liao* are known as typical methods for the former, while *Berenger PML* (*perfectly matched layer*) for the latter. Since the details of these methods have been fully described in books and papers published so far, here we will review the simplest method called *Mur absorbing boundary condition*.

Consider a case where a plane wave propagates to a negative direction along the x axis with the velocity of v and vertically hits a boundary located at $x = 0$, as shown in Fig. 7.5. If we assume that an electric field is directed to the z axis, such a wave will be expressed

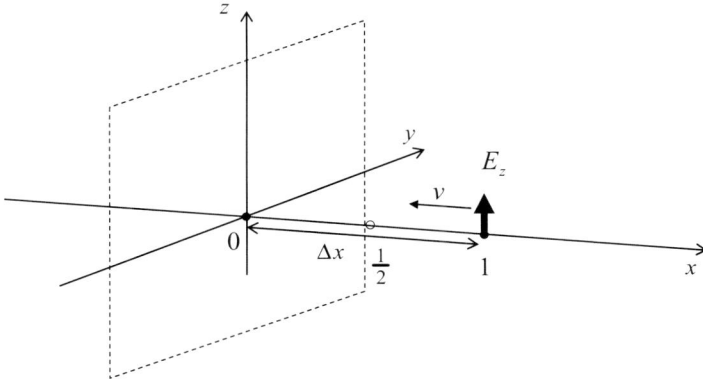

Figure 7.5 Mur first-oder boundary condition.

as $E_z = E_z(x+vt)$. Partially differentiating both sides of this formula with respect to x and t, we obtain the following differential equation for the propagating wave:

$$\frac{\partial E_z}{\partial t} = v \frac{\partial E_z}{\partial x}. \tag{7.21}$$

If no reflection occurs at the boundary, this wave equation should hold even at the boundary of $x = 0$. Thus, we will approximate this using the central difference with respect to t and x. First, the time difference is given as

$$\frac{E_z^n - E_z^{n-1}}{\Delta t} = v \frac{\partial E_z^{n-\frac{1}{2}}}{\partial x}.$$

Then, the partial differential with respect to x is approximated as

$$\frac{\partial E_z^{n-\frac{1}{2}}}{\partial x} \approx \frac{E_z^{n-\frac{1}{2}}(1) - E_z^{n-\frac{1}{2}}(0)}{\Delta x},$$

where we only write an index corresponding to the x coordinate, for simplicity, with an index at the border as 0.

Combining these results, we obtain a difference equation at the boundary as

$$\frac{E_z^n(\frac{1}{2}) - E_z^{n-1}(\frac{1}{2})}{\Delta t} = v \frac{E_z^{n-\frac{1}{2}}(1) - E_z^{n-\frac{1}{2}}(0)}{\Delta x}. \tag{7.22}$$

Since the integer lattice points are to be delivered to E_z both for time and space, the above terms expressed by half-integer lattice points are anyway approximated by averaging as shown below:

$$E_z^n(\tfrac{1}{2}) \approx \frac{E_z^n(1) + E_z^n(0)}{2},$$

$$E_z^{n-\frac{1}{2}}(1) \approx \frac{E_z^n(1) + E_z^{n-1}(1)}{2}.$$

Inserting these relations into Eq. (7.22), we finally obtain a difference equation for E_z at the boundary as

$$E_z^n(0) = E_z^{n-1}(1) + \frac{v\Delta t - \Delta x}{v\Delta t + \Delta x}\left\{E_z^n(1) - E_z^{n-1}(0)\right\}, \qquad (7.23)$$

which is called *Mur first-order boundary condition*.

Since the above condition is only valid for incidence perpendicular to a boundary plane, it is often inappropriate for an ordinary use. Thus, the higher-order condition is usually employed, which allows an expression for oblique incidence at the boundary. The second-order condition is obtained as follows: Consider a plane wave propagating to a direction that is not exactly equal to but is close to $-x$ direction. The z component of the electric field for such a wave is generally expressed as

$$E_z = E_{z0}\, e^{i(\mathbf{k}\cdot\mathbf{r} - \omega t)}, \qquad (7.24)$$

and its derivative with respect to x is given by

$$\frac{\partial E_z}{\partial x} - ik_x E_z = 0. \qquad (7.25)$$

k_x is further approximated according to the following procedure:

$$
\begin{aligned}
-k_x &= \sqrt{k^2 - k_y^2 - k_z^2} = k\sqrt{1 - (k_y^2 + k_z^2)/k^2} \\
&= k\left\{1 - (k_y^2 + k_z^2)/(2k^2) + \cdots\right\} \\
&\approx \frac{\omega}{v} - \frac{v}{2\omega}(k_y^2 + k_z^2),
\end{aligned} \qquad (7.26)
$$

where we take $k_x < 0$ because we consider here a wave propagating to $-x$ direction, and we utilize the relation $k = \omega/v$. Inserting this relation into Eq. (7.25), we obtain

$$\frac{\partial E_z}{\partial x} + i\left\{\frac{\omega}{v} - \frac{v}{2\omega}(k_y^2 + k_z^2)\right\} E_z = 0,$$

which is rewritten as

$$-i\omega\frac{\partial E_z}{\partial x} + \frac{\omega^2}{v}E_z - \frac{v}{2}(k_y^2 + k_z^2)E_z = 0,$$

and then leads to

$$\frac{\partial}{\partial t}\left(\frac{\partial E_z}{\partial x} - \frac{1}{v}\frac{\partial E_z}{\partial t}\right) + \frac{v}{2}\left(\frac{\partial^2 E_z}{\partial y^2} + \frac{\partial^2 E_z}{\partial z^2}\right) = 0, \qquad (7.27)$$

where we have used the following replacements: $-i\omega \to \partial/\partial t$, $ik_y \to \partial/\partial y$ and $ik_z \to \partial/\partial z$.

We take a central difference of the above differential equation around a spatiotemporal lattice point of $(\frac{1}{2}, i_2, i_3 + \frac{1}{2}, n-1)$ using the following relations:

$$\frac{\partial^2 E_z^{n-1}}{\partial t \partial x} \approx \frac{1}{\Delta t \Delta x}\left\{E_z^{n-\frac{1}{2}}(1,,)\right.$$
$$\left. -E_z^{n-\frac{1}{2}}(0,,) - E_z^{n-\frac{3}{2}}(1,,) + E_z^{n-\frac{3}{2}}(0,,)\right\},$$

$$\frac{\partial^2 E_z^{n-1}}{\partial t^2} \approx \frac{1}{\Delta t^2}\left\{E_z^n(,,) - 2E_z^{n-1}(,,) + E_z^{n-2}(,,)\right\},$$

$$\frac{\partial^2 E_z^{n-1}}{\partial y^2} \approx \frac{1}{\Delta y^2}\left\{E_z^{n-1}(,i_2+1,) - 2E_z^{n-1}(,i_2,) + E_z^{n-1}(,i_2-1,)\right\},$$

$$\frac{\partial^2 E_z^{n-1}}{\partial z^2} \approx \frac{1}{\Delta z^2}\left\{E_z^{n-1}(,,i_3+\tfrac{3}{2})\right.$$
$$\left. -2E_z^{n-1}\left(,,i_3+\tfrac{1}{2}\right) + E_z^{n-1}\left(,,i_3-\tfrac{1}{2}\right)\right\},$$

where we have omitted the arguments unless necessary. Inserting these relations into Eq. (7.27) and performing the somewhat complicated calculation, we obtain the *Mur second-order boundary condition* (Exercise (1)) as

$$E_z^n(0, i_2, i_3 + \tfrac{1}{2}) = -E_z^{n-2}(1, i_2, i_3 + \tfrac{1}{2})$$
$$+\frac{v\Delta t - \Delta x}{v\Delta t + \Delta x}\left\{E_z^n\left(1, i_2, i_3 + \tfrac{1}{2}\right) + E_z^{n-2}\left(0, i_2, i_3 + \tfrac{1}{2}\right)\right\}$$
$$+\frac{2\Delta x}{v\Delta t + \Delta x}\left\{E_z^{n-1}\left(0, i_2, i_3 + \tfrac{1}{2}\right) + E_z^{n-1}\left(1, i_2, i_3 + \tfrac{1}{2}\right)\right\}$$
$$+\frac{\Delta x(v\Delta t)^2}{2\Delta y^2(v\Delta t + \Delta x)}\left\{E_z^{n-1}\left(0, i_2 + 1, i_3 + \tfrac{1}{2}\right)\right.$$
$$-2E_z^{n-1}\left(0, i_2, i_3 + \tfrac{1}{2}\right)$$
$$+E_z^{n-1}\left(0, i_2 - 1, i_3 + \tfrac{1}{2}\right) + E_z^{n-1}\left(1, i_2 + 1, i_3 + \tfrac{1}{2}\right)$$
$$\left.-2E_z^{n-1}\left(1, i_2, i_3 + \tfrac{1}{2}\right) + E_z^{n-1}\left(1, i_2 - 1, i_3 + \tfrac{1}{2}\right)\right\}$$
$$+\frac{\Delta x(v\Delta t)^2}{2\Delta z^2(v\Delta t + \Delta x)}\left\{E_z^{n-1}\left(0, i_2, i_3 + \tfrac{3}{2}\right) - 2E_z^{n-1}\left(0, i_2, i_3 + \tfrac{1}{2}\right)\right.$$
$$+E_z^{n-1}\left(0, i_2, i_3 - \tfrac{1}{2}\right) + E_z^{n-1}\left(1, i_2, i_3 + \tfrac{3}{2}\right)$$
$$\left.-2E_z^{n-1}\left(1, i_2, i_3 + \tfrac{1}{2}\right) + E_z^{n-1}\left(1, i_2, i_3 - \tfrac{1}{2}\right)\right\}.$$

$$(7.28)$$

7.1.3 *Incident Field*

The incident field is usually introduced by considering a plane wave propagating in vacuum. Consider a case where an incident wave with the wave vector $\mathbf{k}$ and angular frequency ω propagates within the xy plane and its electric field is directed to the z direction (TM) as shown in Fig. 7.6. We assume that the wave is expressed as

$$E_{iz} = E_{i0}\,p\,(\tau),\qquad(7.29)$$

where a function $p(\tau)$ is defined as $p(\tau) \equiv A(\tau)\cos(\omega\tau)$ with an amplitude modulation $A(\tau)$. τ is defined as $\tau \equiv t - \mathbf{k}\cdot\mathbf{r}/\omega - t_0$ with a phase delay expressed by a time t_0, which is explicitly described as $\tau = t + (x\cos\phi + y\sin\phi)/c - t_0$ in the present case. Using the same expressions, the components of magnetic fields are described as

$$H_{ix} = -\sin\phi\,\frac{E_{i0}}{Z_0}\,p(\tau),$$
$$H_{iy} = \cos\phi\,\frac{E_{i0}}{Z_0}\,p(\tau),$$

where Z_0 is an impedance of free space defined as $Z_0 = \sqrt{\mu_0/\epsilon_0}$.

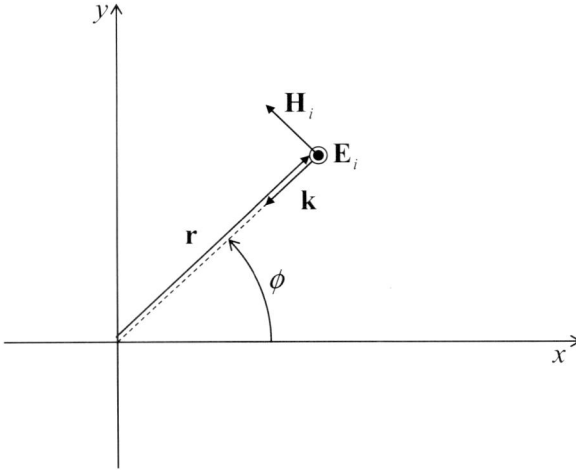

Figure 7.6 Incident field.

Putting $t = n\Delta t$, $x = i_1 \Delta x$ and $y = i_2 \Delta y$, for example, we can convert the above expressions into the FDTD style for the 2D TM mode as follows:

$$E_{iz}^n(i_1, i_2) = E_{i0}\, p\left(n\Delta t + \frac{i_1 \Delta x \cos\phi + i_2 \Delta y \sin\phi}{c} - t_0\right), \quad (7.30)$$

$$H_{ix}^{n+\frac{1}{2}}\left(i_1, i_2 + \tfrac{1}{2}\right) = -\sin\phi\, \frac{E_{i0}}{Z_0}$$
$$\times p\left(\left(n + \tfrac{1}{2}\right)\Delta t + \frac{i_1 \Delta x \cos\phi + \left(i_2 + \tfrac{1}{2}\right)\Delta y \sin\phi}{c} - t_0\right),$$
$$(7.31)$$

$$H_{iy}^{n+\frac{1}{2}}\left(i_1 + \tfrac{1}{2}, i_2\right) = \cos\phi\, \frac{E_{i0}}{Z_0}$$
$$\times p\left(\left(n + \tfrac{1}{2}\right)\Delta t + \frac{\left(i_1 + \tfrac{1}{2}\right)\Delta x \cos\phi + i_2 \Delta y \sin\phi}{c} - t_0\right),$$
$$(7.32)$$

which should be delivered to each lattice point as an incident field.

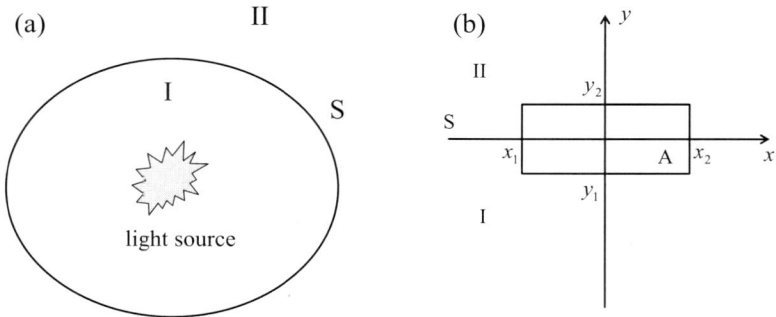

Figure 7.7 (a) Virtual surface S enclosing a light source, which eventually divides a whole space into I and II. (b) Calculation of the continuity condition at the surface S.

7.1.4 *Near-to-Far Transformation*

The electric and magnetic fields thus calculated through a method of finite differences are anyway those in near fields. Thus, if we calculate the reflection and transmission spectra or their angular dependence, it is generally necessary to calculate the far fields only from the information of the near fields. The method of near-to-far transformation was thus devised mainly for designing antennas in electromagnetic wave engineering. Though this method is somewhat complicated, we will investigate its essential principle from a standpoint of physics.

At first, we will prove that the electric and magnetic fields induced by a light source are equivalent to those induced by equivalent electric and magnetic currents assumed to be conducted on a virtual surface enclosing the light source [Schelkunoff (1943, 1951)]. As shown in Fig. 7.7a, we divide a space around the light source into two: One is that enclosed by a virtual surface of S, which we denote as I, while the other is its outside denoted as II. Then, let us consider the continuity condition for the electric and magnetic fields at the surface S, as has been done at an interface between two dielectric materials (see Sec. 2.2.4.1). We consider the following two cases and make them equivalent so as to obtain the equivalent electric and magnetic fields: (1) The electric and magnetic fields exist both in regions of I and II, as if the surface S is absent, and

(2) they exist only in a region of *II*. Instead, electric and magnetic currents are conducted on the surface *S*.

First, we consider a case (1). As has been done previously, we will integrate the Maxwell equations within a region of *A* (see Fig. 7.7b):

$$\int_{x_1}^{x_2} dx \int_{y_1}^{y_2} dy \, \nabla \times \mathbf{E} = -\int_{x_1}^{x_2} dx \int_{y_1}^{y_2} dy \, \frac{\partial \mathbf{B}}{\partial t}, \quad (7.33)$$

$$\int_{x_1}^{x_2} dx \int_{y_1}^{y_2} dy \, \nabla \times \mathbf{H} = \int_{x_1}^{x_2} dx \int_{y_1}^{y_2} dy \, \frac{\partial \mathbf{D}}{\partial t}, \quad (7.34)$$

where we assume that the surface *S* is smooth enough so that we can regard it as flat within a small region of *A*. We further assume that the space is a vacuum except for that corresponding to the light source. Writing Eq. (7.33) in each Cartesian component, we obtain

$$\int_{x_1}^{x_2} dx \int_{y_1}^{y_2} dy \left(\frac{\partial E_z}{\partial y} - \frac{\partial E_y}{\partial z} \right) = -\int_{x_1}^{x_2} dx \int_{y_1}^{y_2} dy \, \frac{\partial B_x}{\partial t},$$

$$\int_{x_1}^{x_2} dx \int_{y_1}^{y_2} dy \left(\frac{\partial E_x}{\partial z} - \frac{\partial E_z}{\partial x} \right) = -\int_{x_1}^{x_2} dx \int_{y_1}^{y_2} dy \, \frac{\partial B_y}{\partial t},$$

$$\int_{x_1}^{x_2} dx \int_{y_1}^{y_2} dy \left(\frac{\partial E_y}{\partial x} - \frac{\partial E_x}{\partial y} \right) = -\int_{x_1}^{x_2} dx \int_{y_1}^{y_2} dy \, \frac{\partial B_z}{\partial t}.$$

We then make $y_2 \to +0$ and $y_1 \to -0$ in these formulas. Since the terms not containing the derivative with respect to *y* will all vanish, the remaining terms give

$$\left(E_z^{II} \right)_{y \to +0} - \left(E_z^{I} \right)_{y \to -0} = 0,$$

$$- \left(E_x^{II} \right)_{y \to +0} + \left(E_x^{I} \right)_{y \to -0} = 0,$$

which lead to

$$\left(\mathbf{E}_{\parallel}^{II} \right)_{y \to +0} = \left(\mathbf{E}_{\parallel}^{I} \right)_{y \to -0}, \quad (7.35)$$

where we put $\mathbf{E}_{\parallel} = (E_x, 0, E_z)$ and assume that the electric field is uniform along the *x* axis within a small range between x_1 and x_2. In a similar way, we obtain

$$\left(\mathbf{H}_{\parallel}^{II} \right)_{y \to +0} = \left(\mathbf{H}_{\parallel}^{I} \right)_{y \to -0}. \quad (7.36)$$

These results guarantee the continuity of the tangential components of the electric and magnetic fields at a virtual surface of *S*.

Then, we will consider a case (2), where the electric and magnetic fields are absent within a region of *I*, while the electric and magnetic currents conduct on the surface S. The electric and magnetic current vectors, **J** and **M**, in our case are defined as follows:

$$\nabla \times \mathbf{E} = -\frac{\partial \mathbf{B}}{\partial t} - \mathbf{M}\,\delta(y), \tag{7.37}$$

$$\nabla \times \mathbf{H} = \frac{\partial \mathbf{D}}{\partial t} + \mathbf{J}\,\delta(y), \tag{7.38}$$

where the presence of a delta function indicates that both currents are conducted on the surface S. The similar integration within a region of A gives

$$\left(E_z^{II}\right)_{y \to +0} - \left(E_z^{I}\right)_{y \to -0} = -M_x,$$
$$0 = -M_y,$$
$$-\left(E_x^{II}\right)_{y \to +0} + \left(E_x^{I}\right)_{y \to -0} = -M_z,$$

which immediately leads to $M_y = 0$. Since we assume that $\mathbf{E}^I = 0$ in the present case, $\left(E_z^{II}\right)_{y \to +0} = -M_x$ and $\left(E_x^{II}\right)_{y \to +0} = M_z$ hold naturally.

Thus, using Eq. (7.35), we can determine the equivalent magnetic current so as to satisfy the relations, $\left(E_z^{I}\right)_{y \to -0} = -M_x$ and $\left(E_x^{I}\right)_{y \to -0} = M_z$, which are expressed in a vectorial form as

$$\mathbf{M} = \mathbf{E}^I \times \mathbf{n}, \tag{7.39}$$

where **n** is a unit vector normal to the surface and directing to *II*. Further, we have omitted $y \to -0$ in the formula, because the surface S is virtual and **E** is naturally a continuous function of space. In a similar manner, we obtain (Exercise (2))

$$\mathbf{J} = \mathbf{n} \times \mathbf{H}^I. \tag{7.40}$$

Next, we will investigate the fields induced by the electric and magnetic currents thus determined [Schelkunoff (1951); Umashankar and Taflove (1982)]. The Maxwell equations containing the source terms of **J** and **M** are given as

$$\nabla \times \mathbf{E} = -\frac{\partial \mathbf{B}}{\partial t} - \mathbf{M}. \tag{7.41}$$

$$\nabla \times \mathbf{H} = \frac{\partial \mathbf{D}}{\partial t} + \mathbf{J}. \tag{7.42}$$

In order to solve these equations, let us divide them into the equations that contain a source term of **J** with the corresponding electric and magnetic field vectors, $\mathbf{E}'$ and $\mathbf{H}'$, and those containing $\mathbf{M}$ with $\mathbf{E}''$ and $\mathbf{H}''$. Thus, the equations are divided into two as follows:

Case (a)

$$\nabla \times \mathbf{E}' = -\frac{\partial \mathbf{B}'}{\partial t}, \tag{7.43}$$

$$\nabla \times \mathbf{H}' = \frac{\partial \mathbf{D}'}{\partial t} + \mathbf{J}. \tag{7.44}$$

Case (b)

$$\nabla \times \mathbf{E}'' = -\frac{\partial \mathbf{B}''}{\partial t} - \mathbf{M}, \tag{7.45}$$

$$\nabla \times \mathbf{H}'' = \frac{\partial \mathbf{D}''}{\partial t}. \tag{7.46}$$

After solving these equations respectively, the calculation of

$$\mathbf{E} = \mathbf{E}' + \mathbf{E}'' \quad \text{and} \quad \mathbf{H} = \mathbf{H}' + \mathbf{H}'' \tag{7.47}$$

will give the solutions for Eqs. (7.41) and (7.42), because adding Eq. (7.43) to Eq. (7.45) gives the original equation of Eq. (7.41), and the same is true for the other.

In Case (a), the similar procedure as has been done in Sec. 6.2 is applicable. Namely, by defining the vector and scalar potentials, $\mathbf{A}'$ and ϕ', the electric field and magnetic flux density are expressed as

$$\mathbf{E}' = -\frac{\partial \mathbf{A}'}{\partial t} - \nabla \phi', \tag{7.48}$$

$$\mathbf{B}' = \nabla \times \mathbf{A}', \tag{7.49}$$

with a Lorentz condition (Eq. (6.5))

$$\nabla \cdot \mathbf{A}' + \frac{1}{c^2} \frac{\partial \phi'}{\partial t} = 0. \tag{7.50}$$

If we assume the light source emits a monochromatic light wave with the angular frequency of ω, all the terms should contain a factor of $\exp[-i\omega t]$. Thus, we replace all the field functions that contain a time variable as in $\mathbf{A}' \rightarrow \mathbf{A}'(\mathbf{r}, \omega) \exp[-i\omega t]$, for example. Therefore, Eq. (7.50) becomes $\nabla \cdot \mathbf{A}' - (i\omega/c^2)\phi' = 0$, which leads to

$$\nabla \phi' = \frac{c^2}{i\omega} \nabla (\nabla \cdot \mathbf{A}'). \tag{7.51}$$

where we have abbreviated $\mathbf{A}' = \mathbf{A}'(\mathbf{r}, \omega)$ and $\phi' = \phi'(\mathbf{r}, \omega)$. Inserting Eq. (7.51) into Eq. (7.48), we obtain

$$\mathbf{E}' = i\omega \mathbf{A}' - \frac{c^2}{i\omega}\nabla(\nabla \cdot \mathbf{A}'), \tag{7.52}$$

$$\mathbf{H}' = \frac{1}{\mu_0}\nabla \times \mathbf{A}', \tag{7.53}$$

where $\mathbf{A}'$ is given, using Eq. (6.22), as

$$\mathbf{A}'(\mathbf{r}, \omega) = \frac{\mu_0}{4\pi}\int_S d\mathbf{r}' \frac{e^{ik|\mathbf{r}-\mathbf{r}'|}}{|\mathbf{r} - \mathbf{r}'|}\mathbf{J}(\mathbf{r}', \omega), \tag{7.54}$$

with $k = \omega/c$. In this expression, the integration should be performed on the whole surface of S.

In Case (b), we define the corresponding vector and scalar potentials so as to satisfy the following relations:

$$\mathbf{D}'' = -\nabla \times \mathbf{A}'', \tag{7.55}$$

$$\mathbf{H}'' = -\frac{\partial \mathbf{A}''}{\partial t} - \nabla\phi''. \tag{7.56}$$

Inserting these relations into Eq. (7.45) and calculating as in Sec. 6.2, we obtain

$$\nabla\left(\nabla \cdot \mathbf{A}'' + \frac{1}{c^2}\frac{\partial \phi''}{\partial t}\right) + \frac{1}{c^2}\frac{\partial^2 \mathbf{A}''}{\partial t^2} - \nabla^2 \mathbf{A}'' = \epsilon_0 \mathbf{M}. \tag{7.57}$$

Thus, imposing the Lorentz condition of

$$\nabla \cdot \mathbf{A}'' + \frac{1}{c^2}\frac{\partial \phi''}{\partial t} = 0, \tag{7.58}$$

we obtain the similar wave equation for $\mathbf{A}''$ with $\epsilon_0 \mathbf{M}$ as a source term. Then, as is similar to Case a), we reach the final expressions for $\mathbf{E}''$ and $\mathbf{H}''$ as

$$\mathbf{E}'' = -\frac{1}{\epsilon_0}\nabla \times \mathbf{A}'', \tag{7.59}$$

$$\mathbf{H}'' = i\omega \mathbf{A}'' - \frac{c^2}{i\omega}\nabla(\nabla \cdot \mathbf{A}''), \tag{7.60}$$

with $\mathbf{A}''$ expressed by

$$\mathbf{A}''(\mathbf{r}, \omega) = \frac{\epsilon_0}{4\pi}\int_S d\mathbf{r}' \frac{e^{ik|\mathbf{r}-\mathbf{r}'|}}{|\mathbf{r} - \mathbf{r}'|}\mathbf{M}(\mathbf{r}', \omega). \tag{7.61}$$

The final solutions are derived by summing the above solutions, which results in

$$\mathbf{E} = i\omega\mathbf{A}' - \frac{c^2}{i\omega}\nabla(\nabla \cdot \mathbf{A}') - \frac{1}{\epsilon_0}\nabla \times \mathbf{A}'', \qquad (7.62)$$

$$\mathbf{H} = \frac{1}{\mu_0}\nabla \times \mathbf{A}' - \frac{c^2}{i\omega}\nabla(\nabla \cdot \mathbf{A}'') + i\omega\mathbf{A}''. \qquad (7.63)$$

To obtain the explicit forms for the electric and magnetic fields in far field, we have to calculate the terms containing the differential operators such that $\nabla \times \mathbf{A}'$ and $\nabla(\nabla \cdot \mathbf{A}')$, and to select the terms that are connected with the electromagnetic wave generation. After somewhat lengthy calculations, the essential parts of these terms are calculated as

$$\nabla \times \frac{e^{ikr}}{r}\mathbf{F} = \mathbf{r} \times \mathbf{F}\frac{ikr - 2}{r^3}e^{ikr}, \qquad (7.64)$$

$$\nabla\left(\nabla \cdot \frac{e^{ikr}}{r}\mathbf{F}\right) = \mathbf{F}\frac{ikr - 2}{r^3}e^{ikr} + (\mathbf{F} \cdot \mathbf{r}) \cdot \mathbf{r}\frac{3 - 3ikr - k^2r^2}{r^5}e^{ikr}, \qquad (7.65)$$

where $\mathbf{F} = \mathbf{J}$ or $\mathbf{M}$, and we have put $\mathbf{r} - \mathbf{r}' \to \mathbf{r}$ with $|\mathbf{r} - \mathbf{r}'| \to r$ (Exercise (3)).

In these formulas, only the terms proportional to r^{-1} contribute to the generation of electromagnetic waves in far field. Thus, we obtain

$$\mathbf{E}(\mathbf{r}, \omega) = \frac{\mu_0}{4\pi}\int_S d\mathbf{r}' \left\{ i\omega\frac{e^{ik|\mathbf{r}-\mathbf{r}'|}}{|\mathbf{r}-\mathbf{r}'|}\mathbf{J}(\mathbf{r}', \omega) \right.$$

$$+ \frac{c^2}{i\omega}\frac{k^2}{}\frac{e^{ik|\mathbf{r}-\mathbf{r}'|}}{|\mathbf{r}-\mathbf{r}'|}\left\{\mathbf{J}(\mathbf{r}', \omega) \cdot \mathbf{e}_{\mathbf{r}-\mathbf{r}'}\right\}\mathbf{e}_{\mathbf{r}-\mathbf{r}'} \bigg\}$$

$$- \frac{1}{4\pi}\int_S d\mathbf{r}'\frac{ik\, e^{ik|\mathbf{r}-\mathbf{r}'|}}{|\mathbf{r}-\mathbf{r}'|}\mathbf{e}_{\mathbf{r}-\mathbf{r}'} \times \mathbf{M}(\mathbf{r}', \omega),$$

which leads to

$$\mathbf{E} = \frac{i\omega\mu_0}{4\pi}\int_S d\mathbf{r}'\frac{\mathbf{J} - (\mathbf{J} \cdot \mathbf{e}_{\mathbf{r}-\mathbf{r}'})\mathbf{e}_{\mathbf{r}-\mathbf{r}'}}{|\mathbf{r}-\mathbf{r}'|}e^{ik|\mathbf{r}-\mathbf{r}'|}$$

$$- \frac{i\omega}{4\pi}\int_S d\mathbf{r}'\frac{\mathbf{e}_{\mathbf{r}-\mathbf{r}'} \times \mathbf{M}}{c|\mathbf{r}-\mathbf{r}'|}e^{ik|\mathbf{r}-\mathbf{r}'|}, \qquad (7.66)$$

where we have put $\mathbf{r} - \mathbf{r}' = |\mathbf{r} - \mathbf{r}'|\mathbf{e}_{\mathbf{r}-\mathbf{r}'}$ with a unit vector $\mathbf{e}_{\mathbf{r}-\mathbf{r}'}$. Further, we have used a relation $k^2c^2/(i\omega) = -i\omega$. In a similar manner, the magnetic field in far field is obtained as

$$\mathbf{H} = \frac{i\omega\epsilon_0}{4\pi} \int_S d\mathbf{r}' \, \frac{\mathbf{M} - (\mathbf{M} \cdot \mathbf{e}_{\mathbf{r}-\mathbf{r}'})\mathbf{e}_{\mathbf{r}-\mathbf{r}'}}{|\mathbf{r} - \mathbf{r}'|} \, e^{ik|\mathbf{r}-\mathbf{r}'|}$$
$$+ \frac{i\omega}{4\pi} \int_S d\mathbf{r}' \, \frac{\mathbf{e}_{\mathbf{r}-\mathbf{r}'} \times \mathbf{J}}{c|\mathbf{r} - \mathbf{r}'|} \, e^{ik|\mathbf{r}-\mathbf{r}'|}. \qquad (7.67)$$

7.1.5 *Non-standard FDTD*

Non-standard finite-difference time-domain (NS-FDTD) method is a kind of generalization of standard FDTD method that utilizes discrete time and spatial differences, and was first proposed by Mickens [Mickens (1989)]. If the functional forms of the fields satisfy the condition, this method will give highly accurate results even when the spatial mesh size is not extremely small, which considerably shortens the computational time and hence greatly enhances the applicability of the FDTD method.

Consider first a 1D wave equation expressed as

$$(\partial_t^2 - v^2\partial_x^2)\psi(x, t) = 0, \qquad (7.68)$$

where we follow the notations employed in the book written by Cole [Cole (2005)] such that $\partial_t^2 = \partial^2/\partial t^2$ and $\partial_x^2 = \partial^2/\partial x^2$. v is a velocity of wave propagating along the x axis. We denote the central time difference as

$$d_t\psi(x, t) = \psi(x, t + \tfrac{1}{2}\Delta t) - \psi(x, t - \tfrac{1}{2}\Delta t), \qquad (7.69)$$

and hence

$$d_t^2 = d_t d_t = \psi(x, t + \Delta t) + \psi(x, t - \Delta t) - 2\psi(x, t). \qquad (7.70)$$

In a similar way, the central spatial difference is expressed by

$$d_x^2 = d_x d_x = \psi(x, t + h) + \psi(x, t - h) - 2\psi(x, t), \qquad (7.71)$$

where Δt and h are time and spatial steps to take differences.

Then, the above wave equation can be approximated using these central differences as

$$\left(\frac{d_t^2}{\Delta t^2} - v^2\frac{d_x^2}{h^2}\right)\psi(x, t) = 0,$$

which leads to

$$(d_t^2 - \bar{v}^2 d_x^2)\,\psi(x, t) = 0, \qquad (7.72)$$

where $\bar{v}^2 = v^2 \Delta t^2 / h^2$. The last expression just corresponds to a result of standard finite difference (SFD) and is called *SFD model*. The time difference is more explicitly written as

$$\psi(x, t + \Delta t) = -\psi(x, t - \Delta t) + 2\psi(x, t) + \bar{v}^2 d_x^2 \psi(x, t), \quad (7.73)$$

which is a *standard FDTD algorithm* for a 1D wave function.

Non-standard FDTD comes from an idea that the derivative and difference are naturally connectable with each other, if a functional form is known beforehand and is adequately chosen. Consider an arbitrary function $f(x)$ and assume that its derivative and difference are combined through a relation

$$f'(x) = \frac{d_x f(x)}{s(h)}. \quad (7.74)$$

where $d_x f(x) = f(x + h/2) - f(x - h/2)$. If we can properly choose $s(h)$ for an arbitrary value of h, we can obtain an exact value of the first derivative of $f(x)$ from the calculation of the first-order difference.

Unfortunately, in the usual case, $f(x)$ is an unknown function so that the above relation cannot be used to obtain $s(h)$. However, if a functional form of $f(x)$ is known beforehand and is expressed, for example, as $f(x) = a_+ \exp[ikx] + a_- \exp[-ikx]$ with $a_\pm$ a complex constant, its derivative and the corresponding difference become

$$f'(x) = ik \left(a_+ e^{ikx} - a_- e^{-ikx} \right), \quad (7.75)$$

and

$$d_x f(x) = a_+ \left(e^{ik(x+\frac{1}{2}h)} - e^{ik(x-\frac{1}{2}h)} \right) + a_- \left(e^{-ik(x+\frac{1}{2}h)} - e^{-ik(x-\frac{1}{2}h)} \right)$$

$$= a_+ e^{ikx} \left(e^{\frac{1}{2}ikh} - e^{-\frac{1}{2}ikh} \right) + a_- e^{-ikx} \left(e^{-\frac{1}{2}ikh} - e^{\frac{1}{2}ikh} \right)$$

$$= 2i \sin(kh/2) \left(a_+ e^{ikx} - a_- e^{-ikx} \right), \quad (7.76)$$

respectively. Comparing these results, we can say that if we put

$$s(h) = \frac{d_x f(x)}{f'(x)} = \frac{2}{k} \sin(kh/2), \quad (7.77)$$

we always obtain the exact value of the derivative from the difference. This is an essence of *non-standard finite difference* (*NSFD*). Similarly, for the second derivative, we obtain the following result (Exercise (4)):

$$f''(x) = \frac{d_x^2 f(x)}{s^2(h)}. \quad (7.78)$$

Using these relations, it is soon proved that a 1D wave equation (7.68) is exactly converted into NSFD as

$$\left(\partial_t^2 - v^2\partial_x^2\right)\psi(x,t) = \left(\frac{d_t^2}{4\sin^2(\omega\Delta t/2)/\omega^2}\right.$$
$$\left. -v^2\frac{d_x^2}{4\sin^2(kh/2)/k^2}\right)\psi(x,t) = 0. \quad (7.79)$$

This is because $\psi(x,t)$ is a general solution for 1D wave equation, if we put $\psi(x,t) = a_+\exp[i(kx - \omega t)] + a_-\exp[i(kx + \omega t)]$. From this result, we can easily derive an exact difference wave equation corresponding to Eq. (7.68) as

$$\left(d_t^2 - u_0^2 d_x^2\right)\psi(x,t) = 0, \quad (7.80)$$

where

$$u_0^2 = \sin^2(\omega\Delta t/2)/\sin^2(kh/2), \quad (7.81)$$

and we have used the relation $k^2 = \omega^2/v^2$.

However, this procedure to obtain an exact difference equation is generally difficult when we extend it to more higher dimensions such as 2D and 3D. Here, we will exemplify a 2D case. In the 2D case, a wave equation is generally expressed as

$$\left(\partial_t^2 - v^2\nabla^2\right)\psi(\mathbf{r},t) = 0, \quad (7.82)$$

where $\nabla^2 = \partial^2/\partial x^2 + \partial^2/\partial y^2$ with $\mathbf{r} = (x,y)$. Corresponding difference equation becomes

$$\left(d_t^2 - \bar{v}^2 D_1^2\right)\psi(\mathbf{r},t) = 0. \quad (7.83)$$

Here, $D_1^2 = d_x^2 + d_y^2$ and $\bar{v}^2 = v^2\Delta t^2/h^2$, where we have employed the same spatial step of h both for the x and y coordinates. According to the above discussion, if $\psi(\mathbf{r},t)$ is expressed as in the form of $\psi(\mathbf{r},t) = a\exp[i(\mathbf{k}\cdot\mathbf{r} - \omega t)]$, then the relations

$$\nabla^2\psi(\mathbf{r},t) = -k^2\psi(\mathbf{r},t),$$

and

$$D_1^2\psi(\mathbf{r},t) = -4\left\{\sin^2(k_x h/2) + \sin^2(k_y h/2)\right\}\psi(\mathbf{r},t),$$

are obtained. Thus,

$$\nabla^2\psi(\mathbf{r},t) = \frac{k^2}{4\left\{\sin^2(k_x h/2) + \sin^2(k_y h/2)\right\}}D_1^2\psi(\mathbf{r},t), \quad (7.84)$$

which leads to

$$\left(\partial_t^2 - v^2\nabla^2\right)\psi(\mathbf{r},t) = \left(\frac{d_t^2}{4\sin^2(\omega\Delta t/2)/\omega^2} -v^2\frac{D_1^2}{4\sin^2(k_xh/2)/k^2 + 4\sin^2(k_yh/2)/k^2}\right)\psi(\mathbf{r},t) = 0.$$

(7.85)

Thus, if we put $u_0^2 = \sin^2(\omega\Delta t/2)/\{\sin^2(k_xh/2)+\sin^2(k_yh/2)\}$, then the exact difference equation should become

$$\left(d_t^2 - u_0^2 D_1^2\right)\psi(\mathbf{r},t) = 0,$$

(7.86)

which is similar to that obtained in the 1D case.

However, we should be careful enough, because u_0^2 contains the components of $\mathbf{k}$ such as k_x and k_y, which means that this relation is valid only for one propagation direction determined by $\mathbf{k}$. In a general case of waves propagating to various directions, a general solution of wave equation should be in the form of $\psi(\mathbf{r},t) = \sum_{\mathbf{k}} a_{\mathbf{k}}\exp[i(\mathbf{k}\cdot\mathbf{r}-\omega t)]$ with a constant value of $|\mathbf{k}|$ for a monochromatic wave, which results in an unseparable form with respect to $\psi(\mathbf{r},t)$:

$$D_1^2\psi(\mathbf{r},t) = 4\sum_{\mathbf{k}} a_{\mathbf{k}}\left\{\sin^2(k_xh/2)+\sin^2(k_yh/2)\right\}e^{i(\mathbf{k}\cdot\mathbf{r}-\omega t)}$$

To solve this problem, Cole proposed an approximate form of difference to minimize the angle-dependent error [Cole (1997, 2005)]. Although a detailed description of his procedure is out of the scope of this book, here we just outline it for the readers' convenience. Cole considered another central difference defined by

$$2d_x^{(2)} f(x,y) = f\left(x+\tfrac12,y+h\right) - f\left(x-\tfrac12,y-h\right) + f\left(x+\tfrac12,y-h\right) - f\left(x-\tfrac12,y+h\right),$$

and instead of applying D_1^2 to the difference wave equation, he utilized $D_\gamma^2 \equiv \gamma D_1^2+(1-\gamma)D_2^2$ with $D_2^2 = (d_x^{(1)}d_x^{(2)}+d_y^{(1)}d_y^{(2)}+D_1^2)/2$. Here, we discriminate $d_{x,y}$ belonging to D_1 from those to D_2 by putting a superscript (1) such as $d_{x,y}^{(1)}$. γ is a parameter determined so as to minimize the angular dependence. He found that

$$\gamma = 1 - \frac{\sin^2(k_x'h/2)+\sin^2(k_y'h/2)-\sin^2(kh/2)}{2\sin^2(k_x'h/2)\sin^2(k_y'h/2)}$$

with $(k'_x, k'_y) = k\left(2^{-1/4}, \sqrt{1 - 2^{-1/2}}\right)$ gave a considerably good result to minimize the error. In fact, he estimated that the error in NSFD model employing this parameter is proportional to $(kh)^6$, whereas that in SFD model is proportional to $(kh)^2$. Since $kh = 2\pi h/\lambda$, it is expected that under a constant wavelength, a reliable result will be obtained even when a mesh size is not so small as compared with the wavelength of light. This promises a very short computational time, even when a system contains a very complicated microstructure. On the contrary, NS-FDTD is applicable only under a condition of constant wavelength and thus essentially monochromatic, which will be somewhat time-consuming when spectral dependence is calculated.

7.2 Reflection due to Multilayer with a Finite Width and Wavelength-Scaled Reflection Spectrum

7.2.1 *Reflection Spectra and Fourier Analyses*

Thin film with a finite width In Secs. 3.2 and 3.3, we have implicitly assumed that the widths of thin film and multilayer are infinite. This assumption is mathematically possible and will make a problem considerably easier. However, actually they are not infinite and moreover the finiteness is one of key structures in bionanophotonics. Hence, we will explicitly consider the effect of finiteness on thin-layer and multilayer interference. In reality, this will be a hard work, because the analytical approach is generally difficult so that only a numerical calculation necessarily plays a central role, the results of which are sometimes difficult to analyze intuitively. In spite of this, here we will numerically calculate the reflection properties of a thin film and multilayer having finite widths and will attempt to analyze the results in an analytical way.

Consider first a case of thin film with a finite width in one direction and an infinite length in the other, as shown in Fig. 7.8. Unlike a thin film with an infinite width, the reflectivity is dependent on the polarization of incident light even under normal incidence. We denote the polarization parallel and perpendicular to a plane of this page as p- and s-polarization, which will be sometimes denoted

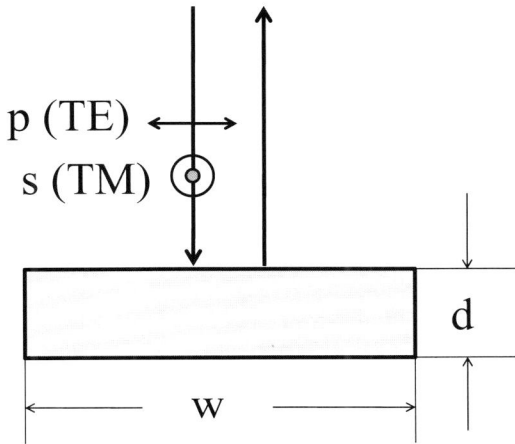

Figure 7.8 Thin layer with a finite width.

as *TE* (*transverse electric*) and *TM* (*transverse magnetic*) modes, respectively, as shown in this figure.

We first show the reflection spectra of thin films with various widths, which are calculated under normal incidence. To calculate them, we have employed the *NS-FDTD* method described in the preceding section. The results are shown in Fig. 7.9. As is seen immediately, for large widths, the reflection spectra are expressed by a smooth function of wavelength and have peaks around 360 nm, which are essentially independent of the polarization. These spectra almost completely agree with thin-film interference for an infinite width, which is expected to show the first-order interference peak at 361 nm under the condition of the thickness of the layer of 55 nm and refractive index of 1.55 with that of the surrounding medium of 1.0.

With decreasing width, the intensity of the reflection spectrum decreases naturally. However, interesting is that the difference between the two polarizations begins to appear below the width of 2000 nm. It is clear that in TM mode, an oscillatory structure seems to overlap on the ordinary reflection spectrum, while in TE mode, such oscillation is hardly seen from the figure. Further, the peak of the reflection spectrum seems to shift to shorter wavelengths.

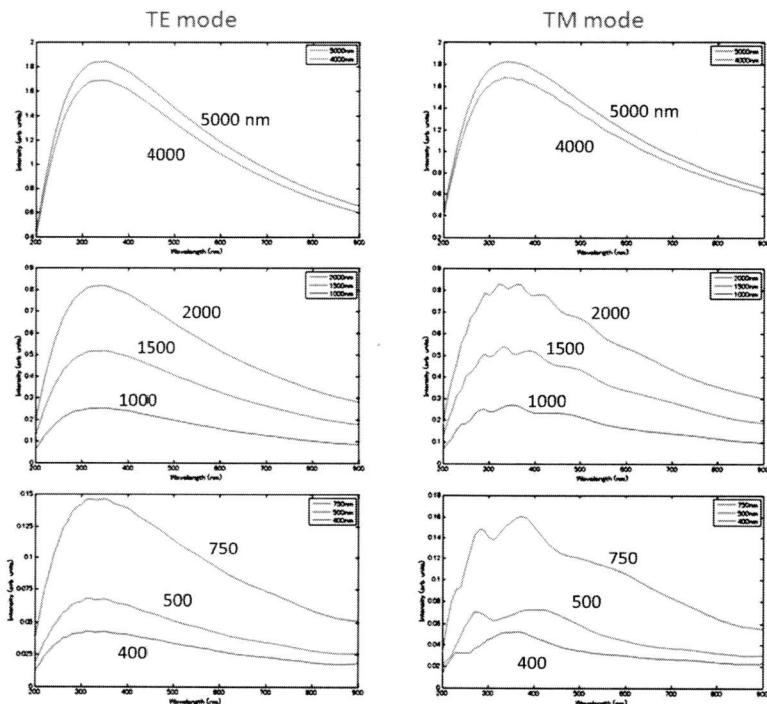

Figure 7.9 Reflection spectra under normal incidence for various widths of thin films. The left and right-hand sides of the figures correspond to p (TE) and s (TM) polarizations. The thickness and refractive index are set at 55 nm and 1.55, respectively. (Courtesy of Dr. D. Zhu.)

These seemingly strange behaviors are intuitively understood if we use the Fourier transform analysis that has been often employed in X-ray scattering problems. First, we will briefly describe the principle of the method. Consider a case where an incident light wave is scattered by a scatterer and produces a scattered light wave. We assume both incident and scattered light waves are expressed by plane waves, whose wave vectors are denoted as $\mathbf{k}_i$ and $\mathbf{k}_s$, respectively. We further consider the simplest case where a single scatterer is placed at a position denoted as S, which is indicated by a position vector $\mathbf{r}$, as shown in Fig. 7.10. Then, the phase factor of the scattered light wave as compared with that scattered hypothetically

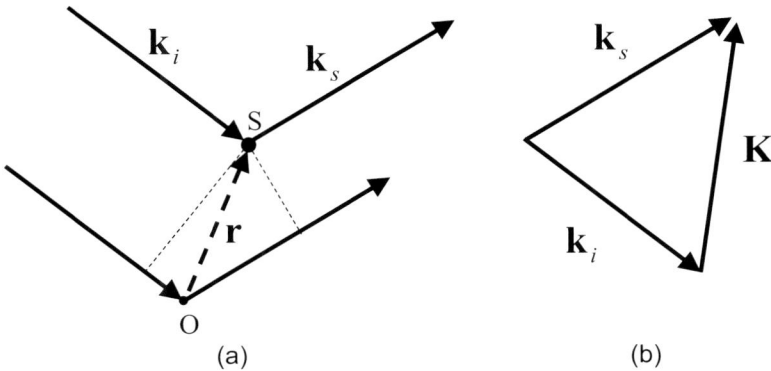

Figure 7.10 (a) Geometry for the scattering of light and (b) Definition of a scattering vector.

at the origin O is expressed as $\exp[-i(\mathbf{k}_s - \mathbf{k}_i) \cdot \mathbf{r}] = \exp[-i\mathbf{K} \cdot \mathbf{r}]$, where $\mathbf{K} \equiv \mathbf{k}_s - \mathbf{k}_i$ is a *scattering vector*.

Consider that many such scatterers are present in space and their distributions are describable in terms of a density distribution function $\rho(\mathbf{r})$ defined as $\rho(\mathbf{r}) = \sum_j \delta(\mathbf{r} - \mathbf{r}_j)$, where $\mathbf{r}_j$ is a position vector of the jth scatterer. Then, the total amplitude of the scattering field will be expressed by a sum of their contributions as

$$\hat{\rho}(\mathbf{K}) = \int d\mathbf{r}\, \rho(\mathbf{r}) e^{-i\mathbf{K} \cdot \mathbf{r}}. \tag{7.87}$$

This is nothing but the Fourier transform of a function $\rho(\mathbf{r})$ in 3D space. Thus, the scattering problem essentially becomes a problem to obtain $\hat{\rho}(\mathbf{K})$ in $\mathbf{K}$ space.

The Fourier transform method will offer an easy way to analyze the characteristics of light scattering from complicated structures, where ordinary analytical methods are not applicable. However, we must be careful enough to use this method, because it does not essentially consider multiple scattering and also does not include the concept of refractive index. Particularly, the latter restriction is serious, because the phase shift during the light propagation within a material and the phase change during the reflection at the interface are all neglected. Therefore, in a strict sense, the Fourier analysis is only applicable in case where the effect of multiple scattering can be ignored, in other words, the difference of the refractive index is small enough. However, in spite of these restrictions, this method offers us the useful information on the scattering problem in a practical use.

Using this method, we perform the Fourier transformation of a thin film with a finite width under the assumption that it consists of an infinitely large number of scatterers. Considering that the density of scatters is uniform, we put $\rho(\mathbf{r}) = 1$ for the inside of a layer, while $\rho(\mathbf{r}) = 0$ for the outside, and obtain

$$
\begin{aligned}
\hat{\rho}(K_x, K_y) &= \int_{-w/2}^{w/2} \int_{-d/2}^{d/2} e^{-i(K_x x + K_y y)} dx dy \\
&= \frac{w \sin(K_x w/2)}{K_x w/2} \cdot \frac{d \sin(K_y d/2)}{K_y d/2},
\end{aligned} \tag{7.88}
$$

where we put the width and thickness of the film as w and d, respectively. Since the intensity of light scattered with the scattering vector of (K_x, K_y) is expressed by a square of the above expression, we plot $|\hat{\rho}(K_x, K_y)|^2$ calculated for $d = 55$ (nm) and $w = 500$ (nm) in Fig. 7.11. The Fourier image consists of periodic structures both in the K_x and K_y directions, whose intensity envelope decreases monotonically with increasing distances from the center. The periodic structure possesses null lines in both directions, whose positions are determined by the conditions of $K_x w/2 = m\pi$ and $K_y d/2 = m'\pi$ in Eq. (7.88), where m and m' are integers.

As shown in Fig. 7.11, under normal incidence and reflection, the scattering vector is always vertical to the film with its magnitude twice the wave vector of incident light. Thus, it is expected that if $|\mathbf{K}|$ coincides null lines of $|\hat{\rho}(K_x, K_y)|^2$, such reflection does not occur basically, otherwise the reflection generally occurs according to the intensity determined by $|\hat{\rho}(K_x, K_y)|^2$. Under normal incidence and reflection, the null lines on the K_y axis are important, which is expressed by $K_y d/2 = m'\pi$ or $d = m'\lambda$. This relation just corresponds to that expected for thin-film interference apart from the refractive index of film and the phase change of reflection at an interface.

Although this scheme explains a general trend of thin-film interference, it does not essentially explain the oscillatory behavior found in the width below 2000 nm for TM mode. We then consider the higher-order scattering process. The second-order scattering

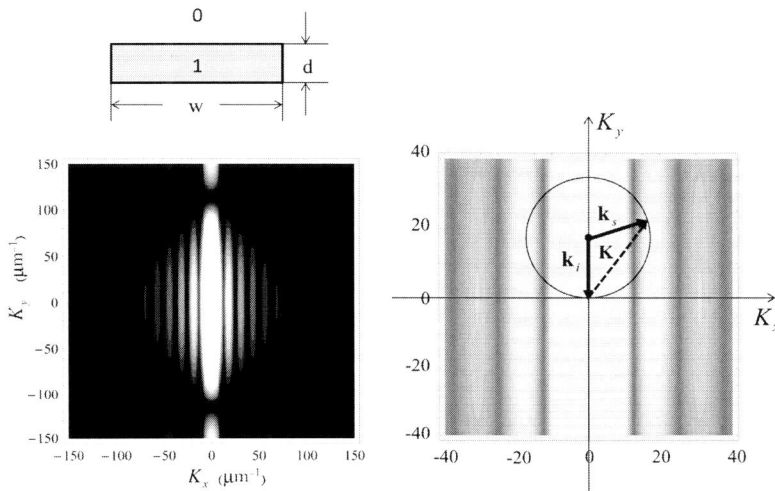

Figure 7.11 Geometry for a thin film with a finite width and its Fourier transformed image. A figure at the lower right shows how to analyze the scattering problem with this image. Under normal incidence and reflection, the scattering vector is vertical to the film with its magnitude twice the wave vector of incident light. The units of the vertical and horizontal axes are μm^{-1}.

process is a process, in which incident light is scattered by a structure to produce the first scattered wave, which is scattered again to produce the second scattering with a different wave vector. The typical process is shown in Fig. 7.12, where we indicate the scattering vector for the first scattering process as $\mathbf{K}_1$ and the second one as $\mathbf{K}_2$.

As is easily understood from this figure, the total scattering intensity depends both on $\mathbf{K}_1$ and $\mathbf{K}_2$. If both of the tips of the vectors agree with the positions of large values of $|\hat{\rho}(\mathbf{K})|^2$, then the second-order scattering will efficiently occur. On the other hand, either tip coincides with a position of $|\hat{\rho}(\mathbf{K})|^2 = 0$, then the scattering will not. We further assume that the wave vector of a light wave produced by the first scattering process has the same magnitude as that of incident one. This assumption guarantees that the second-order scattering is a real process, but not virtual one. Under this

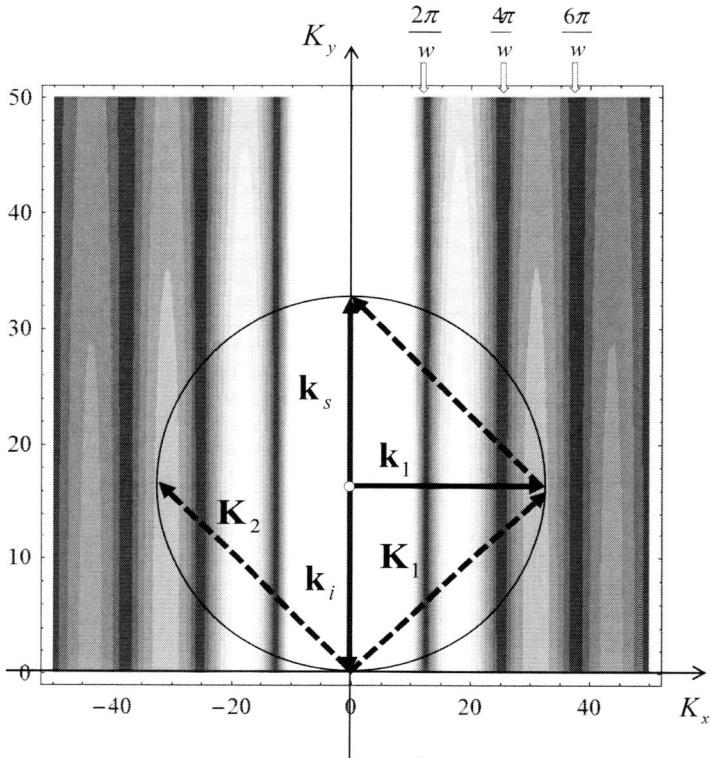

Figure 7.12 Second-order scattering in **K** space for thin-film reflection with a finite width. Dashed line segment with an arrow is a scattering vector corresponding to each scattering process, while solid line segment with arrows indicate the wave vectors for the incident, the first scattering and the reflected waves. The units of the vertical and horizontal axes are μm^{-1}.

assumption, it is expected that the tip of the wave vector of the first scattered wave should lie on a circumference whose radius agrees with $|\mathbf{k}_i|$. Further, under normal incidence and reflection, the first scattering vector should take a direction parallel to a film. Thus, $\mathbf{K}_1$ and $\mathbf{K}_2$ become symmetric with respect to the K_y axis, and the scattering intensity of the second-order scattering process will be the largest when K_{1x} or K_{2x} is nearly equal to $3\pi/w, 5\pi/w, \cdots$, while it will vanish when it agrees with $2\pi/w, 4\pi/w, \cdots$.

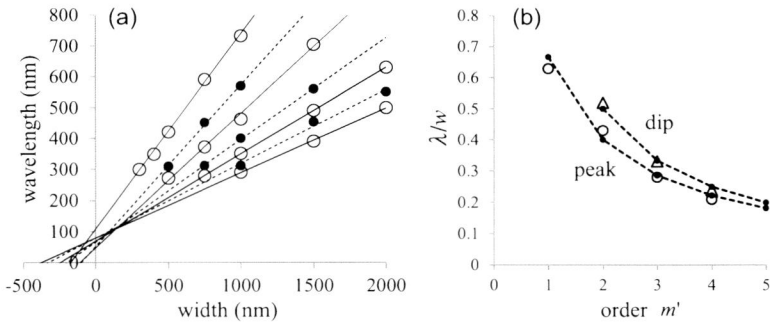

Figure 7.13 (a) Positions of peaks (open circles) and dips (filled circles) of oscillation evaluated from the reflection spectra of finite-width thin films, corresponding to Fig. 7.9. The gradients of peak and dip positions with changing width are evaluated by a linear interpolation, which are plotted in the right figure against the order of lateral stationary waves generated in the thin film. (b) Open circles and triangles are those obtained from peaks and dips in the left figure, while small filled circles combined with dashed line segments are obtained theoretically by the relations of $\lambda/w = 2/(2m' + 1)$ for peak positions and $2/(2m')$ for dip positions, respectively, with $m' = 1, 2, 3 \cdots$. (Courtesy of Dr. D. Zhu.)

Since this analysis does not take the concept of refractive index into account, the quantitative agreement cannot be attained naturally. In spite of this, we find a wonderful linear correlation between them, as shown in Fig. 7.13a, when we plot the wavelengths of peaks and dips of the oscillation against the width of thin film. Although the straight lines thus obtained do not pass the origin, their gradients corresponding to the peaks and dips are generally in good agreement with those deduced from the Fourier analysis, as shown in Fig. 7.13b.

From this analysis, it is strongly suggested that oscillation appearing in the reflection spectrum comes from the generation of laterally propagating waves along the film. For example, in a case of $w = 750$ (nm) of Fig. 7.14a, peaks appear at 560, 360, 280 and 220 nm, which are thought to correspond to $k \approx 3\pi/w$, $5\pi/w$, $7\pi/w$ and $9\pi/w$, where $k = 2\pi/\lambda$. Thus, it is expected that the laterally propagating waves thus generated will have wavelengths of $\lambda \approx 2w/3$, $2w/5$, $2w/7$ and $2w/9$, which strongly suggested that

a)

b)

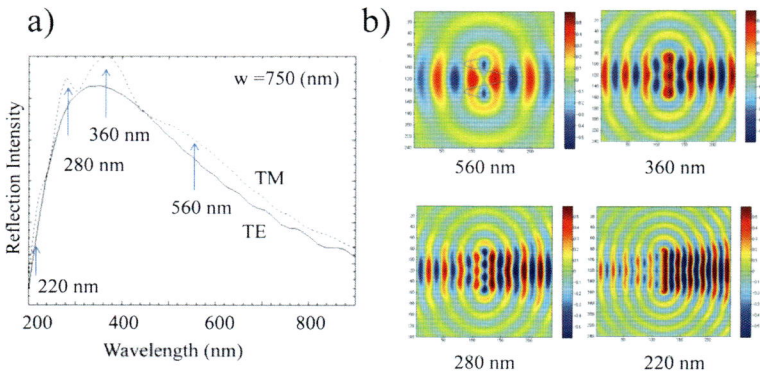

Figure 7.14 Reflection spectrum from a thin film with the width of 750 nm under normal incidence and reflection. The thickness and refractive index are set at 55 nm and 1.55, respectively. The right figures are the near field pattern evaluated by fixing time at various wavelengths corresponding to the peak positions indicated in the left figure. The calculation was made by means of the NS-FDTD method. The red and blue colors correspond to the positive and negative values of electric field. The layer is placed vertically at the center of the figure, while the incident light is injected from the left. We can clearly see stationary waves corresponding to $\lambda = 2w/3, 2w/5, 2w/7$ and $2w/9$ within a layer. (Courtesy of Dr. D. Zhu.)

they exist as stationary waves[a]. The NS-FDTD calculation excellently proves the presence of such waves as shown in Fig. 7.14b. In this figure, we show the near field at a certain time after the injection of an incident wave from the left. The dark regions indicate the electric field with negative sign. We can see clearly two, three and four negative portions at the wavelengths of 560, 360, and 280 nm, indicating the presence of the stationary waves.

Considering the above analysis, we can easily deduce that *s*-polarization is more favorable to produce laterally propagating waves, because the direction of light emission from a dipole induced

---

[a] Actually, the calculations of these wavelengths result in 500, 300, 214, and 167 nm for $w = 750$ (nm), respectively, which differ considerably from the peak positions observed in the reflection spectrum. However, they are rather close to the apparent wavelengths of 533, 300, and 233 nm for the first three, which are read from the patterns shown in Fig. 7.14b. Thus, for the quantitative understanding of this phenomenon, more detailed analysis is anyway necessary, which includes the introduction of the effective refractive index.

within a film is uniform in angle for s-polarization, whereas for p-polarization, it is considerably limited toward the normal direction. Further, when the width of the thin film is too narrow, the central part in Fourier image becomes so wide that the light scattering process cannot cause the oscillation, while when it is too wide, the Fourier image will be concentrated to the central part so that no clear oscillation will appear. Thus, the oscillatory phenomenon found in the reflection spectrum is only limited when the wavelength of light is comparable with the width of the layer. The numerical calculation excellently indicates these expectations.

Multilayer with a finite width   Next, we will apply the similar Fourier transform method to multilayer reflection. At first, we will show the characteristics of the Fourier image of a multilayer with a finite width. In Fig. 7.15b, we show a typical example for a multilayer indicated in Fig. 7.15a. The Fourier image is characterized by strongly bright spots along the K_y axis, which are accompanied by several weak spots between them, while several spots gradually reducing brightness are seen along the K_x axis. The positions of strongly bright spots depend on the periodicity of multilayer, while those of weaker ones along the K_y and K_x axes depend on the number of layers M and the width of multilayer w, respectively.

The positions of bright and weak spots, and dark gaps between these spots are analytically obtainable from the following relation:

$$
\begin{aligned}
\rho(K_x, K_y) &= \int_{-w/2}^{w/2} e^{iK_x x} dx \cdot \sum_{j=0}^{M-1} \int_{-d/2+ja}^{d/2+ja} e^{iK_y y} dy \\
&= \sum_{j=0}^{M-1} e^{iK_y ja} \int_{-w/2}^{w/2} e^{iK_x x} dx \cdot \int_{-d/2}^{d/2} e^{iK_y y} dy \\
&= \frac{1 - e^{iK_y Ma}}{1 - e^{iK_y a}} \cdot \frac{e^{iK_x w/2} - e^{-iK_x w/2}}{iK_x} \cdot \frac{e^{iK_y d/2} - e^{-iK_y d/2}}{iK_y} \\
&= \frac{1 - e^{iK_y Ma}}{1 - e^{iK_y a}} \cdot \frac{w \sin(K_x w/2)}{K_x w/2} \cdot \frac{d \sin(K_y d/2)}{K_y d/2}, \quad (7.89)
\end{aligned}
$$

where a and d denote a period of layers and the thickness of the layer, respectively. Then, we obtain

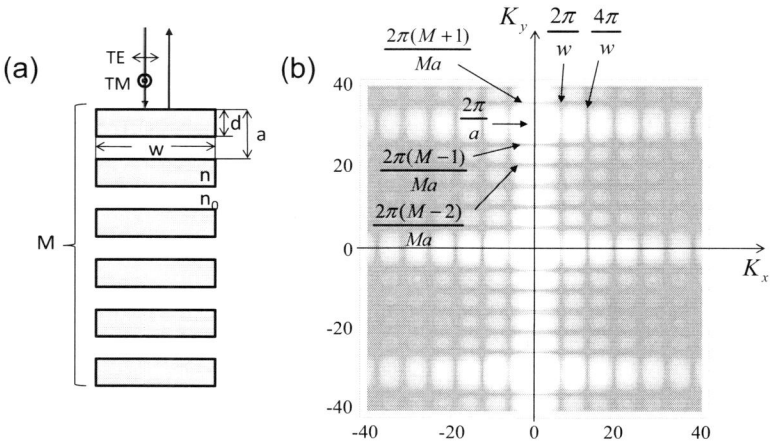

Figure 7.15 (a) Geometry of multilayer reflection under normal incidence and reflection, and (b) Fourier transform of the multilayered structure for $M = 6$. The units of the vertical and horizontal axes are μm^{-1}.

$$|\rho(K_x, K_y)|^2 = \left| \frac{w \sin(K_x w/2)}{K_x w/2} \cdot \frac{d \sin(K_y d/2)}{K_y d/2} \cdot \frac{\sin(K_y M a/2)}{\sin(K_y a/2)} \right|^2.$$
(7.90)

Thus, the positions of strongly bright spots along the K_y axis are obtained from the relation $K_y a/2 = m\pi$ with m an integer, while weaker spots between them are caused by a function of $\sin^2(K_y M a/2)$. Hence, the positions of dark gaps are calculated by the relation $K_y M a/2 = p\pi$ with p an integer. On the other hand, dark gaps between weak spots along the K_x axis are calculated from the relation $K_x w/2 = q\pi$ with q an integer. We indicate typical values of these particular points in Fig. 7.15b.

We will first investigate the light reflection related to bright and weak spots along the K_y axis. For this purpose, we investigate the reflection properties of a multilayer with an infinite width under normal incidence and reflection. In this case, the Fourier image is restricted on the K_y axis and the first-order reflection takes place only on this axis. Thus, we solely consider spots or gaps along the K_y axis, as shown in Fig. 7.16a. It is soon understood that if a tip of a scattering vector **K** coincides with a bright spot, strong reflection

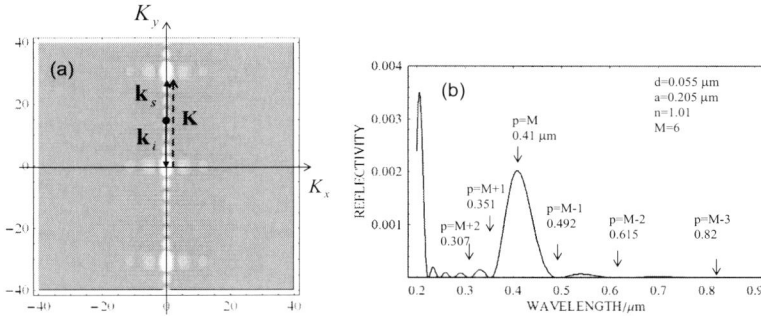

Figure 7.16 (a) First-order scattering expressed in K space, which is due to a multilayer with a sufficient large width under normal incidence and reflection. The units of the vertical and horizontal axes are μm^{-1}. (b) Reflection spectrum calculated for a multilayer with an infinite width. The thickness and refractive index of the layer are set at 55 nm and 1.01, respectively, while the refractive index of the remaining space and layer-to-layer spacing are set at 1.0 and 150 nm respectively. Arrows are expected positions of dips or peaks estimated from (a).

will occur, while if it does with a dark gap, reflection will vanish. Thus, for the first-order reflection, a strongly bright spot is obtained from the relation $K = 2\pi p/(Ma)$ with $p = M$, i.e. $\lambda = 2a$, while vanishing points are obtained from $\lambda = 2(M/p)a$ with $p = M \pm 1$, $M \pm 2, \cdots$.

Since this expectation should be absolutely valid in case where the difference of the refractive indices is sufficiently small, we calculate a reflection spectrum for the refractive index difference of 0.01, and compare it with the above estimations. The result is shown in Fig. 7.16b with the estimations indicated by arrows. The agreement is quite satisfactory. From this figure, it is clear that side bands often seen in multilayer reflection are resulted essentially from a multilayered structure. Thus, it is proved that the Fourier transform method is also effective for the analysis of multilayer reflection.

We will extend this method to a multilayer with a finite width. We first show the reflection spectra for multilayer with various widths under normal incidence and reflection, as shown in Fig. 7.17. Similar to thin-film interference, the reflection spectra resemble to that of a

TE mode

TM mode

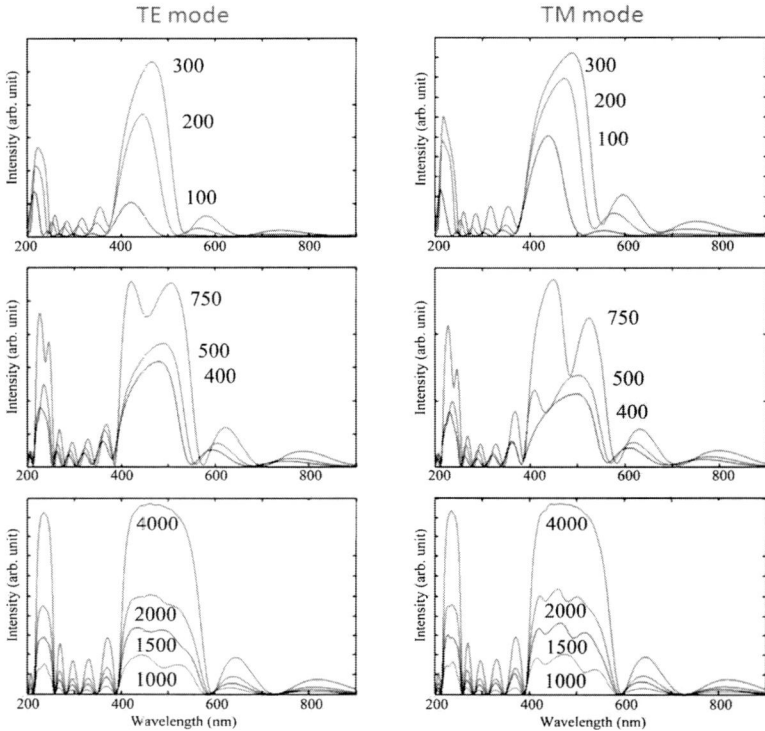

Figure 7.17 Reflection spectra of multilayers with various widths from 100 to 4000 nm for TE and TM modes, calculated under normal incidence and reflection. The multilayer consists of 6 layers with the thickness of 55 nm and the refractive index of 1.55, which are separated by the spacing of 150 nm. The refractive index of the remaining space is set at 1.0. (Courtesy of Dr. D. Zhu.)

multilayer with an infinite width when the width is sufficiently large, while the oscillation is superposed when the width lies in a range of 300–2000 nm. Below this range, the spectrum becomes remarkably sharper and blue-shifted. It is also noticed that as is similar to thin-film interference, the oscillatory structure is prominent in TM mode, while unlike the thin-film case, less prominent structure is also observed in TE mode. In addition, the period of oscillation is larger with decreasing width.

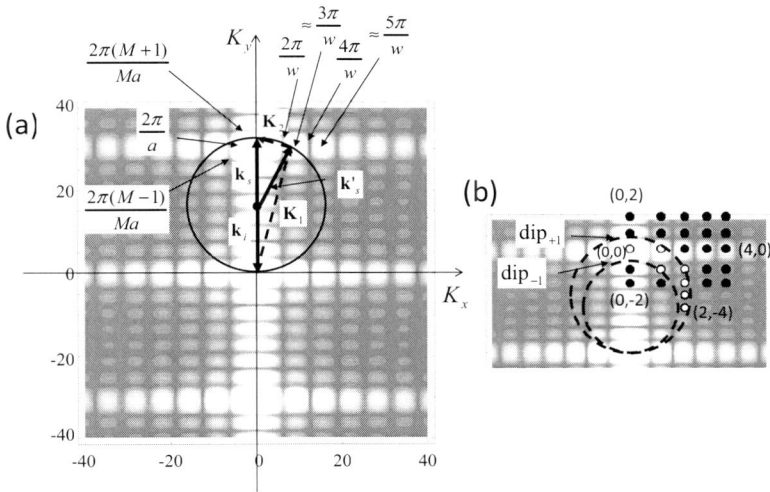

Figure 7.18 (a) Second-order scattering due to a finite-width multilayer, expressed in the K space. The scattering is considered to selectively occur when the tip of the scattering vector is coincident with a bright spot in the **K** space. In (b), we show the bright spots (open circles), which will contribute to an oscillatory structure within the first reflection band. The two circles correspond to the dip positions located at both sides of the reflection band. Each bright spot is characterized by a pair of coordinates shown in this figure. The units of the vertical and horizontal axes are μm^{-1}.

It is reasonable to guess from the result of thin-film interference that a main band and its side bands originate from the first-order scattering process due to a structure along the K_y axis, while an oscillation found within a main band will be due to the second-order one based on the structure along the K_x axis. In Fig. 7.18a, we show spots in the Fourier image, which will possibly contribute to the second-order scattering. Unlike the thin-film interference, the structure along the K_x axis is further structured by that along the K_y axis so that a lot of spots and gaps are found to exist.

The second-oder scattering is, in principle, connected with all these bright spots and dark gaps. However, if we concentrate on the oscillation found within a main reflection band, which is cut off by two spectral dips denoted by dip_{+1} and dip_{-1}, we only concentrate on the spots and dips in the Fourier image within an area enclosed

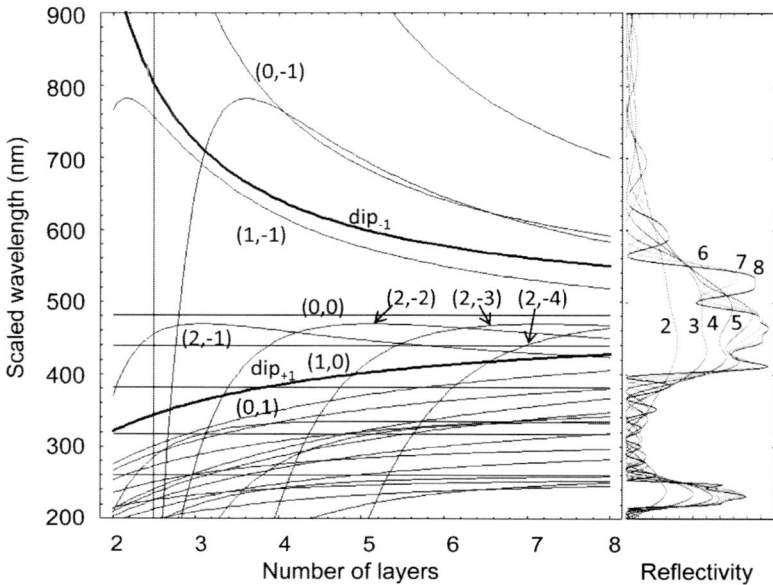

Figure 7.19 Simulation of the wavelengths of the second-order scattering for various numbers of layer M, which are calculated from the positions of bright spots shown in Fig. 7.18b. The curves are obtained by hypothetically considering that the wavelengths are expressed by a continuous function of M. "dip$_{\pm 1}$' indicate the positions of dips appearing at the both sides of the reflection band, which are estimated using the Fourier analysis. A pair of coordinates indicate the position of the bright spots in K space, as shown in Fig. 7.18b. Since the Fourier analysis does not give the true wavelength, the vertical axis is scaled according to a method described in the next section. The right figure shows the reflection spectra for finite-width multilayers with the width of 1.0 μm for various numbers of layers from 2 to 8 under normal incidence and reflection. The thickness of the layer and the spacing between the layers are set at 55 and 150 nm, respectively, whereas the refractive index of the layer is set at 1.55 with that of the surrounding medium of 1.0. (Courtesy of Dr. D. Zhu.)

by two circles, whose diameters are $2\pi(M + 1)/(Ma)$ and $2\pi(M - 1)/(Ma)$, respectively. The spots thus obtained are shown as open circles in Fig. 7.18b. From this figure, it is clear that several bright spots are considered to contribute to the oscillation of the spectrum. We name these spots in order as shown in the figure and estimated their positions with changing number of layers.

The result is shown in Fig. 7.19. As is seen immediately, quite complicated curves are obtained. Although qualitative agreement is obtained when the number of layers is small, it is generally difficult to assign the one-to-one correspondence between the spectral peaks/dips and the spots/dips in the Fourier image. To make thing worse, the spectral modulation is more complicated when the number of layers is increased, because many spots will appear along the K_y axis. Thus, at the present stage, it seems difficult to analyze the shape of the reflection spectrum in an analytical way, particularly when the number of layers is increased.

7.2.2 *Diffraction Grating or Multilayer Interference?*

Multilayer with a finite width can be regarded as a simple multilayer when the width is sufficiently large, while it is regarded as a diffraction grating arranged vertically when the width is extremely small. Thus, it is quite an interesting problem to investigate where a border between them exists. This is also an important problem to consider the characteristics of finite-width multilayer, because the multilayer interference is characterized by specular reflection with high reflectivity within a particular wavelength region, while diffraction grating is characterized by the presence of angularly broadened and wavelength-dependent diffraction spots. Thus, completely different behavior comes out, if only the width of multilayer is varied.

To shed light on this problem, we will focus on the peak position of the reflection spectrum for finite-width multilayer, which is observed under normal incidence and reflection. This is because the peak positions of the above two processes differ considerably as shown in Fig. 7.20, where we show the reflection spectra for typical examples of a multilayer and diffraction grating calculated analytically. We also investigate the reflection spectra for multilayers with a width of 2000 nm and 50 nm using the NS-FDTD method. It is found that these spectra agree fairly well with those obtained analytically.

Actually, in the reflection spectra obtained between the above two extremes, the above-mentioned oscillatory behavior is over-lapped. Thus, we determine the peak position as a middle point of

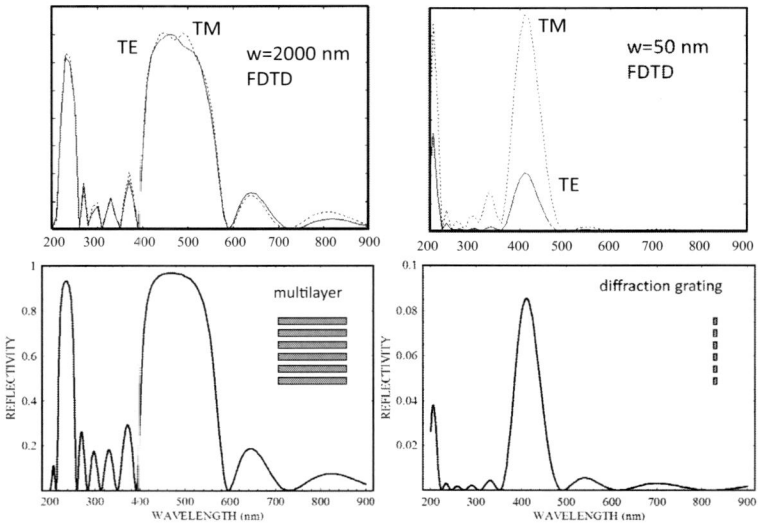

Figure 7.20 Reflection spectra of a multilayer and a vertically aligned diffraction grating under normal incidence and reflection. Upper two figures are calculated for the width of 2000 and 50 nm using the NS-FDTD method for two polarization directions, while the lower ones are analytically calculated for a multilayer with an infinite width and a diffraction grating with an infinitely small width. The number of layers is set at 6 with the refractive index of the layer is 1.55, while that of the surrounding medium is set at 1.0. The layer thickness and layer-to-layer spacing are set at 55 and 150 nm, respectively. (Courtesy of Dr. D. Zhu.)

the reflection band evaluated at a half height. Further, to investigate the generality, we have calculated the reflection spectra with varying layer widths and layer-to-layer spacings. However, it is found that by changing spacing, the spectral positions at both extremes also change. Therefore, we newly define a relative peak position as $\{p(w; a) - p_g(a)\}/\{p_m(a) - p_g(a)\}$, where a and w are the period of the layers and the layer width, respectively. $p(w; a)$, $p_g(a)$ and $p_m(a)$ are the peak position of a finite-width multilayer calculated by NS-FDTD method, and those analytically obtained for a diffraction grating and a multilayer, respectively.

The result is shown in Fig. 7.21. By employing the relative peak position, all the values seem to be well expressed along a smooth

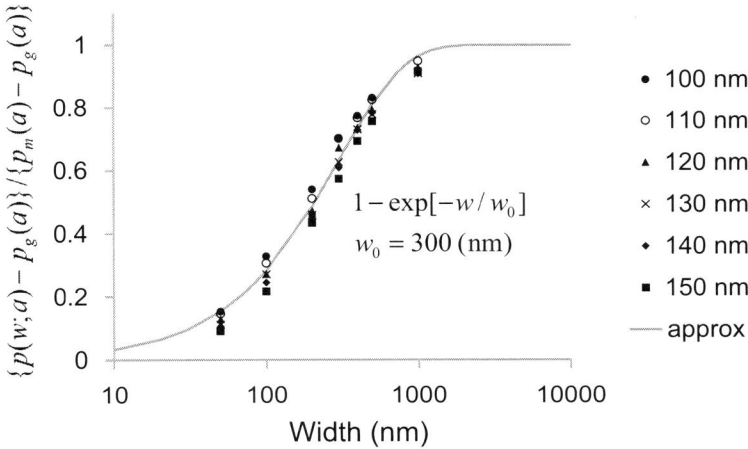

Figure 7.21 Relative peak position of the first reflection band of finite-width multilayer with varying layer width and layer-to-layer spacing. The calculations are performed by NS-FDTD method under normal incidence and reflection for 6 layers with the refractive index of 1.55 and the thickness of 55 nm, which are separated by 100-150 nm spacings with the refractive index of 1.0. Solid curve expresses a fitting curve using a function of $1 - \exp[-w/w_0]$ with $w_0 = 300$ (nm). (Courtesy of Dr. D. Zhu.)

function of multilayer width. In this figure, it should be noticed that when the width becomes large enough, the curve approaches that of a multilayer with an infinite width, $p(w; a) \rightarrow p_m(a)$, while when it becomes sufficiently small, the curve approaches that of a diffraction grating, $p(w; a) \rightarrow p_g(a)$. It is found that the transient is well expressed by a function of $1 - \exp[-w/w_0]$ with $w_0 = 300$ (nm), which shows that the border between multilayer and diffraction grating seems to lie around 300 nm in the present case.

7.2.3 *Reflection Spectrum Scaled by Average Refractive Index*

In the preceding section, we have shown that the relative peak position of multilayer reflection is well expressed by a smooth function of multilayer width, which is eventually independent of periodicity of the layers. This result gives a hint to scale the reflection

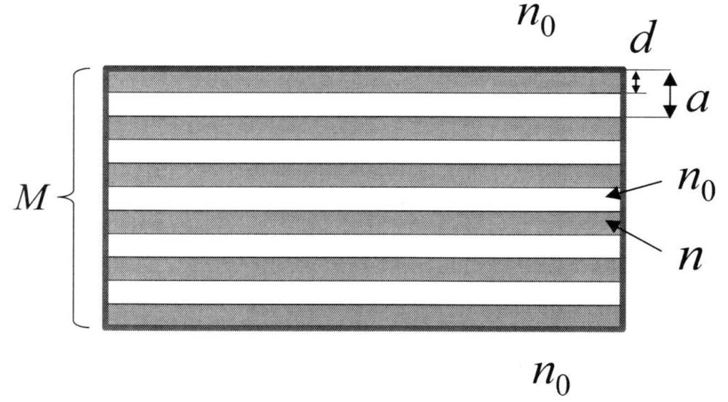

Figure 7.22 Average refractive index calculated for a multilayer having an infinite width. The areal averaging is performed within an enclosed region.

spectrum by appropriately selecting a scaling parameter. We have chosen an average refractive index as a parameter.

Consider first the simplest case of multilayer with an infinite width. Although various methods of averaging can be considered, we have chosen a simple area-averaging method within a region that is enclosed by a rectangle shown in Fig. 7.22. The average refractive index thus obtained is expressed as

$$n_{av} = \frac{Mnd + (M-1)n_0(a-d)}{Md + (M-1)(a-d)}, \tag{7.91}$$

where n and n_0 are refractive indices of layers and surrounding medium, respectively. The effectiveness of this choice of average method can be confirmed by calculating a reflection spectrum whose horizontal axis is scaled by this average refractive index, $\lambda_{sc} \equiv \lambda/n_{av}$. The result is shown in Fig. 7.23a, where we show the reflection spectra of multilayers with various refractive indices of layers. Since the refractive index of layer is changed without changing thickness in this figure, the peak position and band width will necessarily change. However, by adopting the scaled wavelength, the peak positions of reflection spectra seem to agree fairly well with each other.

Although small shifts of dip positions in the side tails are seen, they are mostly due to the broadening of the reflection band with increasing difference of refractive indices. To confirm this, we have

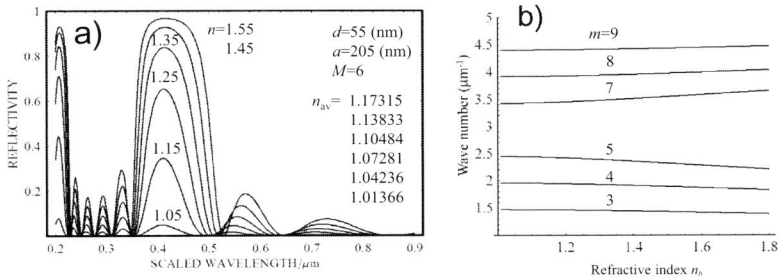

Figure 7.23 (a) Reflection spectra of multilayer, which are scaled by average refractive index. The number of high-refractive index layer is 6 with the thickness of 55 nm and the spacing between layers is set at 150 nm with the refractive index of 1.0. The refractive index of the former layer is varied from 1.05 to 1.55. (b) Variation of the positions of dips appearing in the side lobes of the first reflection band against refractive index of layers, where an ideal multilayer is assumed with the number of 6. The thickness and spacing of the layers are to be kept constant so as to agree with the optical path length determined by 55×1.55 nm.

simulated the positions of dips appearing in side tails with changing refractive index of the layer, which is shown in Fig. 7.23b. In this figure, the dip positions are calculated by putting[a] $\sin^2 p\theta = 0$ in Eq. (3.82). It is clear, below the refractive index of $\sim$1.3, the reflection band width is almost unchanged and hence the dip positions seem to remain constant. However, above 1.3, the band width is remarkably increased, which seems to drive the dips away from the center. Thus, in a strict sense, we should say that the scaling is applicable only for a small refractive-index difference. However, since the peak position of reflection band can be scaled fairly well, one can say that the scaling is also effective even in a high-refractive index region.

Since the scaling of wavelength using an average refractive index is found to work well particularly with respect to the peak position, we will apply this method to multilayer with a finite width. As described above, a multilayer with a finite width has two aspects: One is a typical multilayer, which appears when the width

---

[a]Since the condition $\sin^2 p\theta = 0$ leads to $p\theta = m\pi$ with m an integer and also from Eq. (3.70) together with $\mu_{1,2} = \exp[\mp i\theta]$, a relation $\cos\theta = (\cos 2\phi - r^2)/(1 - r^2)$ with $\phi = kn_b d_b$ is proved to hold. From this relation, a wavelength giving a dip is obtained as $\lambda = 4\pi n_b d_b / \cos^{-1}[(1 - r^2)\cos(m\pi/p) + r^2]$.

(a)

(b)

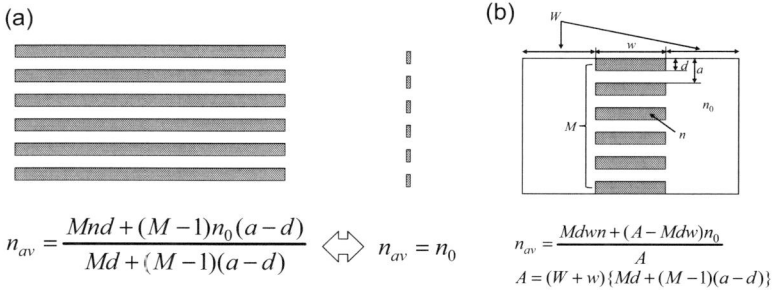

$$n_{av} = \frac{Mnd + (M-1)n_0(a-d)}{Md + (M-1)(a-d)} \quad \Longleftrightarrow \quad n_{av} = n_0 \qquad n_{av} = \frac{Mdwn + (A - Mdw)n_0}{A}$$

$$A = (W + w)\{Md + (M-1)(a-d)\}$$

Figure 7.24 (a) Multilayer and diffraction grating with their expected average refractive indices. (b) Average refractive index proposed for a multilayer with a finite width. An extra space, whose width is expressed by W, is added to a multilayer to calculate an average refractive index.

is sufficiently large, and the other is a diffraction grating, which does when the width is sufficiently small. In the former, the average refractive index should be expressed by Eq. (7.91), while in the latter, it should become $n_{av} = n_0$. If we would employ the same relation of Eq. (7.91), it is clear that even in the limit of $w \to 0$, n_{av} would not change in principle. Thus, it is necessary to construct a new model that is applicable to both of these extremes.

Alternatively, we consider an extra space expressed by the width of W and add it to a multilayer with a finite width, as shown in Fig. 7.24. The average refractive index in this case is simply obtained as

$$n_{av} = \frac{Mdwn + (A - Mdw)n_0}{A}, \qquad (7.92)$$

where $A = (W + w)\{Md + (M - 1)(a - d)\}$. This procedure will eventually make up for the above fault. This is because in the limit of $w \to \infty$, n_{av} becomes that expressed in Eq. (7.91), while in the other limit of $w \to 0$, it becomes n_0. Further, we can determine W empirically so that the peak position of the reflection band of multilayer relative to that of the diffraction grating will be linearly correlated with the average refractive index thus defined. One of such examples is shown in Fig. 7.25, where we show that if $W = 200$ (nm) is employed, the relative peak position is almost linearly correlated with the average reflective index determined by

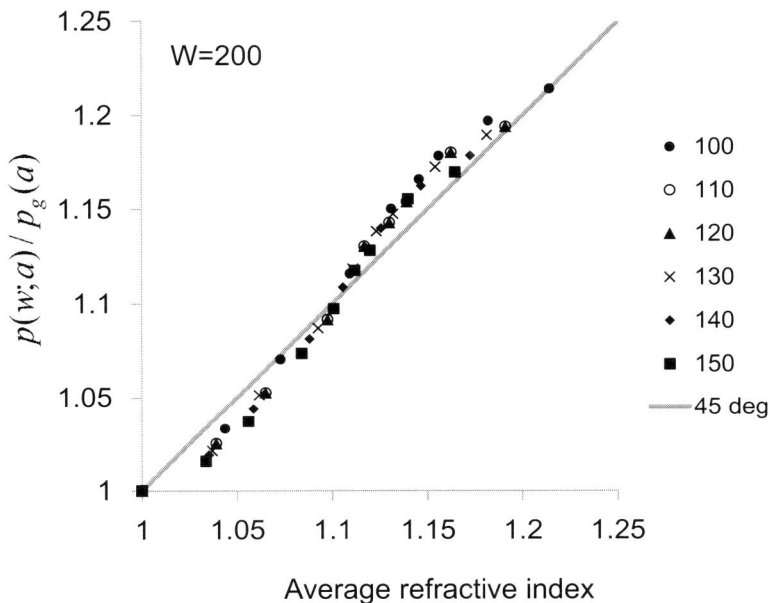

Figure 7.25 The extra space W shown in Fig. 7.24 is empirically determined so that a ratio of the peak position of multilayer reflection to that of a diffraction grating will be coincident with the average refractive index determined by areal averaging. In the figure, an example for $W = 200$ (nm) is shown. Many dots correspond to data for various values of refractive index of layer and various layer-to-layer spacings. (Courtesy of Dr. D. Zhu.)

Eq. (7.92), even when the layer width and spacing between layers are varied.

Using $W = 200$ (nm) and Eq. (7.92), we have performed a scaling of wavelength for the reflection spectra obtained by NS-FDTD method. In Fig. 7.26, we show the results for the refractive indices of 1.25 and 1.5 with various widths of multilayer ranging from 50 to 500 nm. The reflection spectrum is close to that of diffraction grating when the width is small, while a remarkable oscillation appear when the width becomes larger. It is clear that the positions of reflection bands seem to be well scaled, particularly when the difference of refractive indices is small. This attempt shows that a spectral shape and position for multilayer reflection is well

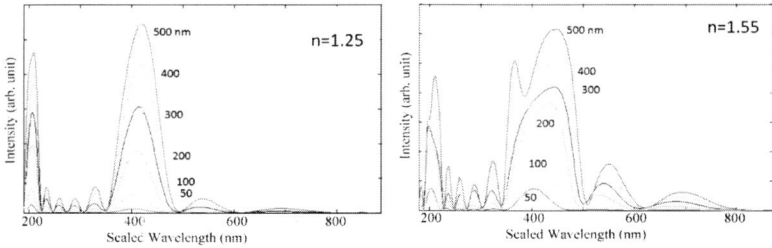

Figure 7.26 Reflection spectra whose wavelengths are scaled by average refractive index using $W = 200$ (nm) for the refractive indices of layers of 1.25 and 1.55, respectively. The reflection spectra with varying layer-to-layer spacing are well scaled so that reflection peaks appear at nearly constant positions. (Courtesy of Dr. D. Zhu.)

predictable by using an average refractive index even when it has a finite width.

7.3 FDTD Method Applied to the *Morpho* Butterfly Scale

The FDTD method demonstrates its full ability when the microstructure is so complicated that any analytical solution cannot be applied. One of such examples is a well-known microstructure equipped on a scale of the *Morpho* butterfly. Since the outline of this butterfly has been fully described in the previous section (Sec. 3.6.3), here we will show its structure and optical properties more in detail through the electron microscopic observations and angle-dependent reflection measurements, and will elucidate the mechanism of its peculiar optical properties, which might not have been clarified without the FDTD method [Zhu *et al.* (2009); Kambe *et al.* (2011)]. Since the microstructure in the *Morpho* butterfly is known to vary from species to species, we will focus on that of *M. rhetenor*, which is known as a typical species of *Morpho* butterflies that display the most glossy blue wing.

In Fig. 7.27, we show the microstructure equipped on a scale of *M. rhetenor*. The microstructure of *M. rhetenor* is characterized by well-developed ridges with a spacing of 600–670 nm. Both sides of a ridge are decorated by shelf structures with an interval of 150–240 nm

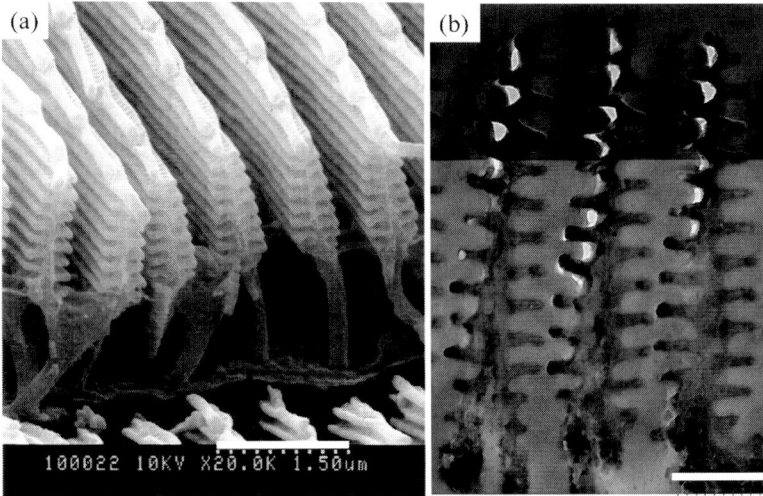

Figure 7.27 (a) Scanning and (b) transmission electron microscopic images at the cross section of a ground scale of the *M. rhetenor*, reproduced from [Kambe *et al.* (2011)] with permission. Scale bars are (a) 1.5 μm and (b) 0.5 μm. (Courtesy of Dr. M. Kambe for the SEM image.)

and the thickness of 50–90 nm. The shelves are alternately sticking to both sides and run equidistantly along the ridge. It is known that the shelves are not parallel to a base plane of scale, but are slightly inclined from the root of scale, which makes the reflection direction to be deflected toward an inner side of the wing. Further, as has been described previously, the ends of shelves observed at the top of ridges seem to be randomly distributed, which disturbs the interference of light reflected among the ridges. Thus, the blue reflection from this type of *Morpho* butterfly originates from the interference of light due to regularly arranged shelves and not due to regularly arranged ridges. In addition, light-absorbing pigments are known to exist at the lower part of the scale.

The microstructure described above was already discovered in 1942, which corresponded to a year just after transmission electron microscope was commercialized. However, since then, the theoretical analysis concerning their reflection properties relied mostly on multilayer interference theory. It was not till the end of the

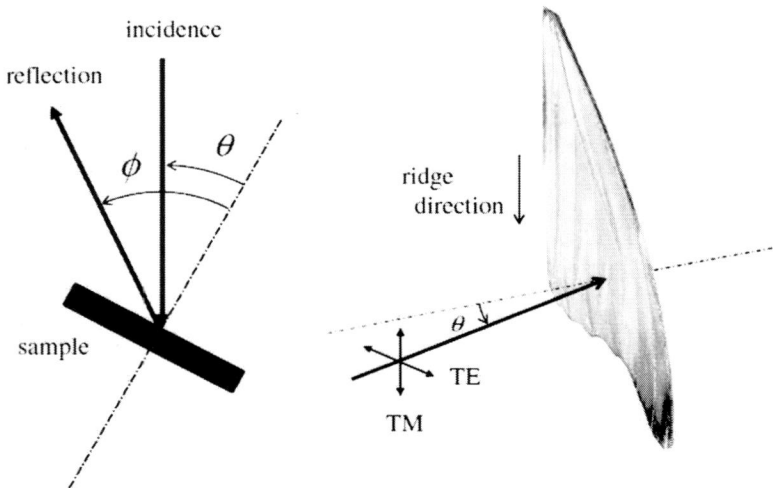

Figure 7.28 Definitions of angular parameters and polarization directions, reproduced from [Kambe *et al.* (2011)] with permission.

20th century that people knew that the essential parts of the *Morpho* blue could not be explained by a simple multilayer interference theory. Owing to recent detailed spectral and angular measurements on a wing or even on a single scale, it is now clarified that the *Morpho* wing shows blue reflection with peculiar angular dependence, which differs completely from that of multilayer reflection.

More recently, our group has reported the detailed angle-resolved reflection spectra on the *M. rhetenor* wing and have clarified that the *Morpho* wing actually possesses strong retrore-flection (or backscattering) properties. In this work, we have compared three typical angle-resolved reflection spectra, which have been frequently employed up to now: (1) detector and (2) sample rotations, and (3) θ–2θ scan. Here, we will define the angular parameters for the angle-resolved measurement, which are summarized in Fig. 7.28. We define the angles of incidence and reflection as θ and ϕ, as shown in the figure. Further, since we adjust the direction of scale to accord with the axis of rotation, the incident light beams with the electric fields vertical and horizontal to the optical table automatically correspond to those parallel and

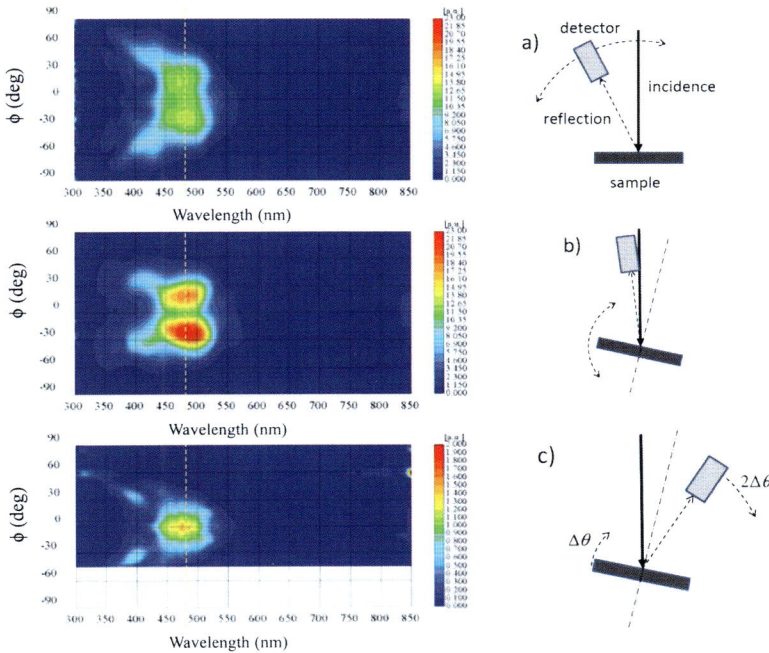

Figure 7.29 Angle-resolved reflection spectra obtained by the methods of (a) detector rotation, (b) sample rotation and (c) θ–2θ scan for the *M. rhetenor* wing under TE polarization. The experimental geometry employed in each measurement is shown on the right side. (Reproduced from [Kambe *et al.* (2011)] with permission.)

perpendicular to the average directions of ridges equipped on the scale, which are called TM and TE mode, respectively.

The detector rotation is a method, in which θ is fixed, while ϕ is varied to obtain the angular dependence of the reflection, and has been most frequently employed in the literatures (see Fig. 7.29a). On the other hand, in the sample rotation method, only a sample is rotated around a rotation axis, while the angle between the incident beam and the detector is fixed so that θ and ϕ will vary simultaneously (see Fig. 7.29b). If we define the difference between their angles as $\eta \equiv \theta - \phi$, $\eta = 0°$ corresponds to a so-called retroreflection (or backscattering) geometry. When a sample is rotated by a step of $\Delta\theta$, while the detector by a step of $2\Delta\theta$, the method is called θ–2θ scan and is well known as the most commonly employed method in X-ray diffractometry (see Fig. 7.29c).

The results of these three measurements are summarized in Fig. 7.29. In the detector rotation method, under normal incidence, the reflection spectrum (Fig. 7.29(a)) is characterized by a broad hump around 480 nm, which extends over a spectral range of 450–530 nm and an angular range of $\pm 45°$. It is noticed that the reflection toward the normal direction is weakly suppressed, and accordingly faint twin peaks appear at $\phi = \sim \pm 20°$. On the other hand, the sample rotation method under almost retroreflection geometry of $\eta \sim 0°$, displays a considerably different pattern (Fig. 7.29(b)). The twin peaks are clearly separated and the suppression to the normal direction is more conspicuous. It is also found that the absolute reflectivity is much higher than that of the detector rotation method. The result of θ–2θ scan is shown in Fig. 7.29(c). In this method, a peak is only observed around $\phi \sim 0°$ in a wavelength range of 430–520 nm, which broadens only within $\pm 20°$. It is found that the absolute reflectivity is much smaller than the results of the other two.

These completely different appearances are well understood, if we employ a comprehensive method of θ–ϕ scan, in which both the sample and detector are rotated. Then, an intensity map for all the angles of θ and ϕ can be obtained. We call this method θ–ϕ scan, which is essentially similar to well-known BRDF (bidirectional reflectance distribution function) measurement. In Fig. 7.30, we show a typical result measured at 480 nm. The horizontal and vertical axes of this map correspond to θ and ϕ, respectively. It is surprising that clear twin spots appear along a 45° inclined line of $\theta = \phi$. The locations of the twin spots are somewhat asymmetric with respect to the origin and located at the angular coordinates of $(17°, 17°)$ and $(-27°, -27°)$, respectively. We can analyze the above three experiments using this map (θ–ϕ map).

Since the detector rotation is a method to scan ϕ, while θ is fixed, it is generally expressed by a vertical line in this map. When $\theta = 0°$, it is coincident with the ϕ axis, which is shown as a dashed line a) in Fig. 7.30. On the other hand, the sample rotation is characterized by a 45° inclined line and in the retroreflection geometry of $\eta \sim 0°$, it is expressed by a straight line of $\theta = \phi$, shown as a dashed line b).

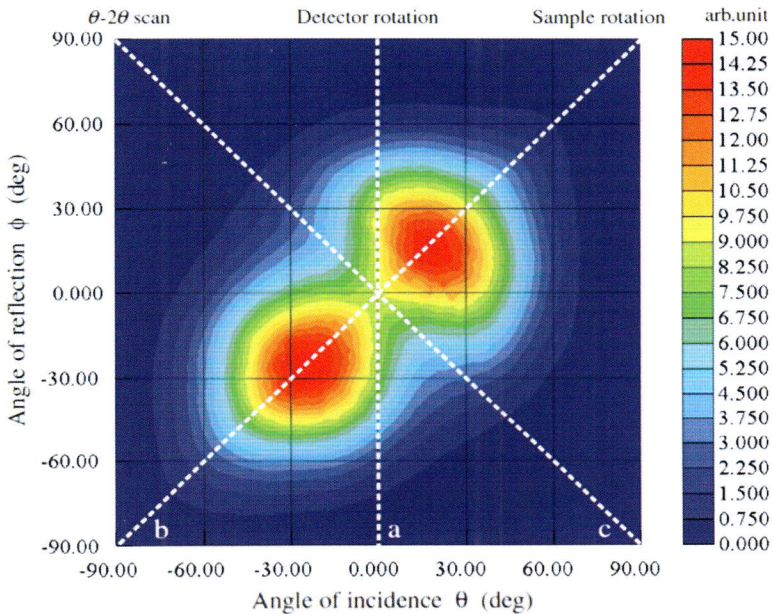

Figure 7.30 θ–ϕ map for the reflection from the *M. rhetenor* wing at the wavelength of 480 nm under TE polarization. Dashed lines a), b) and c) indicate the angular scans corresponding to detector rotation, sample rotation, and θ–2θ scan, respectively. (Reproduced from [Kambe *et al.* (2011)] with permission.)

Further, θ–2θ scan is expressed by $-45°$ inclined line shown as a dashed line c).

Comparing these with the θ–ϕ map measured at 480 nm, it is quite easy to understand their apparent patterns. Since the scattering properties of the *Morpho* butterfly wing is characterized by the twin spots lying on a 45° inclined line, the detector rotation method only probes the foot parts of the twin spots, while the θ–2θ scan does only a valley between the spots. In contrast, the sample rotation method probes the most important parts of the reflection characteristics by passing each center of the twin spots.

The above experiments have clarified a quite peculiar result that the wing of *M. rhetenor* displays strong retroreflection capability in a blue region. To solve this mystery, we re-investigate the electron microscopic image of the *M. rhetenor* scale in detail. It is clear that the retroreflection is somehow related to the alternate shelf structure shown in Fig. 7.27. The presence of alternate shelf structure was already reported in the 1980s [Ghiradella (1984)], and was often taken into account in various computational models [Plattner (2004)]. With inspecting shelf structure more carefully, we have noticed the shelves on the right and left sides has a common root, which is flatted and slightly inclines to form the alternate shelves. Thus, the flattened shelf leans to the left so that the right shelf is always higher than the left one when one faces the root of scale. Thus, the shelf structure is essentially asymmetric in itself and as a whole on a scale level.

In order to investigate the relation between these structural features and the peculiar retroreflection properties, the analytical approach is anyway difficult, because the microstructure is quite complicated and moreover involves a considerable amount of irregularity. It is of a great interest to elucidate which part of the above complicated structure actually contributes to the production of the *Morpho* blue and also to the retroreflection capability. The FDTD method is quite effective to analyze this type of complex problem.

First, we have attempted to investigate the angle-dependent reflection from an intact structure obtained directly from a transmission electron micrograph, as shown in Fig. 7.31a. The result is shown in Fig. 7.31b, where we show the angular dependence obtained under normal incidence at several wavelengths between 400 and 550 nm. Although the overall intensity is actually increased around a blue region of 480 nm, the large irregularity found in the angular dependence disturbs any quantitative interpretation. Since the intact model in Fig. 7.31a seems to assume irregularities both in ridge structure and its arrangement, we select four ridges labeled as 1–4 and calculate the angular dependence for each ridge, which is shown in Fig. 7.31c. It is clear that the angular dependence is largely dependent on ridge employed so that a large amount of irregularity seems to be involved even in each ridge. Thus, even if the FDTD method is believed to be an all-round player, the sampling number is quite important to investigate the microstructure involving the irregularity. In the present case, it is clear that a sampling number

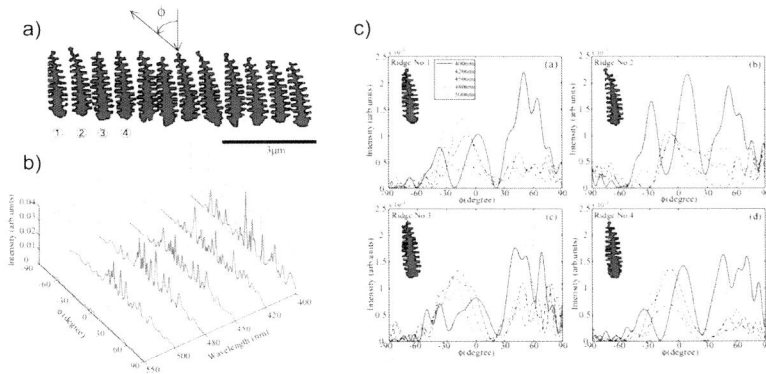

Figure 7.31 (a) Actual microstructure obtained by binarizing the TEM data for the transverse cross section of a *M. rhetenor* ground scale. Scale bar is 3 μm. (b) Angular dependence of the reflection intensities at the wavelengths of 400, 420, 450, 480, 500, and 550 nm calculated for the structure shown in (a) under normal incidence. (c) Angular dependence of the reflection for various wavelengths from 400 to 500 nm in the TM mode. The ridge number shown in each figure corresponds to the ridge in (a). The calculations were performed with a mesh size of 10×10 nm^2. (Reproduced from [Zhu *et al.* (2009)] with permission.)

of ridges is so small that we cannot attain the statistical average on their structures and hence on their optical properties.

However, it is natural that as a sampling number is increased, a computational load will grow extremely as well. Thus, to bring out a correct answer from the *Morpho* wing, it is anyway necessary to make a simple model that does not accompany irregularity and yet contains the essential part of the *Morpho* blue. Inspecting the microstructure minutely, we presume that the retroreflection originates from an alternately stacked shelf structure. To confirm this hypothesis, we consider the following two simple models: One is a flat multilayer and the other is an alternating structure. The former model consists of totally nine flat shelves with a width of 300 nm and a thickness of 55 nm, which are equidistantly arrayed with an interval of 205 nm. The refractive index of the shelf is assumed to be 1.55, while that of the surrounding medium is set at 1.00. In the latter model, each shelf in the former model is cut off at a middle of the shelf and is shifted down by 118 nm to mimic the alternating

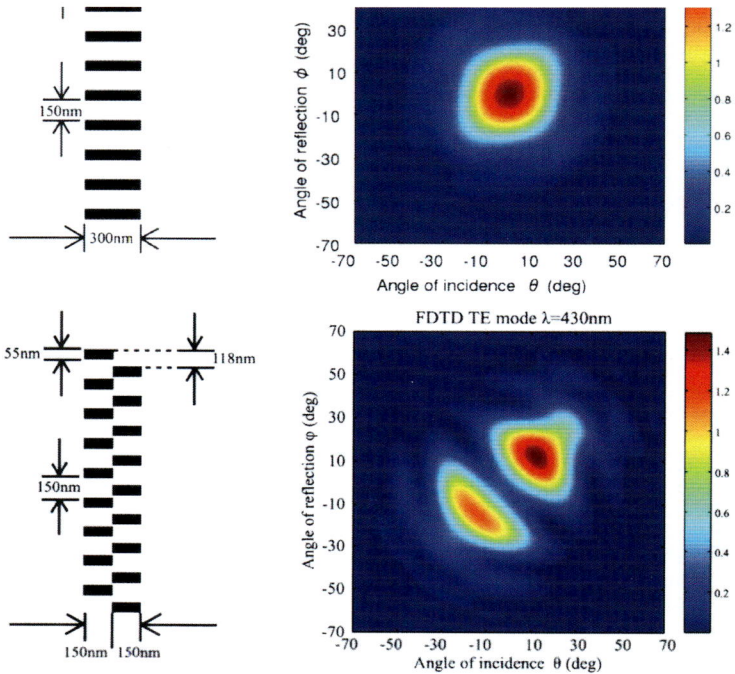

Figure 7.32 Left: Flat (upper) and alternating (lower) multilayer models. Right: θ–ϕ maps at 430 nm under TE polarization, calculated by the NS-FDTD method. (Reproduced from [Kambe *et al.* (2011)] with permission.)

shelf structure, which we call alternating multilayers hereafter. The amount of the shift is chosen to agree with a half of the optical path length of the sum of the shelf and spacing.

The θ–ϕ maps at peak positions of the reflection bands calculated by the NS-FDTD method are shown in Fig. 7.32. It is clear that the flat model shows only a single spot in the θ–ϕ map, which is located at the origin, while in the alternating multilayers, twin spots are clearly seen. The spots are not exactly the same, but the spot in the left side somewhat elongates. This is natural that the alternating structure considered here is not symmetric with respect to a center line of the structure. From the above calculation, it is strongly suggested that the alternating multilayers gives twin spots with strong suppression of reflection toward a normal direction.

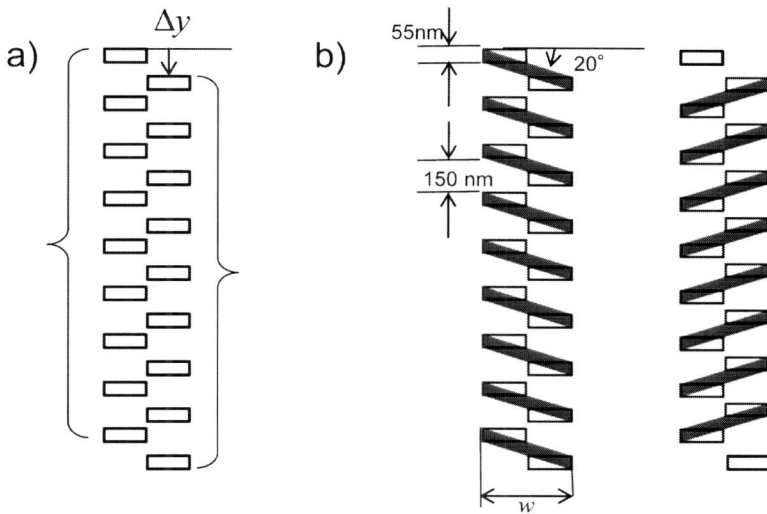

Figure 7.33 Comprehensive models for alternating shelf structure. (a) Alternating multilayers are regarded as two multilayers with a finite width, one of which is shifted vertically by an amount of Δy, and (b) that regarded as an ensemble of oblique shelves made by connecting the shelves at both ends.

We will further investigate the origin of this phenomenon using the following two intuitive models (Fig. 7.33): One is based on the fact that alternating multilayers are regarded as two multilayers with a finite width, one of which is shifted vertically by an amount of Δy. If Δy is equal to $\lambda/4$, where λ is a wavelength of the incident light, this type of structure will strongly suppress the reflection toward a normal direction, because the reflected light from both multilayers will eventually disappear owing to destructive interference. The other model is to regard the alternating multilayers as an ensemble of oblique shelves made by connecting both ends of the right and left shelves.

As for the first model, according to a usual theory of light interference, the scattered/reflected light intensity from two exactly the same structures separated by Δx and Δy in the Cartesian coordinates is generally expressed by (Exercise (5))

$$I(\theta, \phi) = I_s(\theta, \phi) \left| 1 + e^{ik(u\Delta x - v\Delta y)} \right|^2$$

$$= 4I_s(\theta, \phi) \cdot \cos^2 \left[\frac{k(u\Delta x - v\Delta y)}{2} \right], \qquad (7.93)$$

where $u = \sin\theta + \sin\phi$ and $v = \cos\theta + \cos\phi$ with $I_s(\theta, \phi)$ the scattered light intensity due to a single structure and k the wave vector of incident light. From this relation, it is expected that when $k(u\Delta x - v\Delta y)/2 = (2m + 1)\pi/2$, that is, $u\Delta x - v\Delta y = (2m + 1)\lambda/2$, the scattering intensity anyway vanishes, regardless of the functional form of $I_s(\theta, \phi)$. Under normal incidence and reflection, the relations of $u = 0$ and $v = 2$ hold so that the above condition is reduced to $\Delta y = -(2m + 1)\lambda/4$. Thus, under the conditions of $\Delta y = \pm\lambda/4, \pm3\lambda/4, \cdots$, the strong suppression to the normal direction will occur.

From the above consideration, the optical property of alternating multilayers is well expressed by a product of that of a one-sided shelf structure and the interference term expressed by the $\cos^2$ function. Using the parameters of $\Delta x = 300/2$ (nm) and $\Delta y = -118$ (nm), we have calculated each of the terms at the wavelengths of 370, 400 and 430 nm, which are summarized in Fig. 7.34. The first and second columns, (a) and (b), are the results for one-sided shelf structure and the $\cos^2$ term, respectively, while the last column, (d), is that for the alternating multilayers. It seems that the pattern in column (a) is sampled by the pattern (b), which results in the pattern (d). In fact, as shown in the column (c), the product of the columns of (a) and (b) partly reproduces the column (d).

Thus, the $\theta - \phi$ map for alternating multilayers seems to be expressed at least partly by a product of the optical properties of a one-sided multilayer and the $\cos^2$ term. Since this treatment is justified only when multiple scattering between the structural units is neglected, clear discrepancy is also observed, if one sees the pattern at 400 nm. Anyway, the retroreflection in this case comes from an appropriate angular broadening due to a single structure and a condition of $\cos^2[k(u\Delta x - v\Delta y)/2] \sim 1$. The latter condition further leads to $k(u\Delta x - v\Delta y)/2 \sim m\pi$. After all, the conditions of $m = 0$ and $m = 1$ contribute to the pattern in the present case of $\Delta x = 150$ (nm) and $\Delta y = \lambda/4$. Hence, the retroreflection is interpreted in terms of the zeroth- and first-order interference due to two multilayers placed separately.

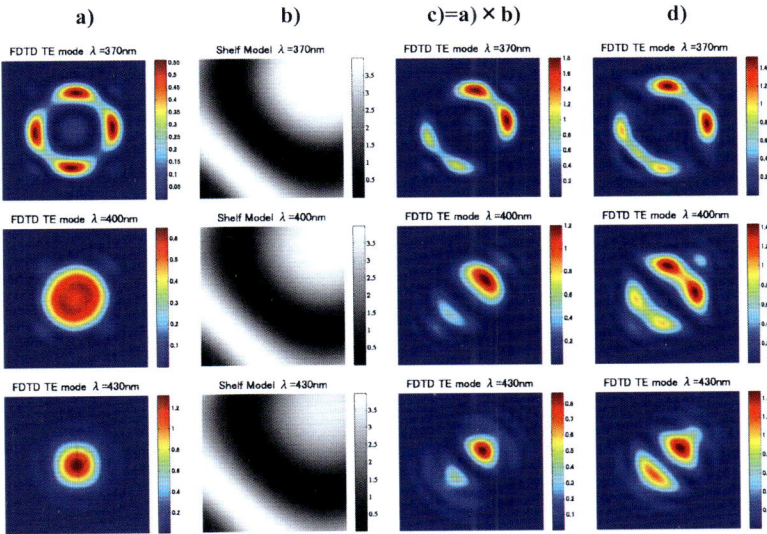

Figure 7.34 θ–ϕ maps calculated for (a) a one-sided shelf structure, (b) $\cos^2[k(u\,\Delta x - v\,\Delta y)/2]$, and (d) full alternating multilayers. The vertical and horizontal axes in each map correspond to ϕ and θ, respectively, with the angular range from -70 to $70°$ for p (TE)-polarization. The third column (c) is obtained by multiplying (a) and (b). The wavelengths employed are 370, 400 and 430 nm from top to bottom. (Reproduced from [Kambe *et al.* (2011)] with permission.)

As for the second model, one may think if both ends of the alternating layers are connected with each other, an obliquely inclined multilayer will be constructed. Further, if both ends with different combinations are connected, two inversely inclined multilayers will be virtually created (see Fig. 7.33b), which will contribute to retroreflection toward two directions normal to the inclined multilayers. However, we should be careful enough to examine this idea, because if the width of the inclined layer is too small, such a multilayer will become a simple diffraction grating aligned vertically. Thus, the direction of reflection normal to the inclined multilayer is no longer justified. It is considered that an obliquely inclined multilayer is one of the limits of an infinitely large width, while a diffraction grating is the other limit of an infinitely small width. Here, we investigate the difference of their reflection

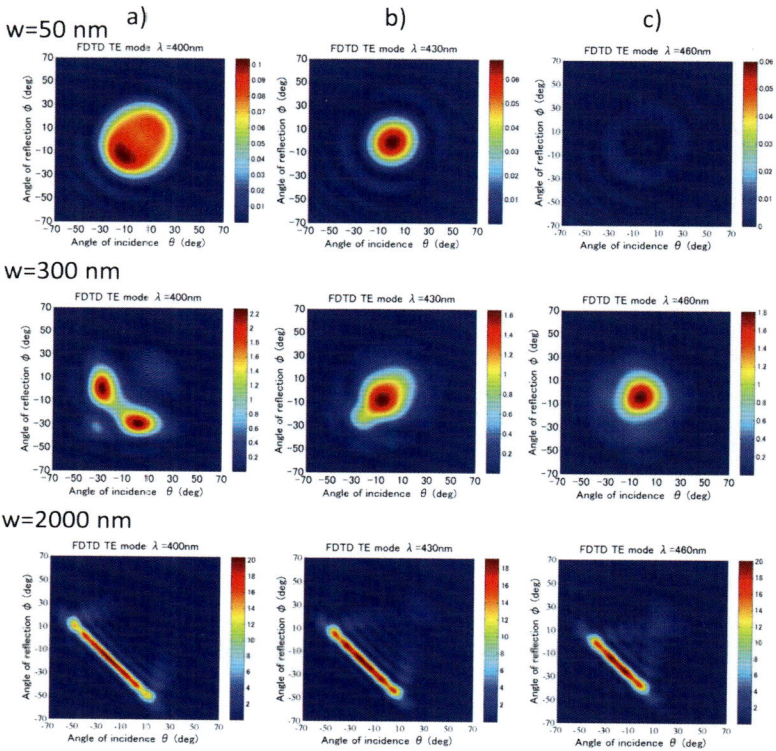

Figure 7.35 θ–ϕ maps calculated for obliquely inclined shelves (the left type in Fig. 7.31b) with the widths of (a) 50, (b) 300, and (c) 2000 nm for p(TE)-polarization. The inclined shelves have a thickness of 55 nm with the interval of 150 nm, both of which are measured vertically. The inclination angle is basically determined by connecting both ends of the alternating shelf structure, which is then approximated by an angle of 20°. The wavelengths are (a) 400, (b) 430 and (c) 460 nm from left to right. The vertical and horizontal axes in each map correspond to ϕ and θ, respectively, with the angular range from −70 to 70°. (Reproduced from [Kambe et al. (2011)] with permission.)

properties by calculating the θ–ϕ map for various widths by means of the NS-FDTD method.

The results for a multilayer leaning to the right are shown in Fig. 7.35. In this figure, the upper three diagrams are the θ–ϕ maps for $w = 50$ (nm), while the middle and lower ones are those for $w =$

300 and 2000 (nm), respectively. The upper one is a representative of a diffraction grating, while the lowest one is that of the inclined multilayer and the middle one comes from our model corresponding to alternating multilayers. It is natural that for a large width of $w = 2000$ (nm), a specular reflection is prominent, which satisfies the interference condition

$$2\left\{nd\cos(\theta_1 + \alpha) + (a - d)\cos(\theta + \alpha)\right\}\cos\alpha = m\lambda, \qquad (7.94)$$

and further that ϕ should satisfy the relation $\phi = -\theta - 2\alpha$, where n, d, a, θ_1 and α are the refractive index and thickness of an inclined layer measured vertically, the interval of layers measured vertically, an angle of refraction and an inclination angle of the layers, respectively, with m an integer. Further, under the conditions of $n = 1.55$, $d = 55$ (nm), $a = 205$ (nm) and $\alpha = 20°$, the first-order multilayer reflection is expected to occur at 442 nm for $\theta = \phi = -\alpha$ and below this wavelength, twin spots indicating the presence of a specular reflection will appear. However, owing to a large number of layers and a large difference in the refractive indices, instead of giving twin spots, it appears as a slender band lying on a straight line of $\phi = -\theta - 2\alpha$. Its maximum angular range is further restricted within a range from $\theta = -77$ to $37°$ owing to the Brewster's law.

On the other hand, if the width becomes too small, the inclined multilayer comes close to a vertically aligned diffraction grating, whose reflection properties are characterized by a factor of $\{\sin(Makv/2)/\sin(kav/2)\}^2$ with M a number of layers. Hence, it shows spots at angles satisfying $kav/2 = m\pi$, that is, $\cos\theta + \cos\phi = m\lambda/a$, which appears as a quasi-circle in θ–ϕ map centered at the origin. In case of a diffraction grating of an infinite size with a period of 205 nm, the longest wavelength to generate the first-order diffraction spot occurs at 410 nm under retroreflection configuration of $\theta = \phi = 0°$. However, when the size of the grating is not infinite, a weak diffraction spot appears around the origin of θ–ϕ map even at longer wavelengths than 410 nm. The calculated result just supports the speculation.

In view of the behavior in each limit case, the reflection properties for the width of $w = 300$ (nm) is considered to be in an intermediate case between these two extremes, because clear twin spots, which are characteristic of multilayer, appear at 400 nm, which are not on a line of $\phi = -\theta - 2\alpha$, but on a line slightly shifted to the origin. Further, the retroreflection takes place around 460 nm and tends to approach the origin when the wavelength becomes longer. Thus, the obliquely inclined multilayer partly explains the presence of retroreflection but is not complete when the layer width is not so large. From a biological viewpoint, however, this half-finished work will be beneficial to the considerable extension of the angular range of reflection, which may be inevitable for a struggle for existence of this butterfly.

So far, the analysis is only confined within a single ridge equipped with an alternating shelf structure and also within a case of s (TM)-polarization as incident light. We will advance the understanding to a more general case where many ridges are arrayed almost equidistantly. However, we will soon encounter the difficulties: If the ridges are arrayed regularly, this structure will give a simple diffraction grating and diffraction spots corresponding to the spacing of ridges are clearly seen in the angular dependence of the reflection. On the other hand, if the ridges are regularly arrayed but their heights are irregular as in an actual scale, an irregular reflection pattern will come out again.

These speculations are easily confirmed by considering a simple model in which 10 sets of alternating multilayers are arrayed equidistantly with an interval of 700 nm, while their heights are distributed randomly within a range of $\pm\Delta y_{\text{max}}$, as shown in Fig. 7.36. The result is shown in Fig. 7.37, which demonstrates that the angular dependence of reflection is expressed as an ensemble

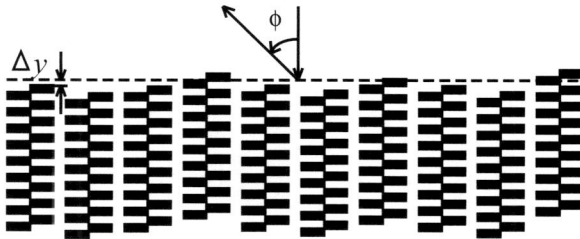

Figure 7.36 Model for an array of alternating multilayers with random heights. The deviation from a mean value of the ridge height is denoted as Δy. (Reproduced from [Zhu *et al.* (2009)] with permission.)

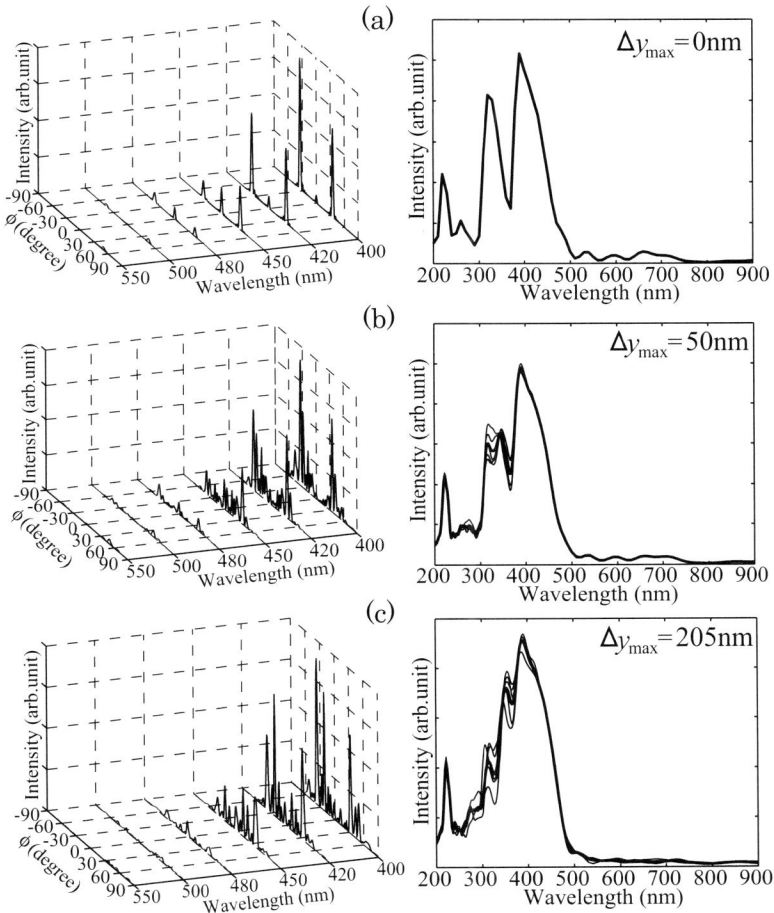

Figure 7.37 Angular dependence of the reflection intensities (left) and angle-integrated reflection spectra (right) of an array of alternating multilayers with irregularity in the ridge height. The irregularity was introduced by giving a uniform random number within a maximum range of $\pm\Delta y_{max}$, where Δy_{max} = (a) 0, (b) 50 and (c) 205 (nm). In (b) and (c), the angle-integrated spectra were calculated for five series. The five curves and their average are plotted as thin and thick curves, respectively. The calculations were performed with a mesh size of 10×10 nm^2. (Reproduced from [Zhu *et al.* (2009)] with permission.)

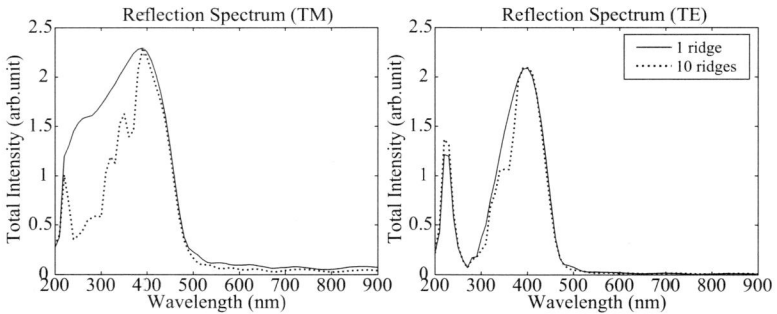

Figure 7.38 Angle-integrated reflection spectra in (a) TM and (b) TE modes for a single alternating multilayer (solid line) and an array of ten alternating multilayers with the irregularity in height with $\Delta y_{max} = 205$ (nm) (dotted line). The reflection spectrum for the array of ten alternating multilayers was obtained by averaging five series of calculations. (Reproduced from [Zhu *et al.* (2009)] with permission.)

of sharp peaks when no height distribution is present, while with increasing irregularity, the angular dependence begins to show severe spike-shaped noises. Thus, it seems to be a tough work to bring out information from the reflection pattern in either case.

To overcome this, we calculate the intensity of reflected light from an array of ridges, which is integrated over all reflection angles. This approach considerably reduces the noise in the spectrum. The results are shown in the right column of Fig. 7.37. These figures are obtained by generating random numbers 5 times in each case, and by being superposed on each figure. In spite of adding the irregularity to the ridge height, the noise appearing in the reflection spectrum is considerably reduced and is confined only within a range of 300–400 nm. It is also noticeable that a peak appearing in a regular array at 340 nm gradually decreases with increasing irregularity.

We compare the reflection spectrum thus obtained for $\Delta y_{max} = 205$ (nm) with that for a single ridge consisting of alternating multilayers. The result is shown in Fig. 7.38. It is quite interesting that a ultraviolet region of 220–380 nm of the reflection spectrum for a single ridge is largely enhanced as compared with an array of ridges. We calculate the case of *p* (TE)-polarization for the comparison. It is surprising that these two spectra almost completely agree with each other and further both spectra measured under *s*- and *p*-polarizations with irregularity included are quite similar with respect to the peak position and the spectral line shape. Thus, the effect to make an array on the optical properties only affects the reflection under *s*-polarization, and is to strongly suppress the intensity of ultraviolet region.

This suppression comes from the scattering of reflected light due to adjacent ridges, which is particularly remarkable for *s*-polarization, since the direction of emission in *s*-polarization is uniformly distributed over a whole angular range, while that in *p*-polarization is restricted within angles around the normal direction. This effect is more effective at shorter wavelengths, because the reflection predominantly leans to large angles when shorter wavelengths are employed under normal incidence. As a consequence, the reflection spectrum for *s*-polarization is considerably affected by the scattering due to adjacent ridges, which prevents the reflection at shorter wavelengths. On the other hand, the reflection spectrum in *p*-polarization is rather unaffected by adjacent ridges, and the reflection spectrum seems to be solely determined by a single alternating multilayer. Thus, a clear difference in mechanisms between the two polarizations is clarified, even though the obtained spectra are similar to each other.

In summary, we have shown the effectiveness of the FDTD method by showing its application to the reflection properties of a multilayer with an finite width and the microstructure in the *Morpho* butterfly as examples. The FDTD method particularly demonstrates its ability in case the electromagnetic field interacting with complicated microstructures is calculated, which is far beyond analytical approach. On the contrary, the major problem concerning this type of calculation method is that it immediately gives an answer to the problem in one hand, but do not provide any physical basis for the answer in the other. Thus, it is absolutely necessary to construct an appropriate model to bring out the mechanism without losing the essence of the structure.

Exercises

(1) Derive the expression for the Mur second-order boundary condition of Eq. (7.28).

(2) Derive the equivalent current expressed by Eq. (7.40).

(3) Derive Eqs. (7.64) and (7.65).

(4) Confirm that Eq. (7.78) actually holds.

(5) Equation (7.93) is a general relation for the interference of light scattered by two exactly the same structures separated by Δx and Δy in the Cartesian coordinate. Derive this relation.

Appendix A

Proofs of Theorems

A.1 Gauss's Theorem

Consider a closed space V enclosed by a surface S, as shown in Fig. A.1. Let us divide V regularly into a lot of infinitely small rectangular parallelepipeds with the side lengths of Δx, Δy and Δz and focus on a certain rectangular parallelepiped among them. At first, we will evaluate a derivative of E_x with respect to x within this small volume, as shown in the right figure. Then, it is proved that the volume integration can be transformed into the surface integration of two faces perpendicular to the x axis as follows:

$$\frac{\partial E_x}{\partial x} \Delta x \Delta y \Delta z \rightarrow \frac{E_x(x + \Delta x, y, z) - E_x(x, y, z)}{\Delta x} \Delta x \Delta y \Delta z$$
$$= \{E_x(x + \Delta x, y, z) - E_x(x, y, z)\} \Delta y \Delta z$$
$$= \{\mathbf{E}(x + \Delta x, y, z) \cdot \mathbf{n}' + \mathbf{E}(x, y, z) \cdot \mathbf{n}\} \Delta y \Delta z,$$

where $\mathbf{n}$ and $\mathbf{n}'$ are unit vectors normal to the faces and are assumed to direct outward. If we also calculate the derivatives with respect to y and z, and sum up all of these results, we obtain

$$\sum \mathbf{E} \cdot \mathbf{n} \Delta S = \left(\frac{\partial E_x}{\partial x} + \frac{\partial E_y}{\partial y} + \frac{\partial E_z}{\partial z} \right) \Delta x \Delta y \Delta z = (\nabla \cdot \mathbf{E}) \Delta V. \quad \text{(A.1)}$$

Similar calculations are performed for all the rectangular parallelepipeds filling the space. Then, the unit vectors $\mathbf{n}$ faced each other for adjacent parallelepipeds are opposite in direction and the two contributions will be canceled, because the direction of a vector $\mathbf{E}$ is the same. Thus, all the contributions from the small surface integrations within the space are canceled out and only the

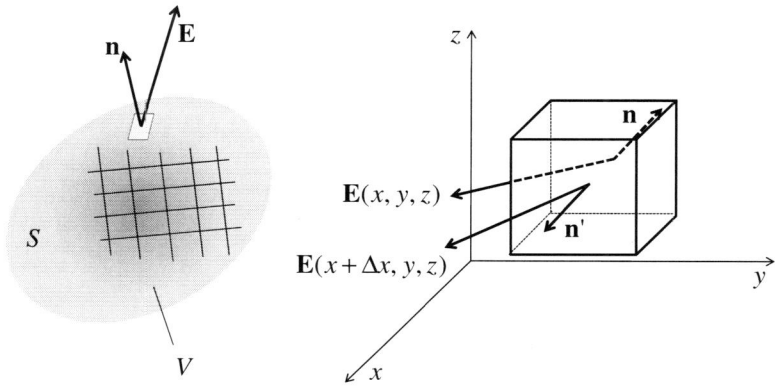

Figure A.1 Gauss's theorem.

contribution of the surface S will remain. On the other hand, the right-hand side of Eq. (A.1) becomes the integration over the total volume V. Therefore, Eq. (A.1) becomes

$$\int_S \mathbf{E} \cdot \mathbf{n} \, dS = \int_V \nabla \cdot \mathbf{E} \, dV, \tag{A.2}$$

which is called *Gauss's theorem*.

A.2 Green's Theorem

The proof of the Green's theorem needs Gauss's theorem described above. Similar to Gauss's theorem, consider a closed space V enclosed by a surface S, as shown in Fig. A.1. In this space, Gauss's theorem states

$$\int_V \nabla \cdot \mathbf{A} \, dV = \int_S \mathbf{A} \cdot \mathbf{n} \, d\sigma, \tag{A.3}$$

where $\mathbf{n}$ is a unit vector normal to the surface S and is assumed to direct outward. $\mathbf{A}$ is an arbitrary vector defined in this space. If $\mathbf{A}$ is expressed by two scalar functions, u and v, as

$$\mathbf{A} = u\nabla v, \tag{A.4}$$

then the following relations are derived:

$$\nabla \cdot \mathbf{A} = \nabla \cdot (u\nabla v) = \nabla u \cdot \nabla v + u\nabla^2 v,$$

$$\mathbf{A} \cdot \mathbf{n} = u\mathbf{n} \cdot \nabla v \equiv u\frac{\partial v}{\partial \mathbf{n}}.$$

By using these relations, Eq. (A.3) is transformed into

$$\int_V u\nabla^2 v\,\mathrm{d}V + \int_V (\nabla u \cdot \nabla v)\mathrm{d}V = \int_S u\frac{\partial v}{\partial \mathbf{n}}\mathrm{d}\sigma. \qquad (A.5)$$

On the other hand, if we put $\mathbf{A} = v\nabla u$, then the following relation holds:

$$\int_V v\nabla^2 u\,\mathrm{d}V + \int_V (\nabla v \cdot \nabla u)\mathrm{d}V = \int_S v\frac{\partial u}{\partial \mathbf{n}}\mathrm{d}\sigma. \qquad (A.6)$$

Subtracting Eq. (A.6) from Eq. (A.5) , we obtain

$$\int_V (u\nabla^2 v - v\nabla^2 u)\mathrm{d}V = \int_S \left(u\frac{\partial v}{\partial \mathbf{n}} - v\frac{\partial u}{\partial \mathbf{n}} \right) \mathrm{d}\sigma, \qquad (A.7)$$

which is called *Green's theorem*.

A.3 Bloch's Theorem

In order to theoretically treat an electromagnetic field in photonic crystal, Bloch's theorem has been often employed as in ordinary crystal whose lattice constant is of an atomic size (hereafter, we call it atomic crystal). Although a lot of works have been devoted to give a proof for this theorem, most of them are concerned with atomic crystal and little is known for photonic crystal. Therefore, here we will derive it again along with a case of photonic crystal.

In photonic crystal, the electric field of light satisfies the following wave equation:

$$\nabla \times (\nabla \times \mathbf{E}(\mathbf{r})) = \epsilon_0\epsilon(\mathbf{r})\mu_0\omega^2\mathbf{E}(\mathbf{r}), \qquad (A.8)$$

where $\epsilon(\mathbf{r})$ is a relative permittivity that is a periodic function of spatial coordinate and can be expanded using reciprocal vectors as

$$\epsilon(\mathbf{r}) = \sum_{\mathbf{G}} \epsilon_{\mathbf{G}}e^{i\mathbf{G}\cdot\mathbf{r}}. \qquad (A.9)$$

We further divide it into two, a homogeneous part and that dependent on the periodicity of crystal, which are expressed as

$$\epsilon(\mathbf{r}) = \epsilon^{(0)} + \sum_{\mathbf{G}\neq 0} \epsilon_{\mathbf{G}}e^{i\mathbf{G}\cdot\mathbf{r}} \equiv \epsilon^{(0)} + \epsilon^{(1)}(\mathbf{r}), \qquad (A.10)$$

where $\epsilon^{(1)}(\mathbf{r})$ is a periodic part of the relative permittivity and satisfies a relation

$$\epsilon^{(1)}(\mathbf{r} + \mathbf{R}) = \epsilon^{(1)}(\mathbf{r}), \qquad (A.11)$$

with **R** being a translation vector of the photonic crystal.

Using this expression, we transform Eq. (A.8) into

$$\left\{ \nabla \times \nabla \times -\epsilon_0 \epsilon^{(1)}(\mathbf{r}) \mu_0 \omega^2 \right\} \mathbf{E}(\mathbf{r}) = \epsilon_0 \epsilon^{(0)} \mu_0 \omega^2 \mathbf{E}(\mathbf{r}), \qquad \text{(A.12)}$$

and put an operation appearing in the left-hand side as

$$\mathcal{L}(\mathbf{r}) \equiv \left\{ \nabla \times \nabla \times -\epsilon_0 \epsilon^{(1)}(\mathbf{r}) \mu_0 \omega^2 \right\}. \qquad \text{(A.13)}$$

Then, the above equation results in the following eigenvalue problem:

$$\mathcal{L}(\mathbf{r})\mathbf{E}(\mathbf{r}) = \epsilon_0 \epsilon^{(0)} \mu_0 \omega^2 \mathbf{E}(\mathbf{r}). \qquad \text{(A.14)}$$

This expression is formally equivalent to the Schödinger equation for an electron in atomic crystal. Further, if **R** is a translation vector of the photonic crystal, then the relation

$$\mathcal{L}(\mathbf{r} + \mathbf{R}) = \mathcal{L}(\mathbf{r}) \qquad \text{(A.15)}$$

holds owing to the periodicity of the permittivity.

Consider a translation operator $\mathcal{T}_\mathbf{R}$ that transfers a function of **r** spatially by an amount expressed by a vector **R**. Thus, by using the translation operator and assuming **R** is a translation vector of the photonic crystal, it is proved that the following relation holds:

$$\mathcal{T}_\mathbf{R}\mathcal{L}(\mathbf{r})\mathbf{E}(\mathbf{r}) = \mathcal{L}(\mathbf{r} + \mathbf{R})\mathbf{E}(\mathbf{r} + \mathbf{R}) = \mathcal{L}(\mathbf{r})\mathbf{E}(\mathbf{r} + \mathbf{R}) = \mathcal{L}(\mathbf{r})\mathcal{T}_\mathbf{R}\mathbf{E}(\mathbf{r}).$$
$$\text{(A.16)}$$

Therefore, a commutation relation $[\mathcal{L}(\mathbf{r}), \mathcal{T}_\mathbf{R}] = 0$ holds between $\mathcal{L}(\mathbf{r})$ and $\mathcal{T}_\mathbf{R}$. Thus, it is generally proved that $\mathcal{L}(\mathbf{r})$ and $\mathcal{T}_\mathbf{R}$ have a common eigenfunction, which we put as **E**(**r**) here. Hence, we can write

$$\mathcal{T}_\mathbf{R}\mathbf{E}(\mathbf{r}) = C(\mathbf{R})\mathbf{E}(\mathbf{r}), \qquad \text{(A.17)}$$

where $C(\mathbf{R})$ is a eigenvalue for a operator $\mathcal{T}_\mathbf{R}$.

If we operate $\mathcal{T}_{\mathbf{R}'}$ from the left of the above equation, then we obtain

$$\mathcal{T}_{\mathbf{R}'}\mathcal{T}_\mathbf{R}\mathbf{E}(\mathbf{r}) = \mathcal{T}_{\mathbf{R}'}C(\mathbf{R})\mathbf{E}(\mathbf{r}) = C(\mathbf{R}')C(\mathbf{R})\mathbf{E}(\mathbf{r}), \qquad \text{(A.18)}$$

where $\mathbf{R}'$ is another translation vector in the photonic crystal. On the other hand, from the feature of the translation operator, the following relation holds naturally:

$$\mathcal{T}_{\mathbf{R}'}\mathcal{T}_\mathbf{R}\mathbf{E}(\mathbf{r}) = \mathcal{T}_{\mathbf{R}'+\mathbf{R}}\mathbf{E}(\mathbf{r}) = C(\mathbf{R}' + \mathbf{R})\mathbf{E}(\mathbf{r}). \qquad \text{(A.19)}$$

Thus, it is proved that the relation

$$C(\mathbf{R}' + \mathbf{R}) = C(\mathbf{R}')C(\mathbf{R}) \tag{A.20}$$

holds generally.

Repeating such a procedure, we can express an arbitrary translation operation in terms of those for three primitive translation vectors. When $\mathbf{R}$ is expressed by the sum of primitive translation vectors as $\mathbf{R} = n_1\mathbf{a}_1 + n_2\mathbf{a}_2 + n_3\mathbf{a}_3$, then the following relation holds generally:

$$C(\mathbf{R}) = C^{n_1}(\mathbf{a}_1)C^{n_2}(\mathbf{a}_2)C^{n_3}(\mathbf{a}_3), \tag{A.21}$$

where $\mathbf{a}_j$ and n_j are a primitive translation vector and an integer, respectively, with $j = 1, 2, 3$. This type of relation holds only when $C(\mathbf{a}_j)$ takes a form of

$$C(\mathbf{a}_j) = \exp[2\pi i x_j], \tag{A.22}$$

where x_j is a real number[a].

Using this expression, we can rewrite Eq. (A.21) as follows:

$$\begin{aligned} C(\mathbf{R}) &= \exp[2\pi i(x_1 n_1 + x_2 n_2 + x_3 n_3)] \\ &= \exp[i(x_1\mathbf{G}_1 + x_2\mathbf{G}_2 + x_3\mathbf{G}_3) \cdot (n_1\mathbf{a}_1 + n_2\mathbf{a}_2 + n_3\mathbf{a}_3)] \\ &= \exp[i\mathbf{k} \cdot \mathbf{R}], \end{aligned} \tag{A.23}$$

where $\mathbf{G}_j$ is a reciprocal vector corresponding to $\mathbf{a}_j$ and we have employed a relation $\mathbf{G}_j \cdot \mathbf{a}_l = 2\pi\delta_{jl}$. We have further put $\mathbf{k} = x_1\mathbf{G}_1 + x_2\mathbf{G}_2 + x_3\mathbf{G}_3$, where $\mathbf{k}$ has a meaning of wave vector. The reason to put $\mathbf{k}$ as above is based on a fact that the factor, Eq. (A.23), defined as an eigenvalue for the translational operator should also become that for $\mathcal{L}(\mathbf{r})$, in which $\mathbf{k}$ corresponds to a wave vector.

Therefore, the following relation holds

$$\mathcal{T}_{\mathbf{R}}\mathbf{E}(\mathbf{r}) = \mathbf{E}(\mathbf{r} + \mathbf{R}) = \exp[i\mathbf{k} \cdot \mathbf{R}]\mathbf{E}(\mathbf{r}), \tag{A.24}$$

which is one of the important expressions for *Bloch's theorem*. Furthermore, if we put

$$\mathbf{u}(\mathbf{r}) = \exp[-i\mathbf{k} \cdot \mathbf{r}]\mathbf{E}(\mathbf{r}), \tag{A.25}$$

then a vector function $\mathbf{u}(\mathbf{r})$ is proved to have the same periodicity, because

$$\mathbf{u}(\mathbf{r} + \mathbf{R}) = \exp[-i\mathbf{k} \cdot (\mathbf{r} + \mathbf{R})]\mathbf{E}(\mathbf{r} + \mathbf{R}) = \exp[-i\mathbf{k} \cdot \mathbf{r}]\mathbf{E}(\mathbf{r}) = \mathbf{u}(\mathbf{r}).$$

---

[a]Actually, it is not necessary to define x_j as a real number. However, to produce a propagating wave without attenuation, the condition of real number is necessary.

Thus, $\mathbf{E}(\mathbf{r})$ is expressed as

$$\mathbf{E}(\mathbf{r}) = \exp[i\mathbf{k} \cdot \mathbf{r}]\mathbf{u}(\mathbf{r}), \qquad (A.26)$$

which means that the electric field in a material having a periodic permittivity is generally expressed as a product of a function with the same periodicity and an exponential factor. This is another expression for Bloch's theorem.

Finally, we expand the function $\mathbf{u}(\mathbf{r})$ in terms of the reciprocal vectors of photonic crystal as

$$\mathbf{u}(\mathbf{r}) = \sum_{\mathbf{G}} \hat{\mathbf{u}}_{\mathbf{G}} \exp[i\mathbf{G} \cdot \mathbf{r}]. \qquad (A.27)$$

Then, inserting this into Eq. (A.26) yields

$$\mathbf{E}(\mathbf{r}) = \exp[i\mathbf{k} \cdot \mathbf{r}]\mathbf{u}(\mathbf{r}) = \sum_{\mathbf{G}} \hat{\mathbf{u}}_{\mathbf{G}} \exp[i(\mathbf{k} + \mathbf{G}) \cdot \mathbf{r}], \qquad (A.28)$$

which is another expression of Bloch's theorem.

Bibliography

Amos, R. M., Rarity, J. G., Tapster, P. R., and Shepherd, T. J. (2000). Fabrication of large-area face-centered-cubic hard-sphere colloidal crystals by shear alignment, *Phys. Rev. E* **61**, 2929–2935.

Anderson, T. F., and Richards, A. G., Jr. (1942). An electron microscope study of some structural colors of insects, *J. Appl. Phys.* **13**, 748–758.

Aydin, C., Zaslavsky, A., Sonek, G. J., and Goldstein, J. (2002). Reduction of reflection losses in $ZnGeP_2$ using motheye antireflection surface relief structures, *Appl. Phys. Lett.* **80**, 2242–2244.

Bernard, G. D., and Miller, W. H. (1968). Interference filters in the corneas of Diptera, *Invest. Ophtalmol.* **7**, 416–434.

Bernhard, C. G., and Miller, W. H. (1962). A corneal nipple pattern in insect compound eyes, *Acta Physiol. Scand.* **56**, 385–386.

Bernhard, C. G., Miller, W. H., and Møller, A. R. (1965). The insect corneal ripple array. A biological broad-band impedance transformer acts as an antireflection coating, *Acta Physiol. Scand.* **63**(suppl. 243), 1–79.

Bernhard, C. G., Gemne, G., and Møller, A. R. (1968). Modification of specular reflexion and light transmission by biological surface structure - To see, to be seen or not to be seen, *Quart. Rev. Biophys.* **1**, 89–105.

Bernhard, C. G., Gemne, G., and Sällström, J. (1970). Comparative ultra-structure of corneal surface topography in insects with aspects on phylogenesis and function, *Z. Vergl. Physiol.* **67**, 1–25.

Berthier, S., Charron, E., and Da Silva, A. (2003). Determination of the cuticle index of the scales of the iridescent butterfly *Morpho Menelaus*, *Opt. Commun.* **228**, 349–356.

Berthier, S., Charron, E., and Boulenguez, J. (2006). Morphological structure and optical properties of the wings of Morphidae, *Insect Sci.* **13**, 145–157.

Berthier, S. (2007). *Iridescences: The Physical Colors of Insects* (Springer Science+Business Media, New York).

Biswas, R., Sigalas, M. M. Subramania, G., and Ho, K.-M. (1998). Photonic band gaps in colloidal systems, *Phys. Rev. B* **57**, 3701–3705.

Blanco, A. López, C., Mayoral, R., Miguez, H., Meseguer, F., Mifsud, A., and Herrero, J. (1998). CdS photoluminescence inhibition by a photonic structure, *Appl. Phys. Lett.* **73**, 1781–1783.

Blanco, A., Chomski, E., Grabtchak, S., Ibisate, M., John, S., Leonard, S. W., Lopez, C., Meseguer, F., Miguez, H., Mondla, J. P., Ozin, G. A., Toader, O., and van Driel, H. M. (2000). Large-scale synthesis of a silicon photonic crystal with a complete three-dimensional band gap near 1.5 micrometres, *Nature* **405**, 437–440.

Boden, S. A., and Bagnall, D. M. (2008). Tunable reflection minima of nanostructured antireflective surfaces, *Appl. Phys. Lett.* **93**, 133108-1-3.

Bogomolov, V. N., Gaponenko, S. V., Germanenko, I. N., Kapitonov, A. M., Petrov, E. P., Gaponenko, N. V., Prokofiev, A. V., Ponyavina, A. N., Silvanovich, N. I., and Samoilovich, S. M. (1997). Photonic band gap phenomenon and optical properties of artificial opals, *Phys. Rev. E* **55**, 7619–7625.

Bohren, C. F., and Huffman, D. R. (1983). *Absorption and Scattering of Light by Small Particles* (John Wiley & Sons, New York).

Bouligand, Y. (1965). Sur une disposition fibrillaire torsadée commune à plusieurs structures biologiques, *C. R. Acad. Sc. Paris* **261**, 4864–4867.

Born, M., and Wolf, E. (1959). *Principles of Optics* (Pergamon Press, London).

Brewster, D. (1845). *A Treatise on Optics* (Lea & Blanchard, Philadelphia).

Brillouin, L. (1922). Diffusion de la lumière et des rayons X par un corps transparent homogène (Diffusion of light and x-rays by a transparent homogeneous body), *Ann. Phys. (Paris)* **17**, 88–122.

Busch, K., and John, S. (1998). Photonic band gap formation in certain self-organizing systems, *Phys. Rev. E* **58**, 3896–3908 .

Caveney, S. (1971). Cuticle reflectivity and optical activity in scarab beetles: The rôle of uric acid, *Proc. R. Soc. Lond. B* **178**, 205–225.

Chen, Q., Hubbard, G., Shields, P. A., Liu, C., Allsopp, D. W. E., Wang, W. N., and Abbott, S. (2009). Broadband moth-eye antireflection coatings fabricated by low-cost nanoimprinting, *Appl. Phys. Lett.* **94**, 263118-1-3.

Clapham, P. B., and Hutley, M. C. (1973). Reduction of lens reflexion by the "moth eye" principle, *Nature* **244**, 281–282.

Cole, J. B. (1997). A high-accuracy realization of the Yee algorithm using non-standard finite differences, *IEEE Trans. Micro. Theory Tech.* **45**, 991–996.

Cole, J. B. (2005). High accuracy nonstandard finite-difference time-domain algorithms for computational electromagnetics: Applications to optics and photonics, in R. E. Mickens (ed.), *Advances in the Applications of Nonstandard Finite Difference Schemes* (World Scientific, Singapore), pp. 89–189.

Darragh, P. J., Gaskin, A. J., Terrell, B. C., and Sanders, J. V. (1966). Origin of precious opal, *Nature* **209**, 13–16.

de Genne, P.-G. (1974). *The Physics of Liquid Crystals* (Claredon Press, Oxford).

Denkov, N. D., Velev, O. D., Kraichevsky, P. A., Ivanov, I. B., Yoshimura, H., and Nagayama, K. (1993). Two-dimensional crystallization, *Nature* **361**, 26.

Denton, E. J., and Land, M. F. (1967). Optical properties of the lamellae causing interference colours in animal reflectors, *J. Physiol.* **191**, 23P–24P.

Denton, E. J., and Land, M. F. (1971). Mechanism of reflexion in silvery layers of fish and cephalopods, *Proc. R. Soc. Lond. A* **178**, 43–61.

Derjaguin, B. V., and Landau, L. (1941). Theory of the stability of strongly charged lyophobic sols and of the adhesion of strongly charged particles in solution of electrolytes, *Acta Physicochim. URSS* **14**, 633–662.

Dufresne, E. R., Noh, H., Saranathan, V., Mochrie, S. G. J., Cao, H., and Prum, R. O. (2009). Self-assembly of amorphous biophotonic nanostructures by phase separation, *Soft Matter* **5**, 1792–1795.

Durrer, H. (1962). Schillerfarben beim Pfau (*Pavo cristatus* L.), *Verhand. Naturf. Ges. Basel* **73**, 204–224.

Durrer, H., and Villiger, W. (1972). Schillerfarben von *Euchroma gigantea* (L.): (Coleoptera: Buprestidae): Elektronenmikroskopische Untersuchung der Elytra, *Int. J. Insect Morphol. Embryol.* **1**, 233–240.

Durrer, H. (1977). Schillerfarben der Vogelfeder als Evolutionsproblem, *Denkschr. Schweiz. Naturforsch. Ges.* **91**, 1–127.

Durrer, H. (1986). Colouration, in J. Bereiter-Hahn, A. G. Matoltsy, and K. S. Richards (eds.), *Biology of the Integument* Vol. 2 *Vertebrates* V. *The Skin of Birds* Chapter 12 (Springer-Verlag, Berlin), pp. 239–247.

Dyck, J. (1971). Structure and colour-production of the blue barbs of *Agapornis roseicollis* and *Cotinga maynana*, *Z. Zellforsch.* **115**, 17–29.

Fox, D. L. (1976). *Animal Biochromes and Structural Colours* (University California Press, Berkeley).

Frank, F., and Ruska, H. (1939). Übermikroskopische Untersuchung der Blaustruktur der Vogelfeder, *Naturwiss.* **27**, 229–230.

Fujii, R. (1993). Cytophysiology of fish chromatophores, *Int. Rev. Cyto.* **143**, 191–255.

Fujimura, Y. (2009). Theoretical study on optical characteristics of photnoc crystal, Thesis (Osaka University).

Gentil, K. (1942). Elektronenmikroskopische Untersuchung des Feinbaues schillernder Leisten von Morpho-Schuppen, *Z. Morph. Ökol. Tiere* **38**, 344–355.

Ghiradella, H., and Radican, W. (1976). Development of butterfly scales II. Struts, lattices and surface tension, *J. Morph.* **150**, 279–298.

Ghiradella, H. (1974). Development of ultraviolet-reflecting butterfly scales: How to make an inteference filter, *J. Morph.* **142**, 395–410.

Ghiradella, H. (1984). Structure of iridescent Lepidopteran scales: Variations on several themes, *Ann. Entomol. Soc. Am.* **77**, 637–645.

Ghiradella, H. (1985). Structure and development of iridescent Lepidopteran scales: The Papilionidae as a showcase family, *Ann. Entomol. Soc. Am.* **78**, 252–264.

Ghiradella, H. (1989). Structure and development of iridescent butterfly scales: Lattices and laminae, *J. Morph.* **202**, 69–88.

Ghiradella, H. (1991). Light and color on the wing: Structural colors in butterflies and moths, *Appl. Opt.* **30**, 3492–3500.

Ghiradella, H. (1994). Structure of butterfly scales: Patterning in an insect cuticle, *Microsc. Res. Tech.* **27**, 429–438.

Ghiradella, H. (1998). Hairs, bristles, and scales, in F. W. Harrison and M. Locke (eds.), *Microscopic Anatomy of Invertebrates* Vol. 11A: *Insecta* (Wiley-Liss, New York), pp. 257–287.

Ghiradella, H. (1999). Shining armor: Structural colors in insects, *Opt. Photonics News* **10**, 46–48.

Ghiradella, H. (2005). Fine structure of basic Lepidopteran scales, in S. Kinoshita and S. Yoshioka (eds.), *Structural Colors in Biological Systems: Principles and Applications* (Osaka University Press, Osaka), pp. 75–93.

Goethe, J. W. (1810). *Zur Farbenlehre (Theory of Colours)* (transl. C. Eastlake, John Murray Publishing, 1840).

Greenewalt, C. H., Brandt, W., and Friel, D. D. (1960a). Iridescent colors of hummingbird feathers, *J. Opt. Soc. Am.* **50**, 1005–1013.

Greenewalt, C. H., Brandt, W., and Friel, D. D. (1960b). The iridescent colors of hummingbird feathers, *Proc. Am. Phil. Soc.* **104**, 249–253.

Greenewalt, C. H. (1960). *Hummingbirds* (Doubleday & Company, New York, republicated by Dover in 1990, New York).

Guild, J. (1932). The colorimetric properties of the spectrum, *Phil. Trans. R. Soc. Lond.* **230**, 149–187.

Hanbury Brown, R., and Twiss, R. Q. (1956). A test of a new type of stellar interferometer on Sirius, *Nature* **178**, 1046–1048.

Hariyama, T., Hironaka, M., Horiguchi, H., and Stavenga, D. G. (2005). The leaf beetle, the jewel beetle, and the damselfly; insects with a multilayered show case, in S. Kinoshita and S. Yoshioka (eds.), *Structural Colors in Biological Systems: Principles and Applications* (Osaka University Press, Osaka), pp. 153–176.

Hertz, H. (1887). Ueber sehr schnelle electrische Schwingungen, *Ann. Phys.* **267**, 421–448.

Hirata, K., and Ohsako, N. (1966). Studies on the structure of scales and hairs of insects. IV. Microstructure of scales of the butterfly, *Morpho menelaus nakaharai* Le Moult, *Sci. Rep. Kagoshima Univ.* **15**, 49–61.

Hongo, T., Hiroshige, T., Toyota, J., and Kumada, M. (2000). *Standard Physiology*, the 5th edition, (Igaku-Shoin, Tokyo), in Japanese.

Hooke, R. (1665). *Micrographia: or some Physiological Descriptions of Minute Bodies made by Magnifying Glasses with Ovservations and Inquiries thereupon* (Fo. Martyn and Fa. Alleftry, London; republicated by Dover Publications, 2003, New York).

Huang, J., Wang, X., and Wang, Z. L. (2006). Controlled replication of butterfly wings for achieving tunable photonic properties, *Nano Lett.* **6**, 2325–2331.

Huxley, A. F. (1968). A theoretical treatment of the reflexion of light by multilayer structures, *J. Exp. Biol.* **48**, 227–245.

Ikeda, M., and Ashizawa, S. (2005). *Doushite Iro wa Mierunoka* (Heibonsha, Tokyo), in Japanese.

Imhof, A., and Pine, D. J. (1997). Ordered macroporous materials by emulsion templating, *Nature* **389**, 948–951.

Imhof, A., Vos, W. L., Sprik, R., and Lagendijk, A. (1999). Large dispersive effects near the band edges of photonic crystals, *Phys. Rev. Lett.* **83**, 2942–2945.

Iohara, K., Yoshimura, M., Tabata, H., and Shimizu, S. (2000). Structurally colored fibers, *Chem. Fibers Int.* **50**, 38–39.

James, R. W. (1948). *The Optical Principles of the Diffraction of X-Rays* (G. Bell and sons, London).

Jiang, P., Bertone, J. F., Hwang, K. S., and Colvin, V. L. (1999). Single-crystal colloidal multilayers of controlled thickness, *Chem. Mater.* **11**, 2132–2140.

Jin, C., Meng, X., Cheng, B., Li, Z., and Zhang, D. (2001). Photonic gap in amorphous photonic materials, *Phys. Rev. B* **63**, 195107-1-5.

John, S. (1987). Strong localization of photons in certain disordered dielectric superlattices, *Phys. Rev. Lett.* **58**, 2486–2489.

Jones, J. B., Sanders, J. V., and Segnit, E. R. (1964). Structure of opal, *Nature* **204**, 990–991.

Kambe, M. (2008). Optical properties of *Morpho* butterfly wings studied by angle-reslolved reflection measurements, Doctorial Thesis, Osaka University.

Kambe, M., Zhu, D., and Kinoshita, S. (2011). Origin of retroreflection from a wing of the *Morpho* butterfly, *J. Phys. Soc. Jpn.* **80**, 054801-1-10.

Kanamori, Y., Sasaki, M., and Hane, K. (1999). Broadeband antireflection gratings fabricated upon silicon substrates, *Opt. Lett.* **24**, 1422–1424.

Kanamori, Y., Roy, E., and Chen, Y. (2005). Antireflection sub-wavelength gratings fabricated by spin-coating replication, *Microelctron. Eng.* **78–79**, 287–293.

Kang, D., Maclennan, J. E., Clark, N. A., Zakhidov, A. A., and Baughman, R. H. (2001). Electro-optic behavior of liquid-crystal-filled silica opal photonic crystals: Effect of liquid-crystal alignment, *Phys. Rev. Lett.* **86**, 4052–4055.

Kawaguti, S. (1965). Electron microscopy on iridophores in the scale of the blue wrasse, *Proc. Japan Acad.* **41**, 610–613.

Kawamura, S. (2010). *The Photobiology of Vision* (Asakura Publishing, Tokyo), in Japanese.

Kerker, M., and Matijević, E. (1961). Scattering of electromagnetic waves from concentric infinite cylinders, *J. Opt. Soc. Am.* **51**, 506–508.

Kerker, M. (1969). *The Scattering of Light and Other Electromagnetic Radiation* (Academic Press, New York).

Kinoshita, S., Yoshioka, S., and Kawagoe, K. (2002a). Mechanisms of structural colour in the *Morpho* butterfly: Cooperation of regularity and irregularity in an iridescent scale, *Proc. R. Soc. Lond. B* **269**, 1417–1421.

Kinoshita, S., Yoshioka, S., Fujii, Y., and Okamoto, N. (2002b). Photophysics of structural color in the *Morpho* butterflies, *Forma* **17**, 103–121.

Kinoshita, S. (2005). *Blue Brilliancy of Morpho butterflies* (Kagaku Dojin, Kyoto), in Japanese.

Kinoshita, S., and Yoshioka, S. (2005a). Phtophysical approach to blue coloring in the *Morpho* butterflies, in S. Kinoshita and S. Yoshioka (eds.),

Structural Colors in Biological Systems: Principles and Applications (Osaka University Press, Osaka), pp. 113–140.

Kinoshita, S., and Yoshioka, S. (2005b). Structural colors in nature: The role of regularity and irregularity in the structure, *ChemPhysChem* **6**, 1442–1459.

Kinoshita, S., and Yoshioka, S. (eds.) (2005c). *Structural Colors in Biological Systems: Principles and Applications* (Osaka University Press, Osaka).

Kinoshita, S. Yoshioka, S., and Miyazaki, J. (2008). Physics of structural colors, *Rep. Prog. Phys.* **53**, 076401-1-30.

Kinoshita, S. (2008). *Structural Colors in the Realm of Nature* (World Scientific Publishing, Singapore).

Kühn, A. (1955). *Vorlesungen über Entwicklungsphysiologie* (Springer-Verlag, Berlin).

Kurachi, M., Takaku, Y., Komiya, Y., and Hariyama, T. (2002). The origin of extensive colour polymorphism in *Plateumaris sericea* (Chrysomelidae, Coleoptera), *Naturwiss.* **89**, 295–298.

Lalanne, P., and Morris, G. M. (1997). Antireflection behavior of silicon subwavelength periodic structures for visible light, *Nanotechnol.* **8**, 53–56.

Land, M. F. (1972). The physics and biology of animal reflectors, *Prog. Biophys. Mol. Biol.* **24**, 75–106.

Larsen, A. E., and Grier, D. G. (1997). Like-charge attractions in metastable colloidal crystallites, *Nature* **385**, 230–233.

Lehmann, O. (1889). Über fliessende Krystalle, *Z. Phys. Chem.* **4**, 462–472.

Leonard, S. W., Mondia, J. P., van Driel, H. M., Toader, O., John, S., Busch, K., Birner, A., Gösele, U., and Lehmann, V. (2000). Tunable two-dimensional photonic crystals using liquid-crystal infiltration, *Phys. Rev. B* **61**, R2389–R2392.

Li, Y., Lu, Z., Yin, H., Yu, X., Liu, X., and Zi, J. (2005). Structural origin of the brown color of barbules in male peacock tail feathers, *Phys. Rev. E* **72**, 010902-1-4.

Liew, S. F., Forster, J., Noh, H., Schreck, C. F., Saranathan, V., Lu, X., Yang, L., Prum, R. O., O'Hern, C. S., Dufresne, E. R., and Cao, H. (2011). Short-range order and near-field effects on optical scattering and structural coloration, *Opt. Express* **19**, 8208–8217.

Lin, S. Y., Fleming, J. G., Hetherington, D. L., Smith, B. K., Biswas, R., Ho, K. M., Sigalas, M. M., Zubrzycki, W., Kurtz, S. R., and Bur, J. (1998). A three-dimensional photonic crystal operating at infrared wavelengths, *Nature* **394**, 251–253.

Linn, N. C., Sun, C.-H., Jiang, P., and Jiang, B. (2007). Self-assembled biomimetic antireflection coatings, *Appl. Phys. Lett.* **91**, 101108-1-3.

Lippert, W., and Gentil, K. (1959). Über lamellare Feinstrukturen bei den Schillerschuppen der Schmetterlinge vom Urania- und Morpho-Typ, *Z. Morph. Ökol Tiere* **48**, 115–122.

Lippert, W., and Gentil, K. (1963). Über den Feinbau der Schillerhaare des Polychaeten Aphrodite aculeata, *L. Z. Morph. Ökol Tiere* **53**, 22–28.

Lythgoe, J. N., and Shand, J. (1982). Changes in spectral reflexions from the irodophores of the neon tetra, *J. Physiol.* **325**, 23–34.

Lythgoe, J. N., Shand, J., and Foster, R. G. (1984). Visual pigment in fish iridocytes, *Nature* **308**, 83–84.

MacAdam, D. L. (1937). Projective transformations of I. C. I. color specifications, *J. Opt. Soc. Am.* **27**, 294–297.

Mallock, A. (1911). Note on the iridescent colours of birds and insects, *Roy. Soc. Proc. A* **85**, 598–604.

Mason, C. W. (1923a). Structural colors in feathers. I, *J. Phys. Chem.* **27**, 201–251.

Mason, C. W. (1923b). Structural colors in feathers. II, *J. Phys. Chem.* **27**, 401–447.

Mason, C. W. (1926). Structural colors in insects. I, *J. Phys. Chem.* **30**, 383–395.

Mason, C. W. (1927a). Structural colors in insects. II, *J. Phys. Chem.* **31**, 321–354.

Mason, C. W. (1927b). Structural colors in insects. III, *J. Phys. Chem.* **31**, 1856–1872.

Maxwell, J. C. (1873). *A Treatise on Electricity and Magnetism* (Claredon Press, Oxford).

McPhedran, R. C., Nicorovici, N. A., McKenzie, D. R., Botten, L. C., Parker, A. R., and Rouse, G. W. (2001). The sea mouse and the photonic crystal, *Aust. J. Chem.* **54**, 241–244.

Meng, Q.-B., Gu, Z.-Z., Sato, O., and Fujishima, A. (2000). Fabrication of highly ordered porous structures, *Appl. Phys. Lett.* **77**, 4313–4315.

Merritt, E. (1925). A spectrophotometric study of certain cases of structural color, *J. Opt. Soc. Am., and Rev. Sci. Instrum.* **11**, 93–98.

Michelson, A. A. (1911). On metallic colouring in birds and insects, *Phil. Mag.* **21**, 554–566.

Michelson, A. A., and Pease, F. G. (1921). Measurement of the diameter of alpha-Orionis with the interferometer, *Astrophys. J.* **53**, 249–259.

Michielsen, K., and Stavenga, D. G. (2008). Gyroid cuticular structures in butterfly wing scales: biological photonic crystals, *J. R. Soc. Interface* **5**, 85–94.

Michielsen, K., De Raedt, H., and Stavenga, D. G. (2010). Reflectivity of the gyroid biophotonic crystals in the ventral wing scales of the green hairstreak butterfly, *Callophrys rubi*, *J. R. Soc. Interface* **7**, 765–771.

Mickens, R. E. (1989). Exact solutions to a finite-difference model of a reaction-advection equation: Implications for numerical analysis, *Numerical Methods for Partial Differential Equations* **5**, 313–325.

Mie, G. (1908). Beiträge zur Optik trüber Medien, speziell kolloidaler Metallösungen, *Ann. Physik* **330**, 377–445.

Miklyaev, Yu. V., Meisel, D. C. Blanco, A., von Freymann, G., Busch, Ko, Koch, W., Enkrich, C., Deubel, M., and Wegener, M. (2003). Three-dimensional face-centered-cubic photonic crystal templates by laser holography: fabrication, optical characterization, and band-structure calculations, *Appl. Phys. Lett.* **82**, 1284–1286.

Moharam, M. G., and Gaylord, T. K. (1981). Rigorous coupled-wave analysis of planar-grating diffraction, *J. Opt. Soc. Am.* **71**, 811–818.

Morris, R. B. (1975). Iridescence from diffraction structures in the wing scales of *Callophrys rubi*, the Green Hairstreak, *J. Ent. (A)* **49**, 149–154.

Munsell, A. H. (1905). *A Color Notation* (Geo. H. Ellis Co., Boston).

Nagaishi, H., and Oshima, N. (1989). Neural control of motile activity of light-sensitive iridophores in the neon tetra, *Pigment Cell Res.* **2**, 485–492.

Nagaishi, H., Oshima, N., and Fujii, R. (1990). Light-reflecting properties of the iridophores of the neon tetra, *Paracheirodon innesi*, *Comp. Biochem. Physiol.* **95A**, 337–341.

Nagaishi, H., and Oshima, N. (1992). Ultrastructure of the motile iridophores of the neon tetra, *Zool. Sci.* **9**, 65–75.

Nakamura, E., Yoshioka, S., and Kinoshita, S. (2008). Structural color of rock dove's neck feather, *J. Phys. Soc. Jpn.* **77**, 124801-1-12.

Neville, A. C., and Luke, B. M. (1969a). A two-system model for chitin-protein complexes in insect cuticles, *Tissue Cell* **1**, 689–707.

Neville, A. C., and Luke, B. M. (1969b). Molecular architecture of adult locust cuticle at the electron microscope level, *Tissue Cell* **1**, 355–366.

Neville, A. C., and Caveney, S. (1969). Scarabaeid beetle exocuticle as an optical analogue of cholesteric liquid crystals, *Biol. Rev.* **44**, 531–562.

Neville, A. C., and Luke, B. M. (1971). A biological system producing a self-assembling cholesteric protein liquid crystal, *J. Cell Sci.* **8**, 93–109.

Neville, A. C. (1975). *Biology of the Arthropod Cuticle* (Springer, New York).

Neville, A. C., Parry, D. A. D., and Woodhead-Galloway, J. (1976). The chitin crystallite in arthropod cuticle, *J. Cell Sci.* **21**, 73–82.

Neville, A. C. (1977). Metallic gold and silver colours in some insect cuticles, *J. Insect Physiol.* **23**, 1267–1274.

Neville, A. C. (1984). Cuticle: organization, in J. Bereiter-Hahn, A. G. Matoltsy and K. S Richards (eds.), *Biology of the Integument. I. Invertebrates* (Springer-Verlag: Berlin), pp. 611–625.

Newton, I. (1704). *Opticks: Or a Treatise of the Reflections, Refractions, Inflections & Colours of Light* (Sam. Smith and Benj. Walford, London; republicated by Dover, New York).

Nicholls, J. G., Martin, A. R., Wallace, B. G., and Fuchs, P. A. (2001). *From Neuron to Brain*, the 4th Edition, Chap. 19–21 (Sinauer Assoc., Sunderland).

Nijhout, H. F. (1991). *The Development and Evolution of Butterfly Wing Patterns* [Smithonian Institution Press, Washington).

Noda, S., Tomoda, K., Yamamoto, N., and Chutinan, A. (2000). Full three-dimensional photonic band gap crystals at near-infrared wavelengths, *Science* **289**, 604–606.

Noh, H., Liew, S. F., Saranathan, V., Mochrie, S. G. J., Prum, R. O., Dufresne, E. R., and Cao, H. (2010a). How noniridescent colors are generated by quasi-ordered structures of bird feathers, *Adv. Mater.* **22**, 2871–2880.

Noh, H., Liew, S. F., Saranathan, V., Prum, R. O., Mochrie, S. G. J., Dufresne, E. R., and Cao, H. (2010b). Contribution of double scattering to structural coloration in quasiordered nanostructures of bird feathers, *Phys. Rev. E* **81**, 051923-1-8.

Ochiai, T., and Sánchez-Dehesa, J. (2001). Superprism effect in opal-based photonic crystals, *Phys. Rev. B* **64**, 245113-1-7.

Onslow, H. (1920). The iridescent colours of insects. II. Diffracton colours, *Nature* **106**, 181–183.

Onslow, H. (1923). On a periodic structure in many insect scales, and the cause of their iridescent colours, *Phil. Trans.* **211**, 1–74.

Orihara, H. (2004). *Physics of Liquid Crystals* (Uchida Rokakuho Pub.: Tokyo), in Japanese.

Parker, A. R., Hegedus, Z., and Watts, R. A. (1998). Solar-absorber antireflector on the eye of an Eocene fly (45 Ma), *Proc. R. Soc. Lond. B* **265**, 811–815.

Parker, A. R. (2000). 515 million years of structural colour, *J. Opt. A: Pure Appl. Opt.* **2**, R15–R28.

Parker, A. R., McPhedran, R. C., McKenzie, D. R., Botten, L. C., and Nicorovici, N.-A. P. (2001). Aphrodite's iridescence, *Nature* **409**, 36–37.

Parker, A. R. (2003). *In the blink of an eye: how vision kick-started the big bang of evolution* (Simon & Schuster, London).

Parker, A. R., Welch, V. L., Driver, D., and Martini, N. (2003). Opal analogue discovered in a weevil, *Nature* **426**, 786–787.

Parker, A. R., and Martini, N. (2006). Structural colour in animals: Simple to complex optics, *Opt. Laser Techol.* **38**, 315–322.

Parretta, A., Sarno, A., Tortora, P., Yakubu, H., Maddalena, P., Zhao, J., and Wang, A. (1999). Angle-dependent reflectance measurements on photovoltaic materials and solar cells, *Opt. Commun.* **172**, 139–151.

Philipse, A. P. (1989). Solid opaline packings of colloidal silica spheres, *J. Mater. Sci. Lett.* **8**, 1371–1373.

Plattner, L. (2004). Optical properties of the scales of *Morpho rhetenor* butterflies: theoretical and experimental investigation of the back-scattering of light in the visible spectrum, *J. R. Soc. Interface* **1**, 49–59.

Poladian, L., Wichham, S., Lee, K., and Large, M. C. J. (2009). Iridescence from photonic crystals and its suppression in butterfly scales, *J. R. Soc. Interface* **6**, S233–S242.

Prum, R. O., Torres, R. H., Williamson, S., and Dyck, J. (1998). Coherent light scattering by blue feather barbs, *Nature* **396**, 28–29.

Prum, R. O., Torres, R., Kovach, C., Williamson, S., and Goodman, S. M. (1999a). Coherent light scattering by nanostructured collagen arrays in the caruncles of the Malagasy asities (Eurylaimidae: Aves), *J. Exp. Biol.* **202**, 3507–3522.

Prum, R. O., Torres, R., Williamson, S., and Dyck, J. (1999b). Two-dimensional Fourier analysis of the spongy medullary keratin of structurally coloured feather barbs, *Proc. R. Soc. Lond. B* **266**, 13–22.

Prum, R. O., and Torres, R. (2003). Structural colouration of avian skin: convergent evolution of coherently scattering dermal collagen arrays, *J. Exp. Biol.* **206**, 2409–2429.

Prum, R. O., Cole, J. A., and Torres, R. H. (2004). Blue integumentary structural colours in dragonflies (Odonata) are not produced by incoherent Tyndall scattering, *J. Exp. Biol.* **207**, 3999–4009.

Prum, R. O. (2006). Anatomy, physics, and evolution of structural colors, in G. E. Hill and K. J. McGraw (eds.), *Bird Coloration*. Vol. I *Mechanisms and Measurements* (Harvard University Press, Cambridge), pp. 295–353.

Raman, C. V. (1928). A new radiation, *Ind. J. Phys.* **2**, 387–398.

Raman, C. V., and Krishnan, K. S. (1928). A new type of secondary radiation, *Nature* **121**, 501–502.

Raman, C. V. (1934a). The orgin of the colours in the plumage of birds, *Proc. Ind. Acad. Sci.* **1**, 1–7.

Raman, C. V. [1934b). On iridescent shells: Part I. Introductory, *Proc. Ind. Acad. Sci. A* **1**, 567–573.

Raman, C. V. (1934c). On iridescent shells: Part II. Colours of laminar diffraction, *Proc. Ind. Acad. Sci. A* **1**, 574–589.

Raman, C. V., and Krishnamurti, D. (1954a). The structure and optical behaviour of iridescent shells, *Proc. Ind. Acad. Sci. A* **39**, 1–13.

Raman, C. V., and Krishnamurti, D. (1954b). The structure and optical behaviour of pearls, *Proc. Ind. Acad. Sci. A* **39**, 215–222.

Lord Rayleigh (1871a). On the scattering of light by small particles, *Phil. Mag.* **41**, 447–454.

Lord Rayleigh (1871b). On the light from the sky, its polarization and colour, *Phil. Mag.* **41**, 107–120.

Lord Rayleigh, F. R. S. (1880). On reflection of vibrations at the confines of two media between which the transition is gradual, *Proc. Lond. Math. Soc.* **s1–11**, 51–56.

Lord Rayleigh, Sec. R. S. (1888a). On the reflection of light at a twin plane of a crystal, *Phil. Mag.* **26**, 241–255.

Lord Rayleigh, Sec. R. S. (1888b). On the remarkable phenomenon of crystalline reflexion described by Prof. Stokes, *Phil. Mag.* **26**, 256–265.

Lord Rayleigh, O. M., F. R. S. (1917). On the reflection of light from a regularly stratified medium, *Proc. R. Soc. Lond. A* **93**, 565–577.

Lord Rayleigh, O. M., F. R. S. (1919). VII. On the optical character of some brilliant animal colours, *Phil. Mag.* **37**, 98–111.

Lord Rayleigh, F. R. S. (1923a). Studies of iridescent colour, and the structure producing it. II. Mother-of-pearl, *Proc. R. Soc. Lond. A* **102**, 674–677.

Lord Rayleigh, F. R. S. (1923b). Studies of iridescent colour, and the structure producing it. III. The colours of labrador felspar, *Proc. Roy. Soc. Lond. A* **103**, 34–45.

Lord Rayleigh, F. R. S. (1923c). Studies of iridescent colour, and the structure producing it. IV. Iridescent beetles, *Proc. Roy. Soc. Lond. A* **103**, 233–239.

Lord Rayleigh, F. R. S. (1923d). Studies of iridescent colour, and the structure producing it. I. The colours of potassium chlorate crystals, *Proc. Roy. Soc. Lond. A* **102**, 668–674.

Lord Rayleigh, For. Sec. R.S. (1930). The iridescent colours of birds and insects, *Proc. R. Soc. Lond. A* **128**, 624–641.

Reinitzer, F. (1888). Biträge zur Kenntnis des Cholesterins, *Monatshefte für Chemie* **9**, 421–41.

Rodieck, R. W. (1998). *The First Steps in Seeing* (Sinauer Assoc., Sunderland).

Romanov, S. G., Fokin, A. V., Alperovich, V. I., Johnson, N. P., and de la Rue, R. M. (1997). The effect of the photonic stop-band upon the photoluminescence of CdS in opal, *phys. stat. sol. (a)* **164**, 169–173.

Saito, A., Yoshioka, S., and Kinoshita, S. (2004). Reproduction of the *Morpho* butterfly's blue: arbitration of contradicting factors, *Proc. SPIE* **5526**, 188–194.

Saito, A., Miyamura, Y., Nakajima, M., Ishikawa, Y., Sogo, K., Kuwahara, Y., and Hirai, Y. (2006). Reproduction of the *Morpho* blue by nanocasting lithography, *J. Vac. Sci. Technol. B* **24**, 3248–3251.

Sakoda, K. (2001). *Optical Properties of Photonic Crystals* (Springer Verlag, Berlin).

Sanders, J. V. (1964). Colour of precious opal, *Nature* **204**, 1151–1153.

Saranathan, V., Osuji, C. O., Mochrie, S. G. J., Noh, H., Narayanan, S., Sandy, A., Dufresne, E. R., and Prum, R. O. (2010). Structure, function, and self-assembly of single network gyroid (I4$_1$32) photonic crystals in butterfly wing scales, *Proc. Natl. Acad. Sci.* **107**, 11676–11681.

Schelkunoff, S. A. (1943). *Electromagnetic Waves* (D. van Nostrand, Toronto).

Schelkunoff, S. A. (1951). Kirchhoff's formula, its vector analogue, and other field equivalence theorems, *Comm. Pure Appl. Math.* **4**, 43–59.

Schmidt, W. J. (1943). Die Mosaikschuppen des *Teinopalpus imperialis* Hope, ein neues Muster schillernder Schmetterlingsschuppen, *Z. Morph. Ökol. Tiere* **39**, 176–216.

Schmidt, W. J., and Ruska, H. (1962). Tyndallblau-Struktur von Federn im Elektronenmikroskop, *Z. Zellforsch.* **56**, 693–708.

Schultz, T. D., and Rankin, M. A. (1985a). The ultrastructure of the epicuticular interference reflectors of tiger beetles (*Cicindela*), *J. exp. Biol.* **117**, 87–110.

Schultz, T. D., and Rankin, M. A. (1985b). Developmental changes in the interference reflectors and colorations of tiger beetles (*Cicindela*), *J. exp. Biol.* **117**, 111–117.

Simon, H. (1971). *The Splendor of Iridescence: Structural Colors in the Animal World* (Dodd, Mead & Company, New York).

Simonis, P., and Vigneron, J. P. (2011). Structural color produced by a three-dimensional photonic polycrystal in the scales of a longhorn beetle: *Pseudomyagrus waterhousei* (Coleoptera: Cerambicidae), *Phys. Rev. E* **83**, 011908-1-8.

Srinivasarao, M. (1999). Nano-optics in the biological world: Beetles, butterflies, birds, and moths, *Chem. Rev.* **99**, 1935–1961.

Stanley, W. M. (1935). Isolation of a crystalline protein possessing the properties of tobacco-mosaic virus, *Science* **81**, 644–645.

Stavenga, D. G., Giraldo, M. A., and Hoenders, B. J. (2006a). Reflectance and transmittance of scattering scales stacked on the wings of pierid butterflies, *Opt. Express* **14**, 4880–4890.

Stavenga, D G., Foletti, S., Palasantzas, G., and Arikawa, K. (2006b). Light on the moth-eye corneal nipple array of butterflies, *Proc. R. Soc. B* **273**, 661–667.

Stöber, W., Fink, A., and Bohn, E. (1968). Controlled growth of monodisperse silica spheres in the micron size range, *J. Colloid Interface Sci.* **26**, 62–69.

Süffert, F. (1924). Morphologie und Optik der Schmetterlingsschuppen, insbesondere die Schillerfarben der Schmetterlinge, *Z. Morph. Ökol. Tiere* **1**, 171–308.

Sun, C.-H., Min, W.-L., Linn, N. C., Jiang, P., and Jiang, B. (2007). Templated fabrication of large area subwavelength antireflection gratings on silicon, *Appl. Phys. Lett.* **91**, 231105-1-3.

Sun, C.-H., Jaing, P., and Jiang, B. (2008). Broadband moth-eye antireflection coatings on silicon, *Appl. Phys. Lett.* **92**, 061112-1-3.

Tabata, H., Kumazawa, K., Funakawa, M., Takimoto, J., and Akimoto, M. (1996). Microstructures and optical properties of scales of butterfly wings, *Opt. Rev.* **3**, 139–145.

Takahashi, K., Yamamoto, H., Onoda, A., Doi, M., Inaba, T., Chiba, M., Kobayashi, A., Taguchi, T., Okamura, T., and Ueyama, N. (2004). Highly oriented aragonite nanocrystal: biopolymer composites in an aragonite brick of the nacreous layer of *Pinctada fucata*, *Chem. Commun.* **2004**, 996–997.

Tilley, R. J. D., and Eliot, J. N. (2002). Scale microstructure and its phylogenetic implications in lycaenid butterflies (Lepidoptera, Lycaenidae), *Trans. lepid. Soc. Japan* **53**, 153–180.

Umashankar, K., and Taflove, A. (1982). A novel method to analyze electromagnetic scattering of complex objects, *IEEE Trans. Electromagnetic Compatibility* **24**, 397–405.

Uno, T. (1998). *Finite Difference Time Domain Method for Electromagnetic Field and Antenna Analyses* (Corona Publishing, Tokyo), in Japanese.

van de Hulst, H. C. (1957). *Light Scattering by Small Particles* (John Wiley & Sons, New York; republicated in 1981 by Dover, New York).

van Driel, H. M., and Vos, W. L. (2000). Multiple Bragg wave coupling in photonic band-gap crystals, *Phys. Rev. B* **62**, 9872–9875.

Velev, O. D., Jede, T. A., Lobo, R. F., and Lenhoff, A. M. (1997). Porous silica via colloidal crystallization, *Nature* **389**, 447–448.

Veron, J. E. N., O'Farrell, A. F., and Dixon, B. (1974). The fine structure of Odonata chromatophores, *Tissue Cell* **6**, 613–626.

Verwey, E. J. W., and Overbeek, J. Th. G. (1948). *Theory of the Stability of Lyophobic Colloids* (Elsevier Publishing; The Dover edition published in 1999).

Vlasov, Yu. A., Petit, S., Klein, G., Hönerlage, B., and Hirlimann, Ch. (1999). Femtosecond measurements of the time of flight of photons in a three-dimensional photonic crystal, *Phys. Rev. E* **60**, 1030–1035.

Vukusic, P., Sambles, J. R., Lawrence, C. R., and Wootton, R. J. (1999). Quantified interference and diffraction in single *Morpho* butterfly scales, *Proc. R. Soc. Lond. B* **266**, 1403–1411.

Vukusic, P., Sambles, J. R., and Lawrence, C. R. (2000). Colour mixing in wing scales of a butterfly, *Nature* **404**, 457.

Vukusic, P., Sambles, R., Lawrence, C., and Wakely, G. (2001). Sculpted-multilayer optical effects in two species of *Papilio* butterfly, *Appl. Opt.* **40**, 1116–1125.

Vukusic, P., and Sambles, J. R. (2003). Photonic structures in biology, *Nature* **424**, 852–855.

Walter, B. (1895). *Die Oberflächen- oder Schillerfarben* (F. Vieweg und Sohn, Braunschweig).

Watanabe, K., Hoshino, T., Kanda, K., Haruyama, Y., Kaito, T., and Matsui, S. (2005a). Optical measurement and fabrication from a *Morpho*-butterfly-scale quasistructure by focused ion beam chemical vapor deposition, *J. Vac. Sci. Technol. B* **23**, 570–574.

Watanabe, K., Hoshino, T., Kanda, K., Haruyama, Y., and Matsui, S. (2005b). Brilliant blue observation from a *Morpho*-butterfly-scale quasi-structure, *Jpn. J. Appl. Phys.* **44**, L48–L50.

Waterhouse, G. I. N., and Waterland, M. R. (2007). Opal and inverse opal photonic crystals: Fabrication and characterization, *Polyhedron* **26**, 356–368.

Watson, G. S., and Watson, J. A. (2004). Natural nano-structures on insects–possible functions of ordered arrays characterized by atomic force microscopy, *Appl. Surf. Sci.* **235**, 139–144.

Wright, W. D. (1929). A re-determination of the trichromatic coefficients of the spectral colours, *Trans. Opt. Soc.* **30**, 141–164.

Xi, J.-Q., Schubert, M. F., Kim, J. K., Schubert, E. F., Chen, M., Lin, S.-Y., Liu, W., and Smart, J. A. (2007). Optical thin-film materials with low refractive index for broadband elimination of Fresnel reflection, *Nature Photonics* **1**, 176–179.

Yablonovitch, E. (1987). Inhibited spontaneous emission in solid-state physics and electronics, *Phys. Rev. Lett.* **58**, 2059–2062.

Yamada, E. (1967). The fine structure of vertebrate retina, *Seitai no Kagaku* **18**, 54–66.

Yang, J.-K., Schreck, C., Noh, H., Liew, S.-F., Guy, M. I., O'Hern, C. S., and Cao, H. (2010). Photonic-band-gap effects in two-dimensional polycrystalline and amorphous structures, *Phys. Rev. A* **82**, 053838-1-8.

Yee, K. S. (1966). Numerical solution of initial boundary value problems involving Maxwell's equations in isotropic media, *IEEE Trans. Ant. Prop.* **14**, 302–307.

Yin, H., Shi, L., Sha, J., Li, Y., Qin, Y., Dong, B., Meyer, S., Liu, X., Zhao, L., and Zi, J. (2006). Iridescence in the neck feathers of domestic pigeons, *Phys. Rev. E* **74**, 051916-1-6.

Yoshida, A., Motoyama, M., Kosaku, A., and Miyamoto, K. (1996). Nanoprotuberance array in the transparent wing of a hawkmoth, *Cephonodes hylas, Zool. Sci.* **13**, 525–526.

Yoshida, A., Motoyama, M., Kosaku, A., and Miyamoto, K. (1997). Antireflective nanoprotuberance array in the transparent wing of a hawkmoth, *Cephonodes hylas, Zool. Sci.* **14**, 737–741.

Yoshino, K., Tada, K., Ozaki, M., Zakhidov, A. A., and Baughman, R. H. (1997). The optical properties of porous opal crystals infiltrated with organic molecules, *Jpn. J. Appl. Phys.* **36**, L714–L717.

Yoshino, K., Lee, S. B., Tatsuhara, S., Kawagishi, Y., Ozaki, M., and Zakhidov, A. A. (1998). Observation of inhibited spontaneous emission and stimulated emission of rhodamine 6G in polymer replica of synthetic opal, *Appl. Phys. Lett.* **73**, 3506–3508.

Yoshino, K., Shimoda, Y., Kawagishi, Y., Nakayama, K., and Ozaki, M. (1999). Temperature tuning of the stop band in trasmission spectra of liquid-crystal infiltrated synthetic opal as tunable photonic crystal, *Appl. Phys. Lett.* **75**, 932–934.

Yoshioka, S., and Kinoshita, S. (2002). Effect of macroscopic structure in iridescent color of the peacock feathers, *Forma* **17**, 169–181.

Yoshioka, S., and Kinoshita, S. (2004). Wavelength-selective and anisotropic light-diffusing scale on the wing of the *Morpho* butterfly, *Proc. R. Soc. Lond. B* **271**, 581–587.

Yoshioka, S., Kinoshita, S., and Saito, A. (2004). Coloration mechanisms of *Morpho* butterflies and production of *Morpho* color plate, *Oyo Butsuri* **73**, 939–942, in Japanese.

Yoshioka, S., and Kinoshita, S. (2005). Structural color of peacock feathers, in S. Kinoshita and S. Yoshioka (eds.), *Structural Colors in Biological Systems: Principles and Applications* (Osaka University Press, Osaka), pp. 195–208.

Yoshioka, S., and Kinoshita, S. (2006). Structural or pigmentary? Origin of the distinctive white stripe on the blue wing of a *Morpho* butterfly, *Proc. R. Soc. Lond. B* **273**, 129–134.

Yoshioka, S., and Kinoshita, S. (2007). Polarization-sensitive color mixing in the wing of the Madagascan sunset moth, *Opt. Express* **15**, 2691–2701.

Yoshioka, S., Nakamura, E., and Kinoshita, S. (2007). Origin of two-color iridescence in rock dove's feather, *J. Phys. Soc. Jpn.* **76**, 013801-1-4.

Yoshioka, S., Nakano, T., Nozue, Y., and Kinoshita, S. (2008). Coloration using higher order optical interference in the wing pattern of the Madagascan sunset moth, *J. R. Soc. Interface* **5**, 457–464.

Yoshioka, S., and Kinoshita, S. (2011). Direct determination of the refractive index of natural multilayer systems, *Phys. Rev. E* **83**, 051917-1-7.

Yoshioka, S., Matsuhana, B., Tanaka, S., Inouye, Y., Oshima, N., and Kinoshita, S. (2011). Mechanism of variable structural colour in the neon tetra: quantitative evaluation of the Venetian blind model, *J. R. Soc. Interface* **8**, 56–66.

Yu, Z., Gao, H., Wu, W., Ge, H., and Chou, S. Y. (2003). Fabrication of large area subwavelength antireflection structures on Si using trilayer resist nanoimprint lithography and liftoff, *J. Vac. Sci. Technol. B* **21**, 2874–2877.

Zhu, D., Kinoshita, S., Cai, D., and Cole, J. B. (2009). Investigation of structural colors in *Morpho* butterflies using the nonstandard-finite-difference time-domain method: Effects of alternately stacked shelves and ridge density, *Phys. Rev. E* **80**, 051924-1-12.

Zi, J., Yu, X., Li, Y., Hu, X., Xu, C., Wang, X., Liu, X., and Fu, R. (2003). Coloration strategies in peacock feathers, *Proc. Natl. Acad. Sci.* **100**, 12576–12578.

Index

DATE DUE

PRINTED IN U.S.A.